This book is dedicated to our wives,
Unni, Ada and Marianne
and to our daughters,
Sonja, Lisa and Julie
Valerie, Elaine and Allison
Maria Raphaela
with appreciation for the support, help and encouragement that we received during the writing of this textbook.

CONVERSIONS BETWEEN U.S. CUSTOMARY UNITS AND SI UNITS (Continued)

U.S. Customary unit		Times conversion factor: Accurate	Times conversion factor: Practical	Equals SI unit	
Moment of inertia (area)					
inch to fourth power	in.^{4}	416,231	416,000	millimeter to fourth power	mm^4
inch to fourth power	in.^{4}	0.416231×10^{-6}	0.416×10^{-6}	meter to fourth power	m^4
Moment of inertia (mass)					
slug foot squared	slug-ft^2	1.35582	1.36	kilogram meter squared	kg·m^2
Power					
foot-pound per second	ft-lb/s	1.35582	1.36	watt (J/s or N·m/s)	W
foot-pound per minute	ft-lb/min	0.0225970	0.0226	watt	W
horsepower (550 ft-lb/s)	hp	745.701	746	watt	W
Pressure; stress					
pound per square foot	psf	47.8803	47.9	pascal (N/m^2)	Pa
pound per square inch	psi	6894.76	6890	pascal	Pa
kip per square foot	ksf	47.8803	47.9	kilopascal	kPa
kip per square inch	ksi	6.89476	6.89	megapascal	MPa
Section modulus					
inch to third power	in.^{3}	16,387.1	16,400	millimeter to third power	mm^3
inch to third power	in.^{3}	16.3871×10^{-6}	16.4×10^{-6}	meter to third power	m^3
Velocity (linear)					
foot per second	ft/s	0.3048*	0.305	meter per second	m/s
inch per second	in./s	0.0254*	0.0254	meter per second	m/s
mile per hour	mph	0.44704*	0.447	meter per second	m/s
mile per hour	mph	1.609344*	1.61	kilometer per hour	km/h
Volume					
cubic foot	ft^3	0.0283168	0.0283	cubic meter	m^3
cubic inch	in.^{3}	16.3871×10^{-6}	16.4×10^{-6}	cubic meter	m^3
cubic inch	in.^{3}	16.3871	16.4	cubic centimeter (cc)	cm^3
gallon (231 in.^{3})	gal.	3.78541	3.79	liter	L
gallon (231 in.^{3})	gal.	0.00378541	0.00379	cubic meter	m^3

*An asterisk denotes an *exact* conversion factor

Note: **To convert from SI units to USCS units, *divide* by the conversion factor**

Temperature Conversion Formulas

$$T(°C) = \frac{5}{9}[T(°F) - 32] = T(K) - 273.15$$

$$T(K) = \frac{5}{9}[T(°F) - 32] + 273.15 = T(°C) + 273.15$$

$$T(°F) = \frac{9}{5}T(°C) + 32 = \frac{9}{5}T(K) - 459.67$$

Transportation Infrastructure Engineering

A Multimodal Integration

SI Edition

Lester A. Hoel

University of Virginia

Nicholas J. Garber

University of Virginia

Adel W. Sadek

University of Vermont

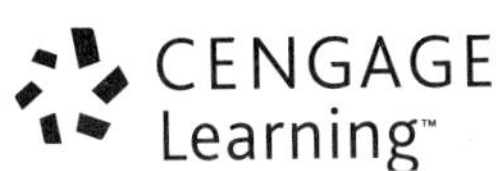

Australia • Brazil • Japan • Korea • Mexico • Singapore • Spain • United Kingdom • United States

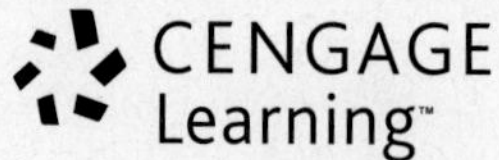

Transportation Infrastructure Engineering: A Multimodal Integration, SI Edition

Publisher, Global Engineering Program: Christopher M. Shortt

Acquisitions Editor, SI Edition: Swati Meherishi

Senior Developmental Editor: Hilda Gowans

Assistant Development Editor: Debarati Roy

Editorial Assistant: Tanya Altieri

Team Assistant: Carly Rizzo

Marketing Manager: Lauren Betsos

Content Project Manager: Jennifer Ziegler

Technical Proofreader: K. K. Chaudhry

Compositor: Glyph International

Rights Acquisition Specialist: John Hill

Senior Art Director: Michelle Kunkler

Cover Designer: Andrew Adams

Cover Images: Plane image © Shutterstock/Mikael Damkier; all other images rest © JupiterImages Corporation.

Text and Image Permissions Researcher: Kristiina Paul

Manufacturing Buyer: Arethea Thomas

Library of Congress Control Number: 2006908593

ISBN-13: 978-0-495-66789-6

ISBN-10: 0-495-66789-7

Cengage Learning
200 First Stamford Place, Suite 400
Stamford, CT 06902
USA

Printed in the United States of America
1 2 3 4 5 6 7 13 12 11 10

Contents

Preface to the SI Edition

This edition of ***Transportation Infrastructure Engineering: A Multimodal Integration*** has been adapted to incorporate the International System of Units (*Le Système International d'Unités* or SI) throughout the book.

Le Système International d'Unités

The United States Customary System (USCS) of units uses FPS (foot-pound-second) units (also called English or Imperial units). SI units are primarily the units of the MKS (meter-kilogram-second) system. However, CGS (centimeter-gram-second) units are often accepted as SI units, especially in textbooks.

Using SI Units in this Book

In this book, we have used both MKS and CGS units. USCS units or FPS units used in the US Edition of the book have been converted to SI units throughout the text and problems. However, in case of data sourced from handbooks, government standards, and product manuals, it is not only extremely difficult to convert all values to SI, it also encroaches upon the intellectual property of the source. Also, some quantities such as the ASTM grain size number and Jominy distances are generally computed in FPS units and would lose their relevance if converted to SI. Some data in figures, tables, examples, and references, therefore, remains in FPS units. For readers unfamiliar with the relationship between the FPS and the SI systems, conversion tables have been provided inside the front and back covers of the book.

To solve problems that require the use of sourced data, the sourced values can be converted from FPS units to SI units just before they are to be used in a calculation. To obtain standardized quantities and manufacturers' data in SI units, the readers may contact the appropriate government agencies or authorities in their countries/regions.

Instructor Resources

A Printed Instructor's Solution Manual in SI units is available on request. An electronic version of the Instructor's Solutions Manual, and PowerPoint slides of the figures from the SI text are available through **www.cengage.com/engineering.**

The readers' feedback on this SI Edition will be highly appreciated and will help us improve subsequent editions.

The Publishers

Preface

This book is intended to serve as a resource for courses in transportation engineering that emphasize transportation in an overall systems perspective. It can serve as a textbook for an introductory course or for upper-level undergraduate and first-year graduate courses. The distinguishing aspect of this book is its multimodal, integrative character that takes a broad view of transportation systems.

This book differs from the widely used textbook, *Traffic and Highway Engineering,* which supports courses that emphasize elements of highway and traffic practice including areas such as planning, operations, design, materials and maintenance, and contains sufficient detail to provide engineering students with a solid background for engineering practice and preparation for engineering professional registration. Other textbooks have been written with a similar "single mode" perspective on topics such as airports, railroads, mass transit, and highways.

Transportation Infrastructure Engineering: A Multimodal Integration serves a different purpose and is intended for a broader audience. Its objective is to provide an overview of transportation from a multimodal viewpoint rather than emphasizing a particular mode in great detail. This book also differs from others that claim the domain of transportation engineering. Some texts include "Transportation Engineering" in their title but emphasize highways with some mention of public transportation. Other texts devote separate chapters or sections to several modes, such as air and mass transit, with little integration of the modes or demonstration of the similarities and differences that may exist from one mode to another. Some fail to provide a context that includes transportation history, its role in society and the transportation profession.

This book places emphasis on explaining the environment in which transportation operates and thus presents the "big picture" to assist students to understand why transportation systems operate as they do and the role they play in a global society. The approach used in this book is to discuss basic concepts in transportation and how they have been applied to various modes. Since every mode of transportation includes vehicles and the route they travel, we refer to that route, whether it is a highway, railroad track, flight path or sea-lane as its "travelway." Thus, for example, the chapter on travelway geometrics describes similarities and differences in design principles for air, rail, and highways and explains how they are used in practice.

The textbook is organized around the fundamentals within the field of transportation engineering. The selection of chapter topics is intended to cover significant professional areas of transportation engineering. These include an overview of transportation in society; transportation systems models; characteristics of the driver, vehicle, and travelway; capacity analysis; planning and evaluation; geometric design of travelways; structural design of pavements; transportation safety; and information technology in transportation. Since this book is an overview of transportation infrastructure engineering it is necessarily less comprehensive than the textbook *Traffic and Highway Engineering*. Students who decide to pursue a career in the field of transportation may also study modes in greater depth in other courses. The pedagogical approach used in this textbook is an extensive use of solved examples in each chapter that illustrate text material, a set of homework problems provided at the end of each chapter, as well as a summary and a list of suggestions for further reading.

Completion of this textbook could not have been possible without the help and support of many individuals and organizations. First, our thanks to those who served as reviewers of manuscript drafts including Murtaza Haider, Stephen P. Mattingly, Carroll J. Messer, and others who preferred to remain anonymous. We also appreciate the help of John Miller and Rod Turochy who provided comments on specific chapters. Special thanks to Jane Carlson and Hilda Gowans who served as our editors and worked closely with us throughout the project. We are also grateful to those professional organizations that permitted us to include material from their manuals and publications thus assuring that emerging transportation professionals will learn the latest in transportation practice as well as theory. These are: The American Association of State Highway and Transportation Officials, The Institute of Transportation Engineers, the Portland Cement Association, The Eno Transportation Foundation, the Transportation Research Board of the National Academies, the American Railway Engineering and Maintenance-of-Way Association, the Association of American Railroads and the U.S. Department of Transportation.

CHAPTER 1

Overview of Transportation

The purpose of this chapter is to describe the context for transportation in terms of its importance to society and the issues raised by the impacts created when new transportation systems and services are provided. The chapter also describes the type of employment opportunities available in the transportation industry with emphasis on the transportation infrastructure sector. Since the popularity and use of transportation modes—such as canals, railroads, air travel, autos, trucking, and ships—will change over time, the history of transportation modes with emphasis on the transportation revolution, which dates from the early 1800s and continues today, is also summarized.

TRANSPORTATION AND SOCIETY

The purpose of transportation is to provide a mechanism for the exchange of goods, people, information, and to support economic improvements for society. Transportation provides the means to travel for purposes of employment, exploration, or personal fulfillment and is a necessary condition for human activities such as commerce, recreation, and defense. Transportation is defined as the movement of people and goods to meet the basic needs of society that require mobility and access. There are many examples of transportation movements that occur daily; a family journeys to another country to seek a better life, a medical emergency requires the immediate transfer of a patient to a hospital, a sales executive travels across the country to attend a management conference, a truckload of fresh produce is delivered to a supermarket, and workers commute from their homes to places of employment.

The quality of transportation affects a society's ability to utilize its natural resources of labor and/or materials. Transportation also influences the competitive position with respect to other regions or nations. Without the ability to transport products easily, a region may be unable to offer goods and services at an attractive price and thus may reduce or lose market share. By providing transportation services safely, reliably, and quickly with sufficient capacity and at a competitive price, a state or nation is able to expand its economic base, enter new markets and import skilled labor.

Every developed nation and region with a strong economic base has invested in high-quality transportation services. In the eighteenth and nineteenth centuries, nations such as England and Spain with a strong maritime presence became the rulers of vast colonial empires and established international commerce with trade routes to North America, India, Africa, and the Far East. In the twentieth century, countries that became leaders in industry and commerce—such as the United States, Canada, Japan, and Germany—relied on modern networks of sea, land, and air transportation. These systems enhanced the capability of their industries to transport manufactured goods, raw materials, and technical expertise and thus maximize the comparative advantage over other competitors. In the twenty-first century, information technology and the integration of land, sea, and air modes have helped to create a global economy. For countries lacking natural resources, reliance on transportation is essential to assure the importation of raw materials needed to manufacture automobiles, electronics, and other export products.

Modern integrated transportation facilities are a necessary but not sufficient condition to assure economic development and prosperity. Without competitive transportation services, the economic potential of a region is limited. To succeed, a region must be endowed with natural or human resources, infrastructure (such as water, power, and sewage disposal facilities), financial capital, adequate housing, and a strong military defense. When these conditions are in place, economic growth will depend on the quality of the internal transportation system, which consists of highways, railroads, airlines, shipping, and ports. It will also depend on the quality of multimodal linkages with the rest of the world, including maritime, rail, trucking, and air services.

Good transportation provides many benefits to society in addition to its role in economic development. Advances in transportation have contributed to the quality of life and have expanded opportunities to engage in the pursuit of happiness, a right of Americans stated by Thomas Jefferson in the Declaration of Independence. Modern transportation systems have provided the world with an unprecedented degree of mobility.

In contrast to earlier times, today we can travel by auto, rail, ship, or airplane to any part of the nation or world to visit friends and relatives and explore new sights. We can also alter our present living conditions by moving elsewhere. Because of good transportation, health care has dramatically improved—for example, drugs, transplants, and medical equipment can be transported in emergency situations to a remote hospital or patients can be moved quickly to specialized medical facilities. Improvements in transportation have contributed to the worldwide decline of hunger as food shortages due to famine, war, or weather have been replenished by air and sea lifts. Other benefits to society include extension of life expectancy, enhanced opportunities for advanced education and technical training, increased incomes and standards of living, broader recreational options, reduction of inequality in education and employment, and wider participation in worldwide multicultural experiences.

The benefits of providing society with improved transportation, whether they are justified on the basis of economic development or mobility, are not achieved without a price. The costs to society are both direct and indirect. Direct costs include capital and operating expenses including right of way, facilities and maintenance. Indirect costs include environmental effects, congestion, property damage, injuries, and deaths. The construction of the 75,140-km National System of Interstate and Defense Highways (named the Dwight D. Eisenhower Interstate System), which began in 1956, took over 40 years to complete with an expenditure of $130 billion. Other major transportation projects include the Panama Canal, completed in 1914 (Figure 1.1), and the transcontinental railroad, completed in 1869 (Figure 1.2). Both projects required the expenditure of

FIGURE 1.1
U.S.S. Arizona in Panama Canal locks, 1921.

FIGURE 1.2
Completion of the Transcontinental Railroad, 1869.

vast sums of money and required the employment of thousands of workers. In more recent times, the "*Big Dig*" in Boston,[1] which replaced an ugly elevated highway with a tunnel system, cost over $14 billion and took 10 years to complete.

Transportation costs are borne by the traveler when crashes or disasters occur. While these events tend to be infrequent, when they do happen, they serve as a reminder of the risks involved. Every mode of transportation brings to mind a major disaster. Examples are the 1912 sinking of the *Titanic* (Figure 1.3), which took 1500 lives—a topic of fascination to this day—or the explosion of the zeppelin *Hindenburg,* which burst into flames while docking after a transatlantic flight from Germany to Lakehurst, New Jersey, in 1936. Modern-day air crashes, though infrequent, are dramatic and catastrophic when they occur, such as United Flight 718 and TWA Flight 2 from Los Angeles that collided over the Grand Canyon in 1956 killing 128 passengers and crew and the explosion and downing of TWA Flight 800 in 1996 while taking off from New York for Paris, in which 230 lives were lost. Air disasters are followed by a thorough investigation by the National Transportation Safety Board of the U.S. Department of Transportation to ascertain the cause and to learn how such

1. The Big Dig is the unofficial name of the Central Artery/Tunnel Project (CA/T), a massive undertaking to route the Central Artery (Interstate 93), the chief controlled-access highway through the heart of Boston, Massachusetts, into a tunnel under the city, replacing a previous elevated roadway. The project also included the construction of the Ted Williams Tunnel (extending Interstate 90 to Logan International Airport) and the Zakim Bunker Hill Bridge over the Charles River.

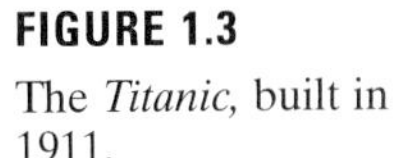

FIGURE 1.3
The *Titanic,* built in 1911.

tragedies can be prevented. Highway crashes are also a significant cost, and in the United States result in the loss of over 40,000 lives each year.

Environmental effects of transportation include noise, air and water pollution, long-term climate effects from carbon monoxide and other pollutants of the internal combustion engine, disturbance of wetlands, desecration of natural beauty, and disruption of natural habitats. These impacts are far reaching and have stimulated environmental legislation to help mitigate the potential damages.

The impact of transportation on society can be illustrated in the following statements:

- Approximately 17.5% of the U.S. Gross Domestic Product (GDP) is accounted for by expenses related to transportation.
- Almost 100% of the energy utilized for propelling transport vehicles is derived from petroleum resources.
- Over 50% of all petroleum products consumed in the United States are for transportation purposes.
- Over 80% of eligible drivers are licensed to operate a motor vehicle.
- Each person in the United States travels an average of 19,300 km each year.
- Over 10% of the U.S. work force is employed in a transportation-related activity.
- In the United States there are over 6 million kilometers of paved roadway, of which about 1.2 million kilometers are used for intercity travel.
- There are approximately 177,000 km of railroads, 10,000 airports, 42,000 km of inland waterways, and 343,000 km of pipelines.

Land use, which is the arrangement of activities in space, is closely interrelated with transportation, since travel takes place from one land use type to another (e.g., from a residence to a work site or from a factory to a warehouse). Several transport options that have been dominant in the past illustrate relationships between land use and transportation and how they have changed over time.

When walking and horsepower was the transport mode, land uses were located in close proximity to each other and walls surrounded many cities. When railroads and rail mass transit dominated, land use patterns assumed a starlike pattern. The central city, with its commercial and industrial activity, was the focal point and residences were located along the radial pathways. Highly concentrated and dense land use patterns emerged in cities such as New York, Philadelphia, Boston, and Chicago since rail transit was readily available in these locations. When the automobile was introduced, land use patterns could be less dense and more diffuse because roads could be built almost anywhere. Suburbs were possible, and with the construction of the National System of Interstate Highways, commercial developments were no longer confined to central city locations. Today, the typical land use pattern is spread out, low density, and homogeneous.

Similarly, cities that were once confined to locations along seacoasts, lakes, rivers, and rail terminals can be located almost anywhere in the country. New forms of transportation such as air and interstate trucking created ubiquitous accessibility and allowed the creation of cities in locations that formerly were not feasible.

In the United States, city and county government or private citizens are responsible for land use decisions at the local level. Decisions to invest in transportation facilities are usually the responsibility of state and federal government and large corporations. Consequently, the failure to coordinate land use and transportation planning often results in inefficiencies in the allocation of resources for both land use and transportation.

CAREER OPPORTUNITIES IN TRANSPORTATION

The four major modes of transportation are air, water, rail, and highway. Each mode has an established market, and the modes compete as well as cooperate with each other. The world has seen profound changes in travel time in the past centuries. In the early nineteenth century, a trip of 500 kilometers took 12 days by stagecoach. With improved transportation technology, travel times were successively reduced to seven days by canal and boat, then eight hours by rail, five hours by automobile, and 50 minutes by air. In the twenty-first century, transportation professionals will face new challenges, including development of new technologies, communications, the search for energy options to replace fossil fuels, and complex issues of environment, financing, and deregulation. Thus, the professional opportunities that will exist in the field of transportation in the twenty-first century are very promising.

The management aspects of freight transportation, known as *business logistics* or operations research, are concerned with the movement and storage of goods between the primary source of raw materials and the location of the finished product. This area of professional specialization, which is considered an element of business administration, has grown in importance as shippers and carriers seek to minimize their transportation costs by utilizing combinations of modes and services that provide the optimal mix of travel attributes including time, cost, reliability, frequency, and security. Typically, logistic managers are trained in a business environment but may also come from academic programs in transportation systems and operations.

A large segment of the transportation industry deals with *vehicle design and manufacture,* including aircraft, cars and trucks, diesel locomotives, transit buses and rail cars, ships, and pipelines. This industry segment is specialized, and several large U.S. companies—such as Boeing Aircraft, General Motors, and Westinghouse—play leading roles. Many other nations, such as France, Japan, Germany, Italy, Great Britain, Sweden and Canada (to name a few), also manufacture transport vehicles. Vehicle design and manufacture involves the application of mechanical systems, electrical systems, and computer engineering skill. It also requires the employment of technically trained mechanics and production workers in other trades. Transportation employs many workers in *service industries*. For passenger modes, jobs include flight attendants, railroad conductors, ship stewards, travel agents, skycaps, maintenance technicians, and ticket agents. In freight modes, jobs include customs agents, truck drivers, railroad yard workers, sailors, stevedores, and security guards. The maintenance and servicing of a vast fleet of vehicles requires a skilled, technical work force to service everything from a personal automobile to a Boeing 747 jet airliner. Fueling for millions of motor vehicles, as well as for airplanes, ships, and railroads, requires a network of storage and distribution facilities and personnel to operate them.

The *transportation infrastructure industry* is also a major source of employment for professionals and deals with all aspects of infrastructure development. Professionals who work in this area are employed by government agencies, consulting firms, construction companies, transportation authorities, and private companies. Professionals who work on transportation issues include engineers, attorneys, economists, social scientists, urban planners, and environmentalists. Among their tasks are to draft legislation, facilitate right-of-way acquisition, monitor the effects of transportation on the economy, prepare environmental impact statements, develop marketing strategies, and develop land use plans and forecasts.

Transportation engineering is the profession responsible for the planning, design, construction, operation, and maintenance of transportation infrastructure. The field includes highways, airports, runways, railroad stations and tracks, bridges and waterways, drainage facilities, ports and harbors, and rail or bus transit systems. Employment opportunities exist in these areas with federal transportation agencies, state government, special transportation authorities, consulting firms, railroad or airline companies, private industry, and professional associations. While this employment sector has been linked to civil engineering, transportation professionals often have academic training in other engineering disciplines, such as mechanical, electrical, aerospace, and information technology. In addition to a basic understanding of transportation principles, the transportation engineer must be broadly educated with knowledge about engineering fundamentals, science, statistics, oral and written communications, computers, economics, history, and social sciences. Typically, the modern transportation engineer has completed a bachelor's degree in engineering and a master's or doctorate in a transportation specialty, as described in the following sections.

Transportation planning involves the process of developing plans and programs that improve present travel conditions. Planners ask questions such as: Should an existing airport be expanded or should a new one be built? Should a freeway be widened? Should a rail transit line be constructed? The process involves defining the problem, setting goals and objectives, collecting travel and facilities data, forecasting future traffic demand, and evaluating options. The planner may also be required to assess the environmental impacts, the effect of the project on land use, and the benefits of the project compared with the cost. Physical feasibility and sources of funding are also considered. The final product is a comparison of various alternatives based on established objectives and criteria and an analysis of how each option will accomplish the desired goals and objectives. A plan is then recommended for consideration and comment by decision makers and the public.

Transportation design involves the specification of features that comprise the transportation facility such that it will function efficiently and in accord with appropriate criteria and mathematical relationships. The final design provides a blueprint for use by the owner and the contractor as it establishes the detailed specifications for the project. The design process involves the selection of dimensions for geometric features of alignment and grade as well as structural elements for bridges and pavements. For runways or highways the pavement thickness is determined. If bridges or drainage structures are required (for example, at a railroad grade crossing or in retrofitting tunnel clearances to accommodate double-stacked railroad cars), a structural design is performed. Provision for drainage facilities, including open channels, culverts, and subsurface elements, is included in the design. Traffic control devices are also specified (for example at railroad grade crossings and within marine terminals).

Traffic control centers for air, rail, or highway systems will require facilities for monitoring and modifying traffic patterns as conditions warrant. Design engineers must be proficient in subjects such as soil mechanics and foundations, hydraulics, land surveying, pavement structure, and geometric design. The design process results in a set of detailed plans that can be used for estimating the facility cost and for carrying out the construction.

Transportation construction involves all aspects of the building process. Typically a construction firm is selected because of its experience, availability of skilled construction workers, and competitive low bid. Some construction firms specialize in a particular aspect of transportation, such as highway construction, airports, seaports, and rail transit. For a very large project, several construction firms usually organize a consortia and subdivide the work into segments. Construction firms also specialize as subcontractors for tasks such as electrical, foundations, pilings, bridges, tunnel borings, framing, plumbing, and earth moving. The transportation engineer's role in construction is to represent the owner to assure the project is being built according to specifications, to approve partial payments, to inspect the work in progress, and to represent the owner in negotiations for changes in work or in disputes that may arise. The transportation engineer may also be employed by the contractor and in this capacity is responsible for estimating costs, managing day-to-day work, dealing with subcontractors, and representing the firm in negotiations with the contracting agency or firm.

Transportation operations and management involves the control of vehicles in real time to ensure that they are traveling in paths that are secure from interference with other vehicles or pedestrians. While each transportation mode has unique traffic control procedures, it is the responsibility of the transportation engineer to devise systems and procedures that will assure both safety and capacity. On highways, individual drivers are in control of vehicles and thus the traffic control system consists of signs, markings, and signals, which are intended to warn and direct motorists. The transportation engineer applies the latest technology to monitor traffic, provide information to motorists, and respond to traffic crashes. Air traffic control is a one-on-one process with a controller monitoring the location of every aircraft and providing directions regarding cruising altitude, speed, take-off, and landing. Rail systems are controlled from a traffic center and by railway signals that automatically assign right-of-way and dictate speed. The locomotive driver may operate under visual or radio control. In each instance the transportation engineer is responsible for developing a control system that is consistent with providing the highest level of safety and service.

Transportation infrastructure maintenance involves the process of assuring that the nation's transportation system remains in excellent working condition. Often maintenance is neglected as a cost-saving tactic, and the result can be catastrophic. Maintenance is not politically attractive, as is new construction, yet the effects of deferred maintenance—if undetected—can result in tragedy and ultimately public investigations of the causes and blame for negligence. Maintenance involves routine replacement of parts, regularly scheduled service, repair of worn surfaces in pavements, and other actions necessary to maintain the vehicle or facility in a serviceable condition. Maintenance also involves data management for work activities and project scheduling as well as the analysis of maintenance activities to assure that they are carried out appropriately and economically. The transportation engineer is responsible for selecting maintenance strategies and schedules, forecasting maintenance cycles, managing risk, handling tort liability, evaluating the economic costs of maintenance programs, testing new products, and scheduling maintenance personnel and equipment.

TRANSPORTATION HISTORY

For thousands of years prior to the 1800s, the means by which people traveled was unchanged. By land, travel was on foot or in vehicles powered by animals. By sea, ships were powered by wind or by humans. Travel was slow, costly, and dangerous. As a result, nations remained relatively isolated and many societies grew, flourished, and declined without the knowledge of other people living elsewhere. In 1790, the year of the first federal census, 4 million people lived in the United States. Poor transportation services kept communities isolated. For example, in 1776, it took almost a month for the citizens in Charleston, South Carolina, to learn that the Declaration of Independence had been ratified in Philadelphia, a distance of less than 1200 kilometers.

At the dawn of the nineteenth century, new technologies were being introduced that had a profound influence on transportation. In 1769, James Watt, a Scottish engineer, patented a revolutionary steam engine design, and in 1807, Robert Fulton, a civil engineer, demonstrated the commercial feasibility of steamboat travel. Since then, books have been written commemorating the story of each of the transportation modes that followed Watt and Fulton. These books describe pioneers, inventors, and entrepreneurs with vision and courage to develop a new technology and thus change society.

Among the principal eras in transportation history are construction of toll roads to accommodate travel on foot and horseback; building of canals and steamships on rivers and inland waterways; expansion of the west made possible by the construction of railroads; development of mass transportation in cities; invention of the airplane and the resulting air transportation system of jet aircraft, airports, and air navigation; introduction of the automobile and highway construction; evolution of intermodal transportation by considering modes as an integrated system; and application of information technology.

In the nineteenth century, *early roads* were primitive and unpaved. Travel was by horseback or in animal-drawn vehicles. In 1808, the Secretary of the

Treasury, Albert Gallatin, who served under President Thomas Jefferson, prepared a report to Congress on the national need for transportation facilities. The report developed a national transportation plan involving roads and canals. Although the plan was not officially adopted, a strong case had been made for the federal government to invest in transportation. Gallatin's report provided impetus for the construction of the first National Road, also known as the Cumberland Road, which connected Cumberland, Maryland, with Vandalia, Illinois. As early as 1827, maintenance on the National Road had become a problem because the stone surface was wearing out and funds for maintenance were lacking.

Road building was not a high priority in the nineteenth century as most traffic was carried by water and later by railroad. The construction of roads and turnpikes was often paid for with private funds, and the roads were maintained by local citizens. Improvements in vehicle design, such as the Conestoga wagon—first built in the mid-1700s—carried most of the freight and people westward over the Alleghenies until about 1850 (Figure 1.4). These covered wagons, drawn by teams of four to six horses, were called the *camels of the prairies.* They were designed with broad-rimmed removable wheels to prevent bogging down in the mud and wagon bottoms were curved to stabilize freight from shifting.

Waterway transportation developed with the introduction of steamboat travel in the United States after the successful voyage of the *North River Steamboat* (also referred to as the *Clermont*) (Figure 1.5). For the first time in history, passengers traveled on the Hudson River from New York City to

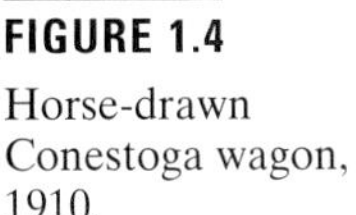

FIGURE 1.4
Horse-drawn Conestoga wagon, 1910.

FIGURE 1.5
Fulton's steamboat, *The Clermont,* 1807.

Albany in a ship not powered by sails. In subsequent years, steamboat transportation flourished on major rivers and lakes and provided passenger service to cities located on Long Island Sound, the Mississippi River, its tributaries, other rivers in the West, and the Great Lakes. To augment the river system, canals were constructed for the purpose of connecting rivers and lakes and opening the West. Water transportation played a key role in the location of cities. Settlements were most likely to take place at locations with access to harbors, rivers, lakes, and streams. Even today, most of the major cities in the United States and the world are located on waterways or on large lakes.

Canals were a dominant mode during the period 1800–1840, when approximately 6400 km of canal were built to connect various waterways in the northeastern portion of the United States. The system of inland waterways and canals served both freight and passengers and provided low-cost transportation between many previously inaccessible locations. One of the most prominent canal projects, the Erie Canal, was completed in 1825 and connected Albany, New York, with Lake Erie in Buffalo (Figure 1.6). This 581-km project spawned a new construction industry as well as the profession of civil engineering. Techniques that were developed in building this project were followed throughout the world in other canal projects, most notably the Suez Canal, which was completed in 1869, and the Panama Canal, which was begun

FIGURE 1.6
The Erie Canal, 1825.

by the French in 1882 and completed by the Americans in 1914. Canals were used to shorten travel distances of circuitous routes by water or horse-drawn wagon. However, travel times on canals themselves were limited by the speed of mules towing the boats or by delays at locks. It was not unusual for long lines to form or for fights to break out between boat crews as to which direction had priority to pass.

Railroad transportation slowly emerged as a new mode during the same period in which canals were being built. The use of rails as a road surface decreased friction forces and enabled horses to pull heavier loads than had been possible in the past. Horse-drawn streetcars were introduced in cities in 1832, and the Baltimore and Ohio (B&O) Railroad inaugurated service in 1830. The introduction of steam engines from England opened a new era of transportation, and the *iron horse* replaced horses as the source of motive power (Figure 1.7). Americans were slow to accept this new technology because they were committed to rivers and canals and the nation had a cheap power source from water. Steam-powered railroads were gradually introduced, first by the South Carolina Canal and Railroad Company in the late 1820s using a steam locomotive named *Best Friend of Charleston*. The B&O Railroad began experimenting with steam when it acquired the *Tom Thumb*. In a race with a horse and carriage in 1830, the *Tom Thumb* lost because a power belt slipped, a story that remains a transportation myth to this day.

By 1850, railroads had proven that they could provide superior service characteristics of time, cost, and reliability when compared to either rivers,

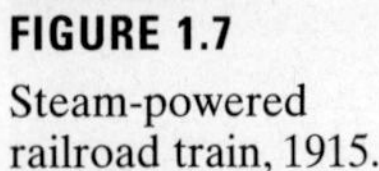

FIGURE 1.7

Steam-powered railroad train, 1915.

canals, or turnpikes. Consequently, funds to build roads for horse-drawn carriages or canals were no longer available and the nation embarked on a massive construction effort of track laying, bridge building, and station construction. In 1840 there were 6400 km of railroad track in the United States whereas in the year 1887 alone, 21,000 km of track were completed (Figure 1.8). The most epic project was the building of the transcontinental railroad, which was completed in 1869 with the driving of the golden spike at Promontory Point, Utah. By the start of the twentieth century, railroads had become the dominant transportation mode for both passengers and freight, with a vast network of rail lines that reached its peak of 416,000 km by 1915.

The United States was transformed by railroads, which opened up the West for settlement. New management tools were developed by railroad companies and adopted by other industries. In 1883, the railroad companies established the system of standard time zones that is still in use today. By the late nineteenth century, railroads had monopoly control of interstate freight commerce and used this power to extract unfair charges from customers, particularly farmers, who rebelled and lobbied Congress for help. As a result, the federal government, through the creation of the Interstate Commerce Commission (ICC) in 1887, began to regulate railroads. Today, these monopoly powers no longer exist and railroads have diminished in importance and no longer have monopoly powers over shippers. Accordingly, the Staggers Act of 1980 deregulated railroads and other transportation modes. In 1996, many of the functions of the ICC were discontinued.

FIGURE 1.8

Workers laying new railroad track, 1881.

The introduction of *containerization* occurred in 1956 when Malcolm McLean modified a tanker vessel to enable the transport of 58 containers. This innovation motivated the railroad industry to become one of the principal modes for the movement of freight. With the growth in container traffic and the construction of large container ports such as Long Beach–Los Angeles, California, railroads have become a vital link for international freight transport by moving goods between seaports and land destinations or as a land bridge serving to link the East and West coasts. The industry expanded its research and development activities in areas such as maintenance, operations, and safety.

In this century, *high-speed passenger rail* projects are being considered for service between heavily traveled city pairs in order to alleviate air traffic congestion on routes of less than 800 km, although passenger rail is no longer a dominant mode as it was in the beginning of the twentieth century.

Urban public transportation serves a different function than intercity modes. City transport is an integral part of urban infrastructure and impacts land use and the quality of urban life. The expansion of city boundaries could only occur with increases in travel speed. In addition to travel factors such as cost, time, and convenience, urban transport modes that are quiet and nonpolluting are preferred. Thus, it is easy to understand why horse-drawn streetcars were replaced by cable cars in the 1870s and later by electrically powered vehicles, which were introduced in the 1880s. In addition to increased speeds and lower costs, the reduction in animal pollution on city streets (with its odor and potential for causing illness and death) was considered a major improvement to the quality of life.

The introduction of the electric *streetcar or trolley* was a revolutionary breakthrough in urban transportation and influenced urban development into the twentieth century. Frank Sprague, who had worked with Thomas Edison in his laboratory in Menlo Park, New Jersey, is credited with creating this new transport mode (Figure 1.9). In 1884, he formed the Sprague Electric Railway and Motor Company and in 1888 electrified a 19-km horsecar line in Richmond, Virginia. Sprague did not invent the electric street railway but was the first to successfully assemble the elements needed for the system to function. These included an overhead wire to collect electricity, an improved control system to facilitate car operation, and a vibration-free suspension system for the motors.

The streetcar proved to be popular and reached a peak of 17.2 billion passengers annually by 1926. Cities everywhere built trolley lines, and by 1916, there were 72,000 km of trolley lines in operation. Cities developed a starlike land use pattern on which lines fanned out from the city center and connected

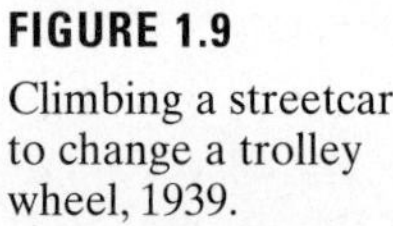

FIGURE 1.9
Climbing a streetcar to change a trolley wheel, 1939.

residential communities and amusement parks located along and at the end of the lines.

The *bus* gradually replaced streetcars as streetcar ridership steadily declined in the 1920s. By 1922 buses carried only about 400 million passengers per year compared with 13.5 billion annual streetcar passengers, but by 1929 bus riders had increased dramatically to 2.6 billion passengers annually. The streetcar industry made a major effort to reverse the declining trend by developing a new systems-engineered vehicle called the President's Conference Committee (PCC) streetcar. Nevertheless, the decline continued and many cities abandoned streetcar service. The first city to do so was San Antonio, Texas, in 1933. After World War II, major cities such as New York, Detroit, Kansas City, and Chicago converted to bus lines. Ironically, today many of these same cities and others, such as Portland, San Jose, and San Diego, have added new streetcar lines, which are now called *light rail.*

The rapid changeover from streetcar to bus created a controversy called the transportation conspiracy. Critics charged that General Motors, the dominant manufacturer of buses in the 1930s, arranged to purchase streetcar companies and then had them replaced with buses. These charges may have some basis in fact, but in reality buses were more economic and flexible than streetcars. Motorists saw trolleys as an impediment to fast and safe driving. Furthermore, the evidence is clear that automobile travel would continue to grow with the inevitable result that streetcar use would decline.

At the end of the 19th century, *rail transit systems* were constructed that were either elevated or in tunnels. Large urban areas required greater capacity and speed than could be provided by street railways or bus lines. The first rapid transit line was opened in London in 1863. By the beginning of the 20th century, rapid transit lines were being constructed in large U.S. cities such as New York, Chicago, Philadelphia, Cleveland, and Boston. After a period of about 50 years in which no new rapid transit systems were constructed, a renewed interest in rail transit occurred. During the 1970s and 1980s, rail transit systems were again constructed in cities such as San Francisco, Washington D.C., Baltimore, and Atlanta, and construction of light rail lines occurred in a large number of cities in the United States.

Air transportation is considered to date from the historic flight of the Wright Brothers on December 17, 1903, when Wilbur and Orville, two bicycle makers from Dayton, Ohio, demonstrated that a self-propelled heavier-than-air machine could be made to fly. The journey of 37 m over the sands of Kitty Hawk, North Carolina, launched a new mode of transportation that would completely change the way people traveled. It was only 24 years later, in 1927, that a young pilot—Charles Lindbergh—would captivate the nation by his solo flight from New York to Paris in 33.5 hours, a distance of more than 5760 km. A nonstop crossing between Tokyo and the West Coast, a distance of 7813 km, was successfully completed in 1933.

These events heralded the beginning of a new age in air travel recognized for its military significance and as a carrier of domestic and international passengers. Prior to World War I (1914–1918), air transportation was in a pioneering stage, with barnstorming pilots demonstrating the new iron bird, while aircraft design and development, particularly in Europe, was making great strides. In World War I, airplanes were used for both combat and reconnaissance, and the postwar period demonstrated the capabilities of air service for delivering mail and passengers. The airline industry was aided by the federal government in the 1920s with contracts to transport mail by air. New airline companies were formed, such as Pan American World Airways (Pan Am) in 1927 and Trans World Airways (TWA) in 1930, and provided international and intercontinental passenger services. During World War II (1939–1945), air power was used extensively and became a key strategic weapon for Germany, Japan, and the United States.

By 1940, propeller aircraft had reached a peak of performance, and although propeller aircraft were still in service in the 1950s, the development of the first jet engine by a British designer, Frank Whittle, in 1938, ushered in a new era in air transportation. The Boeing Aircraft Company delivered the first American-built commercial jet to Pan Am in 1958, and air speeds were increased from 576 to 912 kmph. The first jet flight from New York to Miami took less than three hours, and flight times coast to coast were reduced to less than six hours. This dramatic improvement in transportation service had a profound influence on international travel. Passengers began to shift to the new Boeing 747s introduced in 1970, and this accelerated the decline of intercity passenger travel by rail. Freight traffic by air did not become a major competitor to shipping and railroads and represented a minor fraction of ton-miles carried. However, in terms of the percentage of dollar-kilometers, airborne freight is significant since the goods carried are high-value commodities. Companies like Federal Express and United Parcel Service (UPS), which deliver packages within one day to destinations all over the world, are examples of the importance of air transportation in moving freight.

Highway transportation, the invention of the *automobile*, and the development of mass production techniques created a revolution in transportation in the United States during the twentieth century and a challenge to harness intelligent technologies into the twenty-first century. In 1895, only four automobiles were produced, and this new invention was seen as a toy for the very wealthy (Figure 1.10). In 1903, Henry Ford founded the Ford Motor Company and perfected a process to mass produce automobiles that could be purchased at a price most Americans could afford. In 1901, there were only 8000 registered automobiles in the United States, but by 1910 the number had increased to 450,000. In 1920, more people traveled by private automobile than by railroad and, by 1930, 23 million passenger cars and 3 million trucks were registered.

At the beginning of the twentieth century, *highways* were not capable of servicing the explosive growth in motor vehicle travel. Roads were in such poor

FIGURE 1.10
Packard limousine, 1912.

condition at the beginning of the twentieth century that for many years the League of American Wheelmen, a bicycle federation formed in 1894, had been lobbying Congress and states for better roads. Even the railroad industry promoted road building with its Good Roads trains, which traveled around the country demonstrating the advantages of hard-surfaced roads. Railroad executives believed that roads should be built so farm products could be transported more easily to train stations.

In 1893, the federal government established the U.S. Office of Road Inquiry (with an authorized budget of $10,000) within the Department of Agriculture to investigate and disseminate information about roads. By 1916, the first Federal Aid Road Act was passed and provided federal support for roads and gave states the authority to initiate projects and administer highway construction through their Department of Highways. Thus began a long-term partnership between the states and the federal government to organize, design, and build the nation's highway system.

In 1956, Congress authorized the construction of a 67,200-km system of Interstate and Defense Highways. The idea of a limited access highway network had been developed prior to World War II, and studies conducted during the administration of President Franklin D. Roosevelt (1932–1945) concluded that these roads should not be financed by tolls. It was envisioned that the new roadway system would connect major cities from the Atlantic to the Pacific Ocean and between Mexico and Canada. The Interstate was promoted as a solution to highway congestion since it would be possible, proponents argued, to drive from New York to California without ever stopping at a traffic light. The Interstate was also expected to serve defense needs. Colonel Dwight Eisenhower, who had completed a national highway tour prior to World War II, believed in the military value of a national system of high-quality roads. President Eisenhower signed the legislation authorizing the Interstate System on June 29, 1956, unleashing a massive construction program that ended in the mid-1990s.

The Interstate Highway System has had a profound impact on both passenger and freight transportation in the twentieth century. Bus transit has replaced rail in all but the largest cities, and trucks, which carried less than 1% of ton-km in 1920, now haul almost 25% of ton-km and 75% of freight revenues.

SUMMARY

The explosion of invention, innovation, and construction that occurred during the past 200 years has created a transportation system in the United States that is highly developed. Today there is a complex array of modes, facilities, and service options that provide shippers and the traveling public a wide range of choices for moving goods and passengers. Each mode offers a unique set of service characteristics in terms of travel time, frequency, comfort, reliability, convenience, and safety. The term *level of service* is used to define the user perception of these attributes. The traveler or shipper compares the relative level of service offered by each mode with the trip cost and makes tradeoffs among attributes in selecting a mode. In addition, a shipper or traveler can choose a public carrier or can use personal resources. For example, a manufacturer can decide to hire a trucking firm to move goods or to use company-owned trucks. Similarly, a homeowner can decide to hire a moving company to assist in relocation or rent a truck and utilize friends and family to do the work. The personal auto presents a choice to the commuter or vacation traveler, to drive a car or to travel by bus, rail, or air. Each of these decisions is complex and involves weighing level-of-service factors to reflect personal preferences.

PROBLEMS

1.1 What are the purposes of a transportation system in a region or a nation?

1.2 If you were asked to define *transportation,* what would you say? Provide three examples to illustrate your definition.

1.3 How does the quality or level of service of a transportation system affect the competitive advantage of one geographic area over another (such as a city, state, or nation)? To what extent is a good transportation system sufficient to assure that the economic potential of a region will be minimized?

1.4 What is the characteristic of nations that have good national and international transportation systems? Name three nations with good transportation systems.

1.5 In addition to providing economic benefits to society, list five examples of other advantages provided by the availability of good transportation.

1.6 Explain the statement that "modern transportation is necessary but not sufficient to assure that a region or nation will be prosperous."

1.7 While it is true that good transportation provides enormous benefits to society, there is a price to be paid. What are the direct and indirect costs of transportation?

1.8 List three major U.S. transportation projects that have been completed in the past 150 years.

1.9 Every mode of transportation has experienced a major disaster that cost many lives and property. Use Internet resources to provide one example of such a disaster for air, ships, railroads, and highways.

1.10 List six environmental effects of transportation.

1.11 Provide five examples to convince someone of the importance of transportation in U.S. society, policy, and daily life.

1.12 Does transportation affect land use patterns? Buttress your answer with examples of the influence of walking/animal power, rail transit, water, highways, and air.

1.13 Advances in transportation technology and service can be measured by improvements in travel time between cities. Consider a trip of 450-km between two city pairs. Contrast the travel time by stagecoach, waterways, rail, and automobile. Since these data are provided in the text, do a similar analysis for a 750-km journey between two cities in your state.

1.14 Define the four professional areas in transportation in which employment opportunities exist: business logistics, vehicle design and manufacture, service sector, and infrastructure engineering.

1.15 Define *transportation infrastructure engineering*. Describe the five elements of this professional field.

1.16 Describe the contribution that each of the following individuals made to improved transportation in the United States: Dwight Eisenhower, Henry Ford, Robert Fulton, Albert Gallatin, Charles Lindbergh, Frank Sprague, Harley Staggers, James Watt, Frank Whittle, and Wilbur and Orville Wright.

1.17 What were the dominant transportation modes in the nineteenth and twentieth centuries? What do you think will be the dominant mode in the twenty-first century?

1.18 When was containerization introduced? How did this development alter the movement of freight worldwide?

1.19 In what U.S. cities can you ride on a rail rapid transit system? Which of these systems were constructed in the second half of the twentieth century?

1.20 What is meant by *level of service,* and how does this concept influence the likelihood of new transportation modes being developed in the future?

References

1. *America's Highways: 1776–1976,* U.S. Department of Transportation, Federal Highway Administration, Washington, D.C., 1976
2. Cavendish, Marshall, *The Encyclopedia of Transport* (undated), ISBN 0 85685 1760.
3. Coyle, J. J., Bardi, E. J., and Novack, R. A., *Transportation,* 6th edition, Thompson-Southwestern, Mason, OH, 2006.
4. Davidson, J. F., and Sweeney, M. S., *On the Move: Transportation and the American Story,* National Geographic Society and Smithsonian Institution, 2003.
5. Eno Transportation Foundation, *Transportation in America,* 19th edition, 2002.
6. Eno Transportation Foundation, *National Transportation Organizations,* 2005.
7. *The Interstate Achievement: Getting There and Beyond,* Transportation Research Board of the National Academies, TR News, May–June 2006.
8. Lambert, M., and Insley, J., *Communications and Transport,* Orbis Publishing Limited, London, 1986.
9. Rogers, Taylor G., *The Transportation Revolution, 1815–1860,* Harper Torchbooks, Harper & Row Publishers, New York, 1968.
10. *Transportation History and TRB's 75th Anniversary,* Transportation Research Board, Transportation Research Circular 461, August 1996.
11. U.S. Department of Transportation, *Moving America: New Directions, New Opportunities,* Washington, D.C., 1990.

CHAPTER 2

Transportation Systems Models

This chapter describes the fundamental principles and characteristics of transportation systems and their components and introduces a number of basic analysis tools and models that can be used to address transportation systems problems. The analysis tools and models discussed include (1) fundamental traffic analysis tools; (2) regression techniques; (3) basics of probability theory; (4) queuing theory; and (5) optimization tools. The description of each tool is accompanied by example problems that illustrate how the tool is used in solving transportation systems problems.

SYSTEMS AND THEIR CHARACTERISTICS

A system is defined as a set of interrelated components that perform several functions in order to achieve a common goal and is therefore an entity that maintains its existence and functions as a *whole* through the interaction of its parts. The behavior of different systems depends on how the parts are related rather than on the parts themselves. Systems have several basic characteristics. First, for a system to function properly, all its components must be present and arranged in a specific way. Given this, systems have properties above and beyond the components of which they consist. In addition, when one element in a system is changed, there may be side effects. For example, improving public transportation in a given city may help reduce the number of vehicles on the nearby network, as more people decide to use public transportation instead of driving. Widening a road may relieve congestion for a while but in the long run may result in attracting new drivers and new trips to that road and, in some cases, may even make conditions worse. Second, systems tend to have specific purposes within the larger system in which they are embedded, and this is what gives a system the integrity that holds it together. For transportation systems, the obvious goal is to move people and goods efficiently and safely. Third, systems have *feedback*, which allows for the transmission and return of information. Feedback is crucial to systems operation and to systems thinking. For transportation systems, there is a feedback relation between transportation and land use systems. The land use system drives travel demand since travel demand will depend upon the spatial distribution of the different land use activities (i.e., where people live, work, shop, etc.). Conversely, the transportation system affects the land use pattern since building new roads, transit lines, and airports will often attract development.

Components of Transportation Systems

A transportation system consists of three components: (1) physical elements, (2) human resources, and (3) operating rules.

Physical Elements

Physical elements include (1) infrastructure; (2) vehicles; (3) equipment; and (4) control, communications, and location systems.

Infrastructure refers to the fixed parts of a transportation system (i.e., parts that are static and do not move). These include travelways, terminals, and stations. *Travelways* vary depending upon the transportation vehicle or mode. For example, highways and roads are travelways for automobiles and trucks. Rail transportation requires railroad tracks, and air transportation utilizes specified air corridors. *Terminals* are required for buses, railroads, aircraft, trucks, and ships. Terminals serve dispatching and storage functions by regulating the arrival and departure of vehicles and for storing vehicles and cargo. They represent points where users can enter or leave the system, and they serve as intermodal interchange points for changing from one mode to another. *Stations* serve only a subset of the functions served by terminals. They are primarily points of system exit or entry. Examples include bus, subway, and railway stations. A parking garage or a regional airport also serves as a station.

Vehicles are the elements of a transportation system that move along the travelway. They include automobiles, buses, locomotives, railroad cars, ships, and airplanes. Most vehicles are self-propelled (e.g., automobiles, locomotives, ships, and aircraft) and some are without propulsion (e.g., railroad cars, barges, and truck trailers).

Equipment refers to physical components whose main function is to facilitate the transportation process. Examples include snowplows, railroad track maintenance vehicles, and baggage-handling conveyor belts at airports.

Control involves the elements required to allocate right-of-way. Allocating right-of-way requires air traffic control centers, traffic signals, and travelway signs.

Communications systems link traffic control centers to travelway equipment such as variable message signs, traffic signals, transit vehicles, air traffic controllers, and pilots. *Location systems* identify individual vehicles in real time, using global positioning systems (GPSs) to track vehicles such as transit vehicles, trucks, and emergency vehicles, thus increasing routing efficiency.

Human Resources

Human resources, essential to the operation of transportation systems, include vehicle operators such as automobile truck and bus drivers, railroad engineers, airline pilots, maintenance and construction workers, transportation managers, and professionals who use knowledge and information to advance the transportation enterprise. Among transportation managers are strategic planners, marketing and maintenance management personnel, operations research and information systems analysts, and administrators.

Operating Rules

Operating rules include schedules, crew assignment, connection patterns, cost/level of service tradeoff, and contingency plans.

Schedules define the arrival and departure times of transportation vehicles at the different transportation terminals and stations. In addition, schedule adherence plays an important role in determining the quality of service of a given transportation mode.

Crew assignment involves assigning operators to the different vehicles (e.g., assigning bus drivers to the different buses in a transit agency's fleet, assigning pilots and flight attendants to flights, etc.). It is a challenging task, since a number of constraints need to be satisfied by each assignment. These include the maximum number of continuous hours a person is allowed to work, the need to match vehicle operators with the type of vehicle they are certified to operate, and the need to minimize costs.

Connection patterns refer to how service is organized over the transportation system or network. An example is the "hub-and-spoke" system (Figure 2.1), in which people and cargo are flown from several cities to a hub area, at which point trips are consolidated according to the final destination.

The hub-and-spoke system introduces a number of challenging operational issues. These include the need to consider the time required for transfer from one vehicle to the other, the need for strict schedule adherence, and the sensitivity of the system to external disturbances in the form of incidents or inclement weather.

Cost/level of service tradeoff involves setting operational rules for transportation systems, and doing so involves a tradeoff between cost and level of

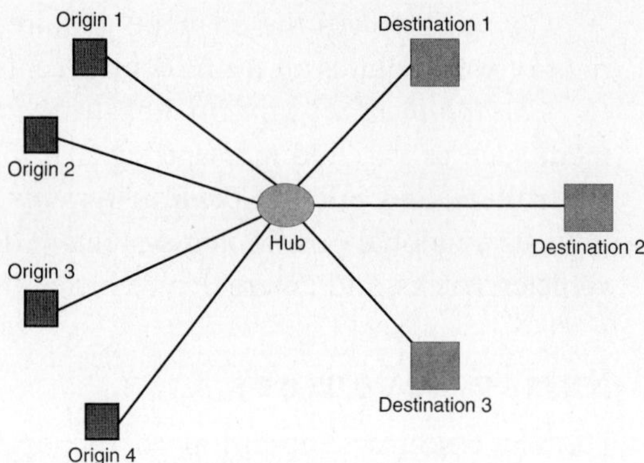

FIGURE 2.1
Hub-and-spoke system.

service. For example, for a transit agency, operating more buses along a route would mean a higher level of service for passengers but a higher operating cost for the agency. For a state highway department, constructing an eight-lane highway versus a four-lane highway would translate into a higher level of service for drivers but a higher cost for the agency and society. For an airline, providing direct service between two cities (as opposed to a connection at an intermediate hub) would mean a higher level of service for travelers but a higher cost for the airline, especially if the demand between the two cities is not large enough to warrant a direct service. The tradeoff between cost and quality of service is a fundamental concept in the operation of transportation systems.

Contingency plans are implemented when something goes wrong with the transportation system. For example, a contingency plan needs to be in place for traffic diversion when a main highway is closed because of an accident or construction, for the evacuation of coastal areas during a hurricane, and for handling surges in traffic demand (such as during special events). Building a good contingency plan often calls for the allocation of additional resources and is another example of the cost/level of service tradeoff.

Tools and Techniques for Analyzing Transportation Systems

The remainder of this chapter is devoted to an introduction of five basic tools and techniques that are widely used for the analysis of transportation systems. These are traffic operations analysis tools, regression analysis, probability, queuing theory, and optimization.

Traffic Operations Analysis Tools

This section describes two traffic operations tools: time–space diagrams and cumulative plots. *Time–space diagrams* are used in cases where many vehicles interact while sharing a common travelway while *cumulative plots* deal with problems involving traffic flow through one or more restrictions along a travelway.

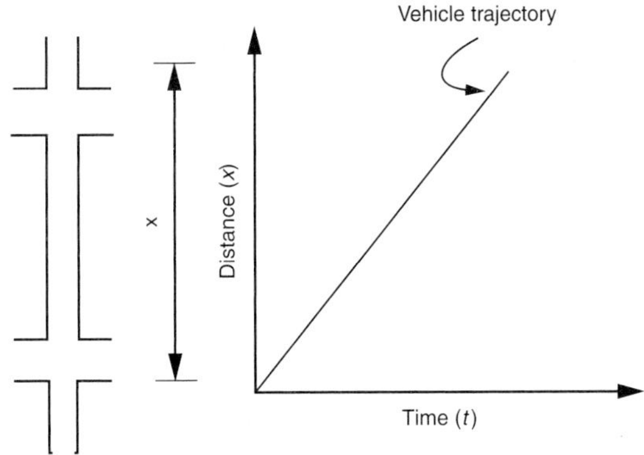

FIGURE 2.2
Time–space diagram.

Time–space diagrams are a simple but effective traffic analysis tool that tracks the position of a single vehicle over time along a one-dimensional travelway. Time–space diagrams may be used to track the position of a vehicle on a freeway, an aircraft on a runway, or a bus on a transit route. Figure 2.2 illustrates an example of a time–space diagram: The vertical axis is the distance (x) along a travelway, and the horizontal axis is the time (t) to traverse that distance. The *trajectory* of one vehicle is a graphical representation of the position of the vehicle (x) as a function of the time (t). Mathematically the trajectory can be represented by a function $x(t)$.

The time–space diagram can also be used to provide a complete summary of the vehicular motion in one dimension and provides information regarding the acceleration and/or deceleration patterns. Because velocity at any time t is given by the slope of the vehicle trajectory, it can be expressed as $u = \frac{dx}{dt}$, which is the first derivative of the function $x(t)$ at time (t).

EXAMPLE 2.1

Describing Vehicle Motion Using a Time–Space Diagram

Figure 2.3 is a time–space trajectory for three vehicles, labeled 1, 2, and 3. Describe the vehicular motion of each vehicle.

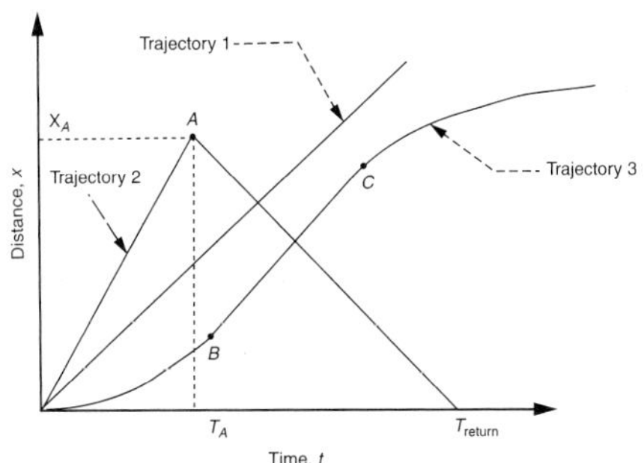

FIGURE 2.3
Time–space trajectories for Example 2.1.

Solution

Trajectory 1 depicts a vehicle moving at a constant speed, since the trajectory is a straight line with a constant slope. Also note that vehicle 1 is traveling forward in one direction.

Trajectory 2 depicts vehicle 2 traveling forward at a constant speed up to point *A*, where it has traveled a distance (X_A) in time (T_A). At point *A*, the vehicle reverses direction, still traveling at a constant speed but more slowly than when going forward since the trajectory slope of the return trip is less than that of the forward trip. At time (T_{return}) the vehicle is back at the starting point.

Trajectory 3 depicts the motion of vehicle 3 as being forward in one direction, but with the velocity varying over time. For the first part of the trip up to point *B*, vehicle 3 is accelerating as indicated by the increase in the slope (velocity) of the trajectory over time. Between points *B* and *C*, the speed is constant. Finally, beyond point *C*, the vehicle decelerates to a stop.

Applications of Time–Space Diagrams

Time–space diagrams are used to analyze situations where vehicles interact with each other while moving on the same travelway. Examples include airplanes with different glide speeds sharing the same runway and subject to minimum separation requirements; scheduling of freight and passenger trains along a single track; and estimating of safe passing sight distances on two-lane roads. In most cases, the analysis can be completed without a time–space diagram. However, as the following examples illustrate, the use of the diagram helps to identify and correct mistakes in the formulation of the problem.

EXAMPLE 2.2

Siding Locations for Passenger and Freight Rail on the Same Track

A freight and passenger train share the same track. The average speed of the freight train is 65 km/h and 130 km/h for the passenger train. The passenger train is scheduled to leave 30 minutes after the freight train departs from the same station. Determine

1. The location of the siding where the freight train will wait so the passenger train can proceed uninterrupted. As a safety precaution, the separation headway of the two trains at the siding should be at least 6 minutes.
2. The time it takes for the freight train to arrive at the siding.

Solution

This problem is solved using the time–space diagram shown in Figure 2.4.

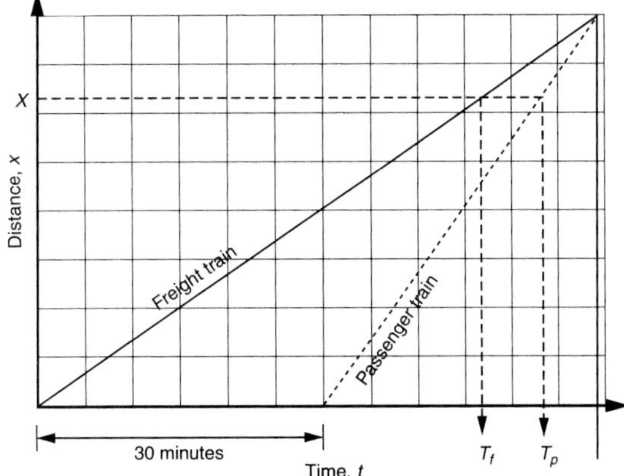

FIGURE 2.4
Time–space diagram for Example 2.2.

Part (1) Figure 2.4 shows the trajectories for the freight and passenger trains. The slope of each trajectory is equal to the average speed for each train (i.e., 65 km/h for the freight train and 130 km/h for the passenger train). The figure also shows that the passenger train departs 30 minutes after the freight train. According to the requirements of the problem, the time headway between the two trains at the siding should be at least 6 minutes.

With reference to Figure 2.4, use X to refer to the location of the siding along the track measured from the location where the trains depart. Also use T_f and T_p to denote the time at which the freight train and the passenger train reach the siding location, respectively. The difference between T_f and T_p equals 6 minutes. Since the speeds of the two trains are given, T_f and T_p can be expressed as follows:

$$T_f = \frac{X}{65}\ \text{h}$$

$$T_p = 0.5 + \frac{X}{130}\ \text{h}$$

The difference between T_p and T_f should be equal to 6 minutes (i.e., 0.10 hr). Therefore,

$$T_p - T_f = 0.10$$

$$0.5 + \frac{X}{130} - \frac{X}{65} = 0.10$$

$$\frac{X}{130} = 0.4$$

$$X = 52 \text{ kilometers}$$

The first siding should be located 52 kilometers from the first station.

Part (2) The time for the freight train to reach the siding is T_f. Therefore,

$$T_f = \frac{X}{65} = \frac{52}{65} = 0.8 \text{ hr} = 48.0 \text{ minutes (answer)}$$

EXAMPLE 2.3

Computing the Average Speed for a Multimodal Journey

A group of three friends (*A*, *B*, and *C*) take a long trip using a tandem bicycle for two persons. Since the bike cannot accommodate a third person, the friends take turns walking. When riding, the average speed is 24 km/h, and when walking the average speed is 6 km/h.

For the following journey scenario, determine the average speed of the group:

- To start the journey, two friends *A* and *B* ride a bicycle, and the third friend, *C*, walks;
- After a while, *B* dismounts from the bike and begins walking, while *A* rides the bicycle alone in the reverse direction to pick up *C*;
- When *A* and *C* meet, they turn the bicycle around and ride forward until they catch up with *B*. When they do, this portion of the trip is completed.

Solution

Solving this problem without the aid of a time–space diagram could be very challenging. Thus, begin by developing a time–space diagram to represent the way the three friends complete this portion of their trip (see Figure 2.5).

Draw a line whose slope is equal to 24 km/h to represent the trajectory of *A* and *B* riding the bicycle. At the same time, *C* is walking and represented by a trajectory whose slope is equal to 6 km/h. Assume that *A* and *B* ride together for time period *X*1.

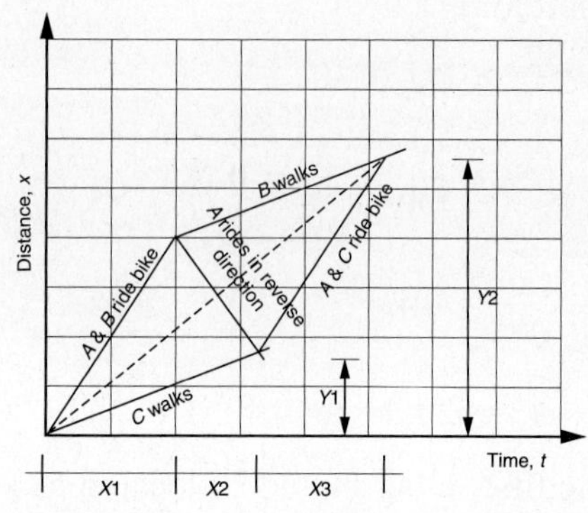

FIGURE 2.5
Time–space diagram for Example 2.3.

After time $X1$, B gets off the bike and starts to walk as represented by a line whose slope is equal to 6 km/h to represent the trajectory "B walks." A then rides the bicycle alone in the reverse direction at a speed of 24 km/h, as shown in the figure, while C is still walking at 6 km/h. A meets C after traveling for a time interval $X2$. The distance, measured from the beginning point of the trip to the point where A and C meet, is denoted by $Y1$.

After A and C meet, they ride together, as depicted by a trajectory whose slope is 24 km/h. Finally, A and C riding together meet B, who is walking. This occurs at a distance $Y2$ from the starting point (see Figure 2.5) and after a time period $X3$ from the moment friends A and C met has elapsed. At that point, this portion of the trip is completed.

The average speed of the group is determined graphically from the slope of a dashed line starting at the intersection of the x- and y-axes and ending at the point $\{(X1 + X2 + X3), (Y2)\}$ shown in Figure 2.5. This slope is equal to $Y2/(X1 + X2 + X3)$.

Alternatively, the problem can be solved analytically with the aid of the time–space diagram by relating the unknown variables $X1$, $X2$, $X3$, $Y1$, and $Y2$ to one another as follows:

With reference to Figure 2.5, the distance $Y1$ can be calculated in two different ways using the expression $D = u \cdot t$:

$$Y1 = 6(X1 + X2) \tag{1}$$

$$Y1 = 24X1 - 24X2 \tag{2}$$

Setting (1) equal to (2) yields

$$X2 = 0.60X1$$

Similarly, the distance $Y2$ can be calculated in two different ways as follows:

$$Y2 = 24X1 + 6(X2 + X3) \tag{3}$$

$$Y2 = 6(X1 + X2) + 24X3 \tag{4}$$

Setting (3) equal to (4) yields

$$X1 = X3$$

Also,

$$\begin{aligned} Y2 &= 24X1 + 6(X2 + X3) \\ &= 24X1 + 6x0.6X1 + 6X1 \\ &= 33.6X1 \end{aligned}$$

The average speed, $S = Y2/(X1 + X2 + X3)$.
Substituting values for $X2$ and $X3$ as previously determined yields

$$S = 33.6X1/(X1 + 0.6X1 + X1) = 12.92 \text{ km/h}$$

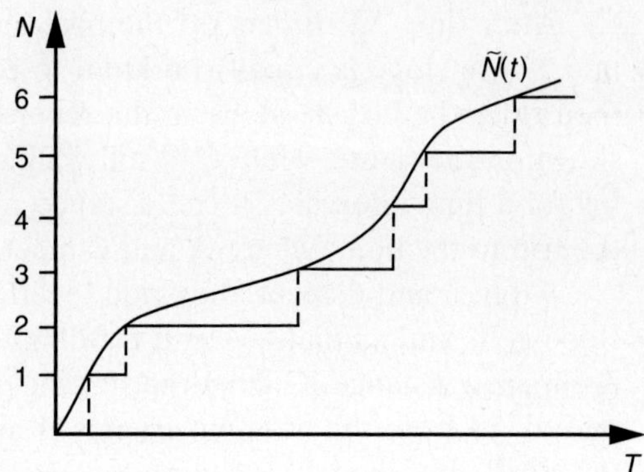

FIGURE 2.6
Cumulative plot approximation.

Cumulative Plots

Cumulative plots are graphs depicting the *cumulative* number of persons or vehicles to pass a given location in time t, expressed as $N(t)$. The cumulative count is typically comprised of discrete units (e.g., cars, buses, people). Therefore, $N(t)$ takes the form of a step function. However, in traffic analysis practice this function is often approximated by assuming a smooth continuous function $\tilde{N}(t)$, most commonly when large numbers of moving objects are involved (see Figure 2.6).

Since $N(t)$ is the number of vehicles or persons during a time interval $(t_1, t_{2,\ldots})$, the number of observations occurring between time t_1 and time t_2 is $(N(t_2) - N(t_1))$. The rate of traffic flow (q), during a given interval (t_1, t_2), is

$$q = \frac{N(t_2) - N(t_1)}{t_2 - t_1} \tag{2.1}$$

Thus the traffic flow (or volume) q is the slope of the function, $N(t)$.

Cumulative plots are useful for analyzing situations involving traffic flow through one or more restrictions along a travelway. Examples include

1. Traffic flow through a bottleneck where there is a reduction in the number of lanes;
2. Traffic flow through a construction zone when one or more lanes are closed;
3. Traffic flow past the location of an accident that is blocking one or more lanes;
4. Traffic flow at a signalized intersection where the traffic signal restricts traffic flow during certain time intervals.

These situations are analyzed using two cumulative plots, one for a point upstream of (or before) the restriction, and another for a point downstream of (or after) the restriction. The upstream cumulative plot represents the arrival pattern of vehicles at the location of the restriction and is called the "arrival" curve $A(t)$, whereas the downstream plot represents the departure pattern and is called the "departure" curve, $D(t)$. The procedure is described in the following example.

Example 2.4

Developing a Cumulative Plot to Depict a Lane Closure

A six-lane freeway (with three lanes in each direction) experiences morning peak traffic volumes of 4800 veh/h; the maximum number of vehicles that a lane can accommodate in an hour is 2000 veh/h/lane. At 8:15 A.M., an accident occurs that completely blocks one lane. At 8:45 A.M., the incident is cleared and the blocked lane is opened to traffic.

Develop a cumulative plot for the situation described, showing both the arrival and departure curves.

Solution

Begin by drawing the arrival curve. Since vehicles arrive at the constant rate of 4800 veh/h, the arrival curve will be a straight line, whose slope is equal to 4800 veh/h (see Figure 2.7). Before 8:15 A.M., all three lanes were open, and each lane could handle 2000 veh/h or a total of 6000 veh/h for three lanes. Thus, prior to the accident, vehicles departed at the same rate they arrived (since 4800 is less than 6000). Thus the arrival and departure curves are identical.

At 8:15 A.M., an accident occurred, closing one lane and reducing the capacity to 4000 veh/h (2 lanes × 2000 veh/h/lane). Thus, as long as the lane is blocked, the available capacity will be less than the number of arriving vehicles (4000 versus 4800). As a result, vehicles will back up, forming a long line waiting to get through the bottleneck. This phenomenon is called *queuing*. The number of vehicles in the queue at any given time, t, is found in Figure 2.7 as the vertical distance between the arrival curve, $A(t)$, and the departure curve, $D(t)$. This is so because the difference between the number of vehicles arriving and the number departing equals the number of vehicles waiting in the queue. At 8:45 A.M. the accident is removed and full capacity is restored. Now vehicles in the queue will

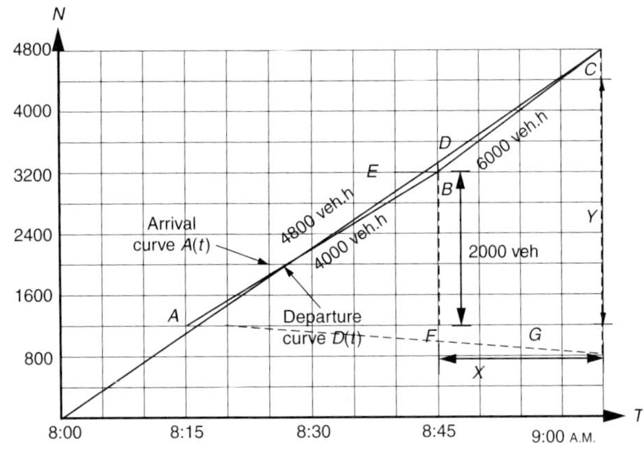

FIGURE 2.7 Cumulative arrival and departure curves for Example 2.4.

begin to depart at the previous rate of 6000 veh/h. Vehicles will continue to discharge from the incident location at the 6000 veh/h rate until all waiting vehicles are served and the queue has dissipated. At that time arriving vehicles will depart as soon as they arrive, and the arrival and departure curves in Figure 2.7 will once again be identical.

EXAMPLE 2.5

Using Cumulative Plots to Evaluate Traffic Queues

Use the cumulative plot developed in Example 2.4 to determine the following:

1. The maximum number of vehicles in the queue;
2. The maximum time a vehicle waits at the incident location;
3. The total vehicle delay resulting from the incident.

Solution

Part (1) The queue length at a given point is given by the vertical distance between the arrival curve and the departure curve. As can be seen in Figure 2.7, the maximum length of the queue is given by the distance BD. This distance can be calculated using the arrival and departure curves:

The total number of arrivals between 8:15 and 8:45 (0.5 h) is 2400 veh

The total number of departures between 8:15 and 8:45 (0.5 h) is 2000 veh

Thus, the maximum number of vehicles in the queue at 8:45 in the morning is

$$BD = 4800 \times 0.50 - 4000 \times 0.5 = 400 \text{ vehicles}$$

Part (2) The time a vehicle waits in the queue is given by the horizontal distance between the arrival and departure curves, since this distance is the time difference between the moment a specified vehicle, n, arrives at the queue and the moment it departs from the congested area. Thus, the maximum delay is the distance BE shown in Figure 2.7. This value represents the delay for the vehicle that departs at the time the incident is cleared.

From the time the accident started at 8:15 A.M. until the lanes are open at 8:45 A.M., a total of 2000 vehicles (calculated as the departure rate of 4000 veh/h multiplied by the duration of the closure of 0.50 h) have departed. The last vehicle to arrive during the lane closure is the 2000th arrival. Since the arrival rate is equal to 4800 veh/h, the 2000th vehicle to arrive did so 2000/4800 or 0.41667 h (or 25 minutes) after the lane closure. However, while the 2000th vehicle joined the queue 25 minutes after the lane closure, it was discharged from the queue 30 minutes after the incident occurred. In other words, the delay for this last vehicle was 5 minutes, which is the maximum delay.

Part (3) The total delay, measured in veh·h, is given by the area of triangle ABC in Figure 2.7. To calculate the area of that triangle, determine the time needed for the queue to dissipate after the incident is cleared. The letter X in Figure 2.7 denotes this time. To calculate X, calculate the distance Y, which represents the number of vehicles that have arrived (or departed) from the moment the incident occurred to the moment traffic conditions returned to normal. Y can be calculated as follows:
From the arrival curve,

$$Y = 4800 \cdot (0.50 + X)$$

From the departure curve,

$$Y = 4000 \cdot 0.50 + 6000 \cdot X$$

Therefore,

$$4800 \cdot (0.50 + X) = 4000 \cdot 0.50 + 6000 \cdot X$$

$$2400 + 4800X = 2000 + 6000X$$

$$1200X = 400, \text{ or}$$

$$X = 400/1200 = 1/3 \text{ h}$$

X, the time for the queue to dissipate once the lanes are open, is 1/3 h or 20 minutes. Therefore,

$$Y = 4000 \cdot 0.50 + 6000 \cdot 1/3 = 4000 \text{ vehicles}$$

The area of triangle ABC can then be calculated as follows:

Area of triangle ACG − area of triangle ABF − area of trapezoid $BCGF$

$$= 0.50 \cdot (0.50 + 0.333) \cdot 4000 - 0.50 \cdot 0.50 \cdot 2000 - 0.50 \cdot (2000 + 4000) \cdot 0.333$$

Total vehicle delay is

$$= 167.66 \text{ veh·h}$$

Regression Analysis Techniques

In many engineering applications, relationships between variables are established based on empirical observations or data collected from controlled experiments or from real-time events observed directly at the site. Typically one

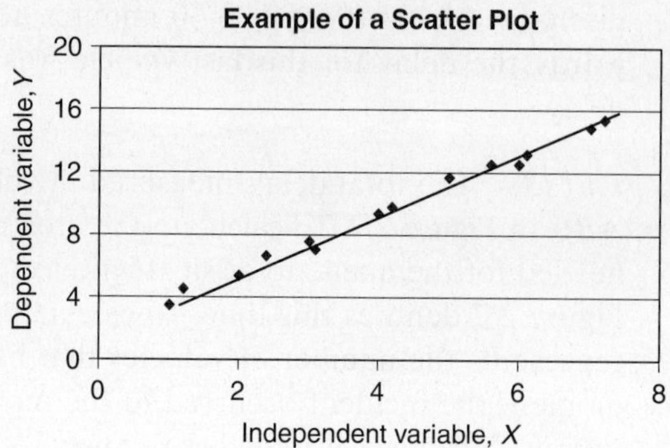

FIGURE 2.8
Scatter plot.

variable is referred to as *dependent* because its value depends on the values of other variables, which are *independent.* Thus, the reason for collecting data is to determine whether or not a relationship exists that can be expressed mathematically between the dependent and independent variables. If the results indicate a relationship that "fits" the data, the mathematical expression can be used in further analysis of transportation problems.

When there is but one *independent variable* to consider, the relationship with the *dependent* variable can be depicted graphically as a scatter plot. The *dependent* variable is plotted along the y-axis, and the *independent* variable is plotted along the x-axis, as illustrated in Figure 2.8. In this figure, it can easily be seen that there appears to be a linear relationship between the variables Y and X of the following form:

$$Y = a + bX \tag{2.2}$$

where

Y = value of the dependent variable
X = value of the independent variable
a = constant that represents the intercept of the fitted line with the y-axis
b = slope of the fitted line

When there are two or more *independent* variables and a greater amount of data, the graphical procedure is replaced by computational techniques. Regression analysis is a useful technique to use when dealing with a number of *independent* variables.

To use regression analysis, assume a mathematical model (i.e., linear, quadratic, exponential, etc.) for the relationship between the *dependent* variable and the *independent* variables. The method of least squares is then used to determine the values of the coefficients for each independent variable so as to minimize the sum of the square of the deviations between the *observed* values of the *dependent* variable Y and the values *estimated* by the assumed mathematical model.

The sum of the square of the deviation between the *observed* and the *estimated* values of the dependent variable, Y, can be expressed as

$$S = \sum_{i=1}^{N} (Y_i - \hat{Y}_i)^2 \tag{2.3}$$

where
Y_i = observed value for Y for observation (i.e., corresponding to the value of X_i)
$\hat{Y}_i$ = estimated value for Y corresponding to the value of X_i

Linear Regression

The simplest case of a linear regression between two variables, a dependent variable, Y, and an independent variable, X, is the relationship

$$Y = a + bX$$

To estimate the values for the two parameters, a and b, use Equations 2.3 and 2.2, by substituting $a + bX_i$ from Equation 2.2 to $\hat{Y}_i$ in Equation 2.3 to obtain Equation 2.4:

$$S = \sum_{i=1}^{N} (Y_i - a - bX_i)^2 \tag{2.4}$$

The partial derivates of S with respect to a and b are determined and set equal to 0, as shown in Equations 2.5 and 2.6:

$$\frac{\partial S}{\partial a} = \sum_{i=1}^{N} \{2(Y_i - a - bX_i)(-1)\} = 0 \tag{2.5}$$

$$\frac{\partial S}{\partial b} = \sum_{i=1}^{N} \{2(Y_i - a - bX_i)(-X_i)\} = 0 \tag{2.6}$$

Solve Equations 2.5 and 2.6 simultaneously to obtain the following expressions for the parameters b and a:

$$b = \frac{\sum_{i=1}^{N} (X_i - \overline{X})(Y_i - \overline{Y})}{\sum_{i=1}^{N} (X_i - \overline{X})^2} \tag{2.7}$$

and

$$a = \overline{Y} - b\overline{X} \tag{2.8}$$

where
$\overline{X}$ and $\overline{Y}$ = mean (average) values for variables X and Y

Determining the values of the numerator and denominator of Equation 2.7 requires that the mean values $\overline{X}$ and $\overline{Y}$ be computed for the observations of the two variables X and Y and then $\overline{X}$ is subtracted from each observation, X_i, to give $(X_i - \overline{X})$. Similarly, $\overline{Y}$ is subtracted from each observation, Y_i, to give $(Y_i - \overline{Y})$. With these calculations complete, the value of b can be determined by using Equation 2.7, and then the value of a determined from Equation 2.8. Example 2.6 illustrates the regression analysis technique using a spreadsheet to make the computations.

EXAMPLE 2.6

Linear Regression with One Independent Variable

One of the most common tasks for transportation engineers is to assess the impact that a proposed new housing or office complex will create for the affected transportation network. The first step in this assessment is to estimate the number of trips the new development will generate. Empirical models developed using data collected at similar sites can be used. Regression analysis techniques are often used to develop these models by relating the dependent variable Y = trips generated by the new complex to one or more independent variables: X_1 = area of the development in square meters, X_2 = the number of employees.

Establish a relationship between the total number of trips generated by an office building and the number of employees. The data consist of the numbers of trips to and from the site observed during the peak hour, and the number of employees. Twenty office buildings were selected for the survey, and the data are shown in Table 2.1.

Develop a regression model that relates the total number of trips generated by an office building (Y) to the number of employees working in the building (X).

Solution

Determine the values for a and b in a linear regression model using Equations 2.7 and 2.8. The dependent variable, Y, is the number of vehicle trips generated, and the independent variable, X, is the number of employees. Compute the mean value, $\overline{Y}$, the mean value, $\overline{X}$, the sum product, $\sum_{i=1}^{n}(Y_i - \overline{Y})(X_i - \overline{X})$, and the sum of the squares, $\sum_{i=1}^{n}(X_i - \overline{X})^2$.

The calculations are performed using Microsoft Excel, as shown in Figure 2.9.

TABLE 2.1 Data for Example 2.6

Building Number	*Vehicle Trips*	*Number of Employees*
1	331	520
2	535	770
3	542	1050
4	261	380
5	702	1150
6	367	380
7	433	820
8	763	1720
9	586	1350
10	1034	1870
11	1038	2260
12	1358	2780
13	890	1760
14	308	580
15	601	1320
16	578	780
17	1310	2320
18	1391	2670
19	1467	3300
20	807	1450

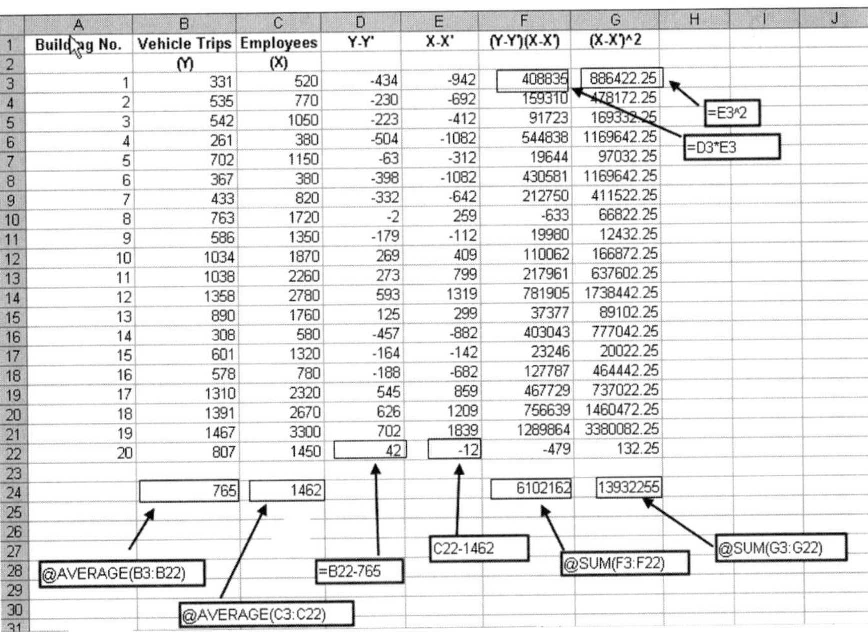

	A	B	C	D	E	F	G
1	Building No.	Vehicle Trips	Employees	Y-Y'	X-X'	(Y-Y')(X-X')	(X-X')^2
2		(Y)	(X)				
3	1	331	520	-434	-942	408835	886422.25
4	2	535	770	-230	-692	159310	478172.25
5	3	542	1050	-223	-412	91723	169332.25
6	4	261	380	-504	-1082	544838	1169642.25
7	5	702	1150	-63	-312	19644	97032.25
8	6	367	380	-398	-1082	430581	1169642.25
9	7	433	820	-332	-642	212750	411522.25
10	8	763	1720	-2	259	-633	66822.25
11	9	586	1350	-179	-112	19980	12432.25
12	10	1034	1870	269	409	110062	166872.25
13	11	1038	2260	273	799	217961	637602.25
14	12	1358	2780	593	1319	781905	1738442.25
15	13	890	1760	125	299	37377	89102.25
16	14	308	580	-457	-882	403043	777042.25
17	15	601	1320	-164	-142	23246	20022.25
18	16	578	780	-188	-682	127787	464442.25
19	17	1310	2320	545	859	467729	737022.25
20	18	1391	2670	626	1209	756639	1460472.25
21	19	1467	3300	702	1839	1289864	3380082.25
22	20	807	1450	42	-12	-479	132.25
23							
24		765	1462			6102162	13932255

FIGURE 2.9 Calculations for Example 2.6.

With the required quantities calculated, the next step is to apply Equations 2.8 and 2.9 to find the parameters, b and a, as follows:

$$b = \frac{\sum_{i=1}^{n}(X_i - \overline{X})(Y_i - \overline{Y})}{\sum_{i=1}^{n}(X_i - \overline{X})^2} = \frac{6102162}{13932255} = 0.438$$

$$a = \overline{Y} - b\overline{X} = 765 - 0.438 \cdot 1462 = 124.9$$

Therefore, the required relationship is

Vehicle Trips = 124.9 + (0.438) (Number of Employees)

Multivariable Linear Regression Using Microsoft Excel

The calculations for Example 2.6 would have been tedious if done manually particularly if the set of observations had been extensive. For this and more complex models containing many variables, software packages for regression analysis are readily available, including Microsoft Excel. For more than one independent variable, Microsoft Excel provides an Add-In feature called Data Analysis ToolPak, and its use is explained in Example 2.7.

EXAMPLE 2.7

Linear Regression Analysis with Two or More Independent Variables

The strength and durability of a pavement section are expressed using a condition index (CI) that ranges from 0 to 100, where 0 is very poor and 100 is excellent. The CI is related to several independent variables:

X_1 = the age in years of the pavement section since it was constructed or resurfaced
X_2 = average daily traffic (ADT)
X_3 = structural number (SN), a measure of pavement ability to carry traffic loads.

The data in Table 2.2 were collected by rating the condition of 20 individual pavement sections. Data also included the number of years since construction or reconstruction, average daily traffic, and the structural number.

Use regression analysis techniques to develop a mathematical model that could be used as a tool to predict the future condition of pavement sections in this region.

Solution

Microsoft Excel provides an Add-In called Data Analysis ToolPak to be used to perform a number of statistical analysis procedures, including regression

TABLE 2.2
Data for Example 2.7

Section Number	*Condition Index (CI)*	*AGE (years)*	*ADT (1000 veh/day)*	*SN*
1	100	1.2	27	4.2
2	93	2.5	15	5.0
3	79	9.2	9	5.1
4	94	2.9	8	5.3
5	79	10.8	12	3.9
6	85	6.3	14	4.3
7	100	0.1	23	4.9
8	97	2.2	17	5.0
9	82	8.1	16	3.1
10	81	9.4	6	5.0
11	88	5.6	27	4.4
12	79	10.0	17	5.2
13	83	7.6	20	4.6
14	76	11.4	13	4.2
15	93	4.0	8	4.0
16	81	9.3	29	5.4
17	100	0.3	5	5.5
18	76	10.4	8	3.8
19	77	10.5	7	3.2
20	84	6.3	17	4.4

analysis. To check that this feature is active, use Excel's Tools menu, and check in the dropdown menu to see if Data Analysis is listed and thus is active. If not, select Add-In from the Tools menu, and check the box next to Analysis ToolPak, as shown in Figure 2.10.

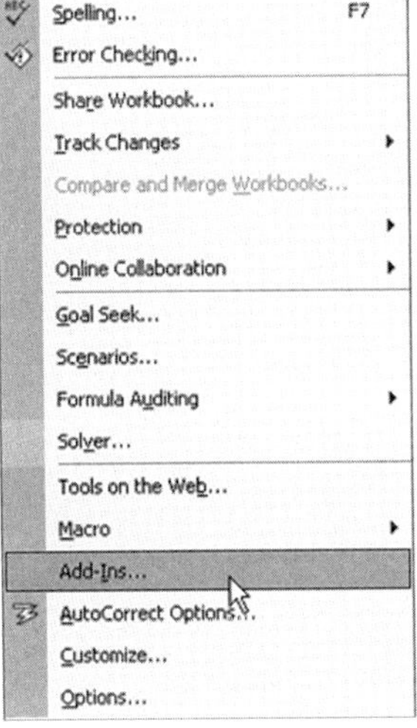

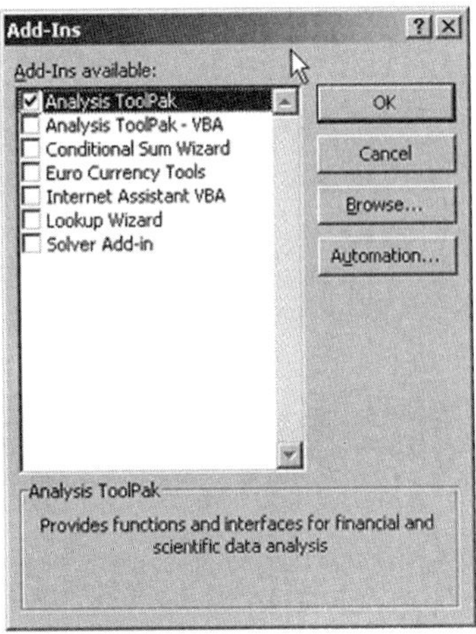

FIGURE 2.10
Tools menu for adding the Analysis ToolPak.

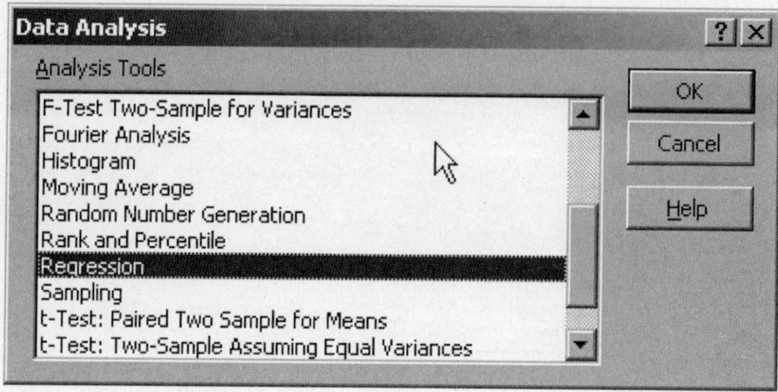

FIGURE 2.11
Selecting the regression analysis option.

Once Analysis ToolPak has been added, enter the data for the problem as shown in the Excel worksheet (Figure 2.12). There are separate columns for the dependent variable Condition Index (CI) and for each of the three independent variables AGE, ADT, and SN. To access the Analysis ToolPak Add-In, select Tools and Data Analysis. A window appears containing the data analysis procedures included in Analysis ToolPak Add-In. Scroll down to Regression and select this option, as shown in Figure 2.11.

Decide which column will contain the dependent variable and which columns will contain the independent variables and label each column appropriately. Specify the input *Y* and *X* ranges, and the output range to display the result, as illustrated in Figure 2.12.

For the "Input Y Range," specify cell range B1 . . . B21, which will contain the values for the dependent variable (CI). (Row 1 is used to label each column.) The "Input X Range" specifies the cells containing the values for the independent variables AGE, ADT, and SN. This range is C1 . . . E21, in columns C, D, and E, cells 1–21. Check the "Labels" box to indicate that the first cell in each column holds the label or name of that variable. Finally, the "Output Range"

B1 SUMMARY OUTPUT

	A	B	C	D	E
1	Section No	CI	AGE	ADT	SN
2	1	100	1.2	27	4.2
3	2	93	2.5	15	5.0
4	3	79	9.2	9	5.1
5	4	94	2.9	8	5.3
6	5	79	10.8	12	3.9
7	6	85	6.3	14	4.3
8	7	100	0.1	23	4.9
9	8	97	2.2	17	5.0
10	9	82	8.1	16	3.1
11	10	81	9.4	6	5.0
12	11	88	5.6	27	4.4
13	12	79	10.0	17	5.2
14	13	83	7.6	20	4.6
15	14	76	11.4	13	4.2
16	15	93	4.0	8	4.0
17	16	81	9.3	29	5.4
18	17	100	0.3	5	5.5
19	18	76	10.4	8	3.8
20	19	77	10.5	7	3.2
21	20	84	6.3	17	4.4
22					

Regression
Input
Input Y Range: $B$1:$B$21
Input X Range: $C$1:$E$21
☑ Labels ☐ Constant is Zero
☐ Confidence Level: 95 %
Output options
◉ Output Range: $I$1
○ New Worksheet Ply:
○ New Workbook
Residuals
☐ Residuals ☐ Residual Plots
☐ Standardized Residuals ☐ Line Fit Plots
Normal Probability
☐ Normal Probability Plots
OK
Cancel
Help

FIGURE 2.12
Using the regression analysis option in Excel.

SUMMARY OUTPUT					
Regression Statistics					
Multiple R	0.98655111				
R Square	0.9732831				
Adjusted R Square	0.96827368				
Standard Error	1.51592037				
Observations	20				
ANOVA					
	df	*SS*	*MS*	*F*	*ignificance*
Regression	3	1339.448016	446.482672	194.2906195	8.56E-13
Residual	16	36.76823292	2.298014557		
Total	19	1376.216249			
	Coefficients	*Standard Error*	*t Stat*	*P-value*	*Lower 95%*
Intercept	98.8696111	2.87293944	34.41409511	1.96898E-16	92.77925
AGE	-2.18471661	0.099702259	-21.91240846	2.33048E-13	-2.39608
ADT	0.0183046	0.047795332	0.38297877	0.706775781	-0.08302
SN	0.27571494	0.543613417	0.507189357	0.618940784	-0.87669

FIGURE 2.13 Regression analysis results for Example 2.7.

shows where the output should go (in this example, the output will begin at cell I1). Additional output can be requested as desired, such as residuals, residual plots, and so on.

The results for this example are illustrated in Figure 2.13.

Two sections of this output are of particular interest, as noted by the box in Figure 2.13. The first section is linked to one of the regression statistics, referred to as "the *R*-square" value or simply as R^2. This statistic has a range of values that is between 0 and 1. It is a measure of how well the model results fit the data. A perfect model is one that exactly fits the data and has an R^2 value of 1 while a model that does not fit the data at all has an R^2 value of 0. Typically values are within the range between 0 and 1, with values closer to 1 indicating a reasonably good fit, as is the case in this example where $R^2 = 0.973$.

The second section noted by the box lists the coefficients of the model. These specify the parameters of the linear model that was fit to the data. In this example, the fitted model describes the deterioration of the pavement sections as follows:

$$\text{CI} = 98.87 - 2.18 \cdot \text{AGE} - 0.02 \cdot \text{ADT} + 0.28\,\text{SN} \tag{2.9}$$

where

CI = condition index
AGE = number of years since construction
ADT = average daily traffic in 1000 veh/day
SN = structural number

EXAMPLE 2.8

Determining Pavement Condition Using a Regression Model

Use the model just developed in Example 2.7 to show how the condition of a pavement section can be expected to change over time. Assume the section has a structural number (SN) of 5.0, and the average daily traffic is 25,000 veh/day.

Solution

Use Equation 2.9, and substitute the values of 5.0 for SN and 25 for ADT. The relationship between CI and age is

$$CI = 98.87 - 2.18 \cdot Age - 0.02 \cdot 25 + 0.28 \cdot 5$$

That is, $CI = 99.77 - 2.18 \cdot Age$

Figure 2.14 plots this relationship to show the deterioration trend of the pavement section over time.

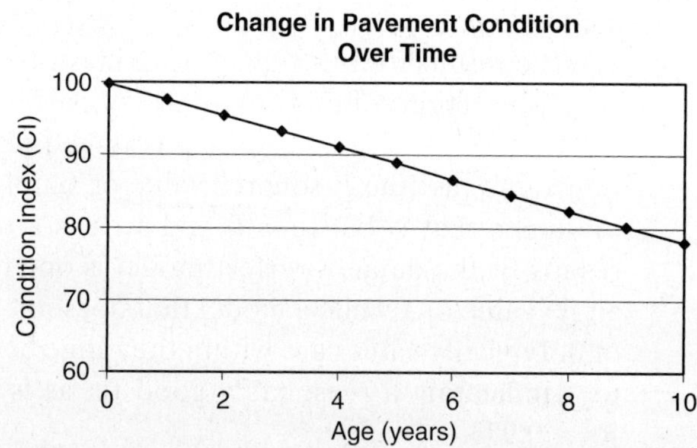

FIGURE 2.14
Deterioration trend for pavement section of Example 2.8.

Regression Using Transformed Variables

A basic assumption of the regression equations considered in the previous sections is that the relationship between the dependent and the independent variables is linear. In some instances a nonlinear relationship may prove to be a better fit to the data and it may still be possible to use linear regression to develop the model by an appropriate transformation of the assumed nonlinear relationship. The following example illustrates how the coefficients of a nonlinear model can be determined using linear regression.

EXAMPLE 2.9

Using Linear Regression to Model the Relationship between Speed and Density

The average speed of traffic on a freeway in km/h (u) and the prevailing traffic density in veh/km (k) is assumed to be described by Equation 2.10.

$$u = ae^{\frac{-k}{b}} \tag{2.10}$$

where

u = average speed in km/h
k = traffic density in veh/km
a, b = model parameters
e = natural log (e = 2.718)

The data shown in Table 2.3 were collected by measuring the average traffic speed at different time points during the day and recording the corresponding density.

Determine the values of the parameters a and b in Equation 2.10.

TABLE 2.3
Data for Example 2.9

Speed (u) km/h	*Density (k) Veh/km*
48	98
96	22
64	71
40	110
64	74
80	40
84	39
104	11
108	10
92	32
84	42
69	68
51	104
76	57
84	39
60	73
113	2
64	73
88	33
93	24

Solution

To convert Equation 2.10 from a nonlinear to a linear form, take the log (base e) to yield

$$\ln u = \ln(ae^{\frac{-k}{b}})$$

Therefore,

$$\ln u = \ln a + \ln(e^{\frac{-k}{b}})$$

$$\ln u = \ln a + \left(\frac{-k}{b}\right)\ln e$$

$$\ln u = \ln a - \frac{k}{b}$$

Equation 2.10 is now a linear relationship between the transformed variable ($\ln u$) and the variable, k. To compute a and b, consider $\ln u$ to be the dependent variable and k the independent variable. Figure 2.15 depicts the Excel formulation using Data Analysis ToolPak to compute a and b. Column A is the speed, B is the density, and C is the log (base e) speed.

Excel is used to run the regression analysis by specifying cells C2 . . . C21 as the dependent variable and cells B2 . . . B21 as the independent. The value

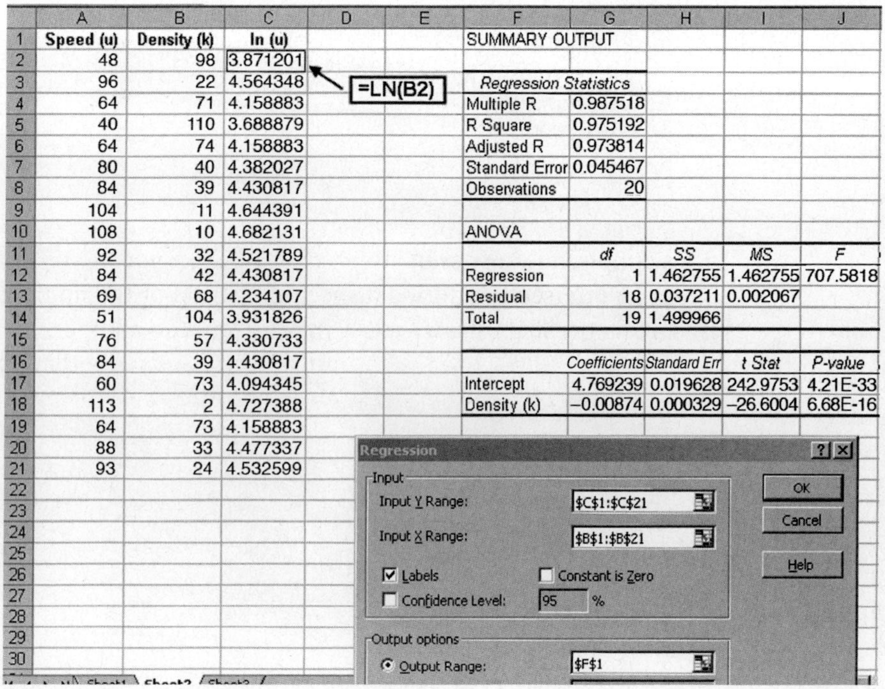

FIGURE 2.15
Solution to Example 2.9.

	A	B	C	D	E	F	G	H	I	J
1	Speed (u)	Density (k)	ln (u)			SUMMARY OUTPUT				
2	48	98	3.871201							
3	96	22	4.564348	=LN(B2)		*Regression Statistics*				
4	64	71	4.158883			Multiple R	0.987518			
5	40	110	3.688879			R Square	0.975192			
6	64	74	4.158883			Adjusted R	0.973814			
7	80	40	4.382027			Standard Error	0.045467			
8	84	39	4.430817			Observations	20			
9	104	11	4.644391							
10	108	10	4.682131			ANOVA				
11	92	32	4.521789				*df*	*SS*	*MS*	*F*
12	84	42	4.430817			Regression	1	1.462755	1.462755	707.5818
13	69	68	4.234107			Residual	18	0.037211	0.002067	
14	51	104	3.931826			Total	19	1.499966		
15	76	57	4.330733							
16	84	39	4.430817				*Coefficients*	*Standard Err*	*t Stat*	*P-value*
17	60	73	4.094345			Intercept	4.769239	0.019628	242.9753	4.21E-33
18	113	2	4.727388			Density (k)	−0.00874	0.000329	−26.6004	6.68E-16
19	64	73	4.158883							
20	88	33	4.477337							
21	93	24	4.532599							

Regression

Input

Input Y Range: $C$1:$C$21

Input X Range: $B$1:$B$21

☑ Labels ☐ Constant is Zero

☐ Confidence Level: 95 %

Output options

◉ Output Range: $F$1

OK | Cancel | Help

of the "intercept" in this case is equivalent to ln a, whereas the "Density" coefficient yielded is equivalent to $(-\frac{1}{b})$. Thus,

$$\ln a = 4.769$$

and

$$a = e^{4.769} = 117.8$$

Also,

$$-\frac{1}{b} = -0.00874$$

and

$$b = 1/0.00874 = 114.4$$

The model can be expressed as follows:

$$u = 117.8e^{\frac{-k}{114.4}}$$

and is illustrated in Figure 2.16.

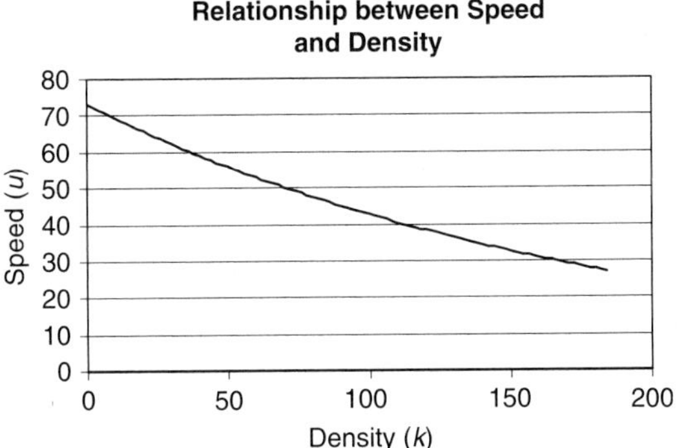

FIGURE 2.16 Developed speed–density relationship.

Probability Theory

In many transportation situations, the outcome is unknown or uncertain. For example, it is impossible to predict the exact number of vehicles that will arrive at an intersection during a specified period or the number of people who will take a particular travel route in preference to another.

Probability theory is a branch of mathematics that deals with the uncertainty of events. It began when the noted French scientist Pascal (1623–1652) invented the theory of probability and predicted the likely outcome of gambling games in order to help friends settle their bets. Since then, probability theory has been applied to a wide range of fields, including traffic and transportation.

A Model of Uncertainty

Probability theory describes uncertainty by referring to *outcomes* and their *probabilities*. The *outcomes* refer to events that might happen, whereas the probabilities specify the likelihood of the occurrence of an outcome.

The outcomes must be *mutually exclusive* and *collectively exhaustive. Mutually exclusive* limits the outcome to only one event. For example, when tossing a coin, *either* a head (H) or a tail (T) occurs, not both at the same time. *Collectively exhaustive* stipulates one of the specified outcomes must occur. For example, when tossing a coin there are only two *outcomes,* a head or a tail. Thus, the probability of an *outcome* is a number between 0 and 1, and the sum of the probabilities of all outcomes equals 1. A *probability model* is basically the enumeration of all possible outcomes and the probability of each outcome.

Examples of Simple Probability Models

Examples of simple probability models are coin tossing and rolling dice. As already noted, tossing a coin has only two outcomes: heads or tails with an equal probability of either outcome equal to 0.50. Another example is throwing a perfectly balanced die. There are six possible outcomes: The die can land with the face showing as 1, 2, 3, 4, 5, or 6. The probability associated with each outcome is 1/6. A probability model is sometimes referred to as an *experiment.* In an *experiment,* the set of all possible outcomes is called the *sample space* and is denoted by the capital Greek letter Ω (Omega).

Events and Their Probabilities

In probability theory, an *event* refers to a set or collection of outcomes. In other words, an event is a subset of the sample space, Ω. For example, in the die experiment, the probability of getting an odd number is the probability of the die showing 1, 3, or 5. The probability of an *event*, A, is defined as the sum of the probabilities of each outcome. In the case of the die toss, the probability of getting an odd number is equal to

$$P[A] = \frac{1}{6} + \frac{1}{6} + \frac{1}{6} = \frac{1}{2} \tag{2.11}$$

The complement of an event, $\overline{A}$, is defined as the subset of Ω that contains all outcomes not belonging to A. In the die example, $\overline{A}$ refers to the event of getting either 2, 4, or 6 on the die. The probability of $\overline{A}$ is equal to $1 - P(A)$.

Given two events A and B, the probability of "A and B" refers to the probability of outcomes that are both in A and in B. This refers to the intersection of two sets and is often written as $A \cap B$, where $\cap$ is the intersection symbol used in set theory. The probability of "A or B" refers to the probability of outcomes that are in A, in B, or in both. This is expressed as $A \cup B$, where $\cup$ is the union operator. The probability of "A or B" is given by the following expression:

$$P(A \cup B) = P(A) + P(B) - P(A \cap B) \tag{2.12}$$

As can be seen from Figure 2.17, the expression $P(A) + P(B)$ includes the probability of each outcome in the event $(A \cap B)$ twice. Thus, $P(A \cap B)$ is subtracted from $P(A) + P(B)$.

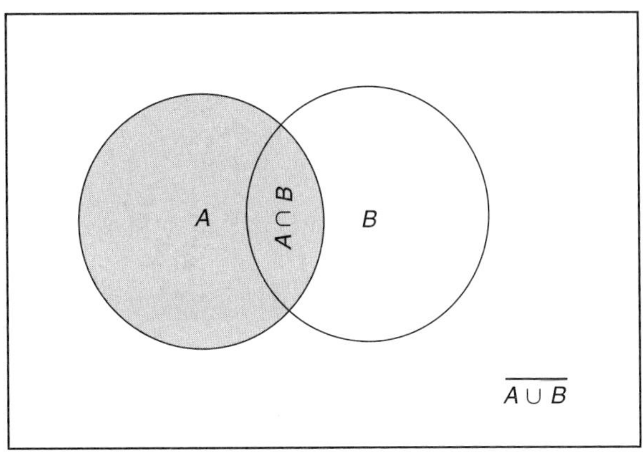

FIGURE 2.17
Computing the probability of "*A* or *B*."

Discrete Random Variables and Their Probability Distributions

A *random variable* is a special type of a probability model that assigns a numerical value to each outcome. The random variable is denoted by an uppercase letter (i.e., X), and the corresponding value that the random variable might take is denoted by a lowercase letter (i.e., x). For example, a random variable, X, could assume as many as n different numerical values $(x_1, x_2, \ldots, x_n)$, with associated probabilities of $(p_1, p_2, \ldots, p_n)$, as shown in the probability tree of Figure 2.18.

The distinguishing feature of a random variable probability model is the fact that the values of $(x_1, x_2, \ldots, x_n)$ are numerical. Random variables can be discrete or they can be continuous. Discrete random variables take specified values with intervals between them whereas continuous variables can take any value without intervals between the values.

Discrete Random Variables

The probability distribution of a discrete random variable lists all possible values for the variable along with their associated probabilities as illustrated in Figure 2.18. The probability distribution of a discrete random variable is often

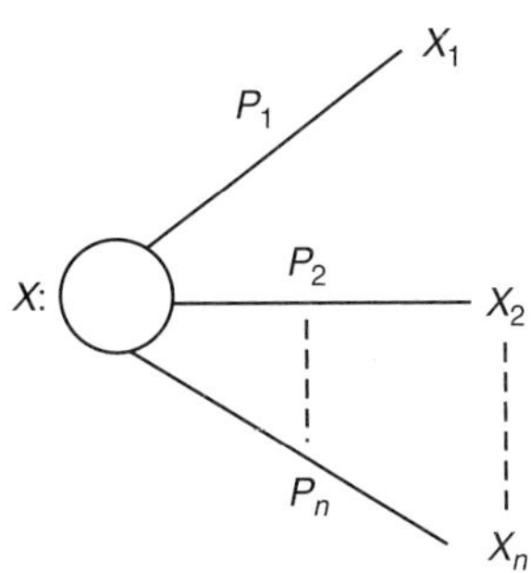

FIGURE 2.18
Probability tree.

referred to as its *probability mass function,* $p(x) = P[X = x]$, which associates each value of a discrete random variable to its probability. The values of $p(x)$ should satisfy the following two conditions:

$$0 \leq p(x) \leq 1 \tag{2.13}$$

$$\sum p(x) = 1 \tag{2.14}$$

In addition to the probability mass function of a discrete random variable, another useful function is its *cumulative distribution function (cdf),* which is defined as

$$F(x) = P[X \leq x] \tag{2.15}$$

In other words, the cumulative distribution function adds those probability values for the random variable, X, that are less than or equal to x. For discrete random variables, the cdf takes the form of a step function with an increase at each of the values that the random variable assumes. The upper and lower limits of this function are 0 and 1.

Summary Measures of Random Variables

A random variable has two types of summary measures. The first measures the center (or mean) of its probability distribution, and the second measures the spread (or variance) in the distribution. The most widely used measure for describing the center of a probability distribution is the mean (μ) or the expectation $E[X]$ defined as follows:

$$E[X] \text{ or } \mu = \sum_{x} x\, p(x) \tag{2.16}$$

The mean or expectation of a random variable does not reveal whether or not the values are similar to each other or widely scattered. For example, the three numbers (10, 20, and 30) and the three numbers (20, 20, and 20) both have identical means, but the spread from the mean is quite different. Thus, a measure of spread is required. The variance is the most commonly used measure of spread for probability distributions and is the expectation of the square of the difference between X and the mean $(X - \mu)^2$. The variance is expressed as

$$\text{Var}[X] = \sum_{x} (X - \mu)^2\, p(x) \tag{2.17}$$

The square root of the variance is called the standard deviation, σ. This term provides a measure of the spread that has the same units as the mean and the random variable. The standard deviation is calculated as Equation 2.18:

$$\sigma(X) = \sqrt{\text{Var}[X]} \tag{2.18}$$

EXAMPLE 2.10

Calculating the Mean and Variance of Walking Speeds

Table 2.4 lists the observed walking speed of a number of pedestrians as they cross at an intersection. Determine the following values for the sample speeds observed:

(a) Average speed
(b) Variance
(c) Standard deviation

Solution

Since each observation is equally likely to occur, $p(x)$ for all recorded values of pedestrian speed x is equal to $1/n$, where n, the number of observations, is 20. The formulae for the mean (μ), the variance, and the standard deviation (σ) in this case can be expressed as follows (the calculations can be performed using Microsoft Excel, as shown in the spreadsheet in Figure 2.19):

$$\mu = \frac{1}{n}\sum_{i=1}^{n} x_i$$

TABLE 2.4
Data for Example 2.10

Pedestrian ID	*Speed (m/s)*
1	1.10
2	1.41
3	1.05
4	1.12
5	1.05
6	1.19
7	1.24
8	1.33
9	1.16
10	1.25
11	1.13
12	1.19
13	1.13
14	1.15
15	1.26
16	1.56
17	1.38
18	1.01
19	1.19
20	1.41

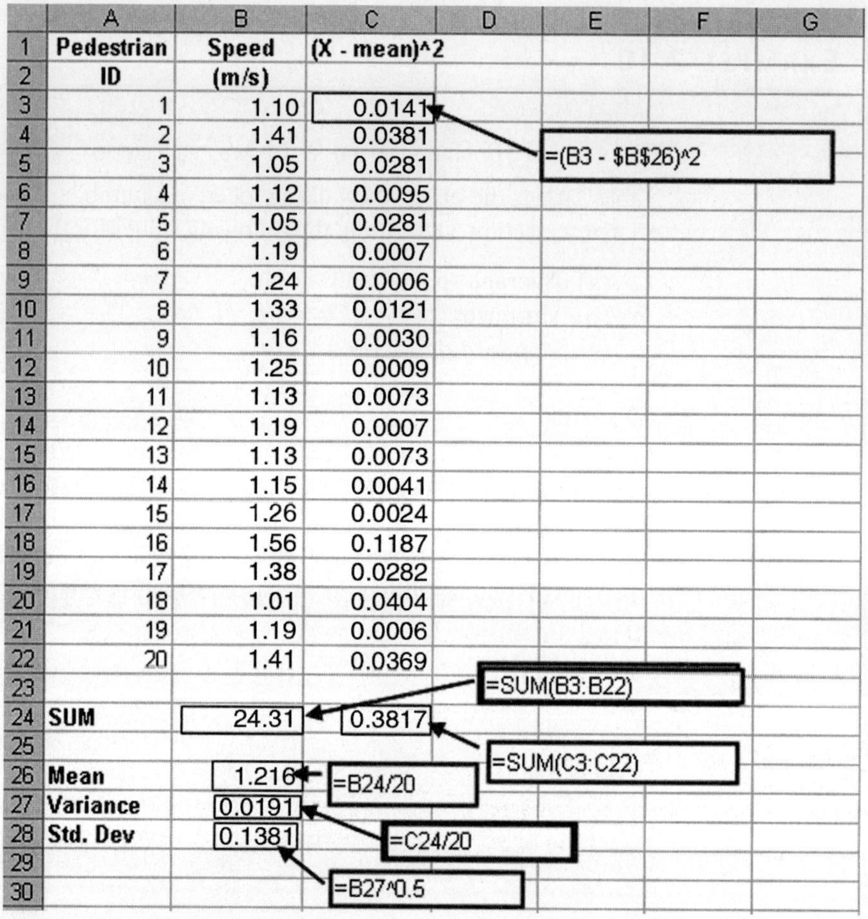

FIGURE 2.19
Calculating the mean, variance, and standard deviation using Excel.

	A	B	C	D	E	F	G
1	**Pedestrian**	**Speed**	**(X - mean)^2**				
2	**ID**	**(m/s)**					
3	1	1.10	0.0141				
4	2	1.41	0.0381				
5	3	1.05	0.0281				
6	4	1.12	0.0095				
7	5	1.05	0.0281				
8	6	1.19	0.0007				
9	7	1.24	0.0006				
10	8	1.33	0.0121				
11	9	1.16	0.0030				
12	10	1.25	0.0009				
13	11	1.13	0.0073				
14	12	1.19	0.0007				
15	13	1.13	0.0073				
16	14	1.15	0.0041				
17	15	1.26	0.0024				
18	16	1.56	0.1187				
19	17	1.38	0.0282				
20	18	1.01	0.0404				
21	19	1.19	0.0006				
22	20	1.41	0.0369				
23							
24	**SUM**	24.31	0.3817				
25							
26	**Mean**	1.216					
27	**Variance**	0.0191					
28	**Std. Dev**	0.1381					
29							
30							

$$\text{Var}[X] = \frac{1}{n}\sum_{i=1}^{n}(X_i - \mu)^2$$

$$\sigma(X) = \sqrt{\text{Var}[X]}$$

(a) To calculate the mean, the observed values are summed in cell B24. This sum is divided by 20 in cell B26 to give the mean value of 1.216 m/s.

(b) To calculate the variance and the standard deviation, a third column C was created to contain the values for $(X - \mu)^2$. The values for $(X - \mu)^2$ were summed in cell C24 to equal 0.3817. The variance was calculated in cell C27 by dividing the value in C24 by 20 (please note that strictly speaking, we should have divided by $(n - 1)$ and not n here, since this is a sample variance, and not a population variance. For simplicity, we will use n here). The standard deviation is calculated in cell B28.

Examples of Discrete Probability Distributions

The Binomial Distribution

The binomial distribution describes an experiment with a sequence of trials with only two outcomes; either the outcome happens or it doesn't (e.g., success or failure). The probability of either outcome is p (success) and $1 - p$ (failure)—the same probability for every trial. A random variable following a *binomial distribution* gives the probability of x successes in n independent trials. For example, in a coin-tossing experiment the variable could describe the number of heads that appear with 10 tosses of a coin. The *binomial distribution* has two parameters: the number of experiments, n; and the probability, p, of success for each trial. The probability distribution of x successes in n trials is given in Equation 2.19:

$$p(x) = P\{X = x\} = \frac{n!}{k!(n - k)!}p^x(1 - p)^{n-x} \tag{2.19}$$

where $n!$ is defined as $n(n - 1)(n - 2) \times, \ldots \times 1$

Microsoft Excel can calculate the binomial distribution. The function = BINOMDIST(x, n, p, 0) gives the probability of $P(X = x)$ for a binomial distribution with parameters n and p. For the same variable, changing the 0 to a 1 (i.e., = BINOMDIST(x, n, p, 1) computes the cumulative distribution function (cdf). To illustrate, Figure 2.20 uses Excel to compute the binomial distribution for a random variable with parameters $n = 10$ and $p = 0.3$. Column B lists the probability density function, whereas Column C lists the cumulative distribution function.

	A	B	C	D	E	F	G	H
1	x	P(X = x)	P(X <= x)					
2	0	0.02825	0.02825		= BINOMDIST(A2,10,0.3,0)			
3	1	0.12106	0.14931					
4	2	0.23347	0.38278					
5	3	0.26683	0.64961		= BINOMDIST(A2,10,0.3,1)			
6	4	0.20012	0.84973					
7	5	0.10292	0.95265					
8	6	0.03676	0.98941					
9	7	0.00900	0.99841					
10	8	0.00145	0.99986					
11	9	0.00014	0.99999					
12	10	0.00001	1.00000					
13								
14								
15								
16								
17								
18								
19								
20								

FIGURE 2.20
The binomial distribution calculations using Excel.

The Geometric Distribution

The geometric distribution also has its basis in a sequence of independent trials. It represents the probability that the first success will occur on the xth trial (for example, in a coin toss series of trials getting a head occurs after $x = 1, 2, 3, \ldots n$). This means that the first $(x - 1)$ trials yield a tail and that the xth trial yields a head. The geometric distribution expresses that probability as follows:

$$p(x) = P[X = x] = (1 - p)^{x-1}p \tag{2.20}$$

EXAMPLE 2.11

Computing the Probability of Aircraft Landings

An airport serves three different types of aircraft: Heavy (H), Large (L), and Small (S). During a typical hour the number of each type of aircraft that land are 30 Heavy, 50 Large, and 120 Small.

Determine the probabilities of the following landing outcomes:

(1) The next aircraft is Heavy.
(2) Exactly 3 out of 10 aircraft are Heavy.
(3) At least 3 out of 10 aircraft are Heavy.
(4) The first Heavy aircraft will be the third one to land.

Solution

Part (1) The probability that a landing aircraft is Heavy can be calculated by dividing the number of Heavy aircraft landing in an hour (30) by the total number of aircraft landing (30 + 50 + 120 = 200):

$$P(\text{landing aircraft is Heavy}) = \frac{30}{200} = 0.15$$

Part (2) The probability that exactly 3 out of 10 landing aircraft are Heavy can be calculated using the binomial distribution. The number of trials n is 10, the number of successes x is 3, and the probability of success p is 0.15. Thus, the probability is calculated using Excel as follows:

$$\begin{aligned} P(\text{3 out of 10 aircraft are Heavy}) &= \text{BINOMDIST}(3, 10, 0.15, 0) \\ &= 0.13 \end{aligned}$$

Part (3) The probability of at least three Heavy aircraft landing out of a total of 10 aircraft is $P(X \geq 3)$. This can occur if three or more landings are

Heavy aircraft. Conversely, it is equal to $1 - P(X \leq 2)$. $P(X \leq 2)$ can be calculated using Excel as follows:

$$P(X \leq 2) = \text{BINOMDIST}(2, 10, 0.15, 1) = 0.82$$
$$P(X \geq 3) = 1 - P(X \leq 2) = 1 - 0.82 = 0.18$$

Part (4) The probability that the first Heavy aircraft landing will be the third aircraft can be computed using the geometric distribution. The probability that the first success will occur on the third trial can be computed as

$$\begin{aligned} P[X = 3] &= (1 - p)^{x-1} p \\ &= (1 - 0.15)^2 \times 0.15 = 0.108 \end{aligned}$$

The Poisson Distribution

A discrete probability distribution with applications in traffic and transportation analysis. It is used to estimate the probability that x number of events occur within a stated time interval, t. For example, the Poisson distribution can be used to describe the arrival pattern of customers at a service facility. In transportation applications, these customers are vehicles in a traffic stream, pedestrians at a signalized crosswalk, or when ships arrive at a sea port. It is formulated as

$$p(x) = \frac{(\lambda t)^x e^{-\lambda t}}{xt} \tag{2.21}$$

where

$p(x)$ = probability that exactly x units will arrive during time interval t
t = duration of the time interval
λ = average arrival rate in passengers or veh/unit time
e = base of the natural logarithm ($e = 2.718$)

The Poisson distribution is most reliable where traffic is free flowing. If traffic is heavily congested or located just downstream of a signalized intersection, it is not accurate.

Microsoft Excel provides a function for calculating Poisson distribution probabilities, which is specified as = POISSON(x, (λt), 0). As was the case with the binomial distribution, when 0 is replaced by 1, a solution to the cumulative distribution function is provided.

EXAMPLE 2.12

Using the Poisson Distribution to Analyze the Arrival of Passengers at an Airport Counter

Travelers arrive at an airport ticketing counter at a rate of 450 passengers/h. What is the probability of 0, 1, 2, 3, and 4 or more travelers arriving over a

15-second time period if the arrival pattern can be described using a Poisson distribution?

Solution

Determine the rate of arrival, λ, in passengers/s. Since the arrival rate is 450 passengers/h, this is equivalent to $450/3600 = 0.125$ passengers/s. During a 15-second interval, λt, equals $0.125 \cdot 15 = 1.875$ passengers per 15-second. To calculate the probability of 0, 1, 2, or 3 persons arriving, use the Excel function = POISSON (x, 1.875, 0), as shown in Figure 2.21.

Calculate the probability of four or more travelers arriving as $1 - P(0, 1, 2, 3)$:

$$P(X \geq 4) = 1.0 - P(X = 0) - P(X = 1) - P(X = 2) - P(X = 3)$$
$$= 1.0 - 0.153 - 0.288 - 0.270 - 0.168 = 0.121$$

	A	B	C	D	E	F	G
1	x	P(X = x)					
2	0	0.153					
3	1	0.288					
4	2	0.270					
5	3	0.168					
6							
7							
8							
9							

POISSON(A2, 1.875, 0)

FIGURE 2.21
Poisson distribution calculations.

EXAMPLE 2.13

Computing the Storage Capacity of a Left-Turn-Only Lane

A left-turn-only lane at an approach to a signalized intersection can store a maximum of five vehicles. The traffic volume is 900 veh/h, and 20% of the vehicles turn left. The time needed to complete one cycle of signal indications is 60 s, and green time allocated for left turns can accommodate a maximum of five vehicles.

Determine the probability that there will be a backup of vehicles waiting to turn left thus blocking the through lane.

Solution

If six or more left-turning vehicles arrive during one 60-second cycle, then one or more will back up into the through lane. The Poisson distribution is assumed to be operable. Calculate λ, the arrival rate of left-turning veh/s:

$$\lambda = \frac{0.20 \times 900}{3600} = 0.05 \text{ left-turning veh/s}$$

Since the cycle length is 60 s and there are 0.05 left-turning veh/s, the number of left-turning veh/cycle, $\lambda t = 0.05 \cdot 60 = 3.0$.

The probability of six or more vehicles arriving in one cycle is equivalent to 1.0 minus the probability of five or less vehicles arriving. Thus

$$P[X \geq 6] = 1.0 - P[X \leq 5]$$

$P[X \leq 5]$ can be calculated using the Excel function = POISSON(5, 3, 1) to find the cumulative distribution function value corresponding to $X = 5$ and $\lambda t = 3$.

(*Note:* The number 1 replaces the zero in the Excel function since the calculation is for the cdf.)

Using the Excel function yields

$$P[X \leq 5] = 0.916 \text{ and } P[X \geq 6] = 1.0 - 0.916 = 0.084$$

The interpretation of the result is that in 8.4% of the cycles, a backup can be expected to occur.

Continuous Distributions

Continuous random variables assume any value within a certain range and are not limited to discrete values. For example, the time interval between the successive arrival of vehicles or pedestrians at an intersection can assume any value within a certain range and therefore is a continuous variable.

For continuous random variables, the probability of the variable assuming a specific value is meaningless. Instead the probability of an outcome is stated for specified intervals. Furthermore, the probability mass functions are replaced by probability density functions $f(x)$. As illustrated in Figure 2.22, the probability of a variable assuming values between a and b is the area under the probability density function between the values a and b.

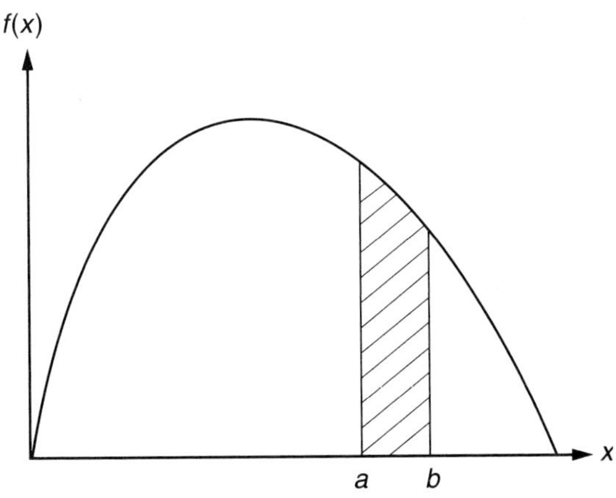

FIGURE 2.22
Continuous random variables and probability calculations.

Mathematically, this is equal to the integral of the function $f(x)$ from a to b:

$$P(a \le X \le b) = \int_{x=a}^{b} f(x)\,dx \tag{2.22}$$

If the cumulative distribution function, $F(x) = P[X \le x]$, is known, the probability that x is between a and b can be calculated as

$$P(a \le X \le b) = F(b) - F(a) \tag{2.23}$$

Normal Distributions

Normal distributions are useful models for describing a wide range of natural phenomena, and they have played a major role in the development of statistical theory. The parameters of the normal distribution are the mean (μ) and the standard deviation (σ). The equation is

$$f(x) = \frac{1}{\sigma\sqrt{2\pi}} \exp\left[-\frac{1}{2}\left(\frac{x-\mu}{\sigma}\right)^2\right] \tag{2.24}$$

The notation for the normal distribution is $N[\mu, \sigma]$, and this distribution is defined by parameters μ and σ. Figure 2.23 illustrates a normal distribution with a mean value of $\mu = 0$ and a standard deviation of $\sigma = 1$ $N[0, 1]$. The normal distribution is bell-shaped such that values near the mean have a higher probability of occurring than those farther from the mean. The area under the curve between values $(\mu - \sigma)$ and $(\mu + \sigma)$ equals 0.6826, indicating that if a random variable is normally distributed, 68% of observations will be within one standard deviation of the mean. Approximately 95% of all observations are within two standard deviations of the mean.

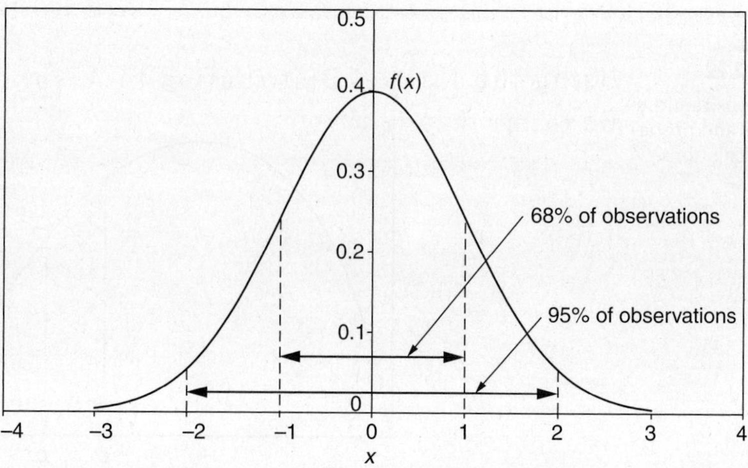

FIGURE 2.23
Normal distribution.

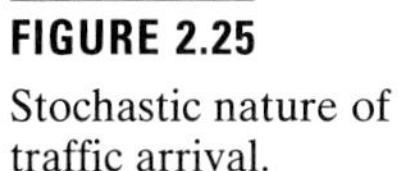

FIGURE 2.25
Stochastic nature of traffic arrival.

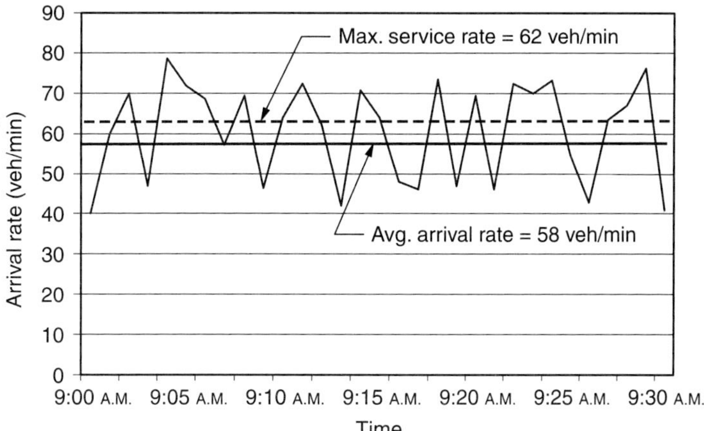

but the work zone can only handle 30 veh/min. Thus, 10 more vehicles will arrive each minute than can pass through. These vehicles will form a queue that keeps getting longer until the arrival rate is less than the departure rate.

Queues always form when the arrival rate exceeds the departure rate. However, queues can also form when the arrival rate is less than the departure rate because of the random nature of the arrival pattern such that there are surges in the arrival rate of vehicles. Figure 2.25 illustrates this phenomenon: The average arrival rate is 58 veh/min, which is less than the maximum service rate of 62 veh/min. However, because of fluctuations in the arrival rate, queues will actually form during the periods when the actual arrival rate is higher than the maximum capacity of 62 veh/min. Eventually, however, these queues would dissipate because the average arrival rate is less than the service rate.

Types of Queues

Queuing systems can be categorized based on several conditions: customer arrival patterns, customer departure or service patterns, and queue disciplines. For the arrival pattern, customers can be assumed to arrive according to one of the following options:

1. At a uniform rate (or at equal time intervals.) This option is a deterministic arrival. Because the arrival rate is deterministic, it is denoted by the letter D.
2. Interarrival times are exponentially distributed (derived from the assumption of Poisson-distributed arrivals) and denoted by the letter M (Markovian).
3. Interarrival times can be assumed to follow a general probability distribution and denoted by the letter G (general).

The service time (or time needed for the vehicle to depart) can also be assumed to follow a uniform distribution (D), a negative exponential distribution (M), or a general probability distribution (G). In queuing theory, queues and queue models are labeled with three alphanumeric values ($x/y/z$), where

x = distribution of the interarrival times (D, M, or G)
y = distribution of the service times (D, M or G)
z = number of servers

For example, a D/D/1 queuing model refers to a system where the arrival rate is uniform, the service time is constant, and there is a single server.

With respect to the queuing discipline, the two most common types are first-in, first-out queues (FIFO) queues and last-in, first-out (LIFO). In FIFO queues, the first person or vehicle to arrive is also the first to depart. For LIFO queues, the last vehicle to arrive is the first to depart. For transportation applications, the FIFO discipline is prevalent.

The following assumptions and notations are used for queuing models:

1. The arrival rate λ is customers/unit time.
2. The number of servers, c, may be one or more working in parallel.
3. The service rate, μ, is the number of customers/unit time.
4. Both the arrival rate, λ, and the service rate, μ, have the same unit of measure (for example, veh/min).
5. The ratio of the arrival rate λ to the service rate, $c\mu$, $\rho = \lambda/c\mu$.
6. $\overline{Q}$ and $\overline{W}$ denote the average number of customers waiting for service and the average waiting time/customer, respectively.
7. The average time spent in the queuing system $\bar{t}$ is equal to the (waiting time + service time).

Two Cases for Queuing Analysis

The first case is when the arrival rate does not exceed the system capacity or $\rho < 1$ and is known as a stable queue. The second case is when the arrival rate exceeds the system capacity, at least temporarily. For this case, cumulative plots described in Example 2.4 are the best analysis tool, especially if the interarrival and service times are deterministic. The following sections describe stable queues and the equations for calculating $\overline{Q}$, $\overline{W}$, and $\bar{t}$ for those problems most often encountered in transportation.

The M/D/1 Queue

Among the queuing models most often used in transportation and traffic analysis. This model assumes that the interarrival times between vehicles are exponentially distributed (i.e., Poisson arrivals), that the service times are deterministic, and that there is one server. For these conditions and assuming that $\rho < 1$, the following equations can be used to calculate $\overline{Q}$, $\overline{W}$, and $\bar{t}$:

$$\overline{Q} = \frac{\rho^2}{2(1 - \rho)} \tag{2.25}$$

$$\overline{W} = \frac{\rho}{2\mu(1 - \rho)} \tag{2.26}$$

$$\bar{t} = \frac{2 - \rho}{2\mu(1 - \rho)} \tag{2.27}$$

where

$\overline{Q}$ = average number of customers waiting for service (excluding the one being served)
$\overline{W}$ = average waiting time/customer
$\bar{t}$ = average time spent in the queuing system
μ = arrival rate, customers/unit time
ρ = ratio of the arrival rate to the service rate

EXAMPLE 2.15

Computing Characteristics for M/D/1 Queues

Travelers arrive at an airport ticket counter at a rate of 70 passengers/h. The average service time is constant and equals 45 s. Time intervals between arrivals are assumed to be exponential.

Determine

1. The average number of customers waiting in the queue
2. The average waiting time in the queue
3. The average time spent in the system

Solution

Calculate the arrival rate, λ, and the service rate, μ. Both λ and μ must have the same units:

$$\lambda = (70 \text{ customers/h})/(60 \text{ min/h}) = 1.16667 \text{ customers/min}$$

$$\mu = (60 \text{ s/min})/(45 \text{ s/customer}) = 1.3333 \text{ customers/min}$$

The ratio of the arrival to the service rate, ρ, is equal to

$$\rho = 1.16667/1.33333 = 0.875$$

Since $\rho < 1$, Equations 2.25 through 2.27 are used in this situation. Calculations are as follows:

(1) Average length of the queue:

$$\overline{Q} = \frac{\rho^2}{2(1 - \rho)} = \frac{0.875^2}{2(1 - 0.875)} = 3.06 \text{ travelers}$$

(2) Average waiting time in the queue:

$$\overline{W} = \frac{\rho}{2\mu(1-\rho)} = \frac{0.875}{2 \times 1.333 \times (1 - 0.875)} = 2.625 \text{ min/travelers}$$

(3) Average time spent in the system:

$$\bar{t} = \frac{2-\rho}{2\mu(1-\rho)} = \frac{2 - 0.875}{2 \times 1.333 \times (1 - 0.875)} = 3.375 \text{ min/travelers}$$

The M/M/1 Queue

Service times follow a negative exponential distribution, as do interarrival times. For some transportation applications, the assumption of an exponential distribution for service times is more realistic than the deterministic case—for example, at a toll booth, service times may vary based upon whether or not a driver has exact change. For an M/M/1 queue, the following equations apply:

$$\overline{Q} = \frac{\rho^2}{(1-\rho)} \tag{2.28}$$

$$\overline{W} = \frac{\lambda}{\mu(\mu - \lambda)} \tag{2.29}$$

$$\bar{t} = \frac{1}{(\mu - \lambda)} \tag{2.30}$$

For an M/M/1 queuing system, the steady-state probability that exactly n customers are in the system can be easily calculated as

$$p^n = (1-\rho)\rho^n \tag{2.31}$$

EXAMPLE 2.16

Computing Characteristics for M/M/1 Queues

For the data in Example 2.15, assume that the service times vary and are replicated by a negative exponential distribution, with a mean value of 45 s/traveler.

Determine

1. The average number of customers waiting in the queue
2. The average waiting time in the queue
3. The average time spent in the system in this case

Solution

The only difference between this example and Example 2.15 is the assumption of an M/M/1 queuing system. Equations 2.28 through 2.30 are applicable. The values for the arrival rate, λ, the service rate, μ, and the ratio of the arrival to service rate, ρ, are the same as Example 2.15).

(1) The average length of the queue:

$$\overline{Q} = \frac{\rho^2}{(1 - \rho)} = \frac{0.875^2}{(1 - 0.875)} = 6.125 \text{ travelers}$$

(2) The average waiting time in the queue:

$$\overline{W} = \frac{\lambda}{\mu(\mu - \lambda)} = \frac{1.166667}{1.333 \times (1.3333 - 1.16667)}$$
$$= 5.25 \text{ min/traveler}$$

(3) The average time spent in the system:

$$\bar{t} = \frac{1}{(\mu - \lambda)} = \frac{1}{1.3333 - 1.16667)} = 6.00 \text{ min/traveler}$$

Optimization and Decision-Making Techniques

Optimization refers to the process that seeks to determine the "best" solution or course of action to a problem. The process provides a way for an objective, systematic analysis of the different decisions encountered when dealing with complex, real-world problems. As one might imagine, optimization problems are ubiquitous in the field of transportation infrastructure engineering. They are often encountered in the planning, design, construction, operations, management, and maintenance of transportation infrastructure.

In the planning process, for example, the transportation planner is typically seeking the *optimal* allocation or use of the available funds. During the design stage, one of the major tasks for the transportation engineer is to identify the optimal alignment for a proposed transportation travelway, an alignment that would minimize the cost of construction and earthwork.

Optimization is also a very powerful tool for traffic operations and management. In traffic signal design, the traffic engineer is typically trying to come up with an *optimal* signal plan (i.e., the *optimal* sequence and duration of the different signal indications) that would minimize the total travel time or travel delay. Finally, within the field of transportation infrastructure maintenance and management, optimization techniques are commonly used to determine optimal treatment strategies and the optimal timing for implementing that strategy.

An optimization process requires a mathematical model appropriate for the problem, which is then solved, yielding an optimal solution. The technique used to solve the model depends upon the model's form—for example, whether the model is linear or contains nonlinear functions. Among the optimization techniques are linear programming, dynamic programming, integer programming, and nonlinear programming.

Regardless of the model form that is used to formulate the mathematical model for the problem under consideration, it involves three basic steps: (1) identification of the decision variables; (2) formulation of the objective function; and (3) formulation of the model's constraints.

Decision variables: These variables represent the decisions to be made. Therefore, for a given problem, if there are n quantifiable decisions, they are represented as the decision variables $(x_1, x_2, x_3, \ldots, x_n)$ whose respective values are determined by solving the optimization model.

Objective function: A mathematical equation that represents the performance measure (e.g., profit or cost) that will be maximized or minimized. It is a function of the decision variables, and the solution of the mathematical program seeks the optimal values for the decision variables that would maximize or minimize the objective function.

Constraints: Restrictions on the values that can be assigned to the decision variables. They are typically expressed in the form of inequalities or equations and an example is $x_1 + x_2$ ″ 10.

An optimization model can therefore be stated as a process for choosing the values of the decision variables so as to maximize or minimize the objective function subject to satisfying the specified constraints.

EXAMPLE 2.17

Using Linear Programming to Maximize Production Strategies

A small manufacturing company produces two microwave switches, A and B. The return/unit of switch A is \$20, while the return/unit of switch B is \$30. Because of contractual requirements, the company manufactures at least 25 units of switch A/week. Given the size of the labor force at the factory, only 250 h of assembly time are available/week. Switch A requires 4 h of assembly, whereas switch B requires three. Formulate a production strategy that will maximize the company's return.

Solution

In this example, formulate the problem mathematically, and then solve using Microsoft Excel.

Mathematical Formulation Formulating a mathematical model involves three basic steps: (1) identification of the decision variables; (2) formulation of the objective function; and (3) formulation of the model's constraints. Each of these steps is described below:

DECISION VARIABLES In this problem, two decisions need to be made, the number of switches of type A to produce/week and the number of switches of type B. Two decision variables are required:

x_A = number of type A switches produced/week and x_B = number of type B switches produced/week.

OBJECTIVE FUNCTION The objective is to maximize the company's return. Since the return on one switch A is \$20 and \$30 for B, the objective function can be formulated mathematically as follows:

$$\text{Maximize } z = 20x_A + 30x_B$$

CONSTRAINTS Constraints represent any restrictions on the values that can be assigned to the decision variables. In this example, there are three groups of restrictions: (1) minimum number of switch A to be produced/week; (2) available hours of assembly; and (3) values of decision variables must be positive. The constraints are formulated as follows:

(1) *Minimum number of switches A:*

$$x_A \geq 25$$

(2) *Assembly hours available:*

$4\,x_A + 3\,x_B \leq 250$ (switch A requires 4 h of assembly, and switch B needs 3 h)

(3) *Allowable values:*

$$x_A, x_B \geq 0$$

The third constraint is called the nonnegativity constraint and is common in most problems.

The complete mathematical model is formulated as follows:

$$\text{Maximize } z = 20x_A + 30x_B$$

subject to

$$x_A \geq 25$$

$$4x_A + 3x_B \leq 250$$

$$x_A, x_B \geq 0$$

Solution Algorithm

The optimization model formulated above is an example of linear programming (LP). It is called *linear* since all the equations for the objective function and constraints are linear in the decision variables. A number of techniques are available for solving LP models. For small problems with two or three decision variables, a simple graphical technique can be used. For larger and more realistic problems with more than three variables, the graphical technique is impractical. For these problems a powerful solution algorithm is the Simplex method, developed by George Dantzig in 1947. It is not necessary to be familiar with this algorithm in order to solve the LP problem since the algorithm has been coded into several commercially available software packages, including Microsoft Excel. The following section illustrates how to use Excel to solve this problem.

Solution Using Microsoft Excel's Solver

Microsoft Excel comes with an Add-In called Solver that can be used for mathematical programming problems, including LP. To check if Solver has been activated, use the Tools menu, and select Solver. If it is not listed, scroll to Add-In in the Tools menu, and check the box next to Solver, as illustrated in Figure 2.26.

With Solver activated, the procedure for solving optimization problems consists of the following four steps:

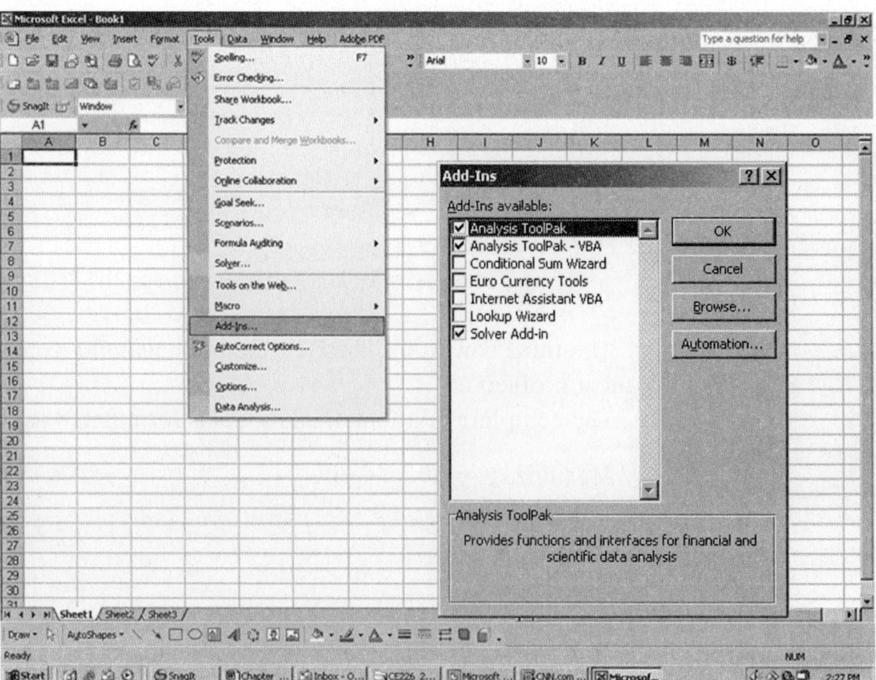

FIGURE 2.26
Adding the Excel Solver Add-In.

FIGURE 2.27
Data input for Example 2.17.

	A	B	C	D	E	F
1		Switch A	Switch B			
2	Min. Number	1			>=	25
3	Assembly Hours	4	3		<=	250
4	Contribution	20	30			
5	Value of Variable	1	1			
6						

Step 1: Data input: Open a new Excel spreadsheet and enter the data of the problem, as shown in Figure 2.27. Label two columns B and C for Switch A and Switch B. Rows 2 and 3 are for values of the parameters for the minimum number and assembly hours' constraints (i.e., the $x_A \geq 25$ constraint, and the $4x_A + 3x_B \leq 250$ constraint, respectively). The parameters to be multiplied by x_A are entered under the Switch A column, and those to be multiplied by x_B are under the Switch B column. Row 4 contains the parameters of the objective function 20 and 30. Row 5 contains the values of the two decision variables. Cells B5 and C5 will ultimately contain the optimal values for x_A and $x_{B,}$ respectively. Initially, this row could be left blank or an estimate could be entered. The ">=" and "<=" signs in cells E2 and E3 are a memory aid and do not affect the solution procedure. Finally, Cells D2, D3, and D4 are intentionally left blank. We will enter a linear expression in each cell in step 2.

Step 2: Enter the linear expressions: The linear expressions of the LP model are formulated and include the objective function and the left-hand side of the constraints. The SUMPRODUCT function in Excel is used to compute these linear expressions. The syntax for this function is = SUMPRODUCT(array 1, array 2, . . .). Array 1 could be a group of cells arranged in a column, a row, or even cells in an $m \times n$ array format (i.e., occupy m rows by n columns). The function proceeds by first multiplying each cell in array 1 by the corresponding cell in array 2, and then adding these products.

Figure 2.28 shows how the SUMPRODUCT function can be used to calculate the linear expressions for the LP model. As noted, arbitrary values have been assumed for the decision variables in cells B5 and C5. In step 4, Solver is used to compute the values of the variables that maximize the objective function.

FIGURE 2.28
Using the SUMPRODUCT function in Excel.

D2 =SUMPRODUCT(B2:C2,B$5:C$5)

	A	B	C	D	E	F	G	H	I
1		Switch A	Switch B						
2	Min. Number	1		1	>=	25			
3	Assembly Hours	4	3	7	<=	250			
4	Contribution	20	30	50					
5	Value of Variable	1	1						
6									
7									
8				=SUMPRODUCT(B3:C3,B$5:C$5)					
9									

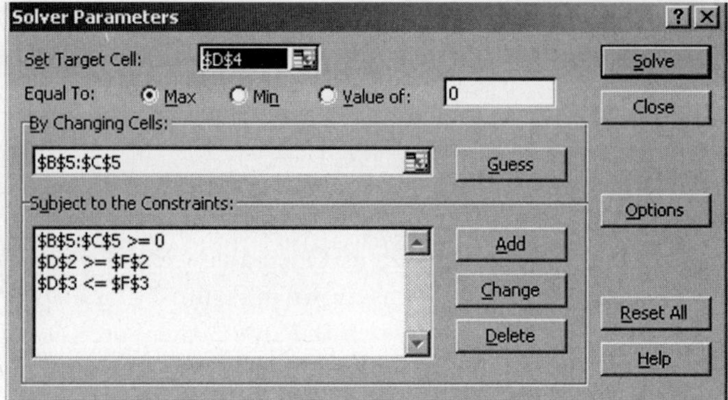

FIGURE 2.29
Formulating the LP model.

The formula for cell D2 is =SUMPRODUCT(B2:C2, B$5, C$5). This formula is then copied into cells D3 and D4. The use of the absolute reference $ for the B$5:C$5 array allows the formula in D2 to be copied into D3 and D4. The assumed values of the decision variables are in cells B5 and C5. Cell D2 contains the number of Type A switches (based upon the assumed values of the decision variables). Cell D3 is the number of assembly hours used to produce one switch A and one switch B in this case. Cell D4 is the total return from the sale of one switch A and one switch B.

Step 3: Use Solver to formulate the LP: Use Solver to direct the program to those cells that contain the decision variables, the objective function, and the constraints. This is achieved by entering the Tools menu and selecting Solver. The window depicted in Figure 2.29 is presented and the user is asked to specify the cells containing the decision variables, the objective function, and the constraints.

Specify the "*Target Cell*" containing the expression to be maximized, which is cell D4 in this example. To *maximize* this value, ask Solver to change the decision variables in cells B5 and C5. Specify the three groups of constraints by clicking on the "*Add*" button, which produces the window shown in Figure 2.30.

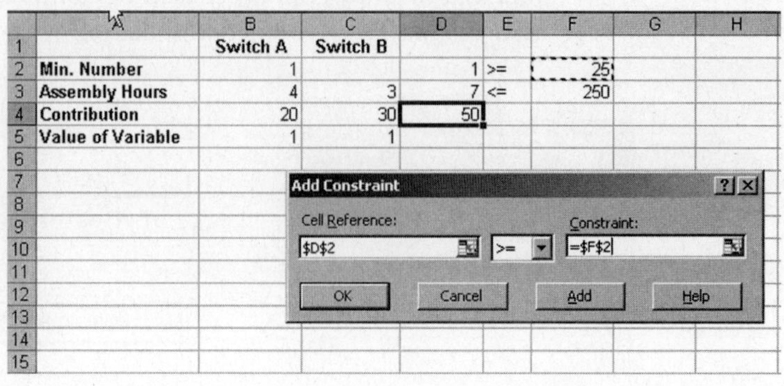

	A	B	C	D	E	F
1		Switch A	Switch B			
2	Min. Number	1		1	>=	25
3	Assembly Hours	4	3	7	<=	250
4	Contribution	20	30	50		
5	Value of Variable	1	1			

FIGURE 2.30
Specifying the constraints.

FIGURE 2.31
Solving the LP model.

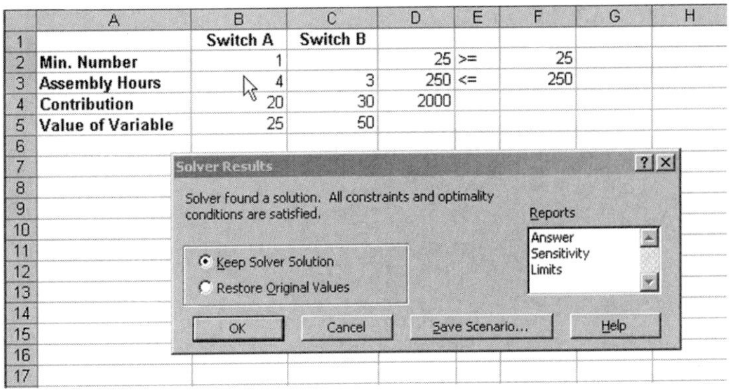

	A	B	C	D	E	F	G	H
1		Switch A	Switch B					
2	Min. Number	1		25	>=	25		
3	Assembly Hours	4	3	250	<=	250		
4	Contribution	20	30	2000				
5	Value of Variable	25	50					

Solver Results

Solver found a solution. All constraints and optimality conditions are satisfied.

Reports: Answer, Sensitivity, Limits

Keep Solver Solution

Restore Original Values

OK | Cancel | Save Scenario... | Help

Clicking on the "Add" button once again would allow the user to enter more constraints.

Step 4: Solve the LP: Solver will solve the model after selecting the Solve button shown in Figure 2.29. The solution will appear on the spreadsheet shown in Figure 2.31. To retain the optimal solution, select the "Keep Solver Solution" option. Additional reports can also be requested.

EXAMPLE 2.18

Optimizing Construction Materials Inventories

A materials supplier provides gravel to various transportation infrastructure construction sites. He purchases the material from three different sources (A, B, and C) for \$140/t, \$180/t, and \$170/t, respectively, and has a contract to meet weekly demand for gravel at four different construction locations. The cost to ship the material to each site, available supply, and the weekly demand are shown in Table 2.5. Note that, the symbol t refers to metric ton or tonne.

To understand Table 2.5, consider that supplier A can provide a maximum of 1000 metric tons of gravel/week at \$140/t. The cost for transporting one metric ton from supplier A to site 1 is \$30. The total demand at site 1 is 600 metric tons.

Determine an optimal policy that would meet the needs at the four sites but at a minimum transportation cost to the supplier.

TABLE 2.5
Data for Example 2.18

Supplier	*Site 1*	*Site 2*	*Site 3*	*Site 4*	*Supply*
A	\$30	\$22	\$37	\$43	1000
B	\$35	\$27	\$34	\$26	1200
C	\$19	\$41	\$33	\$29	800
Demand	600	700	500	800	

Solution

Formulate the problem mathematically, and solve the LP formulation using Excel.

Mathematical Formulation

Let

$c_{i,j}$ = unit cost of transporting one metric ton of gravel from supplier i (where i is 1, 2, or 3) to site j (where j is 1, 2, 3, or 4). For example, $c_{2,3}$ = \$34, which is the unit cost of transporting one metric ton from supplier B to site 3.

s_i = the amount available from supplier i (e.g., 1000 metric tons for $i = 1$)

d_j = demand needed at site j (e.g., 600 metric tons for $j = 1$)

p_i = purchase price of one metric ton of gravel from supplier i (e.g., \$140 for $i = 1$)

Proceed to formulate the LP by identifying the decision variables, the objective function, and the constraints.

DECISION VARIABLES For this problem, the decisions are how many metric tons should be transported from a particular supplier to a given site. Define the following decision variables:

$x_{i,j}$ = metric tons of gravel/week to be transported from supplier i to site j (i=1 . . . 3; j=1, . . . ,4)

There are 12 decision variables in this problem, since we have three suppliers and four sites ($3 \times 4 = 12$). For example, the variable $x_{2,3}$ is the number of metric tons of gravel/week transported from supplier 2 to site 3.

OBJECTIVE FUNCTION The goal here is to minimize the total cost, which includes the purchase price as well as the transportation cost. This could be formulated mathematically as follows:

$$\text{Minimize } z = \sum_{i=1}^{3} p_i \times \left(\sum_{j=1}^{4} x_{i,j} \right) + \sum_{i=1}^{3} \sum_{j=1}^{4} c_{i,j}\, x_{i,j}$$

The first term expresses the purchase cost whereas the second term is the transportation cost.

CONSTRAINTS The constraints of this problem are of three types: The first ensures the available supply at each supplier is not exceeded; the second ensures the required demand arrives at each site; and the third is a nonnegativity constraint for the $x_{i,j}$'s. These constraints are defined as follows:

1. *Supply capacity constraints:*

$$\sum_{j=1}^{4} x_{i,j} \leq s_i \qquad \text{for } i = 1, 2, \text{ and } 3$$

2. *Demand constraints:*

$$\sum_{i=1}^{3} x_{i,j} = d_i \qquad \text{for } j = 1, 2, 3, \text{ and } 4$$

3. *Nonnegativity constraints:*

$$x_{i,j} \geq 0 \qquad \text{for } i = 1, \ldots, 3 \text{ and } j = 1, \ldots, 4$$

The LP model can be summarized as follows:

$$\text{Minimize } z = \sum_{i=1}^{3} p_i \times \left(\sum_{j=1}^{4} x_{i,j} \right) + \sum_{i=1}^{3} \sum_{j=1}^{4} c_{i,j}\, x_{i,j}$$

Subject to

$$\sum_{j=1}^{4} x_{i,j} \leq s_i \qquad \text{for } i = 1, 2, \text{ and } 3$$

$$\sum_{i=1}^{3} x_{i,j} = d_i \qquad \text{for } j = 1, 2, 3, \text{ and } 4$$

$$x_{i,j} \geq 0 \qquad \text{for } i = 1, \ldots, 3 \text{ and } j = 1, \ldots, 4$$

Excel Solution

To complete this problem, use Excel Solver as shown in the following steps:

Step 1: Data input: Enter the data of the problem. As shown in Figure 2.32, the input data table from Table 2.5 is copied into the cell range A1 . . . F5. Then create a 3 × 4 matrix of cells into the cell range B10 . . . E12 to contain the values of the 12 decision variables as shown in Figure 2.32. Initially, the value 1 is assumed for these variables and the cells will contain the optimal solution when the solution process is complete.

Step 2: Enter the linear expressions: The linear expressions for the left-hand side of the constraints and for the objective function are entered. From the supply constraint equations note that the left-hand side is the sum of the decision variables that occupy a particular row in the spreadsheet. For example, for supplier 1, the total number of metric tons of gravel supplied is given by the sum of the values in cells B10 . . . E10. Enter the formula =SUM(B10 . . . E10) in cell F10. Copy it into cells F11 and F12.

Similarly, for the demand constraints, the left-hand side represents the sum of the decision variables in a given column. For site 1, the total number of metric tons transported to that site is given by the sum of the decision variables in cells B10 . . . B12. Enter the formula =SUM (B10 . . . B12) in cell B13 and then copy it into cells C13, D13, and E13.

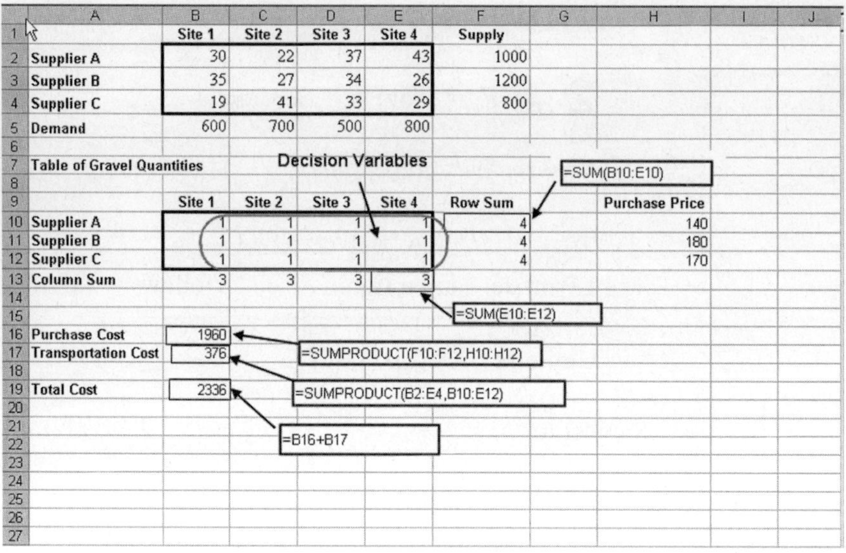

FIGURE 2.32 Data input for Example 2.18.

For the objective function, the purchase cost is in cell B16 and the transportation cost is in cell B17. The expression in cell B16, SUMPRODUCT (F10:F12, H10:H12), multiplies the total quantity supplied (i.e., cells F10 . . . F12) by the unit purchase cost (in cells H10 . . . H12). For the transportation cost, the expression in cell B17, SUMPRODUCT(B2:E4, B10:E12) multiplies the decision variables by the unit transportation cost and sums the product; that is, it computes the quantity $\sum_{i=1}^{3}\sum_{j=1}^{4} c_{i,j} x_{i,j}$. Add the values in cells B16 and B17 to obtain the value for the objective function in cell B19.

Step 3: Use Solver to formulate the LP: Specify the cells containing the decision variables (the changing cells), the objective function (the target cell) that should be maximized or minimized, and the constraints. This is shown in Figure 2.33.

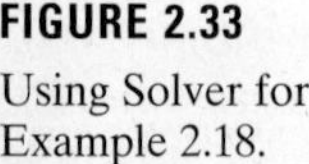

FIGURE 2.33 Using Solver for Example 2.18.

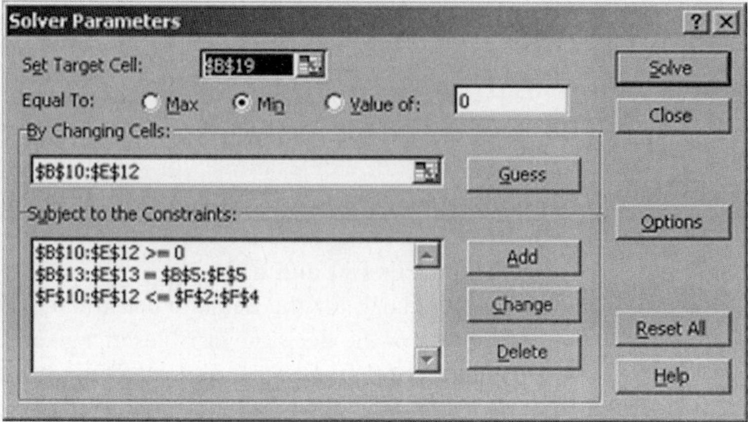

FIGURE 2.34
Final solution for Example 2.18.

	A	B	C	D	E	F	G	H	I
1		Site 1	Site 2	Site 3	Site 4	Supply			
2	Supplier A	30	22	37	43	1000			
3	Supplier B	35	27	34	26	1200			
4	Supplier C	19	41	33	29	800			
5	Demand	600	700	500	800				
6									
7	Table of Gravel Quantities								
8									
9		Site 1	Site 2	Site 3	Site 4	Row Sum		Purchase Price	
10	Supplier A	0	700	300	0	1000		140	
11	Supplier B	0	0	0	800	800		180	
12	Supplier C	600	0	200	0	800		170	
13	Column Sum	600	700	500	800				
14									
15									
16	Purchase Cost	420000							
17	Transportation Cost	65300							
18									
19	Total Cost	485300							
20									

The target cell is B19. Minimize the value to minimize the total cost. The changing cells are B10 . . . E12, and the following constraints are specified: (1) the supply constraints (F10 . . . F12 <= F2 . . . F4); (2) the demand constraints (B13 . . . E13 <= B5 . . . E5); and (3) the nonnegativity constraints (B10 . . . E12 >= 0).

Step 4: Solve the LP model: The LP model is then solved by clicking on the Solve model. The optimal solution obtained is shown in Figure 2.34. The result is 600 metric tons from supplier C to site 1, 700 metric tons from supplier A to site 2, 300 metric tons from supplier A, 200 metric tons from supplier C to site 3, and 800 metric tons from supplier B to site 4.

SUMMARY

This chapter has provided a brief overview of transportation systems and their components and described fundamental mathematical tools and models that can be used to solve transportation systems problems. Among the techniques introduced were (1) traffic analysis tools such as timespace diagrams and cumulative plots; (2) regression analysis techniques that can be used to develop empirical models for transportation applications; (3) probability distributions that can be used for a number of traffic operations and design problems; (4) queuing models; and (5) optimization techniques. In the next chapters of this textbook, these tools will be employed to solve a number of transportation systems and infrastructure engineering problems.

PROBLEMS

2.1 What are some of the basic characteristics of systems?

2.2 Briefly list the different components of transportation systems.

2.3 What types of problems are best addressed using time–space diagrams? Which problems are addressed using cumulative plots?

2.4 A freight and passenger train share the same track. The average speed of the freight train is 72 km/h, whereas that of the passenger train is 112 km/h. The passenger train is scheduled to leave 20 minutes after the freight train departs from the same station. Determine the location where a siding would need to be provided to allow the passenger train to pass the freight train. Also determine the time it takes for the freight train to arrive at the siding. As a safety precaution, the separation headway between the two trains at the siding should be at least equal to 6 minutes.

2.5 Three friends embark on a trip using a tandem bicycle that can carry two of them at a time. To complete the trip, they proceed as follows: first, two friends (friends A and B) ride the bicycle at an average speed of 24 km/h for exactly 15 minutes. While this is taking place, the third friend (friend C) walks at an average speed of 6 km/h. After 15 minutes, friend A drops friend B and then rides back to meet friend C at an average speed of 27 km/h. When friend A meets friend C, they ride together at an average speed of 25.5 km/h, until they meet friend B. The cycle just described is then repeated until the trip is completed. Determine the average speed of the three friends.

2.6 A freight train and a passenger train share the same rail track. The freight train leaves station A at 8:00 A.M. The train travels at a speed of 45 km/h for the first 10 minutes and then continues to travel at a speed of 60 km/h. At 8:35 A.M., the passenger train leaves station A. The passenger train travels first at a speed of 75 km/h for 5 minutes and then continues to travel at a speed of 105 km/h. Determine the location of the siding where the freight train will have to be parked to allow the faster passenger train to pass through. As a safety precaution, it is determined that the time headway between the two trains should not be allowed to fall below 5 minutes.

2.7 Travelers arriving at a certain airline counter at a given airport arrive according to the pattern shown below. It is estimated that on average it takes 45 seconds to serve a customer at the counter. For the first 30 minutes (i.e., from 9:00 to 9:30), the airline has only two counters open. At 9:30, however, a third counter opens and remains open until 10:30.

Time Period	*15-Min Count*	*Cumulative Count*
9:00–9:15	45	45
9:15–9:30	60	105
9:30–9:45	55	160
9:45–10:00	40	200
10:00–10:15	35	235
10:15–10:30	55	290

(a) Draw a cumulative plot showing the arrival and departure patterns for the travelers at the airline counters.
(b) What is the length of the queue at 9:30?
(c) What is the maximum length of the queue?
(d) What is the time at which no one remains in line?
(e) What is the total wait time for all customers in units of customer/minutes?

2.8 An incident occurs on a freeway that has a capacity in the northbound direction, before the incident, of 4400 veh/h and a constant flow rate of 3200 veh/h during the morning commute. At 7:30 A.M., a traffic accident occurs and totally closes the freeway (i.e., reduces its capacity to zero). At 7:50 A.M., the freeway is partially opened with a capacity of 2000 veh/h. Finally, at 8:10 A.M., the wreckage is removed and the freeway is restored to full capacity (i.e., 4400 veh/h).

(a) Draw cumulative vehicles' arrival and departure curves for the scenario described.
(b) Determine the total magnitude of delay, in units of veh.h, from the moment the accident occurs to the moment the queue formed totally dissipates.

2.9 A six-lane freeway (i.e., three lanes in each direction) has a capacity of 6000 veh/h/lane under normal conditions. On a certain day, an accident occurs at 4:00 P.M. The accident initially results in blocking two of the three lanes of the freeway and hence reduces the capacity to only 2000 veh/h for that direction. At 4:30 P.M., the freeway capacity is partially restored to a value of 4000 veh/h. Finally, at 5:00 P.M., the accident is totally cleared, and the full capacity of the freeway is restored. Given that the traffic demand at the accident site is given by the following table, determine

(a) The maximum length of the queue formed at the accident site;
(b) The time the queue dissipates; and
(c) The total delay.

Time Period	*15-Min Volume*
4:00–4:15	700
4:15–4:30	900
4:30–4:45	1100
4:45–5:00	1200
5:00–5:15	800
5:15–5:30	700
5:30–5:45	1100
5:45–6:00	900

2.10 To evaluate the condition of bike paths, a condition index was developed that rates the surface condition of each bike path segment on a scale from 0 to 100, with 100 referring to a segment in perfect condition. The table below shows the inspection data for several bike path segments in terms of the condition index (CI) for each segment, along with its AGE (i.e., the number of years since its construction):

Condition Index (CI)	*Age (years)*
100	0
98	0.5
96	1.2
93	3
100	0
93	2
88	4
86	7
100	0
95	2
90	4
83	8
100	0
92	3
85	6
82	9
81	10

It is postulated that the deterioration of bike path surfaces can be expressed using the following equation:

$$\text{CI} = a + b(\text{AGE}) + c(\text{AGE})^2$$

Using regression, develop a deterioration prediction curve for bike paths. Plot the resulting curve to show the typical deterioration trend for bike paths.

2.11 You are asked to develop a relationship between the total number of trips generated by a wholesale tire store and the gross floor area of the store. To do this, you compile a data set showing the average number of vehicle trips/day to and from several tire stores. The gross floor area of the stores is 100 m^2. The data compiled are as shown below. How well does your model fit the data?

Store ID	*Trip Ends/Day*	*Gross Floor Area*
1	170	9
2	300	14.5
3	250	12
4	350	17
5	340	18
6	200	11
7	230	14
8	250	16
9	100	6
10	400	19
11	150	8
12	380	17.5
13	220	11.5
14	270	12.5
15	280	14

2.12 It is postulated that the relationship between the average speed of a traffic stream, u, in km/h, and the density (which gives the number of vehicles/unit length), k, in veh/km for a given transportation facility can be expressed as follows:

$$u = c \ln \frac{k_j}{k}$$

where c and k_j are parameters. To fit the above equation, average speeds and density were collected from the facility at different times of the day and different usage levels. The data collected are as shown below. Use regression to fit the above equation to the data. What are the values for the two parameters c and k_j?

Speed, u (km/h)	*Density, k (veh/km)*
85	14
65	28
59	32
16	78
40	44
32	53
77	17
72	22
43	40
24	56
21	60
56	36
55	37
59	32

2.13 In the context of probability theory, explain what is meant by a random variable.

2.14 Give some examples of random variables that arise within the context of transportation systems problems and that follow each of the following probability distributions: (1) binomial distribution; (2) geometric distribution; (3) Poisson distribution; and (4) normal distribution.

2.15 Differentiate between the probability distribution function (pdf) and the cumulative distribution function (cdf).

2.16 In the context of probability density functions, what does the p-fractile function compute? Illustrate using a simple diagram.

2.17 The following table lists the observed speeds of a number of trains as they pass a certain point midway between two stations. Determine (1) the average speed; (2) the variance; and (3) the standard deviation.

Train ID	*Speed (km/h)*
1	114
2	111
3	117
4	88
5	100
6	85
7	108
8	72
9	87

Train ID	*Speed (km/h)*
10	101
11	108
12	66
13	121
14	77
15	69
16	88
17	93
18	95
19	101
20	108

2.18 Pedestrians arrive at a signalized intersection crossing at the rate of 600 pedestrians/h. The duration of the red interval for the pedestrians at that intersection is 45 seconds. Assuming that pedestrians' arrival pattern can be described using a Poisson distribution, what is the probability that there will be more than 10 pedestrians waiting to cross at the end of the pedestrian red interval?

2.19 Airplanes arrive at an airport at an average rate of 10 aircraft/h. Assuming that the arrival rate follows a Poisson distribution, calculate the probability that more than four aircraft would land during a given 15 minutes.

2.20 An approach to an intersection carries an average volume of 1000 veh/h with 15% of the vehicles desiring to turn left. The cycle length at the intersection is equal to 75 seconds. The city would like to construct a left-turn bay at the intersection in order to minimize the probability of the left-turning vehicles blocking the through lane. You are asked to determine the minimum length of the left-turning bay at that approach so that the probability of an arriving left-turning vehicle not finding enough room on the left-turn bay is less than 10%. Assume the average vehicle length is equal to 6 m. Assume Poisson arrivals.

2.21 A trucking company has enough capacity to transport 2000 metric tons of a certain material per week. If the weekly demand for transporting such material is normally distributed with a mean of 1750 metric tons and a standard deviation of 300 metric tons, determine

(a) The probability that within a given week, the company would have to turn down requests for transporting the material;

(b) The capacity that the company should maintain so that the probability of it turning down transportation requests is less than 5%.

2.22 A certain company operates small ferryboats between a small island and the mainland. Each ferryboat can carry a maximum of six vehicles. It is estimated that vehicles arrive at the ferryboat dock at the rate of 12 veh/h. The company is interested in determining the frequency at which the ferryboats should be operated so that the probability of a vehicle being left at the dock because the boat does not have enough capacity does not exceed 10%. Assume that vehicles arrive according to a Poisson distribution.

2.23 An airport serves four different types of aircraft: Heavy (H), Large (L), Medium (M), and Small (S). During a typical hour, 40 Heavy aircraft, 50 Large, 60 Medium, and 70 Small aircraft land. Determine the probability that

(a) The next aircraft to land is a Small aircraft;
(b) In a stream of 20 aircraft, at least 5 Small aircraft would land;
(c) The first Medium aircraft would be the fifth aircraft.

2.24 In the context of queuing theory, what is the difference between the time a customer spends in the queue and the time he or she spends in the system?

2.25 Give some examples of queuing systems in transportation systems.

2.26 On what basis are different types of queues distinguished?

2.27 Why do queues form?

2.28 At a given airport, aircraft arrive at an average rate of 8 aircraft/h following a Poisson distribution. The average landing time for an aircraft is 5 minutes. However, this time varies from one aircraft to another. This variation can be assumed to be exponentially distributed. Determine

(a) The average number of aircraft awaiting clearance to land;
(b) The average time an aircraft spends in the system;
(c) The probability that there will be more than five aircraft awaiting clearance to land.

2.29 Travelers arrive at the ticket counter of a particular airport at the rate of 90 customers/h. The average service time per customer is more or less fixed and is equal to 30 seconds. Determine the average queue length, the average waiting time in the queue, and the average time spent in the system.

2.30 Travelers arrive at a ticket counter of an Amtrak train station at the rate of 100 travelers/h. It has been estimated that it takes an average of 30 seconds to serve each customer at the counter. Assuming that arrivals can be described using a Poisson distribution, determine the average wait time in the queue and the average number of customers waiting in the queue.

2.31 Travelers at an airport arrive at a given security check point at the rate of 120 travelers/h. Check-in times for passengers vary according to a negative exponential distribution with an average value of 25 seconds per passenger. Determine

(a) The average number of passengers waiting in line in front of the security check point;
(b) The average waiting time for passengers;
(c) The average time a passenger spends in the system.

2.32 In the previous problem, it is desired to limit the probability that there would be more than seven passengers in line to a value less than 5%. Determine the maximum rate of arrival that should be allowed.

2.33 Give some examples of optimization problems arising within the field of transportation infrastructure engineering.

2.34 What are the three basic steps in formulating optimization models?

2.35 A certain transit authority must repair 120 subway cars/month. At the same time, the authority must refurbish 60 subway cars. Each task can be done in the authority's facility or can be contracted out. Private contracting increases the cost by $1000/car repaired, and by $1500/car refurbished. Car repair and refurbishing take place in three shops—namely, the assembly shop, the machine shop, and the paint shop. Repairing a single car consumes 2% of the assembly shop capacity and 2.5% of the machine shop capacity. On the other hand, refurbishing a single car takes up 1.5% of the assembly shop capacity and 3% of the paint shop capacity. Formulate the problem of minimizing the monthly expense for private contracting as a linear program, and solve it using Microsoft Excel's Solver.

2.36 You are required to come up with the best plan to transport finished products from three production plants to four marketplaces. The production capacities for the three plants are 2000, 3500, and 4000 units, respectively. At the same time, the demand that has to be met at each marketplace is for 3200 units at market 1, 2800 units at market 2, 2000 at market 3, and 1500 at market 4. The unit shipping costs are given in the table below.

	Market 1	*Market 2*	*Market 3*	*Market 4*
Plant 1	4.5	6.5	4	7
Plant 2	11	4	12	3
Plant 3	5	7	8	4

References

1. Dagnazo, C. F., *Fundamentals of Transportation and Traffic Operations,* Elsevier Science Ltd., Oxford, United Kingdom, 1997.
2. Denardo, E. V., *The Science of Decision Making: A Problem-Based Approach Using Excel,* John Wiley & Sons, Inc., New York, 2002.
3. Garber, N. J., and Hoel, L. A., *Traffic and Highway Engineering,* Brooks/Cole, Pacific Grove, California, 2002.
4. Hillier, F. S., and Lieberman, G. J., *Intorduction to Mathematical Programming,* McGraw-Hill, New York, 1995.
5. Ignizio, J. P., and Cavalier, T. M., *Linear Programming,* Prentice Hall, Upper Saddle River, NJ, 1994.
6. Khisty, C. J., and Mohammadi, J., *Fundamentals of Systems Engineering: With Economics, Probability and Statistics,* Prentice Hall, Upper Saddle River, NJ, 2001.
7. Ossenbruggen, P., *Systems Analysis for Civil Engineers,* John Wiley & Sons, Inc., New York, 1984.
8. ReVelle, C. S., Whitlatch, E. E., and Wright, J. R., *Civil and Environmental Systems Analysis,* Prentice Hall, Upper Saddle River, NJ, 1997.
9. Washington, S. P., Karlaftis, M. G., and Mannering, F. L., *Statistical and Econometric Methods for Transportation Data Analysis.* Chapman & Hall/CRC, Boca Raton, FL, 2003.

CHAPTER 3

Human, Vehicle, and Travelway Characteristics

The main components of any mode of transportation are human beings, the vehicle, and the travelway. In the highway mode, human beings are the drivers and pedestrians, the vehicle is the automobile, and the travelway is the roadway. Similarly, in rail transportation, human beings are the locomotive drivers, the train is the vehicle, and the travelway is the railroad. To provide an efficient and safe transportation system, it is essential for the transportation engineer to have adequate knowledge of the characteristics and limitations of those components that are of importance to the operation of the system.

Awareness of the interrelationships among these components is also important in order to determine the effects, if any, they have on each other. These characteristics also become critical when the control of the operation of any transportation system is under consideration. For example, when traffic engineering measures such as control devices are to be used, certain characteristics of drivers—(e.g., how quickly they react to a stimulus), of vehicles (e.g., how far the vehicle travels during a braking maneuver), and of roadway conditions of the roadway (e.g., the grade) are of particular importance.

It should be noted, however, that knowing average limitations might not always be adequate; sometimes it may be necessary to obtain information on the full range of limitations. In the United States, for example, automobile drivers' ages range from 16 to above 70 and can exceed 80, and airplane pilots' ages range from 18 to over 70. Sight and hearing vary considerably across age groups and can vary even among individuals of the same age group.

Similarly, the automobile fleet consists of a wide range of vehicles, varying from compact cars to articulated trucks, just as the airplane fleet varies from single-engine propeller planes to jumbo jets such as the 747 jumbo jets

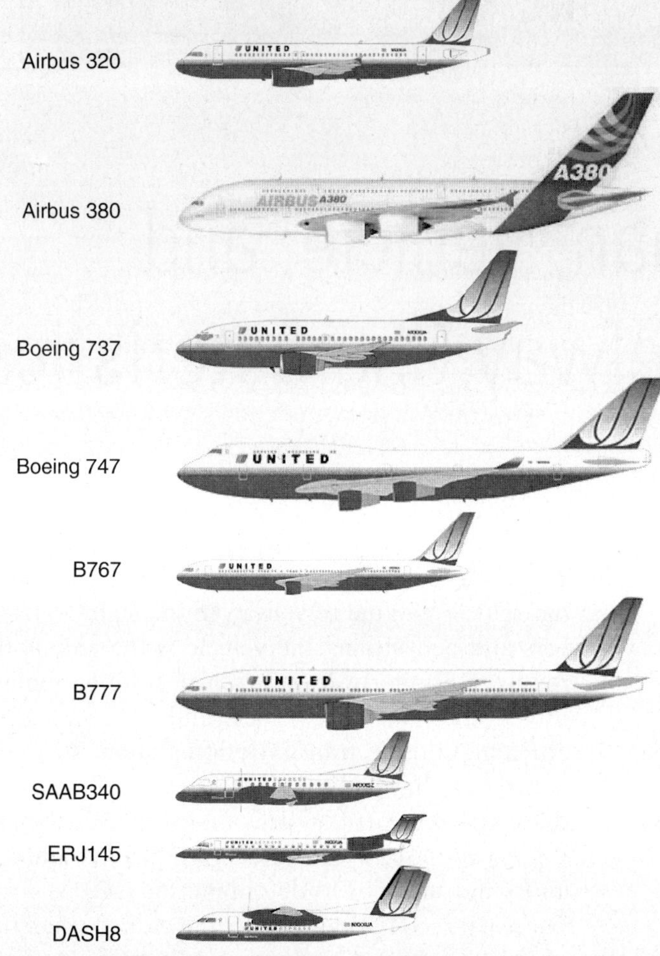

FIGURE 3.1
Different types of Aircraft used in civil aviation.

Source: United Airlines Web site, Airbus Web site.

(Figure 3.1). The characteristics of these different airplanes vary significantly. The maximum acceleration, turning radii, and ability to climb grades differ considerably among the various types of automobiles, just as the take-off and landing distances and the maximum flying heights for different airplanes also vary. Therefore, the road must be designed to accommodate a wide range of characteristics and at the same time allow use by drivers and pedestrians with a wide range of physical and psychological characteristics; an airport runway must be designed to accommodate the take-off and landing requirements of all the planes using it. Similarly, port and harbor facilities should be designed to accommodate the characteristics of the ships expected to use the port.

This chapter discusses characteristics of the human, vehicle, and travelway as they relate to highway, air, rail, and water modes of transportation.

Human Characteristics

A major problem that faces transportation engineers when they consider human characteristics or factors—usually referred to as *ergonomics*—in the design of transportation systems is the varying skills and perceptions of humans using and/or operating the system. This is demonstrated in the wide range of people's abilities to react to information. Studies have shown that these abilities may also vary in an individual under different conditions, such as the influence of alcohol, fatigue, stress, and time of day. Therefore, it is important that the criteria used for design purposes should be compatible with the capabilities of those that use and/or operate the transportation system. Transportation engineers must have some knowledge of how human beings function.

The use of an average value such as mean reaction time may not be adequate for a large number of users or operators of the system. Both the 85th and the 95th percentiles have been used to select design criteria, and, in general, the higher the chosen percentile, the wider the range covered.

The Human Response Process

Actions taken by operators and users of transportation systems result from their evaluation of, and reaction to, information they obtain from certain stimuli that they see or hear.

Visual Reception

The receipt of stimuli by the eye is the most important source of information for both users and operators of any transportation system, and some general knowledge of human vision will therefore aid in the design and operation of most transportation systems. The principal characteristics of the eye are visual acuity, peripheral vision, color vision, glare vision and recovery, and depth perception.

Visual acuity is the ability of an observer to resolve fine details of an object. It can be represented as the reciprocal of the smallest pattern detail in minutes of arc of visual angle that can be resolved. Visual angle (φ) of a given target is given as

$$\varphi = 2 \arctan\left(\frac{L}{2D}\right) \tag{3.1}$$

where

L = diameter of the target (letter or symbol)
D = distance from eye to target in the same units as L

In general, an observer will have the same response for different objects that subtend the same visual angle if all other visual factors are the same. This fundamental premise is a factor used to set legibility standards in transportation engineering, such as those given in the manual on uniform *Traffic Control*

Devices published by the Federal Highway Administration. It should be noted that drivers usually have several other expectancy cues, such as layout or word length, that may alter their visual performance.

Direct extrapolation from letter or symbol legibility/recognition sizes to word signs may therefore be misleading. For example, a study of field and laboratory data by Greene determined that a mean visual angle of 0.00193 (6.6 minutes of arc) was required for a Deer Crossing sign, while a more complex Bicycle Crossing sign required a mean visual angle of 0.00345 rad (11.8 minutes of arc).

Two types of visual acuity—*static acuity* and *dynamic acuity*—are of importance in ground transportation systems. *Static acuity* is the ability of a person to resolve fine details of an object when both the person and the object are stationary. Factors that influence static acuity include the background brightness, contrast, and exposure time which can be defined as the time an observer will take to read and understand a given message. As background brightness increases, and all other factors remain constant, static acuity tends to increase up to a background brightness of about 30 candela (cd)/m^2 ft and then remains constant even with an increase in illumination. When other visual factors are held constant at an acceptable level, the optimal time required for identification of an object with no relative movement is between 0.5 and 1.0 s. The exposure time can, however, vary significantly depending on how complicated and/or unfamiliar the sign is.

The ability of individuals to resolve fine details of an object that has a relative angular movement depends on their *dynamic visual acuity*, an important factor that should be considered in designing transportation systems. For example, signs displayed on highways and railroads should be adequately legible so that drivers or locomotive operators can easily read and understand the infor-mation given. Similarly, in-vehicle displays in automobiles and aircraft cockpits should have a minimum legibility standard. This is of significant importance for older drivers, who tend to have less visual acuity than their younger counterparts.

Variable message signs (VMSs), sometimes referred to as *changeable message signs (CMSs)*, are now commonly used to give real-time information on traffic conditions and parking availability. Factors influencing the legibility of these devices include resolution, luminance, contrast, and glare protection. The interaction among these different factors is complex, and there are many human factors publications for recommended standards. For example, Boff and Lincoln have developed legibility standards for aircraft application and Kimura et al. have developed guidelines for appropriate color, contrast, and luminance levels that can be used for in-vehicle displays.

Peripheral vision is the ability of an individual to see objects beyond the cone of clearest vision. Most people have clear vision within a conical angle of 3–5° and fairly clear vision within a conical angle of 10–12°. Although objects can be seen beyond this zone, details and color are not clear. The cone for peripheral vision could be one subtending up to 160°, but this value is affected by the relative speed of the object.

Peripheral vision is also affected by age—for instance, significant changes occur from about age 60, and this factor should be considered in deciding the location of in-vehicle and off-vehicle displays. For example, if a display is located far from a driver's normal forward field of view, the driver may not be able to understand the information on the display because it is outside his or her peripheral vision. This is of significant importance for in-vehicle displays in automobiles as drivers will tend to switch their vision between the roadway and the display. Dingus et al. determined that this switching occurs about every 1.0 to 1.5 s.

A study conducted by Pop and Faber determined that the driving performance of an individual is better when a display is positioned directly in front of the driver. Also, Weintraub et al. determined that for roadway displays, the switching time increases with the distance the display is away from the roadway. Therefore, the farther away the display is, the less time the driver devotes to the roadway and the display.

Color vision is the ability of an individual to differentiate one color from another, and deficiency in this ability is usually referred to as *color blindness.* Between 4 and 8% of the population suffer from this deficiency. It is therefore not advisable to use color alone to disseminate critical transportation information. Additional means are therefore used to facilitate the recognition of transportation systems information. For example, in order to compensate for color blindness, traffic signs are usually standardized in size, shape, and color. Standardization not only aids in distance estimation but also helps the color-blind individual to identify signs, as shown in Figure 3.2.

Glare vision and recovery is usually classified into two types: direct and specular. *Direct glare* occurs when relatively bright light appears in the individual's field of vision, and *specular glare* occurs as a result of a reflected image by a relatively bright light appearing in an individual's field of vision. Visibility is reduced when either direct or specular glare occurs, and both types of glare cause discomfort in the eyes. The sensitivity to glare increases as we age, with a significant change at about age 40.

Recovery from the effects of glare occurs sometime after an individual passes the light source that causes the glare, a phenomenon usually referred to as *glare recovery*. Studies have shown that glare recovery takes about 3 s when the movement is from dark to light and about 6 s when movement is from light to dark.

Glare vision is of particular importance during night driving, particularly for elderly drivers, who in general tend to see more poorly at night. Attention should therefore be given to the design and location of street lighting so that glare effects are reduced to a minimum, particularly in areas with a relatively high percentage of elderly drivers.

Basic principles that can be used to minimize glare effects include reducing luminary brightness and increasing the background brightness in an individual's field of vision, to obtain minimum interference with the visibility of the

FIGURE 3.2
Traffic signs.

Source: Manual on Uniform Traffic Control Devices, 2003 edition, Federal Highway Administration, U.S. DOT, Washington, D.C., 2001.

driver. For example, specific actions include using higher mounting heights, positioning lighting supports farther away from the travelway, and restricting the light from the luminary.

Depth perception is the ability of an individual to estimate speed and distance. This characteristic is of more importance on two-lane highways during passing maneuvers, when the lack of accurate speed and distance estimation may result in head-on crashes. *Depth perception* also influences the individual's ability to differentiate between objects. The human eye is not reliable for estimating absolute values of speed, distance, size, and acceleration.

Hearing perception occurs when the ear receives sound stimuli; hearing perception is important when warning sounds are given. Loss of some hearing ability is not of significant importance in the design and operation of transportation systems as it can normally be corrected by a hearing aid.

Walking speeds of individuals are important in the design of many transportation systems. For example, a representative walking speed of pedestrians is required in the design of signalized intersections. Similarly, rail and air terminals are designed mainly for pedestrians who are either walking or waiting.

Observations of pedestrian movements have indicated that walking speeds vary between 0.9 and 1.8 m/s. Significant differences have also been observed between male and female walking speeds. At intersections, the mean male walking speed has been determined to be 1.5 m/s and for females 1.4 m/s. The *Manual on Uniform Traffic Control Devices (MUTCD)* suggests the use of a more conservative value of 1.2 m/s for design. Studies have shown that walking speeds tend to be higher at midblock than at intersections and that the speeds of older people will generally be at the lower end of the speed range.

Factors that affect pedestrian speeds include the time of day, air temperature, presence of snow or ice, and the trip purpose. Age is the factor that most commonly results in lower walking speed. The lower end of the speed range (0.9 m/s) is used as the walking speed for design of transportation facilities that will be extensively used by older people.

Perception reaction time is the period of time between the time a driver perceives an obstruction and the time action is taken to avoid the object. The time depends on several factors, including the distance to the object, the driver's visual acuity, the ability of the driver to react, and the type of obstruction and varies considerably from one individual to another. Perception reaction times are significantly higher for elderly drivers. Allowance should be made for the time it takes a driver to read a sign before taking necessary action. Factors that influence this time include the type of text, number of words, sentence structure, and whether the driver is engaged in another activity. Research by Dudeck has shown that a short word of four to eight characters requires a minimum exposure time (reading time) of 1 s, while that for each unit of information is 2 s. Thus, a sign that has 12 to 16 characters per line will require a minimum of 2 s to read. The American Association of State Highway and Transportation Officials (AASHTO) recommends the use of 2.5 s for perception reaction time, which exceeds the 90th percentile of reaction time for all drivers.

Passenger Behavior Characteristics in Transportation Terminals

Transportation terminals serve as an important component of the system in that they provide the facility for passengers to change from one vehicle to another of the same mode or from one mode to another. For example, the passenger terminal of an airport enables passengers to change from a ground transportation mode to an air mode, while a train terminal may enable passengers to change from an automobile to a train. Terminals should be designed to take account of the behavioral characteristics of passengers and should consider the characteristics that have been discussed, such as visual reception and walking speeds. For example, visual reception is of importance in the placement of signs to provide adequate information so passengers can use the terminal facility efficiently. Walking speeds will be used to determine the need for moving walkways.

Physiological characteristics relate mainly to the perceived comfort of the passenger. The temperature in the terminal is a typical example and should range between 21°C–24°C. The noise level should not exceed the speech interference (SPI) level, which is between 60 and 65 dBA. Other factors include the provision of retail shops and toilet facilities.

Psychological characteristics relate to the perceived safety of the passengers, who must feel that it is safe if they are expected to use a transportation terminal. The greatest security is achieved by the presence of uniformed law officers or an environment that provides for easy communication with law officers in case of an emergency.

VEHICLE CHARACTERISTICS

An important component of any transportation system is the vehicle. The transportation engineer must therefore be familiar with the characteristics of the vehicle that will serve the system, which, in turn, will influence the design of the travelway. The characteristics of the design vehicle influence the geometric alignment and pavement structure of a surface travelway. The characteristics of a design airplane influence the taxiways and runway configurations of an airport. These are discussed in terms of static and dynamic characteristics.

Static Characteristics of Automobiles

The physical components of a highway are designed to be compatible with the size of the largest and heaviest vehicle expected. These components include lane width, shoulder width, length and width of parking bays, and vertical curves. The axle weights of the vehicles expected on the highway are important for the design of the pavement structure and when maximum grades are being determined. With the passage of the Surface Transportation Assistance Act of 1982, maximum sizes and weights of trucks on interstates and other qualifying federal-aid highways were established. These include:

- 360 kN gross weight, with axle loads of 90 kN for single axles and 150 kN for double axles
- 259 cm width for all trucks
- 14.6 m length for semitrailers and trailers
- 8.5 m length for each twin trailer

Those states with higher weight limits before the law was enacted are allowed to retain them for intrastate travel. Also, limits on overall truck length are no longer allowed to be set.

Because the static characteristics of the predominant vehicles are used to establish certain geometric parameters of the road, vehicles have been classified so that they represent static characteristics within a particular class. AASHTO has classified automobiles into four general classes of vehicles: passenger cars, buses, trucks, and recreational vehicles. Vehicles included in the passenger-car class are all sizes of passenger cars, sport/utility vehicles, minivans, vans, and pick-up trucks. Those in the bus class include intercity (motor coaches), city transit, school, and articulated buses. Vehicles in the truck class are single-unit trucks, truck tractor–semitrailer combinations, and trucks or truck tractors with semitrailers in combination with full trailers. Vehicles in the recreational-vehicle class are motor homes, cars with camper trailers, cars with boat trailers, motor homes with boat trailers, and motor homes pulling cars. Table 3.1 gives the physical dimensions for 19 design vehicles, and Table 3.2 gives minimum turning radii for design vehicles that represent different vehicles within each of the four general classes. The largest and most frequent vehicle expected to use the facility is selected as the design vehicle, and the following guidelines are provided for selection:

- For parking lots or a series of parking lots, passenger-car class could be considered.
- For intersections on residential streets and park roads, a single-unit truck class could be considered.
- For intersections of state highways with city streets on which buses travel, but on which relatively few large trucks travel, a city transit bus class could be considered.
- For intersections of highways with low-volume county highways and township/local roads with traffic volume under 400 Average Daily Traffic (ADT), a large school bus (84 passengers) or a conventional school bus (65 passengers) could be considered.
- For other intersections of state highways and industrial streets with high volumes of traffic and/or providing large truck access to local facilities, the minimum design vehicle is the WB20 (WB65 or WB67) (see Table 3.1).
- For intersections of freeway ramp terminals with arterial crossroads the minimum design vehicle is the WB20 (WB65 or WB67) (see Table 3.1).

The characteristic of vehicle categories that influence design of intersections when speeds are 15 km/h or less are (1) the minimum centerline radius (CTR); (2) the out-to-out track width; (3) the wheel base; and (4) the path of the inner rear tire of the vehicle as it makes a turn at the intersection. When turns are made at speeds of 10 km/h or less, the turning radius and turning path depend mainly on the size of the vehicle making the turn. These have therefore been established for each design vehicle. For example, the minimum turning paths for the passenger and WB20 (WB65 and WB67) design vehicles are shown in

TABLE 3.1
Design Vehicle Dimensions

		Dimensions (ft)											
		Overall			Overhang								
Design Vehicle Type	Symbol	Height	Width	Length	Front	Rear	WB_1	WB_2	S	T	WB_3	WB_4	Typical Kingpin to Center of Rear Axle
Passenger Car	P	4.25	7	19	3	5	11	—	—	—	—	—	—
Single-Unit Truck	SU	11–13.5	8.0	30	4	6	20	—	—	—	—	—	—
Buses													
Intercity Bus (Motor Coaches)	BUS-40	12.0	8.5	40	6	6.3[a]	24	3.7	—	—	—	—	—
	BUS-45	12.0	8.5	45	6	8.5[a]	26.5	4.0	—	—	—	—	—
City Transit Bus	CITY-BUS	10.5	8.5	40	7	8	25	—	—	—	—	—	—
Conventional School Bus (65 pass.)	S-BUS 36	10.5	8.0	35.8	2.5	12	21.3	—	—	—	—	—	—
Large School Bus (84 pass.)	S-BUS 40	10.5	8.0	40	7	13	20	—	—	—	—	—	—
Articulated Bus	A-BUS	11.0	8.5	60	8.6	10	22.0	19.4	6.2[b]	13.2[b]	—	—	—
Trucks													
Intermediate Semitrailer	WB-40	13.5	8.0	45.5	3	2.5[a]	12.5	27.5	—	—	—	—	27.5
Intermediate Semitrailer	WB-50	13.5	8.5	55	3	2[a]	14.6	35.4	—	—	—	—	37.5
Interstate Semitrailer	WB-62*	13.5	8.5	68.5	4	2.5[a]	21.6	40.4	—	—	—	—	42.5
Interstate Semitrailer	WB-65** or WB-67	13.5	8.5	73.5	4	4.5–2.5[a]	21.6	43.4–45.4	—	—	—	—	45.5–47.5
"Double-Bottom"-Semitrailer/Trailer	WB-67D	13.5	8.5	73.3	2.33	3	11.0	23.0	3.0[c]	7.0[c]	23.0	—	23.0
Triple-Semitrailer/Trailers	WB-100T	13.5	8.5	104.8	2.33	3	11.0	22.5	3.0[d]	7.0[d]	23.0	23.0	23.0
Turnpike Double-Semitrailer/Trailer	WB-109D*	13.5	8.5	114	2.33	2.5[e]	14.3	39.9	2.5[e]	10.0[e]	44.5	—	42.5
Recreational Vehicles													
Motor Home	MH	12	8	30	4	6	20	—	—	—	—	—	—
Car and Camper Trailer	P/T	10	8	48.7	3	10	11	—	5	19	—	—	—
Car and Boat Trailer	P/B	—	8	42	3	8	11	—	5	15	—	—	—
Motor Home and Boat Trailer	MH/B	12	8	53	4	8	20	—	6	15	—	—	—
Farm Tractor[f]	TR	10	8-10	16[g]	—	—	10	9	3	6.5	—	—	—

* = Design vehicle with 48-ft trailer as adopted in 1982 Surface Transportation Assistance Act (STAA).
** = Design vehicle with 53-ft trailer as grandfathered in with 1982 Surface Transportation Assistance Act (STAA).
a = This is overhang from the back axle of the tandem axle assembly.
b = Combined dimension is 19.4 ft and articulating section is 4 ft wide.
c = Combined dimension is typically 10.0 ft.
d = Combined dimension is typically 10.0 ft.
e = Combined dimension is typically 12.5 ft.
f = Dimensions are for a 150–200 hp tractor excluding any wagon length.
g = To obtain the total length of tractor and one wagon, add 18.5 ft to tractor length. Wagon length is measured from front of drawbar to rear of wagon, and drawbar is 6.5 ft long.

- WB_1, WB_2, and WB_4 are the effective vehicle wheelbases, or distances between axle groups, starting at the front and working towards the back of each unit.
- S is the distance from the rear effective axle to the hitch point or point of articulation.
- T is the distance from the hitch point or point of articulation measured back to the center of the next axle or center of tandem axle assembly.

Source: Adapted from *A Policy on Geometric Design of Highways and Streets,* American Association of State Highway and Transportation Officials, Washington, D.C., 2004. Used by permission.

Note: 1 ft = 0.3 m

TABLE 3.2
Minimum Turning Radii of Design Vehicles

Source: A Policy on Geometric Design of Highways and Streets, American Association of State Highway and Transportation Officials, Washington, D.C., 2004. Used by permission.

Design Vehicle Type	Passenger Car	Single-Unit Truck	Intercity Bus (Motor Coach)		City Transit Bus	Conventional School Bus (65 pass.)	Large[2] School Bus (84 pass.)	Articulated Bus	Intermediate Semitrailer	Intermediate Semitrailer
Symbol	P	SU	BUS-40	BUS-45	CITY-BUS	S-BUS36	S-BUS40	A-BUS	WB-40	WB-50
Minimum Design Turning Radius (ft)	24	42	45	45	42.0	38.9	39.4	39.8	40	45
Centerline[1] Turning Radius (CTR) (ft)	21	38	40.8	40.8	37.8	34.9	35.4	35.5	36	41
Minimum Inside Radius (ft)	14.4	28.3	27.6	25.5	24.5	23.8	25.4	21.3	19.3	17.0
Design Vehicle Type	**Interstate Semitrailer**		**"Double Bottom" Combination**	**Triple Semitrailer/trailers**	**Turnpike Double Semitrailer/trailer**	**Motor Home**	**Car and Camper Trailer**	**Car and Boat Trailer**	**Motor Home and Boat Trailer**	**Farm[3] Tractor w/One Wagon**
Symbol	WB-62*	WB-65** or WB-67	WB-67D	WB-100T	WB-109D*	MH	P/T	P/B	MH/B	TR/W
Minimum Design Turning Radius (ft)	45	45	45	45	60	40	33	24	50	18
Centerline[1] Turning Radius (CTR) (ft)	41	41	41	41	56	36	30	21	46	14
Minimum Inside Radius (ft)	7.9	4.4	19.3	9.9	14.9	25.9	17.4	8.0	35.1	10.5

* = Design vehicle with 48-ft trailer as adopted in 1982 Surface Transportation Assistance Act (STAA).
** = Design vehicle with 53-ft trailer as grandfathered in with 1982 Surface Transportation Assistance Act (STAA).
1 = The turning radius assumed by a designer when investigating possible turning paths and is set at the centerline of the front axle of a vehicle. If the minimum turning path is assumed, the CTR approximately equals the minimum design turning radius minus one-half the front width of the vehicle.
2 = School buses are manufactured from 42-passenger to 84-passenger sizes. This corresponds to wheelbase lengths of 11.0 ft to 20.0 ft, respectively. For these different sizes, the minimum design turning radii vary from 28.8 ft to 39.4 ft and the minimum inside radii vary from 14.0 ft to 25.4 ft.
3 = Turning radius is for 150–200 hp tractor with one 18.5 ft long wagon attached to hitch point. Front wheel drive is disengaged and without brakes being applied.

Note: 1 ft = 0.3 m

Figures 3.3 and 3.4. Those for other design vehicles are given in *Policy on Geometric Design of Highways and Streets*. These turn paths are based on a study of the turning paths of scale models of the representative vehicle of each class. Table 3.2 gives the minimum turning radii for different design vehicles. It should be emphasized, however, that these minimum turning radii are for turns made at speeds of 15 km/h or less. When turns are made at higher speeds, the radii depend mainly on the speeds at which the turn is made.

Static Characteristics of Airplanes

The static characteristics of airplanes also vary considerably. Depending on the type of airplane, the maximum take-off weight can be as low as 7 kN for the Cessna-150 and as high as 3800 kN for the Boeing 747-400. These airplanes

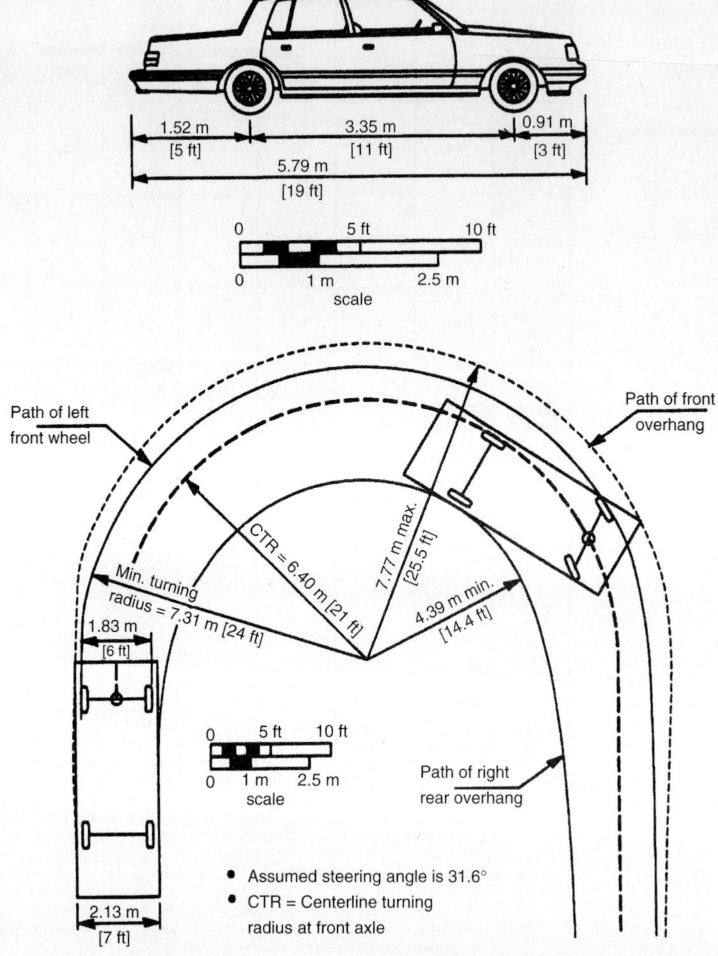

FIGURE 3.3

Minimum turning path for passenger design vehicles.

Source: A Policy on Geometric Design of Highways and Streets, American Association of State Highway and Transportation Officials, Washington, D.C., 2004. Used by permission.

can be classified into two general categories: *transport aircraft* and *general aviation aircraft.*

Aircraft may also be classified based on their airworthiness certification and aircraft operations regulations in Title 14 of the Code of Federal Regulations. These are described by GRA Incorporated in the report *Economic Values for FAA Investment and Regulatory Decisions; A Guide*. The *airworthiness classification* is based on the approval given by the Federal Aviation Administration (FAA) to the design of the aircraft. This approval is required for each type of aircraft to be flown in the United States. Title 14 contains four parts that deal with the airworthiness standards for aircraft. These are:

Part 23: Includes "Normal, Utility, Acrobatic and Commuter Category Airplanes." The normal utility and acrobatic aircraft are limited to a maximum of nine passengers and a maximum take-off weight of 55 kN, while the commuter airplanes have a maximum of 19 passengers and a maximum take-off weight of 85 kN.

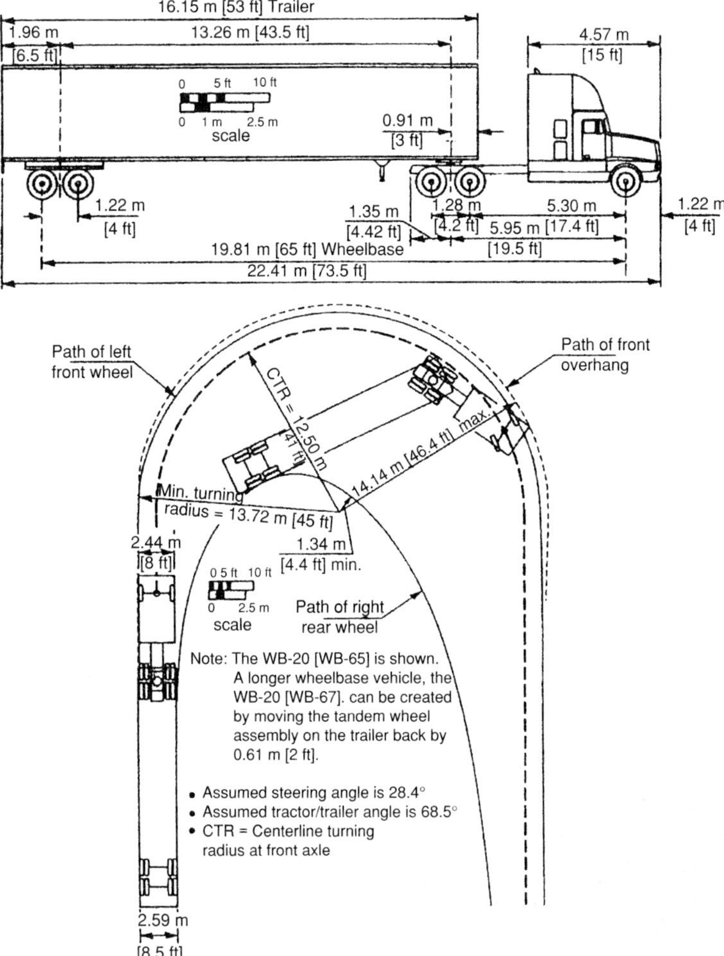

FIGURE 3.4
Minimum turning path for WB-20 (WB-65 and WB-67) design vehicles.

Source: A Policy on Geometric Design of Highways and Streets, American Association of State Highway and Transportation Officials, Washington, D.C., 2004. Used by permission.

Part 25: Includes "Transport Category Airplanes." These are fixed-wing aircraft that do not meet the standards of Part 23. In general, these include piston-powered fixed-wing aircraft and turboprops with less than 20 seats. This group also includes larger turboprops and all jet-powered airplanes.

Part 27: Includes "Normal Category Rotorcraft." These are piston- or turbine-powered rotorcraft that have a maximum take-off weight of 27 kN and a maximum of nine passenger seats.

Part 29: Includes all "Transport Category Rotorcraft" that do not meet the requirements of Part 27.

Title 14 also contains several parts that relate to the standards for civilian aircraft operation in the United States. These are:

Part 91: Referred to as the "General Aviation" operations and includes powered aircraft that do not include activities that require regulations under any of the other parts. This part is therefore the least restrictive.

An operator who intends to conduct commercial operation under this category should also obtain air carrier or some other operating certificate.

Part 121: Gives standards for domestic and flag operations conducted by holders of air carrier or operating certificates. This includes the use of aircraft to transport passengers, with passenger seats higher than nine, or to transport cargo with capacity exceeding 33.5 kN. Most airlines operate under this structure.

Part 125: Contains standards for non-commercial operations that use fixed-wing aircraft with 20 or more seats and do not fit into Parts 135 or 137. Note that these aircraft are not included in Part 91.

Part 133: Contains regulations for operating rotary wing aircraft carrying an external load.

Part 135: Contains regulations for operations such as transportation of mail, certain sightseeing or air tour flights, and commuter flights. Commuter flights are described as "common carriage flights that operate on a regular schedule but are conducted with rotary-wing aircraft or fixed-wing aircraft with nine or fewer passenger seats or with a payload capacity of 33.5 kN or less."

Part 137: Contains regulations governing the aerial application of substances to support activities such as agriculture, firefighting, public health, and cloud seeding.

Based on this classification system, the following categories were obtained:

1. Two-engine narrow body
2. Two-engine wide body
3. Three-engine narrow body
4. Three-engine wide body
5. Four-engine narrow body
6. Four-engine wide body
7. Regional jet under 70 seats
8. Regional jet 70 to 100 seats
9. Turboprops under 20 seats (Part 23)
10. Turboprops under 20 seats (Part 25)
11. Turboprops with 20 or more seats
12. Piston engine (Part 23)
13. Piston engine (Part 25)

The FAA has also classified airplanes into two categories—*airplane design group* and *aircraft approach category*—for the purpose of selecting appropriate airport design standards, as the design criteria for airports are based on the airplanes intended to operate at the airport. The *airplane design group* is based on

TABLE 3.3
Airport Reference Code, Static Characteristics and Approach Speeds of Different Airplanes

Source: Adapted from Advisory Circular AC No: 150/5300-13, U.S. Department of Transportation, Federal Highway Administration, Washington, D.C., 2004.

Aircraft	*Airport Reference Code*	*Approach Speed (knots)*	*Wingspan (ft)*	*Length (ft)*	*Tail Height (ft)*	*Maximum Take-Off (lb)*
Cessna-150	A-I	55	32.7	23.8	8.0	1,600
Beech Bonanza A36	A-I	72	33.5	27.5	8.6	3,650
Beech Baron 58	B-I	96	37.8	29.8	9.8	5,500
Cessna Citation I	B-I	108	47.1	43.5	14.3	11,850
Beech Airliner 1900-C	B-II	120	54.5	57.8	14.9	16,600
Cessna Citation III	B-II	114	53.5	55.5	16.8	22,000
Bae 146-100	B-III	113	86.4	85.8	28.3	74,600
Antonov AN-24	B-III	119	95.8	77.2	27.3	46,305
Airbus A-320-100	C-III	138	111.3	123.3	39.1	145,505
Boeing 727-200	C-III	138	108.0	153.2	34.9	209,500
Boeing 707-320	C-IV	139	142.4	152.9	42.2	312,000
Airbus A-310-300	C-IV	125	144.1	153.2	52.3	330,693
MDC-8-63	D-IV	147	148.4	187.4	43.0	355,000
Boeing 747-200	D-V	152	195.7	231.8	64.7	833,000
Boeing 747-400	D-V	154	213.0	231.8	64.3	870,000

Note: 1 knot = 1.85 km/h; 1 ft = 0.3 m

the wingspan of the aircraft and is indicated by a Roman numeral (I, II, III, IV, or V). The *aircraft approach category*, designated by a letter (A, B, C, or D), is based on the aircraft's approach speed.

A coding system for airports known as the *Airport Reference Code (ARC)* (see Table 3.3) relates the design criteria for categories of aircraft intended to operate at that airport on a regular basis using the airplane design group and aircraft approach category of the aircraft. Table 3.3 gives examples of ARCs for different airplanes. ARCs for other planes can be obtained from the *FAA Advisory Circular 150/5300-13.* For example, if an airport is designed to serve the Boeing 747-200 with a wingspan of 59.65 m and an approach speed of 152 knots, the ARC of the airport is D-V.

Table 3.3 shows that the static characteristics of airplanes vary considerably. Important static characteristics that influence the design of airports are the maximum take-off load and the wingspan of the planes that are expected to use the airport. In general, the higher the maximum take-off load of the plane, the longer the required take-off and landing lengths.

Static Characteristics of Railroad Locomotives

Locomotives can be classified into five general categories based mainly on the type of power used:

- Electric
- Diesel–electric

- Steam
- Magnetic levitation (Maglev)
- Other types (gas, turbine-electric)

Electric Locomotives

The power supply to an electric locomotive is achieved by using either a direct current (dc) system or an alternating current (ac) system. The power is transmitted from an external source of supply, and the capacity of the locomotive is therefore not limited internally. The current is transmitted by the use of either collector shoes that ride on a third rail or through overhead wires. The third-rail system is mainly used when heavy current and low voltage are employed, while overhead wires are used when high voltage is necessary, mainly because of safety reasons. Electric motors can be coupled as multiple units under one controller or they can be used in single units.

Diesel-Electric Locomotives

The power supply to a diesel-electric locomotive consists of a diesel engine prime mover that is directly connected to a dc generator, thus forming a complete power plant. These locomotives are therefore self-contained, with each having its own power plant and traction motor. This gives diesel–electric locomotives an advantage over the electric locomotive in that the extensive power distribution that is required for the electric locomotives is not necessary for the diesel–electric alternative. Diesel–electric locomotives can also be used in single units or in multiple units that are controlled from one cab. Single units are used mainly in railroad yard operations, while the multiple units are used for line-haul operation.

Steam Locomotives

These locomotives receive their power from reciprocating steam engines that are much less efficient than the diesel–electric system and have therefore been widely replaced by diesel–electric locomotives. This system is now mainly used in developing countries because of its relatively lower capital cost per unit of horsepower.

Magnetic Levitation Trains (Maglev)

In this type of locomotive, there is no contact between the bearing structure and the vehicle. The power is derived from sets of magnets and coils that are suitably placed to produce the forces that are required for levitation, propulsion, and guidance. Figure 3.5 illustrates the basic principle of magnetic levitation.

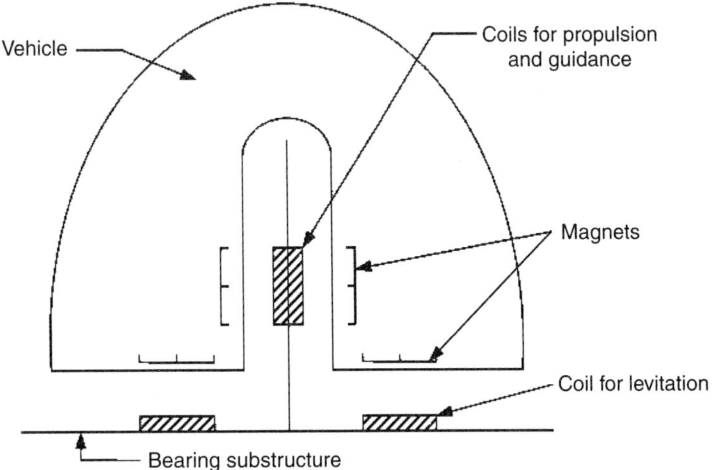

FIGURE 3.5
Basic principles of magnetic levitation.

Source: Railway Engineering, V.A. Profillidis, Avebury Technical, 1995.

Tests have shown that these trains are capable of traveling at very high speeds and can traverse relatively higher longitudinal grades. For example, a magnetic levitation test track was constructed with a 5 m-high superelevated section and a design speed of about 400 km/h. The cars were 54 m long, weighed 108 metric tons, and were capable of carrying 200 passengers. The track had a minimum radius of curvature of over 4000 m and a maximum longitudinal gradient of 10%.

Static Characteristics of Waterborne Vessels

Waterborne vessels can be broadly classified into passenger vessels and cargo vessels. Passenger vessels can be further classified into passenger ferries and passenger liners, while cargo vessels can be further classified into tankers and dry cargo. Figure 3.6 shows examples of waterborne vessels.

Passenger Vessels

A passenger vessel is defined as one that has accommodation for more than 12 passengers. For this reason, it is common for cargo vessels to have accommodation for up to 12 passengers so that they will not be subjected to the more stringent passenger vessel regulations.

The main difference between passenger ferries and passenger liners is that ferries usually transport passengers, automobiles, and some freight for shorter distances, while liners mainly transport passengers over relatively longer distances. There are currently very few passenger liners in operation with the primary purpose of purely transporting passengers. This is mainly due to the competition of air transportation, which is much faster and cheaper for

FIGURE 3.6
Examples of different types of waterborne vehicles.

Passenger Cruise: *Crown of Scandinavia* (*Source:* The United Steamship Company, Denmark)

Freight ship: *Tor Flandria* (*Source:* The United Steamship Company, Denmark)

Ferry ship: *First Ferry,* Hong Kong (*Source:* Hong Kong Vessel Society)

TABLE.3.4
Static Characteristics of Some Cruise Ships

Source: www.en.wikipedia.org/wiki/passenger_ship

Name	*Length (ft)*	*Beam (ft)*	*Gross Tonnage (tons)*	*Passenger Capacity*	*Cruising Speed (knots)*
Grandeur of the Seas	916	106	74,000	1950	22
Rhapsody of the Seas	915	105.6	75,000	2000	22
Splendour of the Seas	867	105	70,000	1,804	24
Majesty of the Seas	880	106	73,941	2,354	19
Nordic Express	692	100	45,563	1,600	19.5

Note: 1 ft = 0.3 m; 1 ton = 0.91 metric ton; 1 knot = 1.85 km/h

passengers. Most of the passenger liners are now serving as cruise ships and are used mainly for vacation purposes. Table 3.4 shows static characteristics of some cruise ships.

Dynamic Characteristics of Transportation Vehicles

The forces that act on a vehicle while it is in motion are the air resistance, the grade resistance, the rolling resistance, and the curve resistance. Techniques for quantitatively estimating these forces are presented in this section.

Air Resistance on Automobiles

The air in front of and around a vehicle in motion causes resistance to the movement of the vehicle, and the force required to overcome this resistance is known as *air resistance.* The magnitude of this force depends on the square of the velocity at which the vehicle is traveling and the cross-sectional area of the vehicle. It is related to the cross-sectional area of the vehicle in a plane that is perpendicular to the direction of motion. It has been shown by Claffey that this force can be estimated from Equation 3.2:

$$F_a = 0.5 \frac{(0.0772 p\, C_D A u^2)}{g} \tag{3.2}$$

where

F_a = air resistance in force (N)
p = density of air (1.227 kg/m^3) at sea level: less at higher elevations
C_D = aerodynamics drag coefficient (current average value for passenger cars is 0.4, for trucks this value ranges from 0.5 to 0.8 but a typical value is 0.5)
A = frontal cross-sectional area (m^2)
u = speed of automobile (km/h)
g = acceleration of gravity (9.81 m/s^2)

EXAMPLE 3.1

Determining the Air Resistance on Moving Automobiles

Determine the difference in air resistance between a passenger car and a single-unit truck if both vehicles are traveling at a speed of 96.5 km/h. Assume that the frontal cross-sectional area of the passenger car is 2.79 m^2 and that for the truck is 10.70 m^2.

Solution

Determine air resistance for the passenger car from Equation 3.2:

$$F_a = 0.5\left(\frac{0.0772\,pC_D A u^2}{g}\right)$$

$$= 0.5\left(\frac{0.0772 \times 1.227 \times 0.4 \times 2.79 \times 96.5 \times 96.5}{9.81}\right) \text{ lb}$$

$$= 501.7 \text{ N}$$

Determine air resistance for the truck from Equation 3.2:

$$F_a = 0.5\left(\frac{0.0772 \times 1.227 \times 0.5 \times 10.7 \times 96.5 \times 96.5}{9.81}\right)$$

$$= 2405.3 \text{ N}$$

Determine difference in air resistances:

Difference in air resistances is (2405.3 – 501.7) = 1903.6 N

Air Resistance on Trains

The equation for the air resistance on trains is similar to that for automobiles, except that since trains are much longer than automobiles, the frictional resistance along the length of the train should also be considered. It is given as

$$F_{at} = C_{t1} A u^2 + C_{t2} p L u^2 \tag{3.3}$$

where

F_{at} = train air resistance (N)
A = front surface cross-sectional area of the train m^2
u = speed of train (km/h)
L = length of train (m)
p = partial perimeter (m) of the rolling stock down to rail level
C_{t1} and C_{t2} = constants

The constant C_{t1} depends on the shape of the front and rear of the train, and C_{t2} depends on the condition of the train surface. Various railway authorities have therefore developed empirical formulae for rolling resistance that also account for air resistance (see Equation 3.7 for rolling resistance of trains).

Grade Resistance

A vehicle traveling on an upgrade is resisted by a force acting in the opposite direction (i.e., downgrade). This force is the component of the vehicle's weight acting downward along the plane of the vehicle's travelway. This force is the *grade resistance*. The grade resistance will tend to reduce the speed of the vehicle if an accelerating force is not applied. The speed achieved at any point along the grade for a given rate of acceleration will depend on the grade and type of vehicle. The grade resistance is given as:

$$\text{Grade resistance} = \text{weight} \times \text{grade in decimal} \tag{3.4}$$

The impact of grade resistance is of more significance in the highway mode than in the railway and air modes. The reason is that grades are much more restricted in the railway and air modes as the weights of vehicles used in these modes are much higher than those for automobiles. For example, as will be discussed in Chapter 6, maximum grades of airports do not exceed 2%, those for rail tracks do not exceed 4%, but highway grades can be as high as 9%.

Rolling Resistance

Forces exist within the vehicle itself that offer resistance to motion. These include forces due mainly to the frictional effect on moving parts and other mechanical resistances and those generated by friction between the wheels of the vehicle and the travelway. The total effect of these forces on motion is known as the *rolling resistance.* Factors that influence this resistance include the speed of the vehicle and the condition of the travelway. For example, a vehicle traveling at 80 km/h on a highway with a badly broken and patched asphalt surface will experience a rolling resistance of 255 N/metric ton of weight, whereas at the same speed on a loose sand surface the rolling resistance is 380 N/metric ton of weight.

Rolling Resistance on Automobiles

Different formulae have been developed for passenger cars and trucks.

The rolling resistance for passenger cars on smooth pavement can be determined from Equation 3.5:

$$F_r = (C_{rs} + 0.0772 C_{rv} u^2) W \tag{3.5}$$

where

F_r = rolling resistance force (N)
C_{rs} = constant (typically 0.012 for passenger cars)

C_{rv} = constant (typically 6.99×10^{-6} s²/m² for passenger cars)
u = vehicle speed (km/h)
W = gross vehicle weight (N)

For trucks the rolling resistance is given as

$$F_{rt} = (C_a + 1.47C_b u^2)W \tag{3.6}$$

where

F_{rt} = rolling resistance force (lb)
C_a = constant (typically 0.2445 for trucks)
C_b = constant (typically 0.00044 sec/ft for trucks)
u = vehicle speed (mph)
W = gross vehicle weight (lb)

EXAMPLE 3.2

Determining the Rolling Resistance on a Passenger Car

Determine the rolling resistance on a passenger car that is traveling at 105 km/h if the weight of the car is 9000 N.

Solution

Use Equation 3.5 to find the rolling resistance:

$$F_r = (C_{rs} + 0.0772C_{rv}u^2)900$$

$$C_{rs} = 0.012$$

$$C_{rv} = 6.99 \times 10^{-6}$$

$$F_r = (0.012 + 0.0772 \times 6.99 \times 10^{-6} \times 105 \times 105)900$$

$$= (0.012 + 0.0059)9000 \text{ N}$$

$$= 0.0179 \times 9000 \text{ N}$$

$$= 161.1 \text{ N}$$

Rolling Resistance on Trains

The American Railway and Engineering and Maintenance-of-Way Association suggests that the rolling resistance for trains could be estimated from Equation 3.7:

$$F_{rT} = 0.3 + \frac{9.07}{m} + 0.0031u + \frac{k}{mn}u^2 \tag{3.7}$$

where

F_{rT} = rolling resistance force (N/metric ton)
m = average load per axle in metric tons

u = speed of vehicle (mph)
n = number of axles
mn = average weight of locomotive or car in tons
k = air resistance coefficient: 0.0123 for conventional equipment; 0.028 for piggyback; 0.0164 for containers

The formula given for rolling resistance in Equation 3.7 is the "modified Davis formula." The original formula was found to give satisfactory results for speeds between 8 km/h and 65 km/h, and the modified formula given in Equation 3.7 was developed to take into consideration modern operations with higher speeds. This equation also accounts for air resistance and is therefore commonly referred to as the *level tangent resistance*.

EXAMPLE 3.3

Resistance on a Train

Determine the rolling resistance on a train with conventional equipment traveling at 130 km/h on a straight and level track section if the load per axle is 18.14 metric tons and the train consists of 16 cars each having four axles.

Solution

Determine resistance: Since the train is on a straight and level track section, the resistance is the level tangent resistance. Use Equation 3.7 to determine level tangent resistance:

$$\begin{aligned} F_{rt} &= 0.3 + \frac{9.07}{m} + 0.003u + \frac{ku^2}{mn} \\ &= 0.3 + \frac{9.07}{18.14} + 0.003 \times 130 + \frac{0.0123 \times 130 \times 130}{18.14 \times 4} \\ &= 4.07 \text{ N/metric ton} \end{aligned}$$

Curve Resistance

When a vehicle travels on a curve section of its travelway, external forces act on the vehicle. Certain components of these forces tend to retard the forward motion of the vehicle. The sum effect of these components is the curve resistance.

Curve Resistance on Automobiles

The radius of the curve, the velocity at which the vehicle is moving, and the gross weight of the vehicle are the factors that determine the magnitude of the curve resistance. Curve resistance can be estimated from Equation 3.8:

$$F_c = 0.5\frac{0.0772u^2W}{gR} \tag{3.8}$$

where

F_c = curve resistance (N)
u = vehicle speed (km/h)
W = gross weight of vehicle (kg)
g = acceleration of gravity
R = radius of curvature (m)

EXAMPLE 3.4

Determining the Curve Resistance on an Automobile

A single-unit three-axle truck traveling on an interstate highway at a speed of 88.5 km/h approaches a horizontal curve with a radius of 274.25 m. Determine the air resistance that acts on the truck as it traverses the curve if the weight on each axle is 22675 N.

Solution

Determine gross vehicle weight:

Gross vehicle weight of the truck = 3 × 22675 = 68025 N

Determine curve resistance using Equation 3.8:

$$F_c = 0.5\frac{0.0772u^2W}{gR}$$

$$F_c = 0.5\frac{0.0772 \times 88.5 \times 88.5 \times 68025 \text{ N}}{9.81 \times 274.25}$$

$$= 7644.1 \text{ N}$$

Curve Resistance on Trains

This resistance depends on the friction between wheel flange and rail, the wheel slippage on the rails, and the radius of curvature. Based on the results of tests performed with actual trains in the United States, the American Railway Engineering Association (AREA) has adopted a recommended value of 4 N/metric ton/° of curve for three-piece trucks without wheel/rail lubrication on standard gage tracks. This is reflected in Equation 3.9, which is recommended by Canadian National Railways and can be used to determine curve resistance on any given track:

$$F_c = 0.279 \times (\text{gage}) \tag{3.9}$$

where

F_c = curve resistance on trains; (N/metric ton) per degree of curvature
gage = gage of track in m

It should be noted that curve resistance developed when starting a train is about twice the value for the train in motion. This should be taken into consideration in the design of a curvature if trains are expected to be stopping at those curves.

Running Resistance

The force that should be applied to overcome the various resistances is the running resistance, which is determined by summing the values for all the resistances derived by computing the appropriate equations.

Power Requirements

The performance capability of a vehicle is measured in terms of the horsepower the engine can produce to overcome the different resistances and put the vehicle in motion. The horsepower is the rate at which work is done, and 1 horsepower is 746 N.m/s. The power delivered by the engine is

$$P = \frac{0.278Fu}{76.04} \tag{3.10}$$

where

P = horsepower delivered (hp)
F = sum of resistances to motion (kg)
u = speed of vehicle (km/h)

Example 3.5

Determining Horsepower Required to Drive a Train around a Curve

Determine the power that is required to operate a train of 16 cars traveling around a curve of 2° at 112.5 km/h on a level track if the total load including that of the locomotive is supported by 64 axles carrying an average of 18.14 metric tons/axle.

Solution

In this problem we need to find the level tangent and curve resistances to obtain the running resistance.

Determine the level tangent resistance from Equation 3.7:

$$F_{rt} = 0.3 + \frac{9.07}{m} + 0.0031u + \frac{ku^2}{mn}$$

Number of axles/car = 64/16 = 4

$$= 0.3 + \frac{9.07}{18.14} + 0.0031 \times 112.5 + \frac{0.0123 \times 112.5 \times 112.5}{18.14 \times 4}$$

$$= 0.3 + 0.5 + 0.35 + 2.15$$

$$= 33 \text{ N/metric ton}$$

Determine the curve resistance using 4 N/metric ton as recommended in the text.

For a 2° curve, resistance = 2 × 4 = 8 N/metric ton

Determine total resistance/metric ton:

Total resistance = level tangent resistance + grade resistance + curve resistance

= (3.3 + 0 + 8) N/metric ton

= 41 N/metric ton

Determine power from Equation 3.10:

$$P = \frac{0.278Fu}{76.04}$$

where

F = total resistance:
= resistance/metric ton × weight of train in metric tons
= 4.1 × 16 × 4 × 18.14
= 47599 N

$$P = \frac{0.278 \times 4759.9 \times 112.5}{76.04}$$

$$= 1957.7 \text{ hp (or 1460.5 kW)}$$

EXAMPLE 3.6

Determining Power Required by Passenger Car to Overcome Resistance

A 13600 N passenger car is traveling at 88.5 km/h on a flat section of road with a horizontal curve of 300 m radius. If the cross-sectional area of the vehicle is 2.7 m^2, determine the horsepower required to overcome the resistance on the vehicle.

Solution

Total resistance = air resistance + rolling resistance + grade resistance + curve resistance

Determine air resistance from Equation 3.2:

$$F_a = 0.5\left(\frac{0.0772pC_DAu^2}{g}\right)$$

$$= 0.5\left(\frac{0.0772 \times 1.227 \times 0.4 \times 2.7 \times 88.5 \times 88.5}{9.81}\right)$$

$$= 408.4 \text{ N}$$

Determine rolling resistance from Equation 3.5:

$$F_r = (C_{rs} + 0.0772C_{rv}u^2)W$$

$$= (0.012 + 0.0772 \times 6.99 \times 10^{-6} \times 88.5 \times 88.5)13600$$

$$= (0.012 + 0.00423)13600$$

$$= 220.7 \text{ N}$$

Determine grade resistance: The road is level; therefore, grade resistance is zero. Determine curve resistance using Equation 3.8:

$$F_c = 0.5\frac{0.0772u^2W}{gR}$$

$$= 0.5\frac{0.0772 \times 88.5 \times 88.5 \times 13600}{9.81 \times 300}$$

$$= 1397.1 \text{ N}$$

Determine total resistance:

$$\text{Total resistance} = 408.4 \text{ N} + 220.7 \text{ N} + 1397.1 \text{ N}$$

$$= 2026.2 \text{ N}$$

Using Equation 3.10, determine horsepower required to overcome resistance:

$$P = \frac{0.278Fu}{76.04}$$

$$P = \frac{0.278 \times 2026.2 \times 88.5}{550}$$

$$= 65.56 \text{ hp (or 48.91 kW)}$$

Braking Distance

The action of the forces on a moving vehicle play an important part in determining the distance required by the vehicle to come to rest from a given speed. Other factors of importance include the deceleration rate, the coefficient of

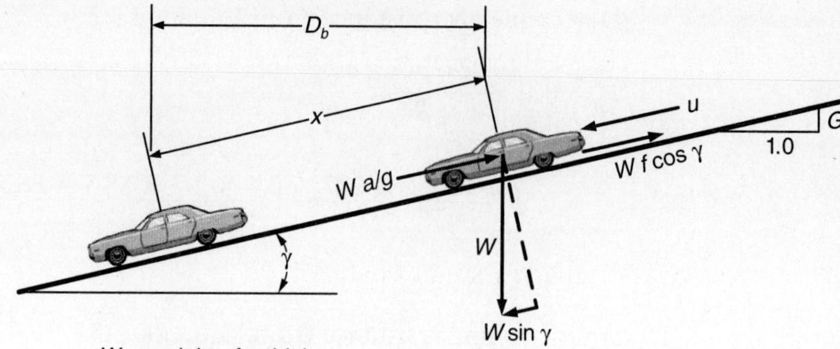

FIGURE 3.7
Forces acting on a vehicle braking on a downgrade.

W = weight of vehicle
f = coefficient of friction
g = acceleration of gravity
a = vehicle deceleration
u = speed when brakes applied
D_b = braking distance
γ = angle of incline
G = tan γ (% grade/100)
x = distance traveled by the vehicle along the road during braking

friction between the tires, and the road pavement in the case of automobiles or between the wheels and tracks in the case of trains.

Braking Distance for Automobiles

Consider a vehicle traveling downhill with an initial velocity of u, in mph, as shown in Figure 3.7. Let

W = weight of vehicle
f = coefficient of friction between road tires and road pavement
γ = angle between the grade and the horizontal
a = deceleration of the vehicle when brakes are applied
D_b = horizontal component of distance traveled during braking (that is, from time brake is applied to time vehicle comes to rest)

Note that the braking distance D_b is the horizontal component of the distance along the incline. The reason for this is that highway distances are measured in the horizontal plane in keeping with the way in which distances are measured in surveying. Consider the following:

$$\text{Frictional force on the vehicle} = Wf\cos\gamma$$

$$\text{Force acting on the vehicle due to acceleration} = W\frac{a}{g} \tag{3.11}$$

where

g = acceleration due to gravity
a = deceleration that brings the vehicle to a stationary position

If u is the initial velocity, then $a = -\frac{u^2}{2x}$ (assuming uniform deceleration), where x = distance traveled along the plane of the grade during braking. The component of the weight of the vehicle = $W \sin \gamma$.

Substituting into $\Sigma F = ma$, we obtain

$$W \sin \gamma - Wf \cos \gamma = W\frac{a}{g} \tag{3.12}$$

Substituting for a in Equation 3.12, we obtain

$$W \sin \gamma - Wf \cos \gamma = W\frac{u^2}{2gx} \tag{3.13}$$

However, $D_b = x \cos \gamma$.

Substituting for x in Equation 3.13, we obtain

$$W\frac{u^2}{2gD_b}\cos\gamma = Wf\cos\gamma - Wf\sin\gamma$$

which gives

$$\frac{u^2}{2gD_b} = f - \tan\gamma \tag{3.14}$$

and

$$D_b = \frac{u^2}{2g(f - \tan\gamma)} \tag{3.15}$$

Note, however, that $\tan \gamma$ is the grade G of the incline (that is, percent of grade/100) as shown in Figure 3.7.

Equation 3.15 can therefore be written as

$$D_b = \frac{u^2}{2g(f - G)}$$

If g is taken as 9.81 m/s^2 and u is expressed in km/h, Equation 3.15 becomes

$$D_b = \frac{u^2}{254.3(f - G)} \tag{3.16}$$

and D_b is given in meters. Also, the friction coefficient f can be represented as a/g, where a is the deceleration rate in m/s^2. AASHTO recommends that a deceleration rate (a) of 3.41 m/s^2 be used as this is a comfortable deceleration rate for drivers. Equation 3.16 then becomes

$$D_b = \frac{u^2}{254.3(0.35 - G)} \tag{3.17}$$

Note that Equation 3.17 is for when the vehicle is traveling down a grade. When the vehicle is traveling up a grade, the equation is

$$D_b = \frac{u^2}{254.3(0.35 + G)} \tag{3.18}$$

A general equation for the braking distance can therefore be written as

$$D_b = \frac{u^2}{254.3(0.35 \pm G)} \tag{3.19}$$

The plus sign is for vehicles traveling uphill, the minus sign is for vehicles traveling downhill, and G is the absolute value of tan γ.

Also, the distance traveled while reducing the speed of an automobile from u_1 to u_2 in km/h is given as

$$D_b = \frac{u_1^2 - u_2^2}{254.3(0.35 \pm G)} \tag{3.20}$$

It should also be noted that the distance traveled between the time the driver observes an object in the vehicle's path and the time the vehicle comes to rest is longer than the braking distance computed from Equation 3.19. The additional distance accounts for the distance traveled during the perception reaction time. The total distance traveled during a braking maneuver is referred to as *the stopping distance* and is given as

$$S(in\ m) = 0.28ut + \frac{u^2}{254.3(0.35 \pm G)} \tag{3.21}$$

The first term of Equation 3.21 computes the distance traveled during the perception reaction time t (s), and u is the velocity in km/h at which the vehicle was traveling when the brakes were applied.

EXAMPLE 3.7

Determining the Stopping Distance for Different Grade Conditions

If the design speed of a rural two-lane highway is 90 km/h, determine the stopping distance of a vehicle that is traveling on the highway at the posted speed limit for the following sections of road:

(i) A level section
(ii) A 5% upgrade section
(iii) A 5% downgrade section

Solution

Use Equation 3.21 to determine minimum stopping sight distance:

$$S = 0.28ut + \frac{u^2}{254.3(.35 + G)}$$

Determine stopping distance for level section:

$$G = 0$$

$$S = 0.28 \times 90 \times 2.5 + \frac{90^2}{254.3(0.35 + 0)}$$

$$= 63 + 91$$

$$= 154 \text{ m}$$

Determine stopping distance for 5% upgrade section:

$$G = 0.05$$

$$S = 0.28 \times 90 \times 2.5 + \frac{90^2}{254.3(0.35 + 0.05)}$$

$$= 63 + 79.63$$

$$= 142.63 \text{ m}$$

Determine stopping distance for 5% downgrade section:

$$S = 0.28 \times 90 \times 2.5 + \frac{90^2}{254.3(0.35 - 0.05)}$$

$$= 63 + 106.17$$

$$= 169.17 \text{ m}$$

Braking Distance for Trains

The braking distance for trains is similar to that for automobiles in that it is the distance traveled by the train to come to rest after the brakes have been applied. It is, however, different from that for automobiles in that the braking distance for a given train may be significantly different from that for another train. Train braking distance is of importance in the design of the signaling system of the railroad, and it may be computed from empirical equations or by conducting dynamic tests using a specific type of train on the rail line of interest.

Several empirical formulae have been developed in Europe for the braking distance of trains. These depend on the type of train and may therefore depend on the type of braking system used. One of two types of brakes is normally used for rail vehicles: *shoe (block) brakes or disc brakes. Shoe brakes* operate by pressure being applied on the metal shoes, which results in a friction force being applied to the wheels. The braking shoes are provided on both wheels of the axle being braked. *Disc brakes* operate by the action of friction on steel discs or cast iron fixed to the axle. A brief description of the methods used to transmit the braking force follows:

- *Air braking:* The air pressure in special conduits is changed by operating a valve in the driver's cab. The disadvantage of this system is that the braking force is not simultaneously applied to all train vehicles.
- *Electropneumatic:* In this system an electric signal is transmitted on line along the train that is used to simultaneously modify air pressure on all wheels through electrically-actuated air valves in each brake.
- *Electromagnetic braking:* In this system the braking force is applied directly to the rails by special electromagnetic shoes, which carry a current during braking. This system may operate independently or in combination with other systems.
- *Electrodynamic braking:* Deceleration is achieved by converting the electric traction motors into electric generators, thereby eliminating the problem of brake shoe wear.

Software packages that can be used to determine the braking distance of trains are available. For example, the braking distance module of the software package RailSim V7 developed by Systra Consulting Inc. can be used to determine the braking distance of a given train for signal design purposes. Incorporated factors include user-specified train compositions such as multiple train lengths, user-specified settings and parameters, and rear wheel to coupler overhang distance, with the ability to process multiple velocities for a single location. Results have indicated that the braking distance of a train could range from about 79 m for an initial speed of 19 km/h to 2900 m for an initial speed of 160 km/h.

German Railways developed two empirical equations, one for passenger trains and the other for freight trains. These are referred to as the Minden formulae and are given as follows:

For passenger trains:

$$L(m) = \frac{3.8u^2}{6.1\psi(1 + \lambda/_{10}) + i} \tag{3.22}$$

For freight trains:

$$L(m) = \frac{3.85u^2}{\left[5.1\psi\sqrt{(\lambda - 5)} + i\right]} \tag{3.23}$$

where

$L(m)$ = braking distance (m)
u = speed of the train (km/hr)
λ = braking percentage (i.e., the ratio of the braking force required for braking 1 metric ton to the total vehicle weight)
ψ = a constant depending on the brake type characteristics. Values range from 0.5 to 1.25

Stopping Distance of Passenger Trains

Belgian Railways have also developed the empirical formula given in Equation 3.24:

$$L(m) = \frac{4.24u^2}{\left[\lambda\left(\frac{57.5u}{u - 20}\right)\right] + 0.05u - i} \tag{3.24}$$

where $L(m)$, λ, and u take the same definitions as those for Equations 3.22 and 3.23.

TRAVELWAY CHARACTERISTICS

The basic characteristics of the travelway of any mode of transportation depend on the associated human and vehicle characteristics for that mode. For example, the minimum sight distance that can be provided on a highway depends on the perception and reaction time of the driver and the forces acting on the braking vehicle. Similarly, railway tracks are designed for relatively lower grades than those for highways because the weight of a train is much higher than that of an automobile, resulting in a much higher grade resistance for trains for any given grade. It should, however, be noted that the important travelway characteristics are different from mode to mode and are therefore discussed separately in this section for the highway, railroad, and airport runways and taxiways.

Roadway Characteristics

The characteristics of the highway that provide for safe stopping and passing maneuvers and highway curvature are presented here, as they have a more direct relationship to those discussed earlier. This material will be referred to in Chapter 6 in the discussion of geometric design of travelways.

Sight Distance

This is the length of the roadway a driver can see ahead at any particular time. There are two types of sight distance: stopping sight distance and passing sight distance.

Stopping Sight Distance (SSD)

This is the minimum sight distance that should be provided on the highway so that when a driver traveling at the design speed of the road observes an obstruction on the road, he or she will be able to stop the vehicle without colliding with the obstruction. It is the sum of the distance traveled during perception reaction time and the distance traveled during braking. It is therefore the same as the stopping distance given in Equation 3.21. The SSD for a vehicle traveling at u km/h is therefore given as

$$SSD = 0.28ut + \frac{u^2}{254.3(0.35 \pm G)} \tag{3.25}$$

where

SSD = stopping sight distance
u = design speed of road, km/h
G = grade of the roadway (i.e., percent of grade/100)

The sight distance at any point on the highway should therefore be at least equal to the SSD. Table 3.5(a) gives values for SSDs for different design speeds on level grades ($G = 0$). The values for upgrades are shorter and longer for downgrades as shown in Table 3.5(b).

TABLE 3.5
Stopping Distances for Different Design Speeds

Source: Adapted from *A Policy on Geometric Design of Highways and Streets,* American Association of State Highway and Transportation Officials, Washington, D.C., 2004. Used by permission.

Design Speed (mph)	Brake Reaction Distance (ft)	Braking Distance on Level (ft)	Stopping Sight Distance Calculated (ft)	Stopping Sight Distance Design (ft)
15	55.1	21.6	76.7	80
20	73.5	38.4	111.9	115
25	91.9	60.0	151.9	155
30	110.3	86.4	196.7	200
35	128.6	117.6	246.2	250
40	147.0	153.6	300.6	305
45	165.4	194.4	359.8	360
50	183.8	240.0	423.8	425
55	202.1	290.3	492.4	495
60	220.5	345.5	566.0	570
65	238.9	405.5	644.4	645
70	257.3	470.3	727.6	730
75	275.6	539.9	815.5	820
80	294.0	614.3	908.3	910

(a) Zero Percent Grades

Design Speed (mph)	Stopping Sight Distance (ft)					
	Downgrades			Upgrades		
	3 %	6 %	9 %	3 %	6 %	9 %
15	80	82	85	75	74	73
20	116	120	126	109	107	104
25	158	165	173	147	143	140
30	205	215	227	200	184	179
35	257	271	287	237	229	222
40	315	333	354	289	278	269
45	378	400	427	344	331	320
50	446	474	507	405	388	375
55	520	553	593	469	450	433
60	598	638	686	538	515	495
65	682	728	785	612	584	561
70	771	825	891	690	658	631
75	866	927	1003	772	736	704
80	965	1035	1121	859	817	782

(b) Different Percent Grades

Note: 1 mph = 1.61 km/h; 1 ft = 0.3 m

Decision Sight Distance

The stopping sight distances obtained from Equation 3.25 are usually adequate for normal conditions when the driver expects the stimulus. These distances may not, however, be adequate for situations when the stimulus is unexpected or when drivers must make unusual maneuvers. In such a case, a longer sight distance is required, and this is usually referred to as *decision sight distance.* This longer sight distance will provide the driver with the option of making evasive maneuvers, which in some cases may be a better option than stopping. In such cases, perception reaction times are longer, resulting in longer sight distances. Examples of locations where *decision sight distances* are preferable include interchanges and intersections that require unusual or unexpected maneuvers, sections of road where there is a change in a cross-section such as toll plazas and lane drops, and sections of road where several sources of competing information are located. Empirical data have been used by AASHTO to determine decision sight distances for different avoidance maneuvers and design speeds, as shown in Table 3.6.

TABLE 3.6

Decision Sight Distance for Different Design Speeds

Source: A Policy on Geometric Design of Highways and Streets, American Association of State Highway and Transportation Officials Washington D.C., 2004.

Design Speed (mph)	Decision Sight Distance (ft)				
	Avoidance Maneuver				
	A	B	C	D	E
30	220	490	450	535	620
35	275	590	525	625	720
40	330	690	600	715	825
45	395	800	675	800	930
50	465	910	750	890	1030
55	535	1030	865	980	1135
60	610	1150	990	1125	1280
65	695	1275	1050	1220	1365
70	780	1410	1105	1275	1445
75	875	1545	1180	1365	1545
80	970	1685	1260	1455	1650

Note: 1 mph = 1.61 km/h; 1 ft = 0.3 m

Avoidance Maneuver A: Stop on rural road—$t = 3.0$ s
Avoidance Maneuver B: Stop on urban road—$t = 9.1$ s
Avoidance Maneuver C: Speed/path/direction change on rural road—t varies between 10.2 and 11.2 s
Avoidance Maneuver D: Speed/path/direction change on suburban road—t varies between 12.1 and 12.9 s
Avoidance Maneuver E: Speed/path/direction change on urban road—t varies between 14.0 and 14.5 s

Passing Sight Distance

This is the minimum sight distance required on a two-lane highway (one lane in each direction) that will permit a driver to complete a passing maneuver without colliding with an opposing vehicle and without cutting off the passed vehicle. The driver should also be able to abort the passing maneuver (that is, return to the right lane behind the vehicle being passed) within this distance if so desired.

Only single passes (that is, a single vehicle passing a single vehicle) are considered in developing the expression for passing distance. Although it is possible for multiple maneuvers (that is, more than one vehicle pass or are passed in one maneuver), it is not practical for minimum criteria to be based on them.

The assumptions made in developing the passing sight distance are as follows:

1. The vehicle being passed (impeder) is traveling at a uniform speed.
2. The passing vehicle's speed has been reduced and it trails the impeding vehicle at the beginning of the passing zone.
3. At the beginning of the passing section, the driver of the passing vehicle uses a short period of time to perceive the section available for passing and decides to start his or her action.
4. If the decision is made to pass, the passing vehicle is accelerated during the passing maneuver, and the average passing speed is about 10 mph more than the impeder vehicle.
5. A suitable clearance exists between the passing vehicle and any opposing vehicle when the passing vehicle reenters the right lane.

A procedure for determining the minimum passing sight distance for two-lane two-way highways was developed by AASHTO using these assumptions. This involves the determination of four component distances, shown in Figure 3.8, that sum to give the passing sight distance.

These are

d_1 = distance traversed during perception reaction time and during the initial acceleration to the point where the passing vehicle just enters the left lane;

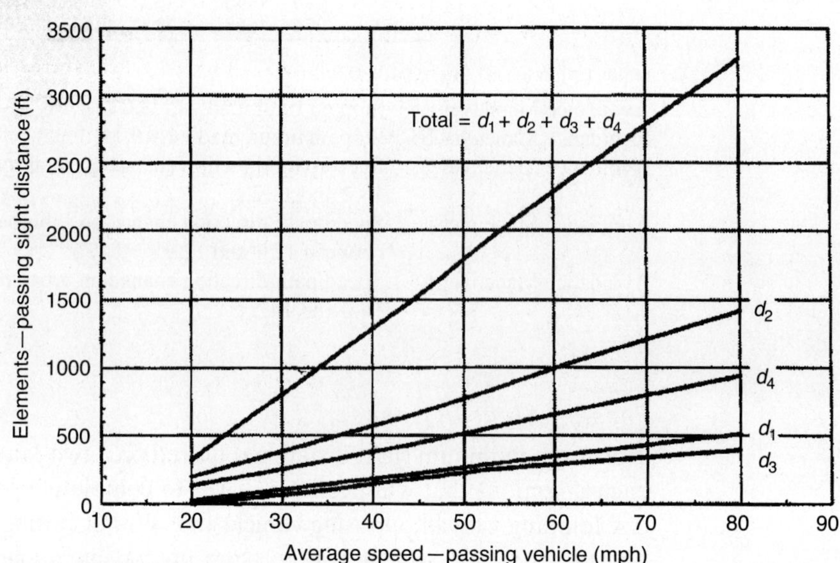

FIGURE 3.8 Elements of and total passing sight distance on two-lane highways.

Source: A Policy on Geometric Design of Highways and Streets, American Association of State Highway and Transportation Officials, Washington D.C., 2004. Used by permission.

d_2 = distance traveled during the time the passing vehicle is traveling in the left lane;
d_3 = distance between the passing and opposing vehicle at the end of the passing maneuver;
d_4 = distance moved by the opposing vehicle during two-thirds of the time the passing vehicle is in the left lane (usually taken as $\frac{2}{3}d_2$).

The distance d_1 is obtained from the expression:

$$d_1 = 0.28t_1\left(u - m + \frac{at_1}{2}\right) \tag{3.26}$$

where

d_1 = distance in m
t_1 = time for initial maneuver in s
a = average acceleration rate (km/h)/s
u = average speed of passing vehicle (km/h)
m = difference in speeds of passing and impeder vehicles in km/h

The distance d_2 is obtained from

$$d_2 = 0.28ut_2$$

where

d_2 = distance in m
t_2 = time passing vehicle is traveling in left lane (s); studies have shown that this time varies between 9.3 and 10.4 s
u = average speed of passing vehicle (km/h)

The clearance distance between the passing and opposing vehicles at the completion of the maneuver has been found to vary between 33.5 and 91.5 m.

Values for these different components computed for different speeds are shown in Table 3.7. It should be noted that these values are for design purposes only and are not used for marking passing and no-passing zones on two-lane highways. Different assumptions are used in the determination of the lengths of passing and no-passing zones on two-lane highways, and these are much shorter. Suggested lengths of passing zones for two-lane highways are shown in Table 3.8.

Minimum Radius of a Highway Circular Curve

The minimum radius of a horizontal curve on a highway can be determined by considering the equilibrium of the dynamic forces acting on the vehicle traveling on the curve. The main forces acting on a vehicle traveling on a curve are an outward radial force (centrifugal) and an inward radial force, which is caused by the frictional effect between the tires and the roadway. If the vehicle is traveling at a high velocity, this frictional force may not be enough to counterbalance the outward radial force, which makes it necessary for the road to be inclined toward the center of the curve. This provides an additional force

TABLE 3.7

Components of Safe Passing Sight Distance on Two-Lane Highways

Source: Adapted from *A Policy on Geometric Design of Highways and Streets,* American Association of State Highway and Transportation Officials, Washington, D.C., 2004. Used by permission.

Component	Speed Range in mph (Average Passing Speed in mph)			
	30–40 (34.9)	40–50 (43.8)	50–60 (52.6)	60–70 (62.0)
Initial maneuver:				
a = average acceleration (mph/sec)[a]	1.40	1.43	1.47	1.50
t_1 = time (sec)[a]	3.6	4.0	4.3	4.5
d_1 = distance traveled (ft)	145	215	290	370
Occupation of left lane:				
t_2 = time (sec)[a]	9.3	10.0	10.7	11.3
d_2 = distance traveled (ft)	475	640	825	1030
Clearance length:				
d_3 = distance traveled (ft)[a]	100	180	250	300
Opposing vehicle:				
d_4 = distance traveled (ft)	315	425	550	680
Total distance, $d_1 + d_2 + d_3 + d_4$ (ft)	1035	1460	1915	2380

[a]For consistent speed relation, observed values are adjusted slightly.

Note: 1 mph = 1.61 km/h; 1 ft = 0.3 m

TABLE 3.8

Suggested Minimum Passing Zone and Passing Sight Distance Requirements for Two-Lane, Two-Way Highways in Mountainous Areas

Source: Adapted from N.J. Garber and M. Saito, *Centerline Pavement Markings on Two-Lane Mountainous Highways,* Research Report No. VHTRC 84-R8, Virginia Highway and Transportation Research Council, Charlottesville, VA, March 1983.

85th-Percentile Speed (mph)	*Available Sight Distance (ft)*	*Minimum Passing Zone*		*Minimum Passing Sight Distance*	
		Suggested (ft)	*MUTCD* (ft)*	*Suggested (ft)*	*MUTCD* (ft)*
	600–800	490		630	
	800–1000	530		690	
30	1000–1200	580	400	750	500
	1200–1400	620		810	
	600–800	520		700	
	800–1000	560		760	
35	1000–1200	610	400	820	550
	1200–1400	650		880	
	600–800	540		770	
	800–1000	590		830	
40	1000–1200	630	400	890	600
	1200–1400	680		950	
	600–800	570		840	
	800–1000	610		900	
45	1000–1200	660	400	960	700
	1200–1400	700		1020	
	600–800	590		910	
	800–1000	630		970	
50	1000–1200	680	400	1030	800
	1200–1400	730		1090	

**Manual on Uniform Traffic Control Devices,* Published by FHWA.

Note: 1 mph = 1.61 km/h; 1 ft = 0.3 m

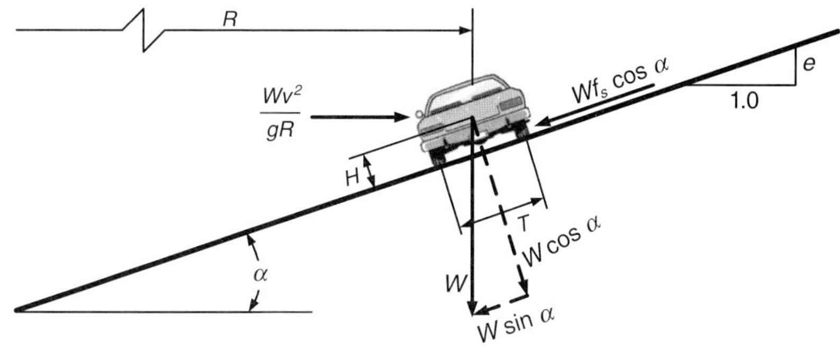

FIGURE 3.9
Forces acting on a vehicle traveling on a horizontal curve section of a road.

W = weight of vehicle
f_s = coefficient of side friction
g = acceleration of gravity
u = speed when brakes applied
R = radius of curve
α = angle of incline
$e = \tan \alpha$ (rate of superelevation)
T = track width
H = height of center of gravity

from the component of weight of the vehicle down the incline (see Figure 3.9). The angle of inclination of the roadway toward the center of the curve is known as the *superelevation.*

Let the minimum radius of the curve be R m and the inclination of the roadway α. The component of the weight down the incline is $W \sin \alpha$, and the frictional force down the incline is $Wf_s \cos \alpha$. The centrifugal force is given as

$$F_c = \frac{Wa_c}{g} \tag{3.27}$$

where

a_c = acceleration for curvilinear motion = u^2/R (R = radius of curve)
W = weight of the vehicle N
g = acceleration of gravity

When the vehicle is in equilibrium with respect to the incline (i.e., the vehicle moves forward but neither up nor down the incline), the three relevant forces may be equated to obtain

$$\frac{Wu^2}{gR} \cos \alpha = W \sin \alpha + Wf_s \cos \alpha$$

where

f_s = coefficient of side friction
$u^2/g = R(\tan \alpha + f_s)$

which gives

$$R = \frac{u^2}{g(\tan \alpha + f_s)} \tag{3.28}$$

TABLE 3.9

Maximum Coefficients of Side Friction for Different Design Speeds

Source: Adapted from *A Policy on Geometric Design of Highways and Streets,* American Association of State Highway and Transportation Officials, Washington, D.C., 2004. Used by permission.

Design Speed (mph)	*Coefficients of Side Friction, f_s*
30	0.20
40	0.16
50	0.14
60	0.12
70	0.10
80	0.08

Note: 1 mph = 1.61 km/h

Tan α is the tangent of the angle of inclination of the roadway and is known as the rate of *superelevation.*

Equation 3.28 can therefore be written as

$$R = \frac{u^2}{g(e + f_s)} \tag{3.29}$$

If g is taken as 9.81 m/s^2, u is measured in km/h, and e is given in percent, the minimum radius R (in m) is given as

$$R = \frac{u^2}{127(0.01e + f_s)} \tag{3.30}$$

It can be seen from Equation 3.30 that to reduce R for a given velocity, either e or f_s or both should be increased. However, there are stipulated maximum values that can be used for either e or f_s. For example, the maximum value for the superelevation (e) depends on whether the highway is located in an urban area, weather conditions (such as the occurrence of snow), and the distribution of slow-moving vehicles in the traffic stream. For highways located in rural highways with no snow or ice, a maximum value for the superelevation is 10%. For values located in areas with snow or ice, maximum values ranging from 8 to 10% are used. For expressways in urban areas, a maximum superelevation rate of 8% is used. Local urban roads are usually not superelevated as speeds on them are relatively low.

The coefficient of side friction f_s varies with the design speed. In general, side friction factors are lower for high-speed design roads than for lower speed design roads. Table 3.9 shows maximum values for f_s recommended by AASHTO for different design speeds.

EXAMPLE 3.8

Determining the Radius of a Horizontal Curve

A horizontal curve is to be designed for a section of an expressway having a design speed of 95 km/h. Determine

(i) The radius of the curve if the superelevation is 6.5%;
(ii) The minimum radius if the expressway is located in an urban area and the maximum superelevation rate can be used.

Solution

Use Equation 3.30 to determine radius of curve for $e = 0.01 \times 6.5$:

$$R = \frac{u^2}{127(0.01e + f_s)}$$

For a design speed of 95 km/h, $f_s = 0.12$ (see Table 3.9):

$$R = \frac{95^2}{127(0.065 + 0.12)}$$

$$= 384 \text{ m}$$

Use Equation 3.30 to find the minimum radius. The minimum radius will be obtained by using the maximum allowable superelevation.

For urban expressway, maximum $e = 8\%$.

$$R = \frac{95^2}{127(0.01 \times 8 + 0.12)}$$

$$\text{R} = 355 \text{ m}$$

Railroad Characteristics

The railroad characteristics that are comparable to those discussed for the highway mode are the stopping distance and the superelevation requirements at horizontal curves. This material will also be referred to in Chapter 6. In general, railroad tracks are not designed to provide a *minimum sight distance* that will allow a train traveling at a high speed to stop if the driver observes an object on the track. The reason is that braking distances of trains can be very high compared to those for automobiles, and it is not feasible to provide sight distances on curves that will allow the train to be stopped before colliding with an object that is observed on the track. Sharp horizontal and vertical curves are therefore avoided in railroad track design, as will be shown in Chapter 6 in the discussion of travelway design. However, at railroad/highway grade crossings with warning devices that allow the driver of an approaching vehicle to determine whether a danger exists because of an approaching train (passive control), the decision to stop or proceed across the crossing is made entirely by the automobile driver. Sufficient sight distance should therefore be provided for drivers of the automobile to safely proceed across the grade crossing when they see the oncoming train.

Sight Distance Requirements at Passive Controlled Railroad Crossings

When drivers of automobiles approach a passive controlled railroad crossing, they have two main options:

- The driver stops at the stop line having seen the approaching train.
- The driver, having seen the train, continues to cross the tracks safely before the train arrives.

Figure 3.10 illustrates the minimum sight distances required for the two options available to the driver of the automobile. The minimum distance (stopping distance) required for the driver to stop at the stop line is given by Equation 3.21 as

$$S = 0.28ut + \frac{u^2}{254.3\,(0.35 + G)}$$

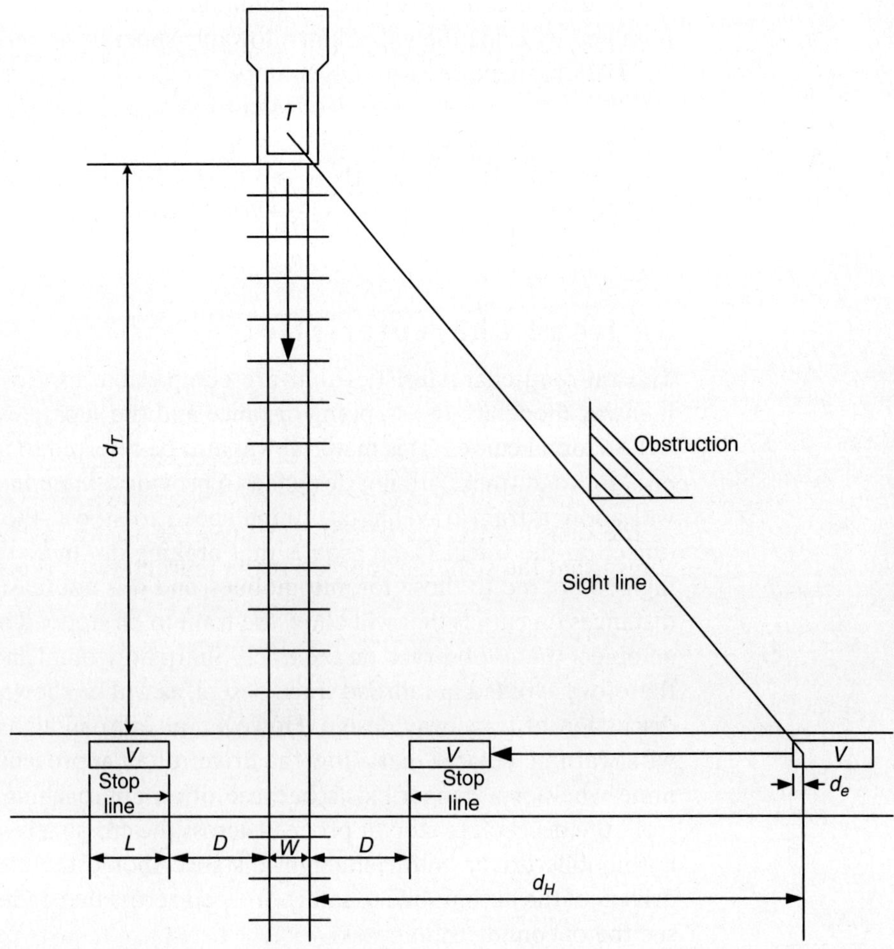

FIGURE 3.10

Conditions for a moving vehicle to safely stop or cross at a railroad/highway crossing.

Source: A Policy on Geometric Design of Highways and Streets, American Association of State Highway and Transportation Officials, Washington, D.C., 2004. Used by permission.

Therefore, the minimum distance (d_H) the driver's eyes should be from the track is the sum of the stopping distance, the distance between the stop line and the tracks, and the distance between the driver's eyes and the front of the vehicle's wheels. This is given as

$$d_H = 0.28u_v t + \frac{u_v^2}{254.3\,(0.35 + G)} + D + d_e \tag{3.31}$$

If it is assumed that the road approaches to the railroad crossing have zero grades, d_H is obtained as

$$d_H = 0.28u_v t + \frac{u_v^2}{89} + D + d_e \tag{3.32}$$

where

u_v = speed of the vehicle (km/h)
t = perception reaction time of the driver
D = distance from the stop line or the front of the vehicle to the nearest rail assumed to be 4.5 m
d_e = distance from the driver to the front of the vehicle, assumed to be 2.4 m

If the driver continues to cross the tracks, it can be seen from Figure 3.10 that the total distance traveled to clear the track is the sum of d_H, the width of the track (W), the distance between the tracks and the stop line on the other side of the tracks (D), and the length of the vehicle (L). The sight distance leg (d_T) on the railroad track is the distance traveled by the train during the time the automobile is traveling this total distance, and is given as

$$d_T = \frac{u_T}{u_v}\left(0.28u_v t + \frac{u_v^2}{89} + 2D + L + W\right) \tag{3.33}$$

Similarly, if the vehicle is stopped at the stop line, a sight distance along the length of the track should be provided to allow the driver to accelerate the vehicle and safely cross the tracks before the arrival of a train that appears just as the driver starts his or her maneuver, as shown in Figure 3.11. It can be shown that the sight distance along the railroad track is given by

$$d_T = 0.28u_T\left[\frac{u_g}{a_1} + \frac{L + 2D + W - d_a}{u_g} + J\right] \tag{3.34}$$

where

d_T = sight distance leg along railroad tracks to permit the vehicle to cross the tracks from a stopped condition
u_T = speed of train, km/h
u_g = maximum speed of vehicle in first gear, assumed to be 2.68 m/s
a_1 = acceleration of vehicle in first gear, assumed to be 0.45 m/s^2
L = length of vehicle, assumed to be 19.8 m

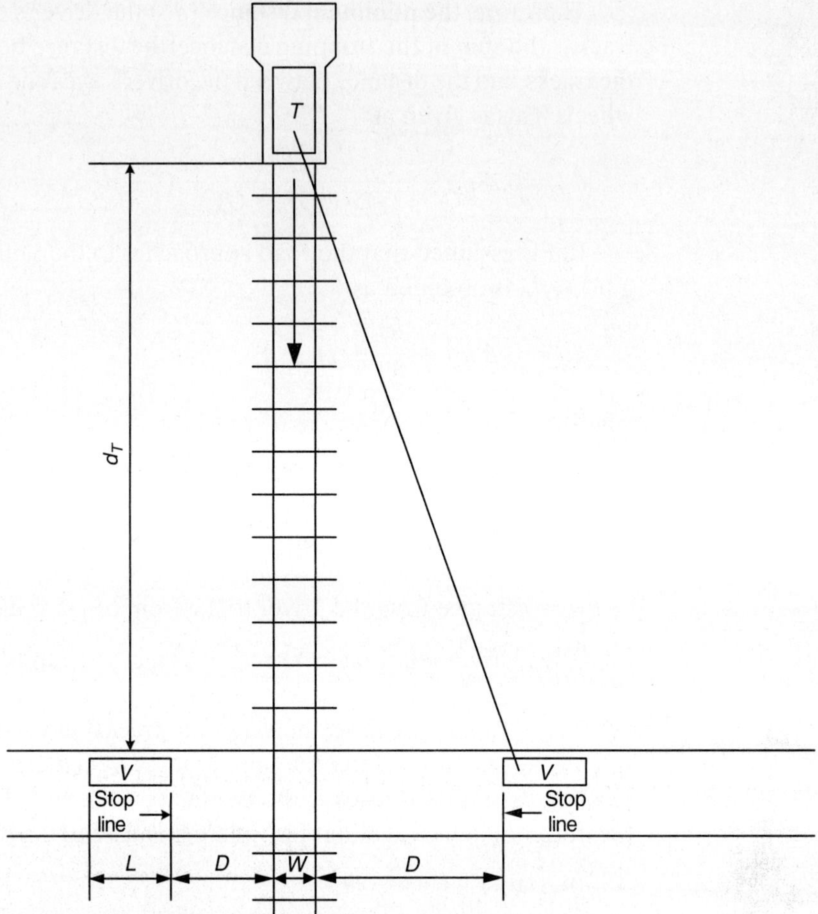

FIGURE 3.11
Conditions for a stopped vehicle to safely depart and cross a single railroad track.

Source: A Policy on Geometric Design of Highways and Streets, American Association of State Highway and Transportation Officials, Washington, D.C., 2004. Used by permission.

D = distance of stop sign to nearest rail, assumed to be 4.5 m
J = sum of perception time and time to activate clutch or automatic shift, assumed to be 2 s
W = distance between outer rails, for a single track; this value is 1.52 m
d_a = distance vehicle travels while accelerating to maximum speed in first gear

$$d_a = \frac{u_g^2}{2a_1} = \frac{2.68^2}{2(0.45)} = 7.98 \text{ m}$$

Table 3.10 gives suggested distances for different approaching speeds of a train and a truck 20 m long that will allow the truck to safely proceed across the grade crossing. Also, the Federal Railroad Administration's program on *Intelligent Grade Crossings* will provide continuous information on train location and speeds. This information will be integrated into the highway traffic management system, with the objective of providing advance warning to automobile drivers of trains approaching the grade crossing. The system will

TABLE 3.10

Required Design Sight Distance for Combination of Highway and Train Vehicle Speeds; 20 m (65 ft) Truck Crossing a Single Set of Tracks at 90 Degrees

Source: A Policy on Geometric Design of Highways and Streets, American Association of State Highway and Transportation Officials, Washington, D.C., 2004. Used by permission.

Train Speed (mph)	Case B Departure from Stop	Case A Moving Vehicle							
		Vehicle Speed (mph)							
	0	10	20	30	40	50	60	70	80
	Distance along railroad from crossing, d_T (ft)								
10	240	146	106	99	100	105	111	118	126
20	480	293	212	198	200	209	222	236	252
30	721	439	318	297	300	314	333	355	378
40	961	585	424	396	401	419	444	473	504
50	1201	732	530	494	501	524	555	591	630
60	1441	878	636	593	601	628	666	709	756
70	1681	1024	742	692	701	733	777	828	882
80	1921	1171	848	791	801	838	888	946	1008
90	2162	1317	954	890	901	943	999	1064	1134
	Distance along highway from crossing, d_H (ft)								
		69	135	220	324	447	589	751	931

Note: 1 mph = 1.61 km/h; 1 ft = 0.3 m

also provide the capability for the railroad engineer to be warned of obstacles or trapped vehicles at the crossings.

EXAMPLE 3.9

Determining the Maximum Safe Speed on a Farm Road/Railroad Crossing

A two-lane road crosses an at-grade single railroad track at 90 degrees. Determine

(i) What maximum speed you will recommend to be posted on the road for vehicles to safely cross the tracks when an approaching train is observed by a driver;
(ii) The maximum distance on the road from the track that the driver should first observe the train.

The following conditions exist:

(i) Speed of trains crossing the highway = 130 km/h
(ii) Design vehicle is a passenger car
(iii) Sight distance on railway tracks when the driver of the vehicle on the road observes a train = 245 m

Solution

Determine speed of passenger car for safe condition (use Equation 3.33):

$$d_T = \frac{u_T}{u_v}\left(0.28u_v t + \frac{u_v^2}{89} + 2D + L + W\right)$$

In this case

$d_T = 245$ m

$D = 4.5$ m

$L = 5.7$ m (see Table 3.1)

W = 1.5 m (for a single track)

$t = 2.5$ s

$$245 = \frac{130}{u_v}\left[\left(0.28u_v(2.5) + \frac{u_v^2}{89} + 2 \times 4.5 + 5.7 + 1.5\right)\right]$$

$$\frac{u_v^2}{89} - 1.185u_v + 16.2 = 0$$

which gives

$$u_v^2 - 105.465u_v + 1441.8$$

Solving the quadratic equation gives

$$u_v = \frac{105.465 \pm \sqrt{105.465^2 - 4 \times 1441.8}}{2}$$

$$u_v = \frac{105.465 \pm \sqrt{11122.87 - 5767.2}}{2}$$

$$= 89.32 \text{ km/h or } 16.14 \text{ km/h}$$

Note that two values are obtained for u_v as they were obtained from a quadratic equation. The reasonable value for this case is 89.32 km/h, and a speed limit of 90 km/h can be used.

Determine maximum distance from the track the driver should first see the train (use Equation 3.32):

$$d_H = 0.28u_v t + \frac{u_v^2}{89} + D + d_e$$

$$d_H = 0.28 \times 90 \times 2.5 + \frac{90^2}{89} + 4.5 + 2.4$$

$$= (63 + 91.01 + 4.5 + 2.4) \text{ ft}$$

$$= 160.91 \text{ m}$$

Railroad Track Characteristics at Horizontal Curves

When a train is moving around a horizontal curve, it is subjected to a centrifugal force acting radially outward, similar to that discussed for highways. It is therefore necessary to raise the elevation of the outer rail of the track by a value E_q, which is the *superelevation* that provides an equilibrium force similar to that on highways. For any given equilibrium elevation, there is an *equilibrium speed.* This is the speed at which the resultant weight and the centrifugal force are perpendicular to the plane of the track. When this occurs, the components of the centrifugal force and the weight in the plane of the track are balanced. If all trains travel around the curve at the equilibrium speed, both smooth riding and minimum wear of the track will be obtained. This is not always the case, as some trains may travel at higher speeds than the equilibrium speed while others may travel at lower speeds.

Trains traveling at a higher speed will cause more than normal wear on the outside rail, while those that travel at lower speeds will cause more than normal wear on the inside rail. Also, when the train is traveling faster than the equilibrium speed, the centrifugal force is not completely balanced by the elevation, which results in the car body tilting toward the outside of the curve. Consequently, under normal conditions, the inclination of the car body from the vertical is less than the inclination of the track from the vertical.

The difference between the inclination of the car from the vertical *(car angle)* and that of the track from the vertical *(track angle)* is known as the *roll angle*. The higher the roll angle, the less comfort is obtained as the train traverses the curve. Full *equilibrium superelevation* (E_q) is, however, rarely used in practice for two main reasons. First, the use of a full equilibrium superelevation may require long curves (spiral curves) connecting the straight portion and circular section of the track. Second, equilibrium superelevation can result in discomfort for passengers on a train traveling at a speed that is much less than the equilibrium speed, or if the train is stopped along a highly superelevated curve. The portion of the *equilibrium superelevation* used in the design of the curve is known as the *actual superelevation* (E_a), and the difference between the actual *superelevation* and the *equilibrium superelevation* is known as the *superelevation unbalanced.*

Equations relating to the superelevation of the curve, the design speed, and the radius of the curve have been developed separately for light rail transit and freight and intercity passenger tracks. These equations are presented in Chapter 6 subsequent to the discussion on classification of travelways.

Airport Characteristics

The specific characteristics discussed in this section are those related to the travelway of the airplane when it is at the airport, such as taxiways and runways. The related airport characteristics are somewhat different from those for the highway and railroad and are dealt with in Chapter 6, as they are directly related to the design of the travelway for different classes of airports. It is, however, necessary for the reader to have a general understanding of how airports are classified to understand the characteristics of taxiways and runways.

Airports are categorized by the type of services they provide and for design purposes by the predominant airplane expected to use the airport. A brief description of airport classification with respect to the services provided is given here, and the classification based on the predominant airplane is given in Chapter 6.

Based on the services provided, airports are generally classified in the following categories:

- Commercial service—primary
- Commercial service—other
- General aviation
  - Basic utility (BU)
  - General utility (GU)
  - Transport
- Reliever airports

Commercial service—primary: Airports having at least 0.01% of the annual U.S. enplanements. Airports in this category must also be served by at least one scheduled passenger service carrier with a minimum of 2500 annual enplanements.

Commercial service—other: Airports that have at least 0.01% of the annual U.S. enplanements but do not satisfy the passenger service criterion.

General aviation: An airport with any of the following characteristics: receiving U.S. mail; considered a significant local, regional, or national interest; having significant military activities; a general aviation heliport serving more than 400 repeating air taxi operations; or serving more than 810 repeating operations.

General aviation basic utility (BU): Airports that accommodate most of the single-engine and many of the smaller twin-engine airplanes.

General aviation general utility (GU): Airports that serve nearly all general aviation airplanes with take-off weights not exceeding 56300 N.

General aviation transport: These airports serve mainly transport-type and business jets and are usually capable of serving turbo jet-powered airplanes. They are usually designed to serve airplanes with an approach speed of 120 knots (note, 1 knot = 1.85 km/h).

Reliever airports: Airports usually located in metropolitan areas with the main objective of relieving congestion at large carrier airports.

Federal Aviation Classification of International Airports

The FAA has also developed a classification system for international airports. International airports are those that serve international air traffic, are designated as ports of entry in the United States from places outside the United States, and provide customs and immigration services. This classification system is in accordance with Article 68 of the *Convention on International Civil Aviation Organization (ICAO),* which requires each signatory government to specify the route an international air service should take within its territory and the airports that may be used by such services. There are four categories of international airports within this classification system:

(a) ***Designated international airports of entry (AOE):*** Airports that are open to all international aircraft for entry and that provide customs services. International flights do not have to obtain prior permission to land, but advance notice of arrival should be made so that inspectors can be made available. An airport under this category should be capable of generating enough international traffic and should provide adequate space and facilities for customs and federal inspections. Examples of these airports include Juneau Harbor SPB in Juneau, Alaska; San Diego International Longfield in San Diego, California; and Houlton International in Houlton, Maine.

(b) ***Landing rights airports (LRAs):*** Permission to land at these airports should have been received by international flights before they arrive. Advance notice of arrival should also be furnished to U.S. Customs. For those airports where Advise Customs Service (ADCUS) is available, the notice of arrival may be transmitted through flight plans, which are considered as applications for permission to land. In some cases customs officers may grant blanket "landing rights" to individuals or companies for a given time period. In such cases only an advance notice of arrival is required. This type of blanket permission is usually given for scheduled airline flights at busy landing rights airports. Airports within the LRA category include many of the major U.S. international airports, such as Los Angeles International and Washington Dulles International Airport.

(c) ***User-fee airports:*** An airport within this category does not meet the customs requirements for clearance services, but a petition has been made for landing rights as a "user service" airport. The costs for inspections at these airports are reimbursable, which may require the aircraft operators to refund to the airport operator the associated costs for providing federal services. Examples of airports within this category are Blue Grass Airfield in Lexington, Kentucky, and Ft. Wayne International in Ft. Wayne, Indiana.

(d) ***ICAO-designated U.S. airports serving international operations:*** These are airports that serve international operations by providing traffic or refueling services. These include airports that regularly serve scheduled and nonscheduled international commercial air transport, those designated as alternates, and those that serve international general aviation flights. It should be noted that this category is not exclusive of the other three categories, as an airport can be in either category (a), (b), or (c) and can then be classified under this category as regular, alternate, or general aviation. For example, Juneau International Airport is in category (a) but is also classified as a regular airport under this category.

Airport Taxiways and Runways Characteristics

Airport taxiways and runways are two main components of the airport that directly serve the aircraft when it is at an airport. Many specific characteristics for taxiways and runways are therefore based on the static characteristics of the aircraft expected to use the airport. Tables that give minimum dimensions for specific design items for airport taxiways and runways are included in Chapter 6 in the discussion of geometric design of travelways.

SUMMARY

Transportation engineers need to study and understand the fundamental elements that are required to design the different components of the transportation mode they are dealing with. This chapter presents the basic characteristics that are of importance in the geometric and structural design of the travelways of the highway, air, and rail modes. It should be noted that extensive research has been conducted on specific aspects of these characteristics, particularly on human characteristics. The material presented in this chapter, however, is confined to that which has direct bearing on the material included for the design sections of this book. Important points of interest include:

Visual reception of humans

Perception reaction time

Walking speeds

Static characteristics of automobiles, airplanes, and railroad locomotives

Power required by automobiles and railroad locomotives in motion to overcome resistance forces

Braking distances for automobiles and railroad locomotives

Superelevation on curves for highways and railroad tracks

Sight distance requirements

Some of the material discussed will be used in Chapter 6 in the discussion of the geometric design of travelways, and in Chapter 7 in the discussion of structural design of travelways.

PROBLEMS

3.1 Describe the two main human characteristics that affect the design of transportation terminals.

3.2 Why is color blindness not of great significance in the operation of an automobile?

3.3 Select at least 10 intersections in your area and determine the following:

(a) The average walking speed of all pedestrians at each intersection;
(b) The average walking speed at each intersection for men and women separately;
(c) The average walking speed of all pedestrians for all intersections combined;
(d) The average walking speed of men and women separately for all intersections combined.

Discuss your results with respect to the value given in the text for average walking speed and any factors you identified that influence the walking speeds of pedestrians.

3.4 Describe the significant differences among the vehicle characteristics for the automobile, locomotive, and airplane that are of significant importance to the transportation engineer.

3.5 Describe how waterborne transportation vehicles are classified and compare the classification system with that for automobiles.

3.6 Describe the system of classification of airplanes by the Federal Aviation Administration (FAA) for the purpose of selecting appropriate airport design standards. Show how this classification system is used in airport design.

3.7 A passenger car being driven on a flat and straight section of a highway at a speed of 105 km/h reaches a curved section of the highway with a grade of 5% and a radius of 450 m.

Determine

(a) The additional force that will be required to maintain the original speed of 105 km/h
(b) The percentage increase in the total force to maintain the original speed of 105 km/h

Assume that the weight of the car is 907 kg, the cross-sectional area is 3.15 m^2, and the car is being driven at sea level.

3.8 Repeat Problem 3.7 for a two-axle truck having a cross-sectional area of 5.76 m^2 and carrying a load of 81630 N/axle.

3.9 Two passenger cars are traveling at 88.5 km/h. The weight of car A is 9000 N, and that of car B is 18000 N. The cross-sectional area for car A is 3.15 m^2, and that of car B is 3.6 m^2. Determine the maximum grade on which passenger car A can travel without its total resistance exceeding that of car B traveling on a straight and level section of road.

3.10 A truck and a passenger car traveling on a section of highway at a speed of 80 km/h enter a short curved section of the road with a grade of 5% and a radius of 270 m. Determine the ratio of the additional force required by the truck to that required by the car for both vehicles to maintain their original speed of 80 km/h. Assume the weight of the car is 11250 N and that of the truck 54000 N.

3.11 Determine the power that is required to operate a train of 32 cars on a level tangent if the total load including the locomotive is supported by 128 axles carrying an average of 22.65 metric tons/axle with conventional equipment traveling at 153 km/h.

3.12 If train conventional equipment, consisting of 10 cars, is traveling at 137 km/h on a curve of 2°, determine the total resistance on the train if the load/axle is 22.65 metric tons with 4 axles/car. Assume that the track gage is 1.37 m.

3.13 A curved segment of an existing road has a radius of 180 m, which restricts the posted speed limit for that segment to 75% of the speed limit of the road. If this segment of the road is to be improved so that the posted speed limit is that of the rest of the road, determine the radius of the improved segment. The maximum superelevation allowed is 8%.

3.14 A train is expected to travel on a railroad having a maximum horizontal curve of 3.5° and a maximum grade of 3%. If the load on each axle is 18.14 tons, with 4 axles/car, determine the maximum number of cars that can be pulled along the track by a single locomotive that has a tractive force of 405000 N, traveling at 105 km/h, with a track gage of 1.37 m.

3.15 A freight train consisting of 75 cars each having 4 axles with each axle carrying a load of 22.5 metric tons on a gage of 1.37 m is expected to travel at a speed of 137 km/h on a section of level track with a maximum horizontal curve of 3°. Determine the number of conventional equipment locomotives that will be required if the available locomotives have a maximum tractive force of 225000 N each.

3.16 Repeat Problem 3.15 if the tractive force of the locomotive is 360000 N, the maximum grade is 4% and the maximum horizontal curve is 3.5°. Discuss your results with respect to that obtained for Problem 3.15.

3.17 An engineer has decided to construct a temporary diversion from a principal arterial due to major rehabilitation work to be carried out on a section of the highway. If the speed on the arterial is 105 km/h, determine the minimum distance from the diversion a road sign should be placed to inform drivers of the speed limit on the diversion.

Design speed of diversion = 55 km/h

Letter height of road sign = 7.5 cm

Perception reaction time = 2.5 s

Grade on arterial leading of the diversion = −3%

Assume that a driver can read a road sign within his or her area of vision at a distance of 4.8 m for each centimeter of letter height.

3.18 A horizontal curve is to be designed for a section of a highway having a design speed of 110 km/h. If the physical conditions restrict the radius of the curve to 285 m, determine

(a) Required minimum superelevation at this curve;
(b) Whether this superelevation obtained is feasible or not? If not, what change would you suggest to execute this design?

3.19 A section of a road has a superelevation of 0.6% and a curve of 180 m radius. What speed limit will you recommend at this section of the highway?

3.20 A temporary diversion has been constructed on a highway of −3% gradient due to major repairs that are being undertaken. Determine the speed limit that should be imposed on the diversion if drivers can see the sign informing them of the diversion at a distance of 120 m from the diversion and the posted speed limit on the highway is 95 km/h.

3.21 An existing section of private road with a negative grade connecting a highway to a housing development is to be improved to provide an expected increase of the posted speed limit from 80 km/h to 90 km/h. By what percentage should the grade be reduced at this section of the highway if the available sight distance of 159 m is just adequate for the speed of 80 km/h and cannot be increased due to existing physical constraints?

3.22 A two-lane highway with a design speed of 65 km/h crosses a single railroad track that serves trains traveling at 150 km/h. The highway serves a new development with a significant percentage of school-age children in the population necessitating the selection of a large school bus (68 passengers) as the design vehicle. If the crossing is controlled by a stop sign, determine

(a) The minimum sight distance along the two-lane highway that will ensure all vehicles stopping at the stop line;
(b) The minimum distance along the railroad track a bus driver should see that will allow him or her to cross the track safely after stopping.

3.23 A railroad crossing with a passive control is formed by a two-lane highway with a design speed of 70 km/h and a railroad track with trains traveling at 150 km/h. Determine the minimum distance a building should be placed from the centerline of the tracks to ensure safe crossing of an approaching vehicle if the building is located 45 m from the center line of the right lane of the highway.

3.24 Describe the four categories of international airports in the federal classification of international airports.

References

1. Advisory Circular AC No. 150/5000-5C, Federal Aviation Administration, U.S. Department of Transportation, December 1996.
2. Advisory Circular AC No. 150/5300-13, Federal Aviation Administration, U.S. Department of Transportation, Incorporating Changes 1 through 8, September 2004.

3. Boff, K. R., and Lincoln, J. E., Guidelines for alerting signals. Engineering Data Compendium: Human Perception and Performance, Vol. 3, Human Systems Information Analysis Center http://iac-dtic.mil/hsirc/ 1988.
4. Carpenter, J. T., Fleishman, R. N., Dingus, T., Szczublewaski, F. E., Krage, M. K., and Means, L.G., *Human Factors Engineering, the TravTek Driver Interface.* Vehicle Navigation and Information Systems Conference, Warrendale, PA: Society of Automotive Engineers, 1991.
5. Claffey, P., *Running Costs of Motor Vehicles as Affected by Road Design and Traffic,* National Cooperative Research Program Report III, Highway Research Board, Washington, D.C., 1971.
6. Dingus, T. A., and Hulse, M. C., Some Human Factors Design Issues and Recommendations for Automobile Navigation Systems, *Transportation Research,* IC(2), 1993.
7. Dudeck, C. L., *Guidelines on the Use of Changeable Message Signs.* FHWA-TS-90-043, Federal Highway Administration, Washington, D.C.
8. Greene, F. A., *A Study of Field and Laboratory Legibility Distances for Warning Symbol Signs.* Unpublished doctoral dissertation, Texas A&M University, 1994.
9. Hulbert, S., "Human Factors in Transportation," in *Transportation and Traffic Engineering Handbook,* 2nd ed. Prentice Hall, Englewood Cliffs, NJ, 1982.
10. *Intelligent Grade Crossings,* Federal Railroad Administration, http/www.fra.dot.gov/us/content/1270, September 2003.
11. Kimura, K., Sugiura, S., Shinkai, H., & Negai, Y. (1988). *Visibility Requirements for Automobile CRT Displays–Color, Contrast, and Luminance.* SAE Technical paper Series (SAE No. 880218, pp. 25–31). Warrendale, PA: Society of Automotive Engineers.
12. *Manual for Railway Engineering,* American Railway Engineering and Maintenance-of-Way Association (AREMA), Washington, D.C., 2005.
13. *Manual on Uniform Traffic Control Devices (MUTCD),* U.S. Department of Transportation Federal Highway Administration, Washington, D.C., 2003.
14. *Peripheral Vision Horizon Display (PHVD),* proceedings of a conference held at NASA Ames Research Center, Dryden Flight Research, March 15–16, 1983, National Aeronautics and Space Administration, Scientific and Technical Information Branch, 1984.
15. *Policy on Geometric Design of Highways and Streets,* American Association of State Highway and Transportation Officials, Washington, D.C., 2004.
16. Popp, M. M., and Faber, B., "Advanced Display Technologies, Route Guidance Systems and the Position of Displays in Cars." *Vision in Vehicles* edited by A.G. Gale, (North Holland Elsevier Science Publishers) 1991.
17. Profillidis, V. A., *Railway Engineering,* Avebury Technical, Aldershot, England, 1995.
18. Roland, G., Moretti, E. S., and Patton, M. L., *Evaluation of Glare from Following Vehicle's Headlight,* prepared for U.S. Department of Transportation, National Highway Traffic Safety Administration, Washington, D.C., 1981.
19. Railsim V7, Systra Consulting and Engineering, http://www.systconsulting.com.
20. *Twin Trailer Trucks Special Report 211,* Transportation Research Board, National Research Council, Washington, D.C., 1986.
21. *Visual Characteristics of Navy Drivers,* Naval Submarine Medical Research Laboratory, Groton, CT, 1981.
22. Weintraub, D. J., Haines, F. F., and Randle, R. J., *Runway to Head-Up Display Transition Monitoring Eye Focus and Decision Times,* proceedings of the Human Factors Society, 29th Annual Meeting, Santa Monica, CA, 1985.
23. International Union of Railways UIC Leaflets 541-5, 4th Edition, Railway Technical Publications, 75015 Paris, France, May 2006.

CHAPTER 4

Transportation Capacity Analysis

The focus of this chapter is on understanding the basic concepts associated with determining the capacity and level of service for several examples of transportation facilities. Capacity analysis is concerned with answering the very important question of how much traffic (e.g., vehicles, pedestrians, aircraft, etc.) a given facility can accommodate under a given operating condition. The basic idea behind capacity analysis is to develop a set of models or analytical equations that relate flow levels, geometrics, environmental conditions, and control strategies, on the one hand, to measures describing the resulting operating or service quality on the other hand. These models or equations allow us to determine the maximum traffic-carrying capacity of a facility and the expected quality or level of service at different flow levels.

In this chapter, we focus on the basic concepts underlying the capacity analysis procedures for a number of transportation facilities, including (1) highways, (2) transit, (3) bike paths, (4) pedestrian facilities, and (5) airport runways. The purpose is to give the reader an understanding of the multimodal and broad nature of the transportation field. Since we will be considering several transportation modes in this chapter, our focus will be on the general procedures and concepts without delving too much into the details of the different analysis procedures. Nevertheless, we will try to direct interested readers to appropriate references, whenever possible, where more details about the different procedures may be found.

The Capacity Concept

The *Highway Capacity Manual* (HCM), one of the most important references for transportation professionals, defines the capacity of a facility as follows:

> The capacity of a facility is the maximum hourly rate at which persons or vehicles can reasonably be expected to traverse a point or a uniform section of a lane or roadway during a given time period under prevailing roadway, traffic, and control conditions.

Three important remarks should be made regarding the HCM's definition of *capacity*. First, it should be noted that the manual defines *capacity* in terms of either vehicles or persons. The capacity for highways, for example, is typically defined in terms of vehicles. For transit or pedestrian facilities, the capacity will need to be expressed in terms of persons. Second, the definition specifies that capacity is defined for a point or for a *uniform* section of a facility. The capacity of a facility varies based upon its geometric characteristics, the mix of vehicles using it, and any control actions applied to it (e.g., traffic signals). Given this, the capacity can only be defined for uniform or homogeneous sections, where the different factors affecting capacity remain unchanged. Finally, the HCM defines capacity as the maximum number of vehicles or persons that a facility can *reasonably* accommodate. The use of the word *reasonably* implies that one should expect the value of the capacity of a given facility to vary slightly from one location to another or from one day to another. This means that capacity values that we typically use in our analysis are not the single highest values ever recorded or expected to occur on a facility, but rather a flow level that can reasonably be achieved repeatedly on a given facility.

The Level of Service Concept

Closely associated with the concept of capacity is the concept of level of service (LOS). For transportation facilities, our interest is not just in determining the maximum number of vehicles, passengers, or pedestrians that a facility can accommodate, but of equal interest is quantifying the quality or level of service (in terms of measures such as delay, convenience, etc.). The quality of operations or level of service for a given facility is a direct function of the flow or usage level on the facility.

Consider the case of a highway—when there are only a few vehicles on the road, drivers are free to choose whatever speed they like, consistent with the conditions of the vehicle and the geometric characteristics of the road. As the flow level or volume increases, vehicles get closer to each other, congestion develops, and the speeds at which drivers can travel are reduced. At the extreme case, gridlock can occur and vehicles' speeds approach zero. Thus the flow levels clearly impact the quality of operations of a transportation facility.

TABLE 4.1
Performance Measures Defining Level of Service

Transportation Mode	*Transportation Facility*	*Performance Measures*
Highway	Freeways	Traffic density (veh/km/lane)
	Multilane highways	Traffic density (veh/km/lane)
	Two-lane highways	Average Travel Speed (km/h)
		Percent time spent following (%)
Transit	Signalized intersections	Control delay (sec/veh)
	Urban streets	Average travel speed (km/h)
	Transit	Service frequency (veh/day)
		Service headway (minutes)
		Passengers/seat
Bicycles	Bike paths	Frequency of conflicting events (events/h)
Pedestrians	Pedestrian facilities	Space (m^2/pedestrian)
Air	Runways	Delay or aircraft waiting time

At low flow levels, operating conditions are favorable. As the flow levels increase, the quality of service deteriorates.

For many transportation facilities, the level of service along a section of the facility is described by assigning the section a letter from A to F, with LOS A referring to the best operating conditions and LOS F the worst. This qualitative description of LOS is typically based on quantitative performance measures such as speed, delay, and traffic density, among others. Table 4.1 lists some of the performance measures that can be used to quantify the LOS for several transportation facilities.

It is important to note, however, that the way the LOS is currently defined is in the form of a step function, as shown in Figure 4.1, where each LOS covers a range of operating conditions. This step-function nature of the LOS definition can lead to some problems. Two similar facilities having the same LOS could

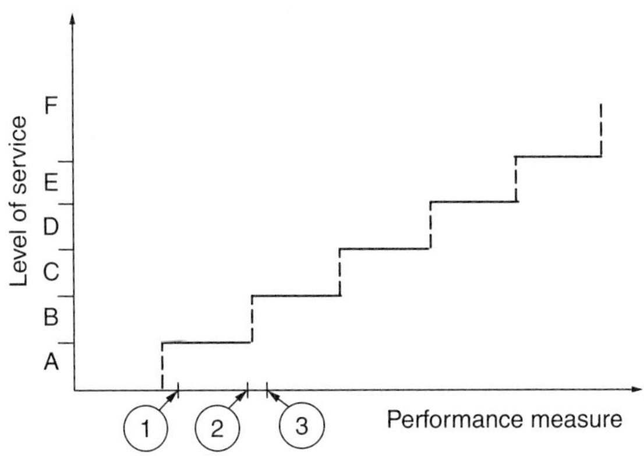

FIGURE 4.1
The step nature of the level of service definition.

vary more than two other facilities having different LOS, depending upon where they are relative to the defined scale. In Figure 4.1, facilities 1 and 2 both have a LOS B, whereas facility C has a LOS C. However, the difference between facilities 1 and 2 is more significant than the difference between facilities 2 and 3. Given this, the qualitative description of the LOS (i.e., the letter designation) should always be used with caution and in conjunction with the actual value of the performance measure upon which the letter designation system is based.

Service Flow Rate

Another concept related to the LOS concept is service flow rate. This represents the *maximum* flow rate that can be accommodated while *maintaining a certain LOS*. We would have a service flow rate for LOS A, another for LOS B, and so on, up to LOS E. The service flow rate corresponding to LOS E is defined to be equal to the capacity of the facility. There is no service flow rate defined for LOS F because it corresponds to unstable flow conditions and to a breakdown in operations.

We will focus on outlining the capacity analysis procedures for highways, transit facilities, bike paths, pedestrian facilities, and airport runways.

Highway Capacity

From a traffic flow analysis standpoint, highway facilities can be divided into two broad categories: (1) *uninterrupted* flow facilities, and (2) *interrupted* flow facilities. *Uninterrupted* flow facilities are those in which there are no external controls interrupting the flow of the traffic stream. A prime example of *uninterrupted* flow facilities is a freeway, where there are no at-grade intersections, no traffic lights, and stop and yield signs. On freeways, the flow conditions are the result of interactions among the vehicles themselves and with the roadway environment.

External controls play a major role in defining the type of traffic operations for *interrupted* flow facilities. Here, traffic flow is regularly interrupted as a result of traffic lights, stop and yield signs, unsignalized intersections, driveways, and other types of interruptions. Almost all urban surface streets fall under the category of *interrupted* flow facilities.

The traffic analysis of *interrupted* flow facilities is more complex and more involved than the analysis of *uninterrupted* flow facilities because the impact of the external controls must be considered. Here, we discuss the basic principles of capacity analysis for both *uninterrupted* as well as *interrupted* flow facilities. Our main focus, however, will be on describing the general procedures for determining the capacity of signalized intersections as a representative example of interrupted flow facilities. Readers interested in learning more about the details of the analysis procedures for both interrupted and uninterrupted flow are referred to standard traffic and highway engineering references and textbooks, including *Traffic and Highway Engineering* by Garber and Hoel. Before describing the capacity analysis procedure for signalized intersections, however, we discuss traffic streams and their basic characteristics, as well as other relevant concepts to capacity analysis in general.

Traffic Stream Characteristics

A highway traffic stream consists of drivers and vehicles interacting with each other and with the roadway environment. To analyze traffic streams, we first need to describe the traffic stream behavior. However, one problem with traffic stream behavior, as opposed to say the behavior of a water stream, is that we are dealing with individual drivers whose exact response or behavior is unpredictable. Nevertheless, there is typically a range of values within which most drivers' behavior would be expected to lie, and this is what is considered in analysis and design.

To describe traffic stream behavior, transportation professionals have devised a set of *macroscopic* parameters and *microscopic* parameters. *Macroscopic* parameters describe the behavior of the traffic stream as a whole, whereas *microscopic* parameters pertain to the behavior of individual vehicles. Among the most important *macroscopic* parameters are (1) flow, (2) speed, and (3) density. Headways and spacing are among the most important *microscopic* parameters. A brief definition of these five important traffic flow parameters follows.

Traffic Flow Parameters

Flow (q)

Flow or volume is defined as the number of vehicles passing a given point on a highway during a given period of time, typically one hour (veh/h). An important flow parameter is the maximum flow value a given facility can reasonably be expected to accommodate. This is often referred to as the capacity (q_m) of a roadway section.

Speed (u)

Speed is the distance traveled by a vehicle during a unit of time. It is usually expressed in km/h, or m/s. The speeds of individual vehicles can be averaged over time (i.e., by averaging the speeds of vehicles as they pass by an observer) or over space (i.e., by averaging the speeds of vehicles occupying a given stretch of a highway at a given point in time). This leads to what we call the time mean speed (u_t) and the space mean speed (u_s), respectively. The space mean speed is typically used for traffic modeling.

Density (k)

Traffic density is defined as the number of vehicles present over a unit length of highway at a given instant in time. Density is typically expressed in veh/km.

Headway (h)

Headway is defined as the difference in time between the moment the front of a vehicle arrives at a point on the highway and the moment the front of the following vehicle arrives at the same point. The time headway is typically expressed in seconds.

The flow of a traffic stream is equal to the inverse of the average time headway:

$$q = l/h_{\text{average}} \tag{4.1}$$

For example, if the average time headway for a given traffic stream is 2 s (i.e., you expect to see a vehicle passing by your point of observation every 2 seconds), the corresponding *hourly* flow value would be equal to 3600/2 = 1800 veh/h (we used 3600 here because there are 3600 seconds in an hour).

Spacing (d)

The space headway (d) is defined as the distance between the front of a vehicle and the front of the following vehicle (in meters). The average spacing of vehicles in a traffic stream is inversely related to the density. If the average spacing between vehicles on a roadway stretch is 100 m, the number of veh/km (i.e., the traffic density) on that stretch is 1000/100 = 10 veh/km. Therefore,

$$k = l/d_{\text{average}} \tag{4.2}$$

Relationships among Macroscopic Traffic Flow Parameters

The three basic macroscopic parameters of a traffic stream (i.e., flow, speed, and density) are related to each other by the following equation:

$$q = uk \tag{4.3}$$

This equation states that the flow or traffic volume is equal to the product of speed and density. So if a 1-km stretch of a roadway contains 15 vehicles (i.e., $k = 15$), and the mean speed of the 15 vehicles is 60 km/h, after 1 hour, 900 vehicles (60×15) would have passed. The value of the flow (q) or traffic volume in this case would be equal to 900 veh/h.

EXAMPLE 4.1

Computing Macroscopic Traffic Parameters

Data obtained from aerial photography showed eight vehicles on a 250-m-long section of road. Traffic data collected at the same time indicated an average time headway of 3 s. Determine (a) the density on the highway, (b) the flow on the road, and (c) the space mean speed.

Solution

From the aerial photographs, the density could be calculated as follows:

Density (k) = 8/250 = 0.032 veh/m = 0.01 × 1000 = 32 veh/km
Flow (q) = 1/average time headway = 1/3 × 3600 = 1200 veh/h

Finally, from Equation 4.3, we have

$$q = uk$$

Therefore,

$$1200 = u \times 32$$

or

u (space mean speed) = 1200/32 = 37.5 km/h

The relationship between the density and flow (Equation 4.3) is generally referred to as the fundamental diagram of traffic flow. The following hypotheses can be made regarding this relationship:

1. At a value for the density equal to 0 (i.e., no vehicles exist on the highway), the flow will also be equal to 0.
2. As the density increases, the flow will also increase.
3. When the density reaches its maximum value (this is typically referred to as the jam density, (k_j), the flow must be zero.
4. It thus follows from (2) and (3) that, as the density increases, the flow initially increases up to a maximum value (q_m). A further increase in density will lead to reduction of the flow, which will eventually become zero when the density is equal to the jam density.

The shape of the relationship between flow and density thus takes the general form shown in Figure 4.2a. The density at which the flow reaches its maximum value (q_m) is commonly referred to as the optimal density (k_o). The value of the optimal density (k_o) can be regarded as dividing the fundamental diagram into two regions. The region to the left of k_o is the stable flow region, where speeds are relatively high and traffic conditions are favorable.

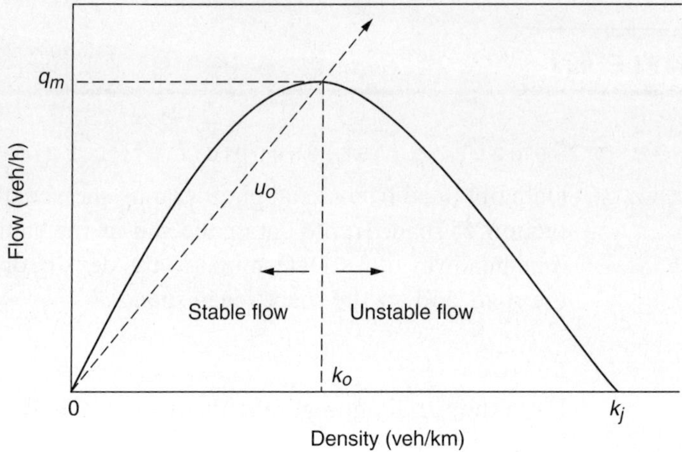

FIGURE 4.2a Relationship between flow and density.

The region to the right of k_o, however, is characterized by unstable conditions, lower volumes, lower speeds, and breakdown in traffic operations. In operating transportation facilities, traffic engineers do their best to make sure facilities operate at densities that are lower than the optimal density (k_o) in order to avoid breakdown in operational conditions.

Since from Equation 4.3 the speed (u) could be expressed as the flow/density (q/k), it follows that speeds at a given point in Figure 4.2a could be represented by radial lines from the origin to that point, as shown in the figure.

Similar hypotheses could be made regarding the relationship between speed and density and the relationship between speed and flow. For the speed–density relationship, when the density approaches zero (i.e., there is little interaction between individual vehicles), drivers are free to select whatever speed they desire, and hence the corresponding speed is what we commonly refer to as the free-flow speed (u_f). As the density increases, the speed decreases until it reaches a value of zero when the road is completely jammed (i.e., when the density is equal to the jam density, k_j). Figure 4.2b depicts this general relationship between the speed and density.

Similarly, for the relationship between speed and flow, one could assume that the speed would be equal to the free-flow speed (u_f) when the density and hence the flow is equal to zero. Continuous increase in flow will then result in a continuous decrease in speed. There will be a point, however, when the further addition of vehicles will result in a reduction in the number of vehicles passing a given point on the highway (i.e., reduction in the flow). The addition of vehicles beyond this point would result in congestion, and both the flow and speed would decrease until they both become zero. The relationship between speed and flow could thus be represented as shown in Figure 4.2c.

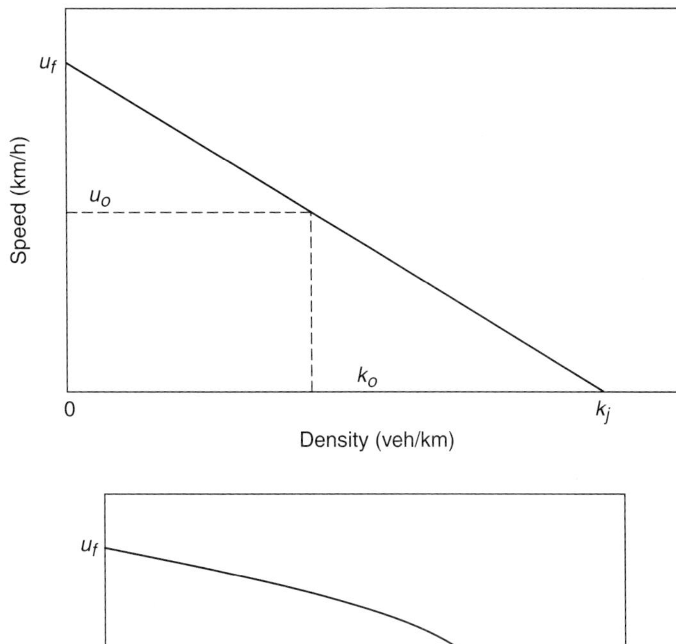

FIGURE 4.2b
Relationship between speed and density.

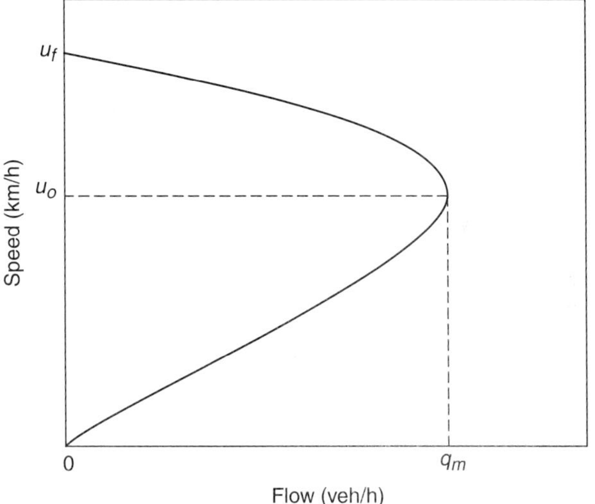

FIGURE 4.2c
Relationship between speed and flow.

From Figures 4.2a through 4.2c, it should be obvious that to avoid congestion it is desirable to operate roadways at densities not exceeding the density at capacity.

Traffic Flow Models

Greenshields was one of the first researchers who attempted to develop models for describing traffic flow. He postulated that a linear relationship exists between speed and density having the following form:

$$u = u_f - \frac{u_f}{k_j} k \tag{4.4}$$

where all the terms are as defined previously. This equation indicates that as the density (k) approaches zero, the speed (u) approaches free-flow speed, u_f.

Also, as the speed (u) approaches zero, the density approaches jam density or k_j.

From Equation 4.3, we know that $q = uk$. Therefore, using the Greenshields equation (Equation 4.4), the flow, q, can be expressed as

$$q = u_f k - \frac{u_f}{k_j} k^2 \tag{4.5}$$

Also, from Equation 4.3, we know that $k = q/u$. Therefore, substituting q/u for k, the relationship between the speed (u) and the flow (q) can be expressed as

$$u^2 = u_f u - \frac{u_f}{k_j} q \tag{4.6}$$

Equations 4.4 through 4.6 describe the three diagrams depicted in Figures 4.2a through 4.2c, respectively. The three equations or diagrams are rather redundant because if only one relation is known, the other two can be derived easily using the basic relation that $q = uk$. Nevertheless, each of the three diagrams has its own use. For theoretical work, the relationship between speed and density is the one typically utilized, since there is only one value for speed for each value of the density. This is not the case with the other two diagrams. The relationship between flow and density is used in freeway and arterial control systems to control the density in an effort to optimize productivity (flow). Finally, the relationship between speed and flow could be used for design to define the tradeoff between the level of service on a road facility, as discussed later in this chapter.

Hourly Volume, Subhourly Volume, and Rates of Flow

For traffic analysis and design purposes, one cannot just use the daily volume to be expected to use the facility by itself. This is because traffic volumes vary considerably over the 24 hours of the day. For example, we typically have a peak traffic flow period or "rush hour" in the morning, when most people are going to work, and another peak in the evening, when people are returning home. If one were to design for the average daily volume, the facility would fail to accommodate the traffic demand during the morning and evening peak periods. Transportation engineers, therefore, typically consider the peak demand for analysis and design purposes.

Moreover, for capacity analysis in particular, the variation of the traffic within a given hour is also of interest. To understand this, let us assume that traffic counts were recorded for each 15-minute period between the hours of 8:00 A.M. and 9:00 A.M. The counts recorded were as shown in Table 4.2. In this case, the actual hourly volume is equal to 120 + 90 + 110 + 80 = 400 veh/h.

TABLE 4.2
Vehicle Counts

Time Period	*Counts (veh/15-min period)*
8:00–8:15 A.M.	120
8:15–8:30 A.M.	90
8:30–8:45 A.M.	110
8:45–9:00 A.M.	80

However, if we only design for the 400 veh/h volume, this would mean that we would experience congestion problems during the first and third 15-minute periods (i.e., between 8:00 and 8:15 A.M., and again between 8:15 and 8:30 A.M.). This is because designing for a volume of 400 vehicles in an hour assumes that the facility would be able to handle not more than 100 vehicles every 15 minutes (400/4 = 100), and, as can easily be seen from Table 4.2, this volume is exceeded during the first and third time periods.

To overcome this problem, it has been the convention for most traffic operational analysis to consider the peak 15-minute period count and to convert that count to the *equivalent hourly rate of flow.* So for the previous example, the peak 15-minute count is 120 veh/15 min, which corresponds to $120 \times 4 = 480$ veh/h. This is the *equivalent hourly rate of flow,* based on the peak 15-minute period, and this volume would be the one used for design and operational analysis purposes.

To facilitate the application of this concept, the transportation community has defined what we call the peak hour factor (PHF). This factor is used to account for the variation of traffic flow within the peak hour itself and is defined as follows:

$$\text{Peak hour factor} = \frac{\text{actual hourly volume}}{\text{max rate of flow}} \tag{4.7}$$

So for the preceding example, the PHF would be computed as 400/480, which is equal to 0.83. The PHF is always less than 1. For a given facility, if the hourly volume (V) and the PHF are known, the maximum rate of flow (v) can be computed easily as follows:

$$v = \frac{V}{\text{PHF}} \tag{4.8}$$

where

v = maximum rate of flow within the hour (veh/h)
V = hourly volume (veh/h)
PHF = peak hour factor

The maximum rate of flow, v, would then be used for design and analysis.

Example 4.2

Computing the Peak Hour Factor (PHF)

Table 4.3 shows the 15-minute counts that have been recorded for a given highway.

(a) Determine the PHF;

(b) The hourly volume for a similar facility was found to be equal to 6000 veh/h. Determine design volume for the facility.

Solution

From Table 4.3, the actual hourly volume (V) is given as

$$V = 1200 + 1400 + 1100 + 1300 = 5000 \text{ veh/h}$$

The maximum rate of flow, v, is computed as follows:

$$v = 1400 \times 4 = 5600 \text{ veh/h}$$

Therefore,

$$\text{PHF} = \frac{5000}{5600} = 0.893$$

For the other facility,

$$V = 6000 \text{ veh/h}$$

Therefore, the design volume or the maximum rate of flow (v) for which the facility needs to be designed can be calculated as follows:

$$v = \frac{V}{\text{PHF}} = \frac{6000}{0.893} = 6720 \text{ veh/h (answer)}$$

The V/C Ratio

Another concept critical to highway capacity analysis is the volume to capacity (v/c) ratio. This is determined by dividing the current or projected demand by the capacity of the facility. The v/c ratio indicates how much of a certain facility's

TABLE 4.3
Vehicle Counts for Example 4.2

Time Period	*Counts (veh/15-min period)*
4:30–4:45 P.M.	1200
4:45–5:00 P.M.	1400
5:00–5:15 P.M.	1100
5:15–5:30 P.M.	1300

capacity is being utilized or used by the demand. The *v/c* concept is closely tied to the concept of service flow rates, previously defined. Dividing the service flow rate for a given LOS by the capacity gives the maximum value for the *v/c* ratio for that particular LOS. If the service flow rate corresponding to a LOS C for a given facility was equal to 1300 veh/h and the capacity of that facility was 2000 veh/h, the maximum *v/c* ratio would be 1300/2000 = 0.65. The maximum *v/c* ratio for a LOS E is always equal to 1 since the service flow rate for LOS E is equal to the capacity of the facility.

Capacity Analysis for Signalized Intersections

In this section, we discuss procedures for the analysis and design of signalized intersections, as one example for *interrupted* flow facilities in which external control plays a major role in defining the traffic flow characteristics. At-grade intersections are locations where different modes of transportation interact, such as cars, trucks, buses, bicycles, and pedestrians. At-grade intersections also represent one of the most complex components of a transportation system. This is because conflicting traffic streams compete for the right-of-way at an intersection.

When traffic volumes are low, traffic at an intersection could be regulated using basic *rules of the road,* or using stop and yield signs. As traffic volumes increase, however, it becomes extremely difficult for drivers to select adequate gaps in conflicting traffic streams for executing their desired maneuvers. When this happens, intersection signalization (i.e., the use of traffic control signals) becomes a must.

Traffic signals play a dramatic role in determining the overall performance level of an arterial system. Poorly designed traffic signals can result in unnecessary and excessive delays. If appropriately designed, a traffic control signal can provide for the orderly movement of traffic and can increase the traffic-handling capacity of an intersection. Traffic signals can generally be divided into two groups: *pretimed* signals and *actuated* signals. Pretimed signals are typically insensitive to current volumes. The cycle length of a pretimed signal is generally fixed, regardless of the current volumes. The operation of actuated controllers varies according to the observed volume. Actuated controllers thus need to be connected to traffic detectors for sensing traffic demand.

While the operation of pretimed signals is not sensitive to current volumes per se, one could still have a range of timing plans for different time periods within a day, using a pretimed controller. Typically, one would have a timing plan for handling the morning peak period or rush hour, another for the evening peak, and a third for the off-peak period. The pretimed controller would be instructed to use the morning signal plan between the hours of 6:30 and 8:30 A.M., the evening plan between 4:30 and 6:30 P.M., and the off-peak plan during the rest of the day. Within each of these periods, the parameters of a pretimed signal plan remain unchanged. Given this, the use of pretimed controllers is most appropriate when traffic conditions do not vary significantly within the different peak periods. In this section, we will focus mainly on pretimed signals, since an understanding of their mode of operation is crucial to understanding more advanced types of control.

Important Definitions

Before discussing the detail of the analysis methodology for signalized intersections, a number of terms need to be defined:

Cycle and cycle length. A traffic signal cycle is a rotation through all the signal indications at a given intersection. Every legal movement would generally receive the "green" indication only once during a given cycle. The time it takes for the signal to go through one cycle of indications is the *cycle length.*

Interval. A time period during which all the signal indications or lights remain unchanged. A cycle generally includes several intervals such as green, change or yellow, clearance or all-red, and the red interval.

Phase. A set of indications (i.e., green and yellow intervals) during which a given set of movements is assigned the right of way. The number of phases for signalized intersections typically ranges between two and four phases. For a two-phase signal, one would typically have one phase dedicated to traffic movements from the east and west approaches and another for the north and south movements.

Offset. A term used in conjunction with coordinated systems of signals. It refers to the time difference between the initiation of green on two adjacent signals. Typically, the offset is measured in terms of the green initiation time of the downstream (t_d) signal relative to the upstream signal (t_u)—that is, the offset is equal to $t_d - t_u$.

Signal Timing Principles

In order to appreciate the basis behind the methodology for the analysis and design of signalized intersections, we need first to discuss the following: (1) the mechanism by which vehicles discharge from a queue waiting at a traffic signal; (2) the time lost in the process; and (3) the concept of the capacity of a given intersection approach. Each of these issues is briefly discussed below.

Discharge Headway and Saturation Flow Rate

Observations of the way vehicles discharge from a queue (i.e., a line of vehicles) have revealed that if the recorded discharge headways (i.e., the interval between the time one vehicle crosses the stop bar and the time the following vehicle crosses) are plotted against the vehicle position in the queue, a graph similar to Figure 4.3 is obtained. This figure explains the mechanism by which vehicles depart from a signalized intersection when it turns green. The first few

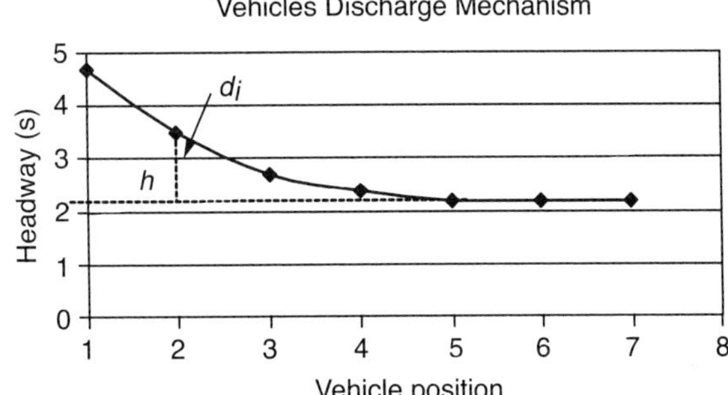

FIGURE 4.3
Discharge headways at signalized intersections.

headways are relatively long, then after the fourth or fifth vehicle, the discharge headways typically level out toward a constant value. This value is known as the *saturation headway* and represents the average headway that can be achieved by a saturated, stable moving queue of vehicles, or the maximum rate at which vehicles can depart from a stop bar, provided that there are vehicles waiting in the queue. The *saturation headway* is often denoted by h and is typically in the range of 2–3 s/veh.

If we assume that the signal remains green all the time and that we have enough vehicles waiting in the queue and that each one consumes h seconds to enter the intersection (i.e., the saturation discharge headway), the total number of vehicles entering in an hour (i.e., the saturation flow rate) can be computed easily as

$$s = \frac{3600}{h} \tag{4.9}$$

where

s = saturation flow rate in units of veh/hour green/lane (veh/hg/ln)
h = saturation headway (s/veh)

We will discuss later how to account for the fact that at a signalized intersection, each approach gets the green only during a fraction of the total cycle length.

Total Lost Time and Effective Green Time

In Equation 4.9, we have assumed that vehicles enter the intersection every h seconds. In reality, however, the average headway is greater than h. As Figure 4.3 shows, for the first four or five vehicles, the headway is actually greater than h, since the drivers of those first four or five vehicles typically require more reaction time in order to accelerate. Let us denote the difference between the actual headway for those first vehicles and the saturation headway by d_i (see Figure 4.3). The sum of these d_i's would give us what we call the start-up lost time l_1. This represents the time lost at the beginning of each phase (i.e., when the signal turns green) as a result of the additional reaction time required by the first four or five vehicles in a queue. We could compute the total time required for a queue consisting of n vehicles to discharge from a signalized intersection as

$$T_n = l_1 + nh \tag{4.10}$$

where

T_n = green time required for the discharge of n vehicles (seconds)
l_1 = start-up lost time (s/phase)
h = saturation headway (s/veh)

In addition to the start-up lost time, which occurs each time a queue starts moving, there is some time lost toward the end of the phase (when the signal is about to turn red). This time is called the clearance lost time and is denoted by l_2. To understand why we need to account for l_2, let us review what typically happens when a phase is about to end. Typically, a signal for a given approach goes through the following sequence of intervals (1) green; (2) yellow; (3) all-red (this is typically a 1-second interval in which the signal indications for all approaches at an intersection are red to ensure the clearance of the intersection before the initiation of the green for a second approach); and (4) red. Vehicles from a given intersection approach would therefore normally move during all of the green and part of the yellow or the clearance interval. The part of the yellow that is not utilized by vehicles plus the all-red interval where everybody is stopped represents time lost, and this is the time that l_2 is designed to capture. Therefore, the total lost time/phase (t_L) is thus equal to the start-up lost time plus the clearance lost time, as follows:

$$t_L = l_1 + l_2 \tag{4.11}$$

To facilitate accounting for the lost time/phase in signal analysis and modeling, the *effective green time* (g_i) was defined. This represents the time during which vehicles are effectively moving at the rate of one every h seconds. This time is given as

$$g_i = G_i + Y_i - t_{Li} \tag{4.12}$$

where

g_i = effective green time for phase i
G_i = actual green time for phase i
Y_i = duration of the yellow interval
t_{Li} = total lost time during phase i

Capacity of a Given Lane

The saturation flow rate (s) as defined in Equation 4.9 gives us the capacity of a single lane on a given intersection approach, assuming that that approach gets a green indication all the time. For a signalized intersection, each approach typically gets green only during a certain fraction of the total cycle length. Therefore, if a given approach has an *effective green time* period equal to g_i, and if the total cycle length is C sec, the capacity of that approach is equal to

$$c_i = s_i \frac{g_i}{C} \tag{4.13}$$

where

c_i = capacity of lane i (veh/h)
s_i = saturation flow rate for lane i (veh/hg)
g_i = effective green time for lane i (s)
C = cycle length (s)

The lane capacity computed could then be multiplied by the number of lanes to yield the capacity for the whole lane group (i.e., group of lanes that move together during a given phase and have similar operating characteristics). While Equation 4.13 is useful for computing the capacity of a given approach or a lane group, it does not address how to compute the capacity for a signalized intersection as a whole. The issue of signalized intersection capacity will be addressed in the next section.

Example 4.3

Computing the Capacity of a Signalized Approach

The eastbound approach of a signalized intersection with a cycle length of 80 s gets 37 s of green. Studies show that the saturation headway for that approach is equal to 2.2 s, the start-up lost time is equal to 2 s, and the clearance lost time is equal to 1 s. If the duration of the yellow or clearance interval is 3.5 s, determine the capacity for that approach, assuming that it consists of two through-moving traffic lanes.

Solution

The saturation flow rate for that approach is first computed from Equation 4.9 as

$$s = \frac{3600}{2.2} = 1636 \text{ veh/h/lane}$$

From Equation 4.11, the total lost time/phase for that approach is computed as

$$t_L = l_1 + l_2 = 2.0 + 1.0 = 3 \text{ s}$$

Next, the effective green time for the approach is computed from Equation 4.12 as

$$g_i = G_i + Y_i - t_{Li} = 37.0 + 3.5 - 3.0 = 37.5 \text{ s}$$

Finally, the capacity for the approach can be computed from Equation 4.13 as follows:

$$c_i = s_i \frac{g_i}{C}$$

$$= 1636 \times \frac{37.5}{80} = 767 \text{ veh/h/lane}$$

Therefore, the capacity of the approach or lane group is given by multiplying the preceding figure by 2, since the approach had two lanes, as follows:

$$c = 767 \times 2 = 1534 \text{ veh/h}$$

The Time Budget and the Critical Lane Concepts

Developing signal timing plans is based upon two concepts, namely the time budget and the critical lane concepts. The time budget concept is concerned with allocating the available time among the competing vehicular and pedestrian streams at an intersection. The critical lane concept states that during any given phase, while several traffic approaches are allowed to move, one particular movement will require the largest amount of time. That particular movement is referred to as the critical lane for that given phase. Satisfying the needs of the critical lane movement would automatically satisfy the needs of all other accompanying movements.

Consider the signalized intersection shown in Figure 4.4 and assume that we have a phase dedicated for the eastbound (EB) and westbound (WB) traffic movements (i.e., these six movements take place at the same time) and a second phase dedicated for the northbound (NB) and southbound (SB) traffic movements. Also assume that we have one lane available for each of these 12 movements (i.e., one lane for the left-turning movement, another for the through, and a third for the right from each approach) and that these three lanes are similar in terms of their ability to accommodate traffic volumes. In this case, the phase serving the EB and WB movements needs to be designed so as to satisfy the needs of the heaviest or most intense movement (i.e., it should be long enough to accommodate that volume), which in this case is the 550 veh/h volume. By doing this, the needs of the other lesser volumes would be automatically satisfied. The NB and SB phase needs to be designed for the 450 veh/h lane, since this is the most intense volume moving in this second phase. The critical lane for phase 1 is the one carrying the 550 veh/h volume, and the critical lane for phase 2 is the lane with the 450 veh/h volume. These two movements are shown in bold in Figure 4.4.

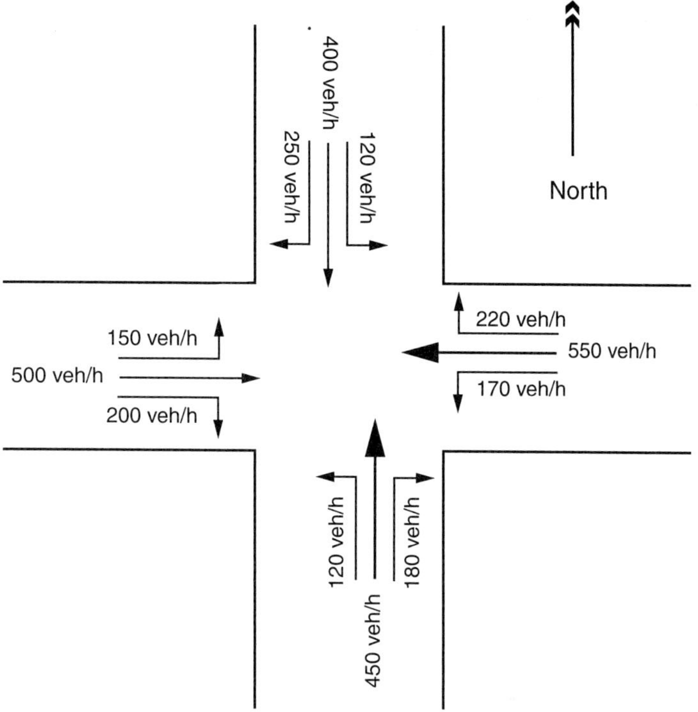

FIGURE 4.4
The critical lane concept.

Therefore, for the intersection shown in Figure 4.4, the available time should be allocated to the vehicles in the critical lane for phase 1 and those in the critical lane for phase 2 while accounting for the time lost in each phase.

Signalized Intersection Capacity

With the concepts of time budget and critical volumes defined, we turn our attention to presenting a simplified version of the concept of signalized intersection capacity. In one sense, the maximum sum of critical lane volumes that a signalized intersection can accommodate can be regarded as a measure for the intersection capacity. This is a simplified measure compared to the more elaborate methods described in the HCM, but nevertheless it is quite useful.

To determine the maximum sum of the critical lane volumes that a signalized intersection can accommodate, first we must determine the time available to allocate for these movements in an hour since there is some time lost in each signal cycle that is not utilized by vehicles. After the available time for allocation is determined, dividing this time by the discharge headway immediately yields the maximum sum of critical lane volumes. The calculation proceeds as follows: First, the total lost time/cycle is computed as

$$L = N \times t_L \tag{4.14}$$

where

L = lost time/cycle (s/cycle)
N = number of phases in the cycle
t_L = total lost time/phase as defined previously (s/phase)

The number of cycles in one hour is given as $3600/C$ cycles, where C is the cycle length in seconds. Therefore, the total lost time in one hour, L_H, is given by

$$L_H = N \times t_L \times \left(\frac{3600}{C}\right) \tag{4.15}$$

Therefore, the time available for allocation, T_G, is equal to

$$T_G = 3600 - N \times t_L \times \left(\frac{3600}{C}\right) \tag{4.16}$$

Dividing this time by the saturation headway, h, would thus give us the maximum sum of critical volumes that the intersection can accommodate, V_c, which can be expressed as

$$V_c = \frac{1}{h}\left[3600 - N \times t_L \times \left(\frac{3600}{C}\right)\right] \tag{4.17}$$

Applications

Equation 4.17 can be used in a number of different ways. First, it can be used to determine the maximum sum of critical lane volumes that a given intersection with a given cycle length can accommodate. Second, it can be used to determine the number of lanes/intersection approaches required if a certain cycle length is desired. Finally, it can be used to determine the minimum cycle length required to accommodate a given set of volumes at a particular intersection. The following examples illustrate these applications of the methodology.

EXAMPLE 4.4

Determining the Maximum Sum of Critical Volumes at an Intersection

A signalized intersection having three phases has a cycle length equal to 90 s. Determine the maximum sum of critical volumes that the intersection can accommodate, given that the lost time/phase is 3.50 s and that the saturation headway is 2 s.

Solution

From Equation 4.17, the maximum sum of critical volumes can be computed as

$$V_c = \frac{1}{h}\left[3600 - N \times t_L \times \left(\frac{3600}{C}\right)\right]$$

$$V_c = \frac{1}{2.0}\left[3600 - 3 \times 3.5 \times \left(\frac{3600}{90}\right)\right] = 1590 \text{ veh/h}$$

EXAMPLE 4.5

Determining the Number of Lanes at an Intersection

Consider the two-phase intersection shown in Figure 4.5. The intersection has a 60-second cycle length, and the time lost/phase is equal to 4 s. The critical volumes are shown in the figure. Determine the number of lanes required for each critical movement. Assume a saturation headway of 2.20 s.

Solution

The first step is to determine the maximum sum of critical volumes that the intersection with its current cycle length can accommodate. This value is then compared to the observed critical volumes shown in Figure 4.5 to determine the number of lanes. The solution proceeds as follows:

$$V_c = \frac{1}{h}\left[3600 - N \times t_L \times \left(\frac{3600}{C}\right)\right]$$

$$V_c = \frac{1}{2.20}\left[3600 - 2 \times 4 \times \left(\frac{3600}{60}\right)\right] = 1418 \text{ veh/h}$$

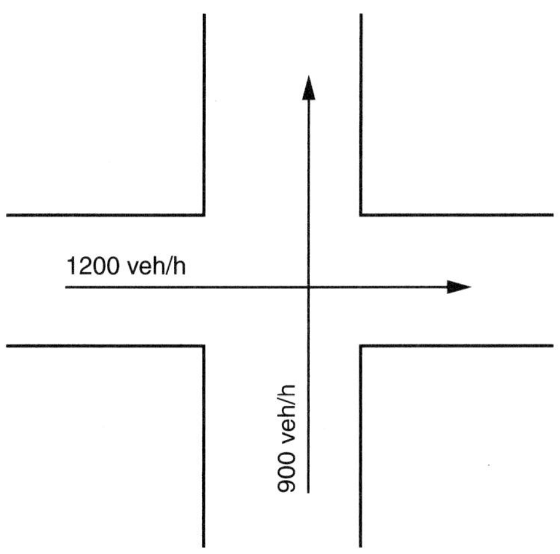

FIGURE 4.5
Intersection for Example 4.5.

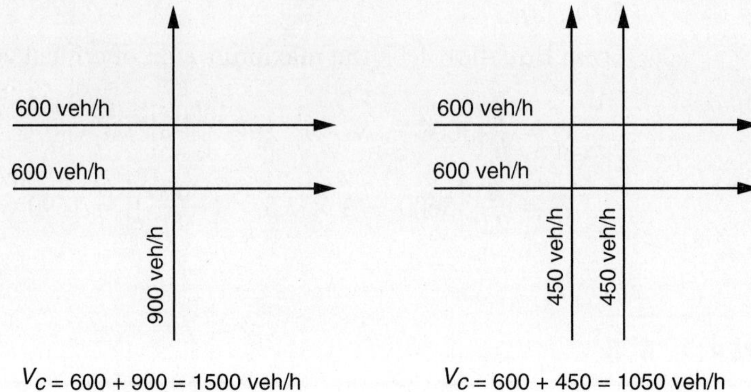

FIGURE 4.6 Solution for Example 4.5.

According to Figure 4.5, the sum of critical volumes is equal to 1200 + 900 = 2100 veh/h. This means that these volumes have to be distributed over a number of lanes so that the sum of critical volumes observed is less than the maximum sum of critical volume that the intersection can handle as determined from Equation 4.17.

We start with two lanes/direction for the east–west street, and one lane/direction for the north–south street, as shown in Figure 4.6. This gives us a sum of critical volumes of 1500 veh/h, which is still more than the intersection can handle (i.e., 1418 veh/h). We then proceed to divide the north–south critical volume as well over two lanes, which gives 1050 veh/h as the sum of critical volumes. Since 1050 veh/h is less than 1418 veh/h, the design is acceptable.

EXAMPLE 4.6

Determining the Minimum Cycle Length

Determine the minimum cycle length for the two-phase intersection shown in Figure 4.7. The intersection's critical volumes are shown in the figure. Assume a saturation headway of 2.10 s and a lost time/phase of 3.50 s.

Solution

For this problem, Equation 4.17 is rearranged in order to solve the cycle length, C. This gives the following equation:

$$C_{\min} = \frac{Nt_L}{1 - \left(\frac{Vc}{3600/h}\right)}$$

where all notations are as used before.

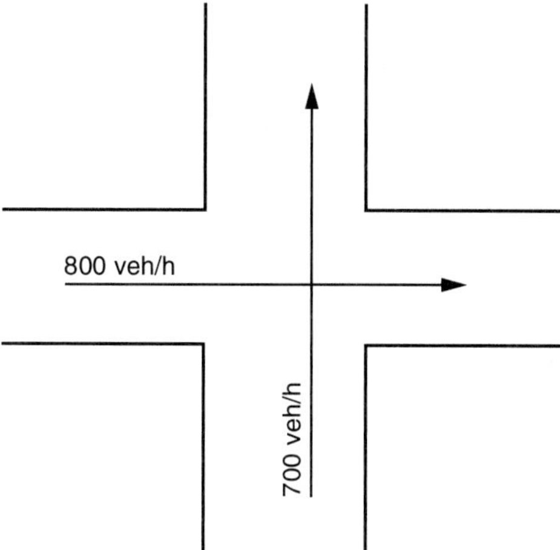

FIGURE 4.7
Problem for Example 4.6.

The sum of critical volumes, V_c, for the intersection shown in Figure 4.6 is 800 + 700 = 1500 veh/h. Therefore, the minimum cycle length is given as

$$C_{\min} = \frac{2 \times 3.5}{1 - \left(\frac{1500}{3600/2.1}\right)} = 56 \text{ s}$$

Desirable Cycle Length

In Example 4.6, we solved for the minimum cycle length needed to accommodate a given sum of critical volumes using Equation 4.17. This equation, however, does not account for the within-the-hour variations of traffic volume (which is accounted for using the peak hour factor). In addition, most signals are timed so that between 80 and 95% of the capacity available is utilized (Equation 4.17 assumes that 100% of the capacity is utilized). Providing some excess capacity is important for transportation systems because of the uncertainty associated with predicting traffic demand.

To account for these two factors (namely, within-the-hour variations and the percent utilization of capacity), Equation 4.17 is modified as follows:

$$C_{\text{des}} = \frac{Nt_L}{1 - \left(\frac{Vc}{(3600/h) \times \text{PHF} \times (v/c)}\right)} \tag{4.18}$$

where

C_{des} = desirable cycle length as opposed to the minimum cycle length of Equation 4.17
PHF = peak hour factor
v/c = desired volume to capacity ratio

Webster Delay Model

When designing a signal plan, a traffic engineer typically attempts to meet the needs of the critical lane movement for each phase while maximizing the effectiveness of the intersection. For isolated intersections, *delay* is commonly the measure of effectiveness used to characterize how well the intersection is performing. Coordinated systems typically try to minimize a "penalty" function, which represents a weighted combination of the number of stops and the total delay.

One of the first transportation researchers to develop a model for delay at signalized intersections was Webster in 1958. Webster's model is based upon developing a cumulative plot (as was discussed in Chapter 2) for the way vehicles arrive and depart at the intersection as shown in Figure 4.8. The figure plots the cumulative number of vehicles arriving and departing at the intersection against time. The time axis is divided into periods of effective green (where vehicles are allowed to move) and effective red (when all vehicles are stopped). It is assumed that vehicles arrive at a uniform rate of flow, namely v vehicles/unit time. This gives a straight line with a slope v for the vehicles' arrival curve.

For departure, during the effective red period, no vehicles can depart, and therefore the departure curve during the red period takes the form of a horizontal line (0 vehicles departing). As soon as the signal turns green, the queue of vehicles that was formed during the red period starts discharging at a rate equal to the discharge headway or saturation flow rate, s veh/h. Discharge at the saturation flow rate continues until the queue is dissipated

FIGURE 4.8
Webster uniform delay model.

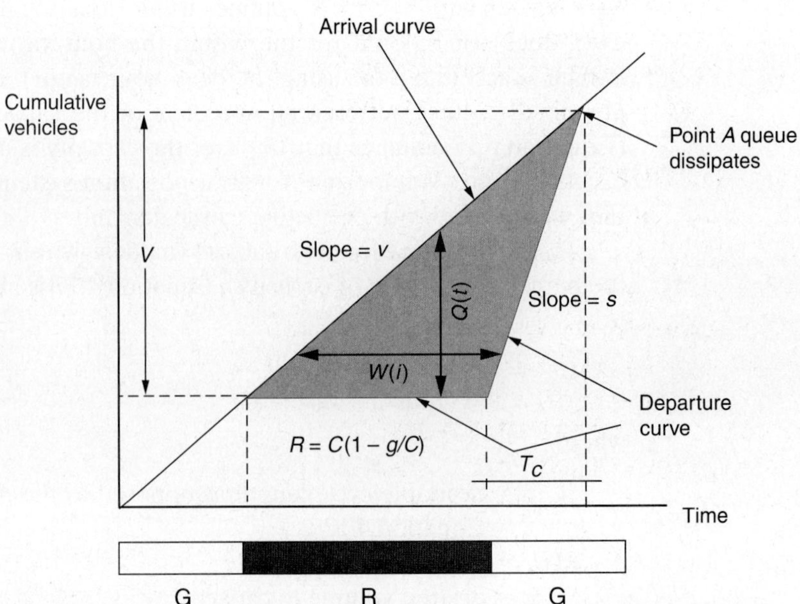

(i.e., the point where the arrival curve meets with the departure curve, point A in Figure 4.8). After point A, vehicles start discharging at a rate equal to their arrival rate.

It should be obvious from Figure 4.8 that the vertical vehicle-scale difference between the arrival and departure curves at time t, $Q(t)$, gives us the number of vehicles waiting in the queue at the intersection, and that the horizontal timescale difference between the two curves, $W(i)$, gives us the time that a vehicle i spends waiting in the queue. Given this, the total or aggregate delay for all vehicles processed by the signal is given by the area of the shaded triangle in Figure 4.8.

Therefore, to determine the aggregate or average delay, the first step is to determine the area of the shaded triangle of Figure 4.8 as follows:

The area of the triangle is equal to one-half the base multiplied by the height, or

$$\text{Aggregate delay} = \frac{1}{2}RV$$

where R is the length of the effective red period and V is the total number of vehicles delayed at the intersection. As a matter of convention, traffic models are typically expressed in terms of the effective green and not the red. In terms of the green, the red period can be expressed as

$$R = C[1 - g/C]$$

where C is the cycle length and g is the length of the effective green period.

In order to find the number of vehicles in the queue, V, we first need to determine the time that elapses between the moment the signal turns green and the time the queue dissipates (i.e., T_c in Figure 4.8). Once T_c is determined, the number of vehicles, V, can be easily computed by multiplying the saturation flow rate, s, by T_c. Determining T_c proceeds as follows:

From Figure 4.8, we have

$$V = v(R + T_c) = sT_c$$

Therefore,

$$R + T_c = (s/v)T_c$$

$$R = T_c(s/v - 1)$$

$$T_c = \frac{R}{(\frac{s}{v} - 1)}$$

Therefore,

$$V = sT_c = \frac{sR}{(\frac{s}{v} - 1)} = R\left[\frac{vs}{s - v}\right] = C\left[1 - \left(\frac{g}{C}\right)\right]\left[\frac{vs}{s - v}\right]$$

The aggregate delay is thus given as

$$\text{Aggregate delay} = \frac{1}{2}RV = \frac{1}{2}C^2\left[1 - \left(\frac{g}{C}\right)\right]^2\left[\frac{vs}{s - v}\right] \tag{4.19}$$

The average delay/vehicle can then be computed from Equation 4.19 by dividing the aggregate delay by the number of vehicles processed/cycle (i.e., $v \cdot C$). This gives the following:

$$\text{Average delay} = \frac{1}{2}C\frac{\left[1 - \frac{g}{C}\right]^2}{\left[1 - \frac{v}{s}\right]} \tag{4.20}$$

Equation 4.20 can also be expressed in terms of the capacity of an intersection approach, c, instead of the saturation flow rate, s, by noting that $c = s \cdot (g/C)$ (Equation 4.20).

This gives us the following expression:

$$\text{Average delay} = \frac{1}{2}C\frac{\left[1 - \frac{g}{C}\right]^2}{\left[1 - \left(\frac{g}{C}\right)\left(\frac{v}{c}\right)\right]} \tag{4.21}$$

Note that in the above equation, the uppercase C refers to the cycle length in seconds, whereas the lowercase c refers to the approach capacity in veh/h.

As listed in Table 4.1, calculating the average delay of the different traffic movements at a signalized intersection forms the basis of determining the LOS for the different movements, as well as for the whole intersection. Table 4.4 shows how the HCM manual defines the different LOS for signalized intersections. LOS C, for example, corresponds to delay values in the range of 20 to 35 seconds.

TABLE 4.4
LOS for Signalized Intersections

LOS	*Control Delay (s/veh)*
A	0–10
B	10–20
C	20–35
D	35–55
E	55–80
F	>80

EXAMPLE 4.7

Calculating the Average Delay for an Intersection Approach

Determine the average delay in s/veh for an intersection approach that gets 40 s of green out of a cycle length of 90 s. The approach carries an hourly volume of 600 veh/h and has a saturation flow rate equal to 1700 veh/h. Also determine the corresponding LOS.

Solution

The average delay for the approach can be directly determined from either Equation 4.20 or Equation 4.21. Using Equation 4.20, the average delay is given by

$$\text{Average delay} = \frac{1}{2}C\frac{\left[1-\frac{g}{C}\right]^2}{\left[1-\frac{v}{s}\right]}$$

$$= \frac{1}{2}90\frac{\left[1-\frac{40}{90}\right]^2}{\left[1-\frac{600}{1700}\right]} = 21.50 \text{ s/veh (answer)}$$

From Table 4.4, this corresponds to LOS C.

Webster Formula for the Optimal Cycle Length

Based on minimizing the approach delay at a signalized intersection, Webster derived a formula for determining the optimal cycle length, C_o, which minimizes the delay at the intersection. The equation for determining the optimal cycle length, C_o, is as follows:

$$C_o = \frac{1.5L + 5}{1 - \sum_{i=1}^{N} Y_i} \tag{4.22}$$

where

C_o = optimal cycle length in seconds
N = number of phases
L = total lost time/cycle, which is equal to the number of phases (N) multiplied by the lost time/phase (t_L)

Y_i = maximum value of the ratios of the approach flows to the saturation flow rates for all lane groups using phase *i*. Therefore, Y_i gives the ratios of the critical lane volume to the saturation flow rate and is computed from

$$Y_i = \max\{v_{ij}/s_j\} \tag{4.23}$$

where

v_{ij} = traffic volume on lane group *j* having the right-of-way during phase *i*
s_j = saturation flow rate on lane group *j*

With the cycle length determined, the available green time (i.e., $C - L$) is allocated among the different phases in proportion to their respective Y_i ratios. Therefore, the effective green time for phase *i* is computed as follows:

$$g_i = \frac{Y_i}{\sum_{i=1}^{N} Y_i} \cdot (C - L) \tag{4.24}$$

where all notations are as defined before.

EXAMPLE 4.8

Signalized Intersection Design

The intersection shown in Figure 4.9 has three phases.

Phase A serves east–west left turns only; phase B serves east–west through and right traffic; and phase C serves north–south left, through, and right movements.

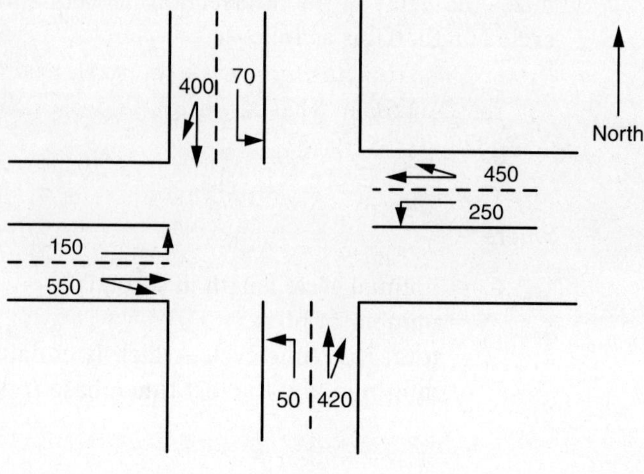

FIGURE 4.9
Equivalent hourly flows at intersection.

Assume lost times of 3.5 s/phase, a yellow interval of 3 s, and the following saturation flow rates:

$$s(TH + RT) = 1700 \text{ veh/h}$$
$$s(LT) = 1600 \text{ veh/h}$$

(a) Using the Webster model, determine the optimal cycle length for the intersection.

(b) Determine the effective green time for phase A.

Solution

(a) The first step is to compute the ratios of the approach flows to saturation flow rates for each of the three phases by dividing the approach flows by the saturation flow rates for all lane groups using each phase. The calculations are organized as follows:

	Phase A		Phase B			Phase C		
v_{ij}	150	250	550	450	50	420	70	400
s_j	1600	1600	1700	1700	1600	1700	1600	1700
v_{ij}/s_j	0.09	***0.16***	***0.32***	0.26	0.03	***0.25***	0.04	0.24

From these calculations, the Y_i for each phase is determined as follows:

$$Y_1 = 0.16$$
$$Y_2 = 0.32$$
$$Y_3 = 0.25$$
$$\sum Y = 0.73$$

The total lost time/cycle, L, in this example is computed by multiplying the number of phases (i.e., three phases) by the lost time/phase (i.e., 3.50 s/phase) as follows:

$$L = 3 \times 3.50 = 10.50 \text{ s/cycle}$$

The optimal cycle length, C_o, can then be determined from Equation 4.29 as follows:

$$C_o = \frac{1.5L + 5}{1 - \sum_{i=1}^{N} Y_i} = \frac{1.5 \times 10.5 + 5}{1 - 0.73}$$

$= 76.9$ s, which is rounded up to 80 s

(b) The effective green for phase A is then computed using Equation 4.30 as follows:

$$g_A = \frac{Y_A}{\sum_{i=1}^{3} Y_i} \cdot (C - L) = \frac{0.16}{0.73} \times (80 - 10.50) = 15.2 \text{ s}$$

A Brief Introduction to the HCM Method

In the preceding presentation, we made a number of simplifications since our main goal was to present the broader framework for the capacity analysis at signalized intersections. Among the main simplifications we made was that we neglected to talk about left-turning movements at intersections and the major role they play in impacting intersection capacity. The HCM contains a detailed methodology that takes that into account, as well as several other factors that we have ignored in the preceding discussion. The interested reader is directed to standard traffic engineering references, including *Highway and Traffic Engineering* by Garber and Hoel, for more details.

TRANSIT CAPACITY

From a capacity standpoint, there are some basic differences between transit and the automobile. For example, while roadway capacity is generally available 24 hours a day, seven days a week, transit capacity and availability depends upon the operating policy of the transit agency (e.g., number of vehicles, hours of operation, etc.). In addition, transit capacity deals with the movement of both *people* and *vehicles*. This means that, for transit, we have to deal with both *vehicle capacity* as well as with *person capacity*.

Vehicle capacity refers to the number of transit units (buses or trains) that can be served by a given transit facility. Typically, *vehicle capacity* is defined for three locations: (1) loading areas or berths, (2) transit stops and stations, and (3) bus lanes and transit routes. As is discussed later, starting with the loading areas, each of these locations directly affects the next location. The vehicle capacity of a transit station, for example, is a direct function of the vehicle capacities of the loading areas of that station. In addition, the capacity of a transit route is controlled by the capacity of the critical stops along that route. One of the most important factors affecting *vehicle capacity* is the vehicle's dwell time, which refers to the time required to serve passengers plus the time required to open and close the doors of the transit vehicle.

Person capacity refers to the number of people that can be carried past a particular location during a given period of time, under specified operating conditions and without unreasonable delay, hazard, or restriction. Person capacity is typically defined for transit stops and stations, as well as for the maximum load point along a transit route or a bus lane. The following three basic factors control person capacity:

Operator policy: The transit agency's policy has a direct impact on transit *person capacity.* For example, the agency's policy regarding whether standing passengers are allowed or not would directly impact the number of passengers that a given transit vehicle can carry.

Passenger demand characteristics: The spatial and temporal distribution of passenger demand directly impacts the number of boarding passengers that can be carried. Because of the spatial distribution of passenger demand, *passenger capacity* is typically defined for the maximum load point along a transit route, and not for the route as a whole. In addition, a transit system should be designed to provide for adequate capacity during peak demand periods. In analysis, this is typically accounted for using the PHF.

Vehicle capacity: This also has a direct impact on the passenger capacity, since it establishes an upper limit to the number of passengers that can use a transit stop or that can be carried past the maximum load point.

The following sections will briefly describe the analysis procedures that can be used to determine the capacity (both *vehicle* as well as *person* capacity) and level of service for transit. While there are several transit modes (for example, buses, streetcars, light rail, rapid rail, and automated guideways), these different modes from a capacity analysis standpoint can be broadly categorized into (1) buses; (2) on-street rail modes, such as streetcars and light rail; and (3) off-street or grade-separated rail modes. The following sections will describe the capacity analysis procedures for these different groups. We will start, however, by discussing some general concepts that are applicable to all three groups.

Transit Capacity Concepts

Vehicle capacity for transit is typically defined for three locations: (1) loading areas or berths, (2) transit stops and terminals, and (3) bus lanes or rail segments. Each of these three locations is briefly described next.

Loading Areas

For buses, a loading area (sometimes called bus berth) refers to the space dedicated for the bus to stop in order to load and unload passengers. A bus stop, as will be discussed later, consists of one or more loading areas. Bus stops along street curbs are the most common type of loading areas, and these could either be in the travel lane itself (i.e., on-line) or in the form of a pullout out of the travel lane (i.e., off-line).

Three primary factors determine the capacity of loading areas: (1) dwell time, (2) dwell-time variability, and (3) clearance time. Dwell time is the time required to serve passengers plus the time required to open and close the doors of the vehicle. *Dwell time* is itself a function of a number of factors, including (1) the number of passengers boarding and alighting the vehicle; (2) the distance between stops (longer distances would result in a large number of passengers at each stop, which in turn will increase dwell time); (3) the fare payment procedures (i.e., whether payment is by coins, tokens, passes, or smart cards); (4) the transit vehicle type (for low-floor buses, for example, the time needed for passengers boarding and alighting is reduced and particularly for the elderly and persons with disability); (5) on-board passenger circulation; and (6) wheelchair and bicycle boarding.

Dwell-time variability accounts for the fact that the dwell time at a given stop is likely to vary depending upon the actual passenger demand present. In analysis, this variability is accounted for using the *coefficient of variation* of dwell time. This is computed by dividing the standard deviation of the observed dwell times at the stop by the dwell time mean value.

Clearance time is the time that elapses after the moment the vehicle closes its doors until it clears the stop. During that time, the loading area is not available for use by another vehicle. For on-line bus stops and rail transit stations, the clearance time is equal to the time needed by the vehicle to start up and travel its own length, thus clearing the stop. For off-line bus facilities, an additional component is needed that is equal to the time required for the stopped bus to find an adequate gap in the adjacent lane's traffic stream, allowing it to reenter traffic.

Stations and Terminals

These are the second locations where vehicle capacity is determined and, for buses, they typically consist of one or more loading areas. Given this, the capacity of a bus stop is directly related to the capacities of the individual loading areas making up the stop. Bus stops can be generally divided into two groups: (1) bus terminals, and (2) on-street bus stops. Bus terminals are typically located off-street, whereas on-street buses are located curbside in one of three locations: (a) nearside (i.e., buses stop immediately before the intersection); (b) farside (buses stop after the intersection); and (3) midblock. From a capacity standpoint, farside stops have the least negative impact on capacity, followed by midblock stops and nearside stops.

Bus Lanes and Rail Segments

Bus lanes refer to any lane on a roadway in which buses operate. These lanes can be for the exclusive use of buses, or buses may have to share them with other traffic. Rail segments are dedicated to the sole use of a transit vehicle. Typically, the vehicle capacity of a bus lane or rail segment is determined by the capacity of the critical bus stop or station located along the lane or rail segment. Bus lanes are divided into three types: Type 1; Type 2; and Type 3. For

Type 1, buses do not make use of the adjacent lane while Type 2 buses have partial use of the adjacent lane, which they would typically share with other traffic. For Type 3, two lanes are provided for the exclusive use of buses. For Types 1 and 2, buses may or may not share the curb lane with other traffic.

In addition to the bus lane type, there are other factors that affect the capacity of bus lanes. For example, bus lane capacity can be increased by dispersing bus stops so that only a subset of the buses in the bus lane uses a certain set of steps. This is often referred to as *skip-stop* operation and can help increase capacity as well as allow for faster trips. The effectiveness of the skip-stop operation pattern is maximized when buses are gathered into platoons, and each platoon is assigned a group of stops. Moreover, the location of the bus stop can impact bus lane capacity—farside stops provide for the highest bus lane capacity, followed by midblock and finally nearside stops. These two issues (i.e., skip-stop operations and bus stop location) are not applicable to grade-separated rail.

Quality of Service Concepts

Different measures are available for assessing the performance of transit. These could reflect the standpoint of the operator or the passengers. The operator point is typically assessed using what is commonly referred to as productivity measures, which include annual ridership, passenger trips/revenue mile, vehicle operating expenses/revenue mile, and so on. The productivity measures, however, do not directly measure passenger satisfaction with the transit quality of service.

Transit quality of service is defined so as to reflect the perceived performance from the passenger's standpoint. In general, measures of transit quality of service can be divided into two main categories: (1) assessing transit service availability, and (2) assessing transit comfort and convenience. In addition, these measures would depend upon the particular element of the transit system being evaluated. As previously mentioned, a transit system can be regarded as consisting of the following three basic elements: (1) transit stops, (2) route segments, and (3) systems. For transit stops, the quality of service measures need to evaluate transit availability and convenience at a single location. For route segments, the measures should address availability and convenience along a route segment, which would consist of two or more stops. Finally, measures are needed for describing availability and convenience of the whole transit system, which would typically consist of several routes covering a specified geographic region.

Table 4.5 shows the transit quality of service framework and lists the different measures that are used for evaluating availability and convenience for the three different elements of a transit system. The measures with a superscript are those used for defining LOS. There are four transit service measures that are used to define the LOS: frequency, hours of service, passenger loads, and reliability.

TABLE 4.5
Transit Quality of Service Framework

Source: Adapted from HCM 2000.

	Service & Performance Measures		
Category	***Transit Stop***	***Route Segment***	***System***
Availability	• Frequency[a] • Accessibility • Passenger loads	• Hours of service[a] • Accessibility	• Service coverage • % person-minutes served
Comfort and Convenience	• Passenger loads[a] • Amenities • Reliability	• Reliability[a] • Travel speed • Transit/auto travel time	• Transit/auto travel time • Travel time • Safety

Note:
a. Service measure that defines the corresponding LOS.

With the general concepts relevant to transit capacity and LOS discussed, we now discuss the details of the analysis procedures for the three groups of transit modes: buses, on-street rail modes, and grade-separated modes.

Bus Capacity Analysis Methodology

Transit capacity for buses is calculated for three locations: loading areas, bus stops and bus lanes. Several factors affect the capacity of transit facilities: (1) dwell time, (2) the coefficient of variation of dwell time, (3) clearance time, (4) failure rate, (5) passenger loads, and (6) skip-stop operation. A discussion of these factors is presented first before the details of the capacity analysis procedures for different types of transit facilities.

Dwell Time

Dwell time refers to the time that elapses while a bus is stopped at a bus stop serving passengers. Specifically, it is the time required to serve passengers at the busiest door plus the time of opening and closing the doors. The HCM recommends a value ranging between 2 and 5 sec for door opening and closing under normal operations.

The best way to determine the dwell time is to directly measure it from the field. This method, however, is only applicable when one is interested in determining the capacity and LOS for a bus route that is already in operation. If the dwell time cannot be measured from the field (e.g., when we need to evaluate the capacity of a proposed new route), typical values based upon common practice are assumed. For example, for bus stops in a city's central business district (CBD) or for major transfer points, a dwell time value of 60 sec may be assumed. For major outlying stops, a value of 30 sec is assumed, or 15 sec for typical outlying stops. Equation 4.25 can also be used to compute dwell time:

$$t_d = P_a t_a + P_b t_b + t_{oc} \tag{4.25}$$

where

t_d = dwell time in seconds
P_a = alighting passengers/bus through the busiest door during peak 15 minutes
t_a = passengers' alighting time (seconds/person)
P_b = boarding passengers/bus through the busiest door during peak 15 minutes
t_b = passengers' boarding time (seconds/person)
t_{oc} = door opening and closing time (seconds)

Note that Equation 4.25 assumes that boarding and alighting passengers use the same door, which is why alighting and boarding times are added. Table 4.6 gives typical values for alighting and boarding times that can be used in conjunction with Equation 4.25. However, it should be noted that the times shown should be increased by 0.50 s if standing passengers are present.

The dwell time should be adjusted if wheelchair users regularly use a bus stop, since the bus door is typically blocked from other use when a wheelchair

TABLE 4.6

Boarding and Alighting Service Times for Transit

Source: Adapted from HCM 2000.

	Availability Doors or Channels		*Typical Boarding Service Times (s/p)*		
Bus Type	***Number***	***Location***	***Prepayment***[b]	***Single Coin Fare***	***Typical Alighting Service Times (s/p)***
Conventional	1	Front	2.0	2.6 to 3.0	1.7 to 2.0
(rigid body)	1	Rear	2.0	NA	1.7 to 2.0
	2	Front	1.2	1.8 to 2.0	1.0 to 1.2
	2	Rear	1.2	NA	1.0 to 1.2
	2	Front, rear[c]	1.2	NA	0.9
	4	Front, rear[d]	0.7	NA	0.6
Articulated	3	Front, rear, center	0.9[d]	NA	0.8
	2	Rear	1.2[e]	NA	—
	2	Front, center[c]	—	—	0.6
	6	Front, rear, center[c]	0.5	NA	0.4
Special single unit	6	3 double doors[f]	0.5	NA	0.4

Note:
NA: data not available.
a. Typical interval in seconds between successive boarding and alighting passengers. Does not allow for clearance times between successive buses or dead time at stop. If standers are present, 0.5 sec should be added to the boarding times.
b. Also applies to pay-on-leave or free transfer solution.
c. One each.
d. Less use of separated doors for simultaneous loading and unloading.
e. Double-door near loading with single exits, typical European design. Provides one-way flow within vehicle, reducing internal congestion. Desirable for line-haul, especially if two-person operation is feasible. May not be best configuration for busway operation.
f. Examples: Denver 16th Street Mall shuttle, airport buses used to shuttle passengers to planes. Typically low-floor buses with few seats serving short, high-volume passenger trips.

lift is in use. In this case, the wheelchair lift time (between 60 and 200 s) should be added to the dwell time. Dwell time should also be adjusted if the transit systems allow for bicycle loading (typically using a folding bus rack on the bus).

EXAMPLE 4.9

Determining Dwell Time and Maximum Load Point

Plans are underway for a bus route serving the central business district and having 10 stops. The route will use 42-seat buses and will require exact fare on boarding. The door opening and closing time is 4 s, and all passengers will be required to board the bus through the front door and alight through the back door. The potential ridership for the route is predicted to follow the pattern shown below:

Stop Number	1	2	3	4	5	6	7	8	9	10
Alighting Passengers	0	5	8	10	12	9	14	17	15	5
Boarding Passengers	25	20	15	16	10	4	3	2	0	0

Studies have shown that the boarding time is 3 s/passenger when no standees are present and that the presence of standees increases the boarding time to 3.50 s/passenger. Alighting time is estimated to be equal to 2 s/passenger. Determine the dwell time at the maximum load point.

Solution

The first step in the solution procedure is to compute the bus load as it arrives at each bus stop in order to determine those bus stops where some people would be standing. This is needed since the presence of standees increases the boarding time by 0.50 sec. The bus loads at the different stops are shown below:

Stop Number	1	2	3	4	5	6	7	8	9	10
Number of Passengers as Bus Arrives at Stop *n*	0	25	40	47	53	51	46	35	20	5

Since the bus can only seat 42 passengers, the bus will have standees when arriving at stops 4, 5, 6, and 7. With this determined, we proceed to determine the alighting and boarding times by multiplying the number of passengers by the given alighting and boarding time/passengers. For stops 4 through 7, the boarding time for the case where standees are present is used. Finally, the bus dwelling time is computed by adding the door opening and closing time (4 s) to the *larger* of the alighting and boarding times, since the bus has two doors—one dedicated for alighting and the other for boarding. The results are shown below:

Stop Number	1	2	3	4	5	6	7	8	9	10
Alighting Time (s)	0	10	16	20	24	***18***	***28***	***34***	***30***	***10***
Boarding Time (s)	***75***	***60***	***45***	***56***	***35***	14	10.5	6	0	0
Dwell Time (s)	79	64	49	60	39	22	32	38	34	14

Boarding times govern for stops 1 through 5 and alighting times govern for stops 6 through 10. Stop 1 is the critical bus stop requiring the longest dwell time.

Coefficient of Variation of Dwell Time

Coefficient of Variation of Dwell Time is computed by dividing the standard deviation of the observed dwell times at a bus stop by the mean dwell time. Experience has shown that this coefficient ranges between 40 and 80%. A value equal to 60% can be assumed in the absence of field observations.

Clearance Time

The clearance time consists of two components: (1) the time needed for the bus to start up and travel its own length, thus exiting the bus stop; and (2) the time needed by the bus to reenter traffic in case of off-line bus stops. Studies have shown that the start-up time for buses is typically in the range of 2 to 5 s and that the time needed by the bus to travel its length ranges between 5 and 10 s. Therefore, for on-line bus stops, the clearance time can be assumed to be equal to 10 s.

For off-line stops, the time needed by the bus to reenter the traffic stream needs to be added to the start-up time and time required by the bus to travel its own length. The reentry delay will depend upon the traffic volume in the adjacent lane. In the absence of other information, Table 4.7 can be used to estimate the reentry time.

Failure Rate

Failure rate refers to the probability that a queue of buses will form behind a bus stop. This probability can be derived from basic statistics. In the analysis procedure, this probability is accounted for using the one-tail normal variate, Z_a, which represents the area under one tail of the normal distribution curve beyond the acceptable levels of probability that a queue will form (Figure 4.10). Table 4.8 gives some of the typical values for Z_a for different failure rates. Values for Z_a can also be determined using Microsoft Excel's function, NORMINV, described in Chapter 2.

TABLE 4.7
Transit Average Reentry Delay

Source: Adapted from HCM 2000.

Adjacent-Lane Mixed-Traffic Volume (veh/h)	*Average Reentry Delay (s)*
100	0
200	1
300	2
400	3
500	4
600	5
700	7
800	9
900	11
1000	14

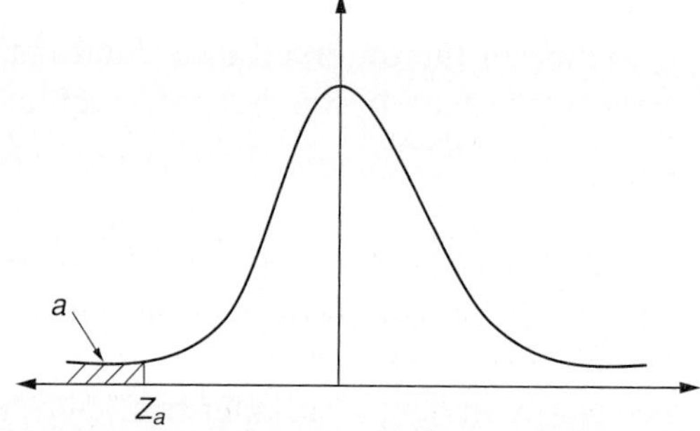

FIGURE 4.10
Normal curve distribution.

In general, for CBD bus stops, a value of Z_a between 1.04 and 1.44 is selected (this corresponds to a failure rate between 7.5 and 15%). For outlying stops, a value of 1.96 is assumed, which corresponds to a failure rate of less than 2.50%.

TABLE 4.8
Values of Percent Failure Associated with Z_a

Source: Adapted from HCM 2000.

Failure Rate (%)	Z_a
1.0	2.330
2.5	1.960
5.0	1.645
7.5	1.440
10.0	1.280
15.0	1.040
20.0	0.840
25.0	0.675
30.0	0.525
50.0	0.000

Passenger Loads

Passenger loads refer to the number of passengers in a single transit unit. Passenger loads are often expressed in terms of what is known as a load factor, which gives the ratio of the number of passengers to the number of seats on the transit vehicle (i.e., a load factor value of 1 would mean that all seats are occupied). In general, for long distances, one should attempt to keep the load factor below 1. For inner-city service, however, load factors are allowed to approach 1.50 or even 2 (which means that there are twice as many passengers on the bus as there are seats).

In transit, we often refer to what is called the maximum scheduled load and the crush load. The maximum scheduled load represents an upper limit for scheduling purposes and is equivalent to the capacity of the vehicle, assuming a reasonable number of standing passengers (load factors typically range between 1.25 and 1.50). Crush loads correspond to load factors exceeding 1.50. Under these circumstances, standing passengers are subject to unreasonable discomfort, and circulation becomes difficult, which in turn increases dwell time and reduces vehicle capacity.

Skip-Stop Operation

Skip-stop operation refers to a type of operation in which the stops are spread and an alternating route stop pattern is implemented (e.g., two- or three-block stop patterns). As previously mentioned, skip-stop operation reduces the total trip time and the number of buses stopping at each bus stop. With an alternating block stopping pattern, the capacity of a bus lane is nearly equal to the sum of the capacities of the two critical bus stops.

Loading Area Capacity

The first location where transit capacity needs to be determined is the loading area or berth. The maximum number of buses/berth/h, B_{bb}, can be determined from Equation 4.26:

$$B_{bb} = \frac{3600\left(\frac{g}{C}\right)}{t_c + \left(\frac{g}{C}\right)t_d + Z_a c_v t_d} \tag{4.26}$$

where

B_{bb} = maximum number of buses/berth/h (buses/h)
g/C = effective signal green time divided by the cycle length
t_c = clearance time between successive buses (seconds)
t_d = average dwell time (seconds)
Z_a = one-tail normal variate corresponding to the allowable failure rate for queue formation
c_v = coefficient of variation of dwell times

Bus Stop Capacity

The capacity of a bus stop is a direct function of the individual capacities of the bus loading areas that it contains. However, the increase in the bus stop capacity is not a linear function of the number of loading areas (i.e., doubling the number of loading areas will not result in doubling the capacity). This is because the loading areas of multiple berths are not used equally, which means that the efficiency of multiple loading areas decreases with the increase in the number of loading areas.

The capacity of a bus stop is therefore computed by multiplying the individual capacity of the loading area by the number of *effective* loading areas, N_{eb}. The *effective* number will always be less than the actual number of loading areas to reflect the reduced efficiency effect mentioned previously. Table 4.9 gives the value of N_{eb} for different multiple linear loading areas. The capacity of the bus stop is then given by

$$B_s = N_{eb} B_{bb} \tag{4.27}$$

where

B_s = maximum number of buses/bus stop/h
N_{eb} = effective number of loading areas

Bus Lane Capacity

For bus lanes, we consider two cases: (1) the case of exclusive urban bus lanes, and (2) the case of mixed-traffic bus lanes.

Exclusive Urban Bus Lanes

In general, the vehicle capacity of an exclusive bus lane is equal to the capacity of the critical bus stop along that lane. However, several factors that affect the vehicle capacity of bus lanes have to be accounted for in the calculation. These factors include (1) the type of bus lane (i.e., whether it is Type 1, 2, or 3); (2) whether skip operation is implemented; (3) the volume to capacity ratio of

TABLE 4.9 Effective Number of Loading Areas for Multiple Linear Loading Bus Stops, N_{eb}

Source: Adapted from HCM 2000.

	On-Line Loading Areas		*Off-Line Loading Areas*	
Loading Area No.	*Efficiency, %*	*No. of Cumulative Effective Loading Areas*	*Efficiency, %*	*No. of Cumulative Effective Loading Areas*
1	100	1.00	100	1.00
2	85	1.85	85	1.85
3	60	2.45	75	2.60
4	20	2.65	65	3.25
5	5	2.70	50	3.75

traffic in the adjacent lane for Type 2 bus lanes; and (4) the bus stop location and right-turning vehicles from the bus lane. Typically, these factors are accounted for through the introduction of appropriate adjustment factors.

Adjustment Factor for Right Turns: Vehicles turning right at a road intersection physically compete with buses in the bus lane for space, since these vehicles typically turn from the bus lane. In addition, turning vehicles may queue and may therefore block the bus from reaching the bus stop. The bus stop location at the intersection (i.e., whether nearside, farside, or midblock) plays a major role in determining the impact of right-turn vehicles on bus operation as well as the impact of buses on right-turning vehicles operation.

The impact of right turns on the bus lane capacity is determined by multiplying the bus lane vehicle capacity without right turns by the right-turn adjustment factor, which is given by

$$f_r = 1 - f_l\left(\frac{v_r}{c_r}\right) \tag{4.28}$$

where

f_r = right-turn adjustment factor
f_l = bus stop location factor
v_r = volume of right turns at specific intersection (veh/h)
c_r = capacity of right turns at specific intersection (veh/h)

Values for the bus location factor, f_l, can be obtained from Table 4.10. The bus location factor is a function of the bus stop location (i.e., whether it is nearside, midblock, or farside) and the bus lane type (Type 1, 2, or 3).

Adjustment for Skip-Stop Operation: The number of buses that can be accommodated by a series of skip stops should theoretically be equal to the sum of the capacities of the bus routes using each stop. An impedance factor, f_k, however, is introduced to reflect the effects of nonplatooned bus arrivals, as well as to reflect the impact of traffic in the adjacent lane on the skip-stop bus operation. The impedance factor, f_k, is computed as follows:

$$f_k = \frac{1 + Ka\,(N_s - 1)}{N_s} \tag{4.29}$$

TABLE 4.10
Bus Stop Location Factor, f_l

Source: Adapted from HCM 2000.

	Bus Lane Type		
Bus Stop Location	***Type 1***	***Type 2***	***Type 3***
Nearside	1.0	0.9	0.0
Midblock	0.9	0.7	0.0
Farside	0.8	0.5	0.0

where

f_k = capacity adjustment factor for skip-stop operations
K = adjustment factor for ability to fully utilize bus stops in a skip-stop operation
a = adjacent-lane impedance factor
N_s = number of alternating skip stops in sequence

Skip-stop operation is most efficient when buses arrive in platoons. To account for this, the adjustment factor, K, is introduced in Equation 4.29. This factor depends upon the arrival pattern of buses. A value of 0.50 is assumed for random arrivals, 0.75 for typical arrivals, and 1.00 for platooned arrivals. The value of the adjacent-lane impedance factor, a, can be computed from Equation 4.30:

$$a = 1 - 0.8\left(\frac{v}{c}\right)^3 \tag{4.30}$$

where (v/c) is the volume to capacity ratio in the adjacent lane.

With the adjustment factors calculated, the exclusive urban bus lane vehicle capacity can be computed from Equation 4.31 for non-skip-stop operation, and Equation 4.32 for skip-stop operation:

Non-skip-stop operation:

$$B = B_1 = B_{bb} \cdot N_{eb} \cdot f_r \tag{4.31}$$

where

B = bus lane vehicle capacity (buses/h)
B_{bb} = bus loading area vehicle capacity at critical bus stop (buses/h)
N_{eb} = number of effective loading areas at critical bus stop (Table 4.9)
f_r = right-turn adjustment factor (Equation 4.28)

Skip-stop operation:

$$B = f_k(B_1 + B_2 + \cdots + B_n) \tag{4.32}$$

where

$B_1 \ldots, B_n$ = vehicle capacities of each set of routes at their respective critical bus stops that use the same alternating skip-stop pattern (buses/h)
f_k = capacity adjustment factor for skip-stop operations (Equation 4.29)

Mixed-Traffic Bus Lanes

With the exception of large cities with high transit demand, mixed-traffic bus lanes are the more common type compared to exclusive lanes. The capacity is computed in essentially the same way for mixed-traffic bus lanes as for exclusive lanes; however, the interference of other traffic on bus operations must be considered. This interference is most obvious when off-line bus stops are used and buses have to wait for adequate gaps in the adjacent lane traffic stream to reenter traffic.

For mixed-traffic bus lanes, the different bus lane types (i.e., Type 1, 2, or 3) describe different lane configurations compared to the lane types of exclusive bus lanes. Type 1 mixed-traffic lane has one traffic lane in the direction of bus travel. Type 2 has two or more lanes, and there are no Type 3 lanes for mixed-traffic lanes.

In general, the impact of traffic on bus operations on mixed-traffic bus lanes can take one of two forms. First, it may interfere with the bus operations especially close to an intersection where queued vehicles may actually prevent a bus from reaching its stop. Second, for off-line bus stops, a stopped bus would have to wait until it found adequate gaps in the traffic stream before joining traffic. This second form of interference (i.e., the reentry delay) is accounted for by including the bus clearance time in the dwell time calculation (see Table 4.7).

To account for the first form of interference, a mixed-traffic adjustment factor, f_m, is used. This is computed in a manner very similar to computing the right-turn vehicle adjustment factor. Specifically, f_m is given by the following equation:

$$f_m = 1 - f_l\left(\frac{v}{c}\right) \tag{4.33}$$

where

f_m = mixed traffic adjustment factor
f_l = bus stop location factor from Table 4.10
v = curb lane traffic volume at critical bus stop (veh/h)
c = curb lane capacity at critical bus stop (veh/h)

The mixed-traffic bus lane vehicle capacity is then computed as

$$B = B_{bb} N_{eb} f_m \tag{4.34}$$

where B is the vehicle capacity for the mixed-traffic bus lanes, and all other notations are as defined before.

EXAMPLE 4.10

Calculating Bus Lane Capacity

A transit route has its transit vehicles operate in mixed-traffic lanes. The route has a total of eight bus stops. The critical bus stop, constraining vehicle capacity, is bus stop 3, which is an on-line stop located at the nearside of a signalized intersection. The following information regarding the operation of the bus route has been compiled:

- Dwell time at bus stop 3 = 40 s
- Curb lane volume = 450 equivalent passenger cars/h (pc/h)
- Capacity of the right curb lane = 700 pc/h
- The signal at bus stop 3 has a cycle length of 90 s, and the bus approach gets the green for 40 s out of the 90-second cycle
- The number of loading areas at bus stop 3 is limited to two berths
- Buses are allowed to use the adjacent lane

Determine the bus lane capacity, given that it is desired that the probability that a queue of buses will form behind a bus stop does not exceed 7.50%.

Solution

Step 1: Calculate the loading area capacity, B_{bb}. In order to do this, we use Equation 4.26 as follows:

$$B_{bb} = \frac{3600\left(\frac{g}{C}\right)}{t_c + \left(\frac{g}{C}\right)t_d + Z_a c_v t_d}$$

In our case,

- $g/C = 40/90 = 0.444$
- For on-line stops, the clearance time, t_c, is assumed to be equal to 10 sec
- The dwell time $t_d = 40$ s
- Since the failure rate is required not to exceed 7.50%, the value of Z_a (from Table 4.8) is found to be equal to 1.44
- The coefficient of variation of dwell time, c_v, is assumed to be equal to 0.60 in the absence of field observations

Therefore, the loading area capacity, B_{bb}, is equal to

$$B_{bb} = \frac{3600 \cdot 0.444}{10 + 0.444 \cdot 40 + 1.44 \cdot 0.60 \cdot 40} = 25 \text{ buses}$$

Step 2: Calculate the bus stop capacity. From Equation 4.27, the bus stop capacity, B_s, is given as

$$B_s = N_{eb} B_{bb}$$

For two linear on-line loading areas, the effective number of loading areas, N_{eb}, from Table 4.9 is equal to 1.85. Therefore,

$$B_s = 1.85 \cdot 25 = 46 \text{ buses}$$

Step 3: Calculate the mixed traffic bus lane capacity. For mixed-traffic bus lanes, the first step is to compute the mixed-traffic adjustment factor, f_m, using Equation 4.33:

$$f_m = 1 - f_l\left(\frac{v}{c}\right)$$

The bus stop location factor, f_l, can be found from Table 4.10. For Type 2 bus lanes (since buses are allowed to use the adjacent lane) and for nearside bus stops, $f_l = 0.90$.

Therefore,

$$f_m = 1 - 0.90 \cdot \left(\frac{450}{700}\right) = 0.42$$

The mixed-traffic bus lane vehicle capacity, B, is then computed from Equation 4.34 by multiplying the bus stop capacity by the mixed-traffic adjustment factor, f_m.

Therefore,

$$B = 0.42 \cdot 46 = 19 \text{ buses/h}$$

EXAMPLE 4.11

Assessing the Impact of using Farside Bus Stops

In Example 4.10, what impact would using farside bus stops, instead of nearside stops, have on the bus lane vehicle capacity?

Solution

The main difference in the calculations for this example compared to the previous example would be in the value for the bus location factor, f_l, used in calculating the mixed-traffic adjustment factor, f_m, in Equation 4.33.

For farside stops and Type 2 bus lanes, the bus location factor, f_l, is equal to 0.50 (Table 4.10).

Therefore,

$$f_m = 1 - 0.50 \cdot \left(\frac{450}{700}\right) = 0.68$$

The mixed-traffic bus lane vehicle capacity, B, is then computed as before by multiplying the bus stop capacity, which was previously determined to be equal to 46 buses/h by the mixed-traffic adjustment factor, f_m, as follows:

$$B = 0.68 \cdot 46 = 31 \text{ buses/h}$$

EXAMPLE 4.12

Impact of Skip-Stop Operation

For the problem described in Example 4.10, the transit agency would like to try skip-stop operations whereby buses would stop at every other stop. Determine the increase in the bus lane vehicle capacity that would result from adopting skip-stop operations. The adjacent lane carries a total of 600 pc/h and has a capacity of 1100 pc/h. Assume random arrivals for the transit vehicles at the bus stop, and farside stops as considered in Example 4.11.

For skip-stop operations, the first step in order to compute the capacity is to determine the impedance factor, f_k, using Equation 4.29:

$$f_k = \frac{1 + Ka(N_s - 1)}{N_s}$$

In our case,

$$K = 0.50 \text{ (random arrivals)}$$
$$N_s = 2.0 \text{ (every other stop)}$$

Using Equation 4.30, a is computed as

$$a = 1 - 0.8\left(\frac{v}{c}\right)^3$$

where (v/c) is the volume to capacity ratio in the adjacent lane. Therefore,

$$a = 1 - 0.8\left(\frac{600}{1100}\right)^3 = 0.87$$

Therefore, the impedance factor, f_k, is given as

$$f_k = \frac{1 + 0.50 \cdot 0.87 \cdot (2 - 1)}{2} = 0.72$$

Finally, the bus lane capacity is computed from Equation 4.32 as follows:

$$B = f_k(B1 + B2)$$

where $B1 = B2 = 31$ buses/h as determined from Example 4.11. Therefore,

$$B = 0.72 \cdot (31 + 31) = 41 \text{ buses/h}$$

On-Street Rail Capacity Analysis Procedure

On-street rail transit includes streetcars and light rail. These modes operate on urban streets, sharing the right-of-way with cars and automobiles. Streetcars often operate in mixed traffic and therefore share a lot of characteristics with buses. Modern light rail typically uses a combination of right-of-way types, which may include on-street operation (often in reserved lanes) as well as private right-of-way with grade crossings.

Similar to buses, the first step in determining the vehicle capacity of streetcars and light rail is to compute the minimum headway between vehicles. However, while in the case of buses the minimum headway was largely a function of the dwell time at the critical bus stop plus the clearance time, for light rail the situation is complicated by the fact that most light rail transit lines use a combination of right-of-way types. In such cases, the line capacity is determined by the weakest link. In some cases, the weakest link could be the on-street segment, especially if there is a traffic signal with an exceptionally long cycle length. In other cases, the capacity could be constrained by the block signal separation requirements of the off-street segment, (rail block signal systems are safety systems designed to prevent trains from colliding with one another). Moreover, the capacity could be constrained by the headway requirements on single-track sections in a third case.

The train headway used for calculating the capacity is therefore the largest of the following three potential controlling headways:

1. The on-street segment headway, which, similar to buses, is primarily a function of the dwell time of vehicles at stations;
2. The block-signaled segment headway; and
3. The single-track headway.

The following sections will describe how each of these three headway types can be calculated for light rail.

On-Street Segment Headway

Similar to buses, the minimum headway for the on-street segment of light rail or streetcars is primarily a function of the dwell time at stations. The dwell time is equal to the sum of (1) the time needed to serve passengers at the busiest door divided by the number of the available channels per door (typically two channels/door); and (2) the time required to open and close the doors, which is typically assumed to be equal to 5 s for modern light rail vehicles. The dwell time can therefore be expressed as follows:

$$t_d = \frac{P_d t_{pf}}{N_{cd}} + t_{oc} \tag{4.35}$$

where

t_d = dwell time in seconds
N_{cd} = number of channels per door for moving passengers
t_{oc} = door opening and closing time in seconds

TABLE 4.11
Passenger Flow Time

Source: Adapted from HCM 2000.

Car Entry	*Passenger Flow Time t_{pf} for Flow Type (s/p)*		
	Mainly Boarding	*Mainly Alighting*	*Mixed Flow*
Level	2.0	1.5	2.5
Steps	3.2	3.7	5.2

P_d = alighting passengers per rail through busiest door during the peak 15 minutes

t_{pf} = passenger flow time (seconds/passenger), as given by Table 4.11

It should be clear, however, that the dwell-time calculation described here cannot account for every variable that is likely to impact dwell time. For example, passenger volumes may vary within the 15-min peak period or trains may run faster or slower than expected, resulting in more passengers per train than estimated. To account for these variations, it is a common practice to add some extra time (commonly referred to as the operating margin) to the transit line's headway to allow for irregular operation and ensure that one train does not delay the following one. The operating margin typically ranges between 15 and 25 s.

With the dwell time appropriately determined, the minimum headway can then be calculated from the following formula:

$$h_{os} = (t_c + (g/C)t_d + Z_a c_v t_d)/(g/C) \tag{4.36}$$

where

h_{os} = minimum headway for the on-street segment (s)

g = effective green time for the signal at the stop with highest dwell time (s)

C = cycle length for the signal at the stop with highest dwell time (s)

t_d = dwell time at critical bus stop (s)

t_c = clearance time between successive trains, which is equal to the sum of the minimum clear spacing between trains plus the time for the train to clear a station. The minimum clear spacing typically ranges between 15 and 20 s, whereas the time needed to clear a station is usually around 5 s. The time required for the train to clear the station can also be calculated from knowledge of the train length and acceleration. It should be noted that some transit agents use the signal cycle length (C) as the minimum clearance time

Z_a = one-tail normal variate corresponding to the probability that queues of trains will form (from Table 4.8 or using Excel's function NORMINV)

c_v = coefficient of variation of dwell times (typically assumed to be equal to 0.40 for light rail operating in an exclusive lane, and 0.60 for streetcar operation in mixed traffic)

The preceding equation is exactly similar to the one used to calculate the capacity of loading areas for buses (Equation 4.26). For light rail, however, where the length of two trains exceeds one city block, the headway should not be less than twice the length of the longest traffic signal cycle (C_{max}). Such headway would minimize the risk of two adjacent trains blocking an intersection.

The Block-Signaled Segment Headway

The headway for the off-street segment is primarily determined by the block signal system. Train block signaling systems are safety systems designed to prevent trains from colliding with one another, as will be described in detail in the next section, which discusses the capacity of grade-separated rail. Generally speaking, light rail lines are not signaled with the minimum possible headway, but instead with the minimum planned headway, which is typically around 3.0 minutes. This can easily make signaled segments the dominant capacity restraint.

Single-Track Sections

Short single-track sections are sometimes used on light rail as a cost-saving measure. In such cases, these sections could impose severe restrictions on light rail capacity, especially if these sections are more than 0.4 km long. Computing the minimum headway, in such cases, amounts to first calculating the time required to travel the single-track section plus one train length. The minimum headway is then determined as twice the travel time.

Calculation of the travel time should account for the time lost during acceleration, deceleration, and station stops. It should also include a speed margin to account for equipment not operating to performance or drivers not driving at the maximum permitted speed. The following equation can be used to calculate the time needed to travel the single-track section:

$$t_{st} = SM\left[\frac{(N_s + 1)}{2}\left(\frac{3S_{max}}{d_s} + t_{jl} + t_{br}\right) + \frac{L_{st} + L}{S_{max}}\right] + N_s t_d + t_{om} \qquad (4.37)$$

where

t_{st} = time to cover the single-track section (s)
L_{st} = length of the single-track section (m)
L = train length (m)
N_s = number of stations on the single-track section
t_d = station dwell time (s)
S_{max} = maximum speed reached (m/s)
d_s = deceleration rate (default value = 1.3 m/s^2)
t_{jl} = jerk-limiting time (default value = 0.5 s)
t_{br} = operator and braking system reaction time (default value = 1.5 s)
SM = speed margin (commonly assumed to be equal to 1.10)
t_{om} = operating margin time (s)

The minimum headway is then taken as equal to twice the single-track travel time calculated previously, as follows:

$$h_{st} = 2 \cdot t_{st} \tag{4.38}$$

Vehicle Capacity

With the minimum headway for each of the on-street, the off-street, and the single-track segments calculated, the controlling headway (i.e., the maximum value out of the three headways calculated) is determined. This value is then used to calculate the vehicle capacity as follows:

$$T = 3600/h_{\min} \tag{4.39}$$

where

T = maximum number of trains/h
$h_{\min}$ = controlling minimum headway in seconds

EXAMPLE 4.13

Calculating the Capacity of a Light Rail Line

A light rail line in the city has two types of right of way. First, the light rail operates within the median of an arterial street with a speed of 55 km/h and goes through the signalized intersections of that arterial. This is followed by a single-track segment, which is 0.6 km long with one intermediate stop. The dwell time can be assumed to be equal to 35 s for all stations. The train is 27 m long and has an initial service acceleration equal to 1 m/s^2. The city blocks are 120 m long, and the g/C ratio for the signal at the critical stop is 0.50. The maximum cycle length along the on-street segment is 90 s. Determine the vehicle capacity of the light rail line.

Solution

To determine the vehicle capacity in this problem, we first need to compute the minimum headway for (1) the on-street segment, and (2) the single-track section. Since no information was provided in the problem statement regarding any restrictions caused by a block-signaled segment, we will have to assume that this is not an issue for this example.

Minimum Headway for On-Street Segment: To compute the minimum headway for the on-street segment, we need to use Equation 4.36, as defined previously:

$$h_{os} = (t_c + (g/C)t_d + Z_a c_v t_d)/(g/C)$$

As specified in the problem,

$$(g/C) = 0.50$$

$$t_d = 35 \text{ s}$$

$$C_{\max} = 90 \text{ s}$$

For light rail operating in an exclusive lane, c_v can be assumed to be equal to 0.40. We also assume that the probability of queues of trains forming is to be limited to 10%. Given this, Z_a, from Table 4.8, is equal to 1.28. What is left, therefore, is to determine the clearance time, t_c.

As discussed previously, the clearance time, t_c, is equal to the sum of (1) the minimum clear spacing between trains, and (2) the time for the train to clear a station. The minimum clear spacing will be assumed to be equal to 20 s in this case. The time needed for the train to clear the station is equal to the time needed by the train to travel a distance equal to its length (i.e., 27 m), starting from rest and accelerating at a rate of 1 m/s^2, as specified in the problem. To calculate that time, we use the well-known formula

$$x = \frac{1}{2}at^2 + u_o t$$

where

x = distance traveled
a = acceleration rate
u_o = initial velocity
t = travel time

Therefore,

$$27 = \frac{1}{2} \cdot 1 \cdot t^2$$

and

$$t = (27 \cdot 2)^{1/2} = 7.35 \text{ s}$$

The clearance time, t_c, therefore is equal to 20 + 7.35 = 27.35 s. Substituting in Equation 4.36,

$$h_{os} = (27.35 + 0.5 \cdot 35 + 1.28 \cdot 0.4 \cdot 35)/0.5 = 125.54 \text{ s}$$

We will round this up to 140 s to include an appropriate operating margin.

It should be noted that because the city blocks can accommodate more than two trains, using a headway that is at least equal to twice the longest cycle length is not an issue here.

Minimum Headway for the Single-Track Section: We next compute the minimum headway for the single-track section, by using Equation 4.37 to calculate the travel time for the single-track section as follows:

$$t_{st} = SM\left[\frac{(N_s + 1)}{2}\left(\frac{3S_{max}}{d_s} + t_{jl} + t_{br}\right) + \frac{L_{st} + L}{S_{max}}\right] + N_s t_d + t_{om}$$

In this example, we use

$SM = 1.1$ (default value)
$N_s = 1.0$ station
$S_{max} = 55$ km/h or 15.3 m/s
$d_s = 1.3$ m/s^2 (default value)
$t_{jl} = 0.5$ s (default value)
$t_{br} = 1.5$ s (default value)
$L_{st} = 0.55$ km or 550 m
$L = 27$ m
$t_d = 35$ s
$t_{om} = 20$ s (assumed operating margin)

Substituting in Equation 4.37 gives

$$t_{st} = 137.5 \text{ s}$$

The minimum headway for the single track section is equal to twice the single-track travel time, t_{st}. Therefore,

$$h_{st}(\text{min}) = 2 \cdot 137.5 = 275 \text{ s}$$

The minimum headway for the single-track controls, and therefore the light rail line vehicle capacity, can be estimated from Equation 4.39 as

$$T = 3600/275 = 13 \text{ trains/h}$$

Grade-Separated Rail Systems

Grade-separated rail transit refers to electric multiple-unit trains that run on fully separated, signaled, double-track rail. For grade-separated rail, the block signal control system plays a major role in determining the capacity of the rail system. This section, therefore, starts with a brief introduction to block control systems and their different characteristics. Following this, the capacity analysis procedure is described.

Block Signal Control Systems

Trains, as opposed to automobiles, run on fixed tracks. Given this, there is always great potential for collisions because trains cannot steer away from dangerous situations as can automobiles. In addition, the deceleration rate for trains is much smaller than that for automobiles, and hence the stopping or braking distance for trains is much longer than that for automobiles. By the time a train engineer sees an obstacle, the engineer would typically not have enough time to stop the train before a collision occurs.

For all these reasons, block signaling systems were introduced as early as the 1850s. The basic idea behind these systems is to divide the rail network into sections known as blocks. Two trains are not allowed to be in the same block at the same time. Moreover, a train cannot enter a block until permitted to do so by a signal that the block ahead can be occupied. For capacity analysis purposes, rail transit control systems can be classified into (1) fixed block, (2) cab signaling, and (3) moving block. Rail capacity increases from fixed-block to cab-signaling to moving-block systems.

Fixed-Block Systems

These consist of electrically-insulated sections of track, known as blocks. The presence of a train within a given block is detected by that train's wheels shorting a low-voltage current. These systems can only indicate that a train is occupying a particular block but cannot tell where exactly the train is along the block. Moreover, at block boundaries, a single train will occupy two blocks for a short period of time.

Fixed-block signal systems can be classified based on the minimum number of blocks downstream a train must remain unoccupied, as well as on the number of different signal indications or lights employed (commonly called aspects in the context of block signal systems). In the simplest two-aspect block system, only two indications are used—red for stop and green for go. In this case, a minimum of two blocks should be left unoccupied ahead of the train, and each block should be at least equal to the braking distance plus a safety distance. This could significantly limit the capacity of the rail line.

To achieve higher capacities and/or safer operations, more complex fixed-block systems with more aspects could be employed. For example, a three-aspect, three-block system would employ three indications—red for stop, yellow for reduce speed and be prepared to stop at the next signal, and green for proceed at full speed—and would use three blocks to separate trains. The addition of an extra block allows for deploying an automatic train stop device as an additional safety feature. This device would automatically activate the brakes of a train at the second red signal behind a train, if the train driver, for some reason, fails to start braking at the first red signal. With this system, safety is improved, but capacity is reduced due to the increase in the train separation distance (Figure 4.11a).

In addition, it is possible to deploy a four-aspect, four-block system. The four aspects or indications can be devised by using double lights as shown in Figure 4.11b. According to that system, a double red indication is for stop, double green is for proceed at full speed, a red-yellow signal is for prepare to stop, and a yellow-green is for proceed at medium speed. In these systems, four blocks would separate trains, but the braking distance plus the safety distance

FIGURE 4.11

Fixed-block transit control systems.

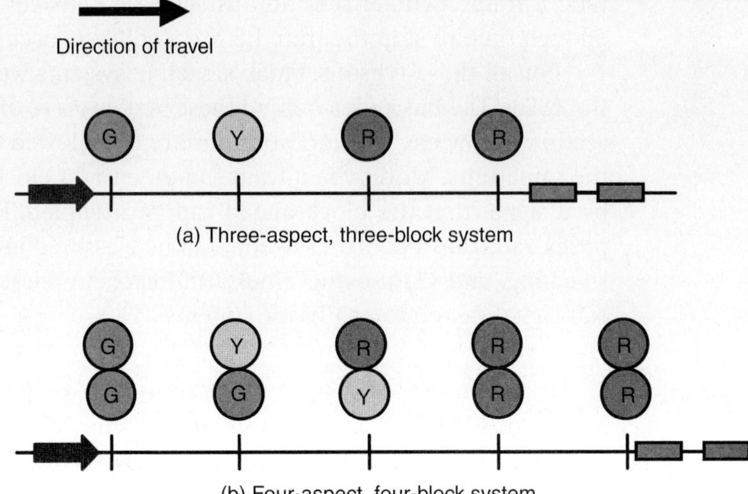

would have to be less than or equal to the distance of two blocks and not just one. This can help increase the capacity compared to the three-aspect, three-block system.

Cab-Signaling Systems

These systems use codes embedded into each track circuit, which can be detected and read by an antenna on each train. The codes communicate the maximum allowable speed for the block to the train. This speed, which is commonly called the reference speed, is often displayed in the train driver's cab (hence the name *cab signaling*).

The reference speed can be changed while a train is in a block depending upon the location of the train ahead. This allows for achieving close to optimum speeds. In addition, cab signaling allows for mitigating the problems of external signal visibility, especially around curves and during inclement weather. It also allows for increasing the number of aspects over what is typical for fixed-block signals. Cab-signaling systems often implement the equivalent of a five-aspect system with the following reference speeds: 80, 65, 50, 35, and 0 km/h.

Moving-Block Systems

These systems use computers to calculate a safe zone behind each moving train that no other train may enter. The system is based upon continuous calculations of the safe zone distance based upon trains' locations, and communication of the appropriate speed, acceleration, or deceleration to each train. The system can thus be regarded as a fixed-block system with very small blocks and a large number of aspects, although physically the system does not have either blocks or aspects. For the system to operate, precise knowledge of each train location and speed, as well as continuous two-way communication with the trains, is required. The computers controlling a moving-block system can be located on each train, at a central location, or they can be dispersed along the wayside. Moving-block signals have the advantage of increasing track capacity and allowing trains to run much closer together.

Capacity Analysis Procedure for Grade-Separated Rail

Assuming that the capacity is not limited by junctions or turnbacks, which is virtually always the case in most modern systems, the combination of the close-in time at stations, the dwell time, and the operating margin will be the capacity constraint. The capacity analysis procedure therefore consists of the following three steps:

1. Determining the close-in time at the maximum load point station;
2. Determining the dwell time at that station; and
3. Selecting a suitable operating margin.

Close-In Time at Maximum Load Point Station

The maximum load point station is usually the main downtown station. However, if a regional transportation planning model is available, with ridership data by station, it can be used to identify more accurately the maximum load point station. The close-in time is defined as the time between a train pulling out of a station and the next train entering the station. The close-in time is sometimes referred to as the safe separation time and is basically a function of the train control system, the train length, approach speed, and the train vehicle performance. It should be noted, however, that sharp curves or downgrades on station approaches would tend to reduce the approach speed and hence would lead to an increase in the close-in time and a corresponding reduction in capacity.

The best approach for determining the close-in time is from existing experience of operating at or close to capacity, or from a computer simulation model. However, if operating data or a simulation model are not available, analytical equations could be used to calculate the close-in time. The analytical procedure will differ depending upon the type of transit control system in place. However, the basic idea is first to determine the approach speed for the longest train that will result in the minimum close-in or separation time. Next, the analyst checks for any speed restrictions (e.g., curves or turnouts) that are within the approach distance of the train. If there are speed restrictions, the most restrictive speed is used along with its corresponding separation time. The following paragraphs will present the details of the procedure for the three types of control systems described: (1) three-aspect, fixed block signaling; (2) cab-signaling; and (3) moving-block systems.

Three-Aspect, Fixed-Block, and Cab-Signaling Systems: Equation 4.40 can be used to calculate the minimum train control separation time for both fixed-block as well as cab-signaling systems:

$$t_{cs} = \sqrt{\frac{2(L_t + d_{eb})}{a}} + \frac{L_t}{v_a} + \left(\frac{100}{f_{br}} + b\right)\left(\frac{v_a}{2d}\right) + \frac{at_{os}^2}{2v_a}\left(1 - \frac{v_a}{v_{max}}\right) + t_{os} + t_{jl} + t_{br} \tag{4.40}$$

where

t_{cs} = train control separation in seconds (to be calculated)
L_t = longest train length (default value = 200 m)
d_{eb} = distance from the front of stopped train to start of station exit block (default value = 10.5 m)
v_a = station approach speed in m/s (the approach speed that corresponds to the minimum separation time is to be calculated)
v_{max} = maximum line speed (default value = 27 m/s or 97 km/h)
f_{br} = braking safety factor expressed as a percentage (default value = 75% of normal rate)
b = separation safety factor which is equal to the number of blocks that separate trains (2.4 for three-aspect, fixed block and 1.2 for cab signaling)
a = initial service acceleration rate (default = 1.3 m/s^2)
d = service deceleration rate (default = 1.3 m/s^2)
t_{os} = time for overspeed governor to operate on automatic systems, or driver perception and reaction time on manual systems (default = 3 s)
t_{jl} = time lost to braking jerk limitation (default = 0.5 s)
t_{br} = brake system reaction time (default = 1.5 s)

Equation 4.40 should be solved for the minimum value of t_{cs}. The easiest approach to do that is to assume a range of values for the approach speed, v_a, and to calculate t_{cs} that corresponds to each of the assumed values of v_a. The calculations can be best done on Microsoft Excel or a similar spreadsheet, as will be illustrated in the upcoming examples.

Moving Block: Equation 4.41 can be used to calculate the safe separation time for a moving-block signal system:

$$t_{cs} = \frac{L_t + P_e}{v_a} + \left(\frac{100}{f_{br}} + b\right)\left(\frac{v_a}{2d}\right) + \frac{at_{os}^2}{2v_a}\left(1 - \frac{v_a}{v_{max}}\right) + t_{os} + t_{jl} + t_{br} \tag{4.41}$$

Equation 4.41 introduces one new parameter compared to Equation 4.40, namely, the positioning error, P_e, whose default value is 6.25 m. It should also be noted that for moving-block systems, the parameter, b, of the equation that refers to the separation safety factor or the number of braking distances or blocks that separate the trains is equal to 1.0, as opposed to 2.4 for fixed block and 1.2 for cab signaling. All other variables are as defined in relation to Equation 4.40.

Dwell Time at Maximum Load Station

Because the train's close-in time is primarily a function of the train's physical performance, along with other fixed characteristics, it was possible to develop analytical models to calculate its value with some precision. Station dwell time is a function of variables that are subject to a great deal of uncertainty. As previously discussed in relation to buses and light rail, the dwell time is a function of the number of travelers waiting at the station and their flow time. Given this, it is very hard to estimate dwell time to the same level of precision as was the case with station close-in time. For grade-separated rail, the common practice is simply to assign a set value to the station dwell time. Experience has shown that the average station dwell time for rail transit systems operating at or close to capacity during the peak hour ranges between 30 and 50 seconds. Values in that range can be used in conjunction with the station close-in time previously determined.

Operating Margin

The last component in calculating the minimum headway for grade-separated rail is the operating margin, which is included in order to account for erratic service situations. The operating margin typically ranges between 15 and 25 second.

Vehicle Capacity

With the station's close-in time and dwell time determined and with an appropriate operating margin selected, the minimum headway, h_{gs}, for a grade-separated rail line is calculated as the sum of these three values, as follows:

$$h_{gs} = t_{cs} + t_d + t_{om} \tag{4.42}$$

The vehicle capacity of the line, in terms of the maximum number of trains/h, can then be calculated easily as

$$T = 3600/h_{gs} \tag{4.43}$$

EXAMPLE 4.14

Vehicle Capacity for a Heavy Rail System with Cab Signaling

A transit agency is planning to undertake a heavy rail transit system project. The agency is interested in determining the vehicle capacity of a transit line for a cab-signaling system. The longest train is expected to be about 200 m and will operate at a maximum speed of 100 km/h. The distance from the front of a stopped train to the station exit block is 10.5 m. No speed restrictions limit approach speed to suboptimal levels.

Solution

In order to determine the vehicle capacity for a grade-separated rail, we need to determine (1) the station close-in time or control separation time, t_{cs}; (2) the dwell time, t_d; and (3) the operating margin, t_{om}.

Close-In Time: For cab signaling, Equation 4.40 should be solved for the minimum separation time. This will be done with the aid of Microsoft Excel:

$$t_{cs} = \sqrt{\frac{2(L_t + d_{eb})}{a}} + \frac{L_t}{v_a} + \left(\frac{100}{f_{br}} + b\right)\left(\frac{v_a}{2d}\right)$$

$$+ \frac{at_{os}^2}{2v_a}\left(1 - \frac{v_a}{v_{\max}}\right) + t_{os} + t_{jl} + t_{br}$$

For this example,

$L_t = 200$ m
$d_{eb} = 10.5$ m
$v_{\max} = 100$ km/h $= 27.8$ m/s
$f_{br} = 75\%$
$b = 1.2$ for cab signaling
$a = 1.3$ m/s^2 (assumed)
$d = 1.3$ m/s^2 (assumed)
$t_{os} = 3$ s
$t_{jl} = 0.5$ s
$t_{br} = 1.5$ s

Equation 4.40 is programmed in Excel, a range of values for the approach speed, v_a, is assumed, and the corresponding control separation time is determined, as shown in Table 4.12.

The values of the separation time are then plotted against the approach speed, as shown in Figure 4.12, and the approach speed, resulting in the minimum separation time is determined as shown below. The approach speed that results in the minimum separation time is around 14 m/s (50 km/h), and the corresponding separation time is about 51 s.

Dwell Time: The dwell time typically ranges between 30 and 50 seconds. We will assume a value of 40 s.

Operating Margin: We assume a value of 20 s for the operating margin.

Vehicle Capacity: Given the preceding information, the minimum headway = 51 + 40 + 20 = 111 s. The vehicle capacity is therefore given by

$T = 3600/111 = 32$ trains/h

TABLE 4.12
Approach Speed versus Control Separation Time for Example 4.14

Approach Speed, v_a (m/s)	*Separation Time, t_{cs} (s)*
2	127.66
4	78.15
6	62.94
8	56.31
10	53.11
12	51.63
14	51.13
16	51.24
18	51.76
20	52.56
22	53.58
24	54.75
26	56.04

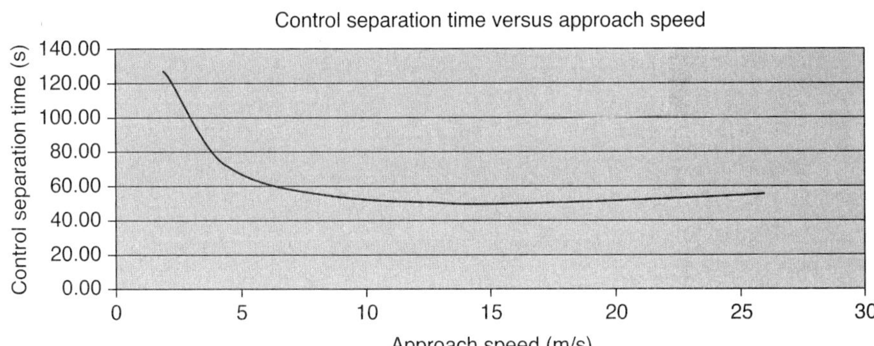

FIGURE 4.12
Control Separation Time Versus Approach Speed.

EXAMPLE 4.15

Vehicle Capacity for a Moving-Block Signal Control System

For the transit line described in Example 4.14, determine the vehicle capacity for a moving-block signal control system instead of cab signaling.

Solution

For a moving-block control system, Equation 4.41 should be solved for the minimum separation time using a similar procedure to that described in relation to Example 4.14:

$$t_{cs} = \frac{L_t + P_e}{v_a} + \left(\frac{100}{f_{br}} + b\right)\left(\frac{v_a}{2d}\right) + \frac{at_{os}^2}{2v_a}\left(1 - \frac{v_a}{v_{\max}}\right) + t_{os} + t_{jl} + t_{br}$$

For moving-block control, the parameter b is equal to 1.0, and the positioning error can be assumed to be equal to 6.25 m. Table 4.13 finds the control separation time for a range of approach speeds, and Figure 4.13 plots the time

TABLE 4.13
Approach Speed versus Control Separation Time for Example 4.15

Approach Speed, v_a (m/s)	*Separation Time, t_{cs} (s)*
2	112.63
4	61.40
6	45.52
8	38.48
10	34.97
12	33.23
14	32.50
16	32.40
18	32.73
20	33.34
22	34.17
24	35.17
26	36.28

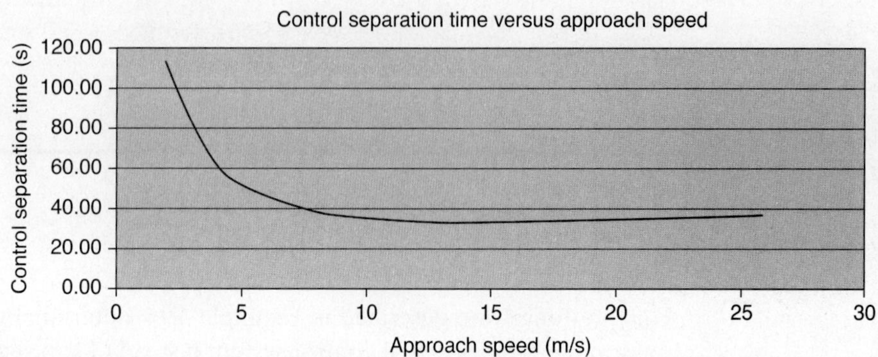

FIGURE 4.13
Control Separation Time Versus Approach Speed.

against the speed. From Figure 4.13, it can be seen that the approach speed that results in the minimum time is around 16 m/s, and the corresponding minimum control separation time is 32.4 s.

Assuming a dwell time of 40 s and an operating margin of 20 s, the minimum headway in this case is 32.4 + 40 + 20 = 92.4 s. The corresponding vehicle capacity is equal to 3600/92.4 = 39 trains/h. This is a significant increase over the capacity of Example 4.14.

Transit Person Capacity

Our focus so far has been on computing the vehicle capacity of transit facilities, which involves determining the maximum number of buses or trains that can be accommodated at a station or along a transit line/h. For transit facilities, in addition to determining the *vehicle capacity,* we are also interested in determining the *person capacity*. For buses, this can be calculated easily by multiplying the bus lane vehicle capacity at the maximum load point, by the allowed passenger loads on board an individual bus, by the peak hour factor (typically assumed to be equal to 0.75 for buses).

For light rail or grade-separated rail, the maximum person capacity, P, is typically calculated by multiplying the vehicle capacity in terms of the maximum number of trains/h, T, by the length of the train, L, by the linear passenger load factor, P, which gives the number of passengers/m of length set by the transit agency policy, by the peak hour factor, PHF. This can be expressed as

$$P = TLP_m(PHF) \tag{4.44}$$

where

P = person capacity (person/h)
T = vehicle capacity (trains/h)
L = length of train (m)
P_m = linear passenger loading level (person/m)
PHF = peak hour factor

The peak hour factor is typically assumed to be equal to 0.80 for heavy rail, 0.75 for light rail, and 0.60 for commuter rail. The linear passenger loading level is around 5.9 passengers/m length for heavy rail and 4.9 passengers/m length for light rail.

Example 4.16

Calculating Transit Person Capacity

Determine the person capacity of the transit route described in Example 4.12. All buses have a capacity of 43 passengers. The transit agency has 10 express buses on which standees are not allowed. For the remaining fleet, up to 50% standees are allowed. Assume a PHF = 0.75.

Solution

As determined in Example 4.12, the bus route has a capacity equal to 41 buses/h. Out of these 41 buses, standees are not allowed on 10 buses (the express buses) whereas they are allowed on the remaining 31 buses.

Therefore,

$$\text{Passenger capacity} = [(10 \cdot 43) + (31 \cdot 43 \cdot 1.50)] \cdot 0.75$$
$$= 1822 \text{ passengers/h}$$

Quality of Service Measures

As discussed previously, measures of transit quality of service can be divided into two main categories: (1) assessing transit service availability; and (2) assessing transit comfort and convenience. Table 4.5 also showed that for LOS designation, four quality of service measures are employed: (1) frequency, (2) hours of service, (3) passenger loads, and (4) reliability. This section defines these four measures and describes how they can be used to determine the LOS for transit stops and routes.

Frequency

Service frequency is the measure used to evaluate the transit availability LOS at transit stops. It determines the number of times per hour a user has access to the transit mode (assuming that the transit stop is within an acceptable walking distance for the user). Table 4.14 shows the different thresholds for the service frequency that are used to define the different LOS. For example, LOS A corresponds to a frequency of more than 6 veh/h or to headways that are less than 10 minutes. It should be noted that a transit agency could decide to operate its vehicles at different LOS throughout the day. For example, during the peak hours, the service may operate at LOS B, whereas it could operate at LOS D at midday.

TABLE 4.14

Service Frequency LOS

Source: Adapted from HCM 2000.

LOS	*Headway (min)*	*Veh/h*	*Comments*
A	<10	>6	Passengers don't need schedules
B	≥10–14	5–6	Frequent service; passengers consult schedules
C	>14–20	3–4	Maximum desirable time to wait if bus/train missed
D	>20–30	2	Service unattractive to choice riders
E	>30–60	1	Service available during hour
F	>60	< 1	Service unattractive to all riders

TABLE 4.15
Hours of Service LOS

Source: Adapted from HCM 2000.

LOS	*Hours per Day*	*Comments*
A	>18–24	Night or owl service provided
B	>16–18	Late evening service provided
C	>13–16	Early evening service provided
D	>11–13	Daytime service provided
E	>3–11	Peak hour service/limited midday service
F	0–3	Very limited or no service

Note:
Fixed route: number of hours per day when service is provided at least once an hour.
Paratransit: number of hours per day when service is offered.

Hours of Service

This measure defines the number of hours during the day when the transit service is available along a route, and hence it is the availability measure for transit routes. Table 4.15 shows how the measure can be used to determine the LOS for a transit route. As with frequency, hours of service LOS can vary over the day.

Passenger Loads

From the passenger's standpoint, passenger loads help determine the comfort level in terms of being able to find a seat or to stand comfortably. The measure uses the area available for each passenger as a measure for the LOS. The space thresholds corresponding to the different LOSs are shown in Table 4.16.

Route Segment Reliability

A number of measures can be used for the service reliability of a transit route segment including (1) on-time performance, (2) headway adherence, (3) missed

TABLE 4.16
Passenger Load LOS

Source: Adapted from HCM 2000.

	Bus		*Rail*		
LOS	*ft²/p (m²/p)*	*p/seat*[a]	*ft²/p (m²/p)*	*p/seat*[a]	*Comments*
A	>12.90 (>1.16)	0.00–0.50	<19.90 (<1.79)	0.00–0.50	No passenger need sit next to another
B	8.60–12.89 (0.77–1.16)	0.51–0.75	14.00–19.90 (1.26–1.79)	0.51–0.75	Passenger can choose where to sit
C	6.50–8.59 (0.59–0.77)	0.76–1.00	10.20–13.99 (0.92–1.26)	0.76–1.00	All passengers can sit
D	5.40–6.49 (0.49–0.59)	1.01–1.25	5.40–10.19 (0.49–0.92)	1.01–2.00	Comfortable loading for standees
E	4.30–5.39 (0.39–0.49)	1.26–1.50	3.20–5.39 (0.29–0.49)	2.01–3.00	Maximum schedule load
F	<4.30 (<0.39)	>1.50	<3.20 (<0.29)	>3.00	Crush loads

Note:
a. Approximate values for comparison. LOS is based on area per passenger.
b. 1 ft^2/p = 0.093 m^2/p

TABLE 4.17

Reliability LOS for On-time Performance

Source: Adapted from HCM 2000.

LOS	*On-Time Percentage*	*Comments*[a]
A	97.5–100.0	1 late bus per month
B	95.0–97.4	2 late buses per month
C	90.0–94.9	1 late bus per week
D	85.0–89.9	
E	80.0–84.9	1 late bus per direction per week
F	<80.0	

Note:
Applies to routes with frequencies of fewer than 6 buses/h scheduled.
a. User perspective, based on 5 round trips/week of their travel on a particular transit route with no transfers.
On-time = 0–5 min late departing published time point (fixed route)
arrival within 10 min of scheduled pickup time (deviated fixed route)
arrival within 20 min of scheduled pickup time (paratransit)

trips, and (4) distance traveled between mechanical breakdowns. From a passenger's standpoint, the on-time performance is the measure that most accurately reflects the user's perception of service reliability. However, when vehicles run at frequent headways, headway adherence becomes more important.

For on-time performance, the common practice is to define a transit vehicle as late when it is more than 5 min behind schedule. Early departures are generally considered as being equivalent to a vehicle being late by the amount of one headway, since passengers would have to wait for the next vehicle. Table 4.17 lists the reliability LOS thresholds for transit vehicles operating with frequencies of fewer than 6 veh/h.

For transit service at frequencies higher than 6 veh/hr, the LOS is defined in terms of headway adherence (or more specifically on the coefficient of variation of headways, c_v) as shown in Table 4.18. The coefficient of variation, c_v, is computed by dividing the standard deviation of the headways by the mean headway.

TABLE 4.18

Reliability LOS for Schedule Adherence

Source: Adapted from HCM 2000.

LOS	*Coefficient of Variation*
A	0.00–0.10
B	0.11–0.20
C	0.21–0.30
D	0.31–0.40
E	0.41–0.50
F	>0.50

Note:
Applies to routes with frequencies greater than or equal to 6 buses/h scheduled.

Pedestrian Facilities

In this section, we discuss the capacity analysis and LOS determination procedures for pedestrian facilities, which include walkways and sidewalks, shared off-street paths, pedestrian crosswalks, and pedestrian facilities along urban streets.

Pedestrian Flow Characteristics

Similar to the traffic flow parameters used in conjunction with vehicle traffic flow, the following parameters are defined for studying the capacity and LOS for pedestrian facilities:

Pedestrian speed: This is the average walking speed for pedestrians, which is typically about 1.2 m/s but varies with age and purpose of the walking trip.

Pedestrian flow: This refers to the number of pedestrians crossing a line of sight across the width of the pedestrian facility perpendicular to the pedestrian path in unit time (p/min). The *pedestrian flow/unit width* is equal to the pedestrian flow divided by the effective width of the pedestrian facility, in units of pedestrians/min/m (p/mm/m).

Pedestrian density: This is computed as the average number of pedestrians/unit area of the pedestrian facility (p/m^2).

Pedestrian space: This refers to the average area provided for each pedestrian. It is equal to the inverse of the density and is expressed in units of square meter/pedestrian (m^2/p).

Flow–Speed–Density Relationships for Pedestrian Traffic

The parameters just defined are related to one another in a manner that is similar to the fundamental relationships among the vehicle traffic flow parameters (Figure 4.2). Similar to vehicle traffic flow, pedestrian flow, density, and speed are related to one another by the following equation:

$$v_{\text{ped}} = S_{\text{ped}} \cdot D_{\text{ped}} \tag{4.45}$$

where

v_{ped} = pedestrian flow rate (p/min/m)
S_{ped} = pedestrian speed (m/min)
D_{ped} = pedestrian density (p/m^2)

This equation can also be expressed in terms of the pedestrian space (M), which, as defined previously, is equal to the inverse of the density as follows:

$$v_{\text{ped}} = \frac{S_{\text{ped}}}{M} \tag{4.46}$$

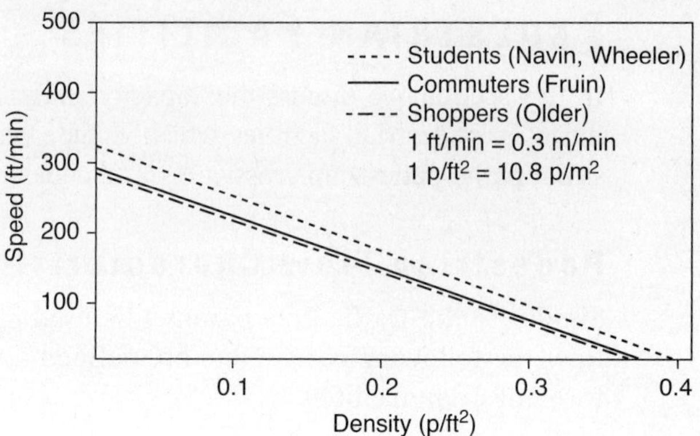

FIGURE 4.14
Relationship between pedestrian speed and density.

Source: Adapted from HCM 2000.

where

M = pedestrian space (m^2/p)

Figure 4.14 shows the relationship between pedestrian speed and density. Similar to vehicle traffic, pedestrian speed can be seen to decrease with increases in density. As density increases, the space available to each pedestrian decreases and, in turn, the degree of mobility afforded decreases.

Figure 4.15 shows the relationship between pedestrian flow and space. The relationship is similar to the flow–density relationship previously developed for vehicles (Figure 4.2).

The maximum flow shown in the above figure corresponds to the capacity of the pedestrian facility. This capacity appears to correspond to densities or space in the range of 0.5 to 0.8 m^2/p (or 5 to 9 ft^2/p). With the aid of Figure 4.15, one can define ranges for the flow rates or space that correspond to the different LOS.

Finally, Figure 4.16 shows the relationship between speed and flow, which once again is very similar to the vehicle traffic relationship. As flow increases (i.e., more pedestrians on the walkways), the speed decreases because pedestrians are getting closer to one another and their ability to choose high walking

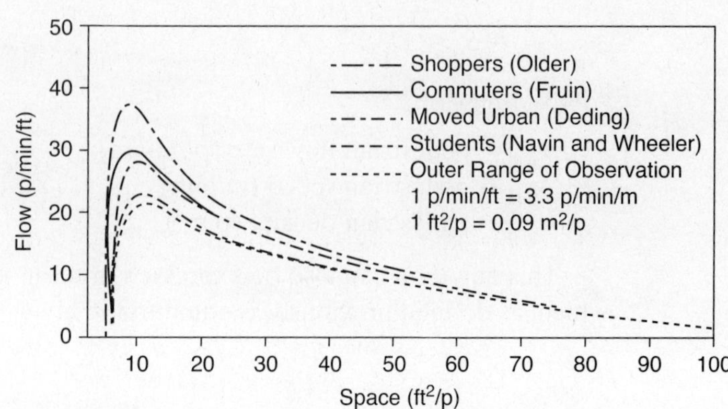

FIGURE 4.15
Relationship between pedestrian flow and space.

Source: Adapted from HCM 2000.

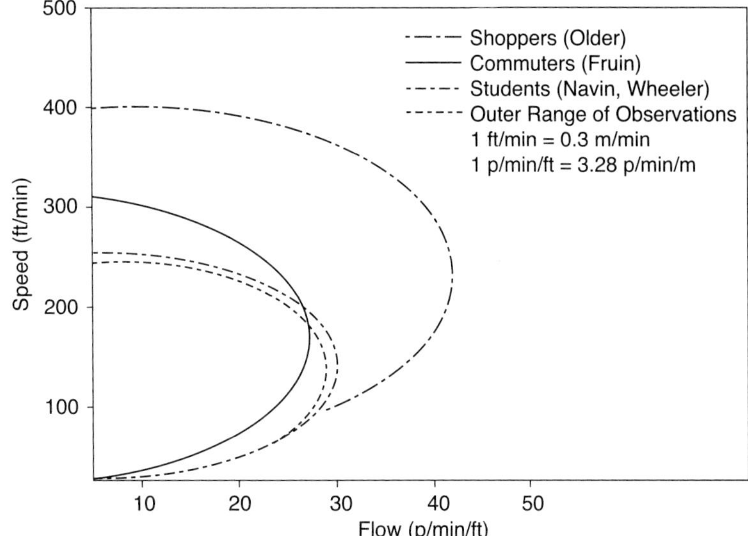

FIGURE 4.16 Relationship between speed and flow.

Source: Adapted from HCM 2000.

speeds is reduced. After reaching the critical crowding level, which corresponds to the capacity of the facility, both speed and flow are reduced.

Capacity Analysis and LOS Concepts

In the absence of other data, the capacity of a pedestrian facility or walkway can reasonably be assumed to be equal to 75 p/min/m or 4530 p/h/m. At capacity, however, pedestrians' speed is severely impaired and would generally be around 0.75 m/s, which is way below the typical average value of 1.2 m/s or 1.5 m/s. Given this, pedestrian facilities are typically designed to operate well below capacity-level operation.

For LOS determination, the basic idea is to define ranges for the space/pedestrian flow rates, and/or speeds that correspond to different LOS. Speed is an important measure since it can be easily measured in the field. The speed at capacity is generally around 0.75 m/s or 45 m/min. Figure 4.17 shows a graphic illustration and a description of the different LOS for a walkway and the range of values for pedestrian space and flow rates that correspond to each LOS.

The LOS criteria shown in Figure 4.17 are based on average flow conditions and do not consider platoon conditions. However, for pedestrian facilities such as sidewalks, for example, the interruption of flow and queue formation at traffic signals can result in surges in traffic demand and in the formation of platoons of pedestrians. Within the platoon, the LOS will generally be one level lower than the LOS that is based on average conditions. The decision as to whether to design the facility for average or for platoon conditions depends upon available space, cost, and policy.

FIGURE 4.17
Pedestrian walkway LOS.

Source: Adapted from HCM 2000.

LOS A
Pedestrian Space > 60 ft²/p *Flow Rate* ≤ 5 p/min/ft

At a walkway **LOS A**, pedestrians move in desired paths without altering their movements in response to other pedestrians. Walking speeds are freely selected, and conflicts between pedestrians are unlikely.

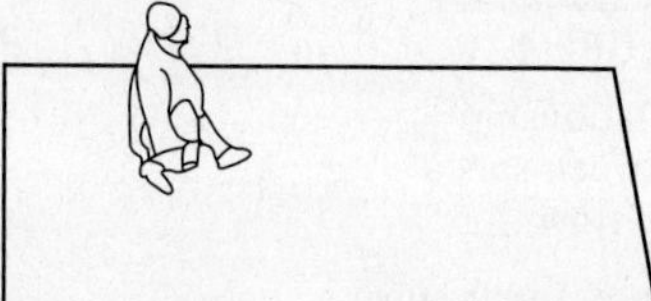

LOS B
Pedestrian Space > 40–60 ft²/p *Flow Rate* > 5–7 p/min/ft

At **LOS B**, there is sufficient area for pedestrians to select walking speeds freely, to bypass other pedestrians, and to avoid crossing conflicts. At this level, pedestrians begin to be aware of other pedestrians, and to respond to their presence when selecting a walking path.

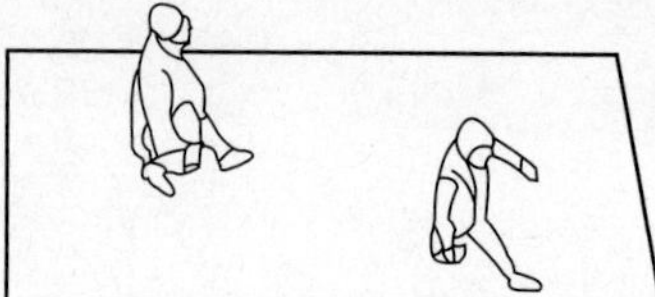

LOS C
Pedestrian Space > 24–40 ft²/p *Flow Rate* > 7–10 p/min/ft

At **LOS C**, space is sufficient for normal walking speeds, and for bypassing other pedestrians in primarily unidirectional streams. Reverse-direction or crossing movements can cause minor conflicts, and speeds and flow rate are somewhat lower.

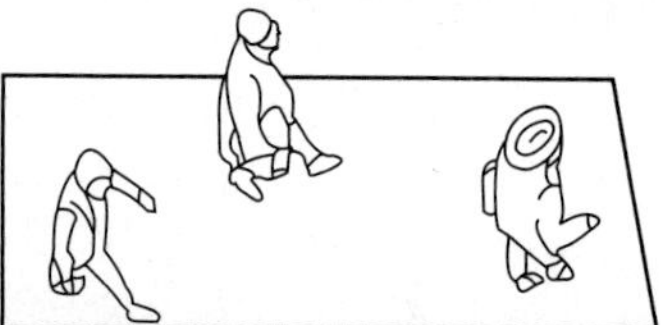

LOS D
Pedestrian Space > 15–24 ft²/p *Flow Rate* > 10–15 p/min/ft

At **LOS D**, freedom to select individual walking speed and to bypass other pedestrians is restricted. Cross- or reverse-flow movements face a high probability of conflict, requiring frequent changes in speed and position. The LOS provides reasonably fluid flow, but friction and interaction between pedestrians is likely.

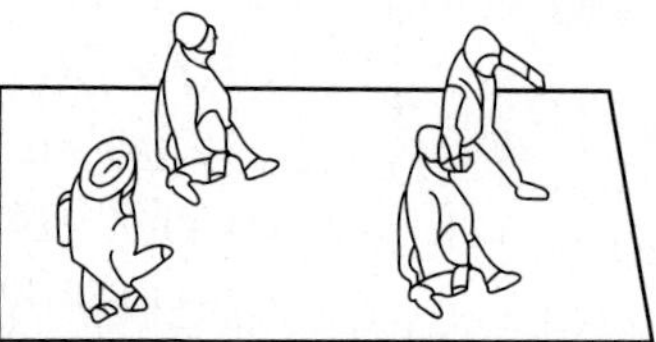

LOS E
Pedestrian Space > 8–15 ft²/p *Flow Rate* > 15–23 p/min/ft

At **LOS E**, virtually all pedestrians restrict their normal walking speed, frequently adjusting their gait. At the lower range, forward movement is possible only by shuffling. Space is not sufficient for passing slower pedestrians. Cross- or reverse-flow movements are possible only with extreme difficulty. Design volumes approach the limit of walkway capacity, with stoppages and interruptions to flow.

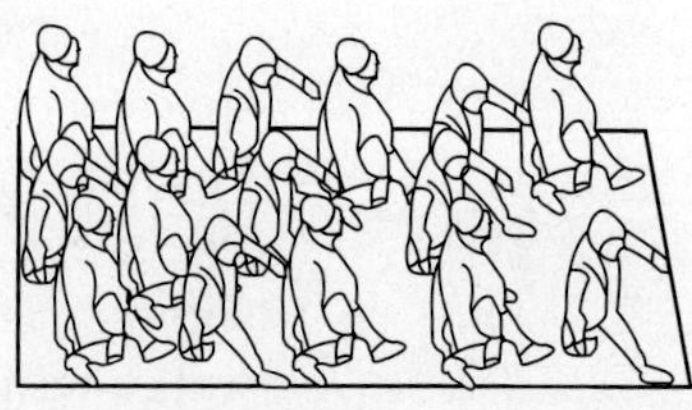

LOS F
Pedestrian Space ≤ 8 ft²/p *Flow Rate* varies p/min/ft

At **LOS F,** all walking speeds are severely restricted, and forward progress is made only by shuffling. There is frequent, unavoidable contact with other pedestrians. Cross- and reverse-flow movements are virtually impossible. Flow is sporadic and unstable. Space is more characteristic of queued pedestrians than of moving pedestrian streams.

Note: 1 ft²/p = 0.09 m²/p; 1 p/min/ft = 3.3 p/min/m

Analysis Methodology

Similar to highways, pedestrian facilities, from a traffic flow point, can also be divided into uninterrupted and interrupted flow facilities. In addition, several pedestrian facilities can be distinguished, including walkways and sidewalks, shared off-street paths, pedestrian facilities at signalized intersections, and pedestrian facilities along urban streets. The analysis procedures and the LOS thresholds for these different facilities vary, and therefore each is treated separately in this section.

Walkways and Sidewalks

These are pedestrian facilities that are separated from motor vehicle traffic. They are for the exclusive use of pedestrians, and their use by bicycles and other users is typically not allowed. Segments of walkways and sidewalks far away from signalized or unsignalized intersections can be regarded as uninterrupted flow pedestrian facilities.

As shown in Figure 4.17, the space available/pedestrian is the primary measure for evaluating the LOS of a walkway or a sidewalk. This can be determined in the field by dividing the number of pedestrians occupying a given area of the facility at a given time by the area. Pedestrian speed can also be observed in the field and used as a supplementary performance measure. To facilitate the determination of LOS, the analysis methodology also allows for using the pedestrian unit flow rate (which can be determined easily from field observations) as a performance measure. To determine the pedestrian unit flow rate, one would need to measure the peak 15-min pedestrian count and the effective width of the walkway (i.e., width excluding widths and shy distances from obstructions on the walkway). With these measured, the pedestrian unit flow rate can be determined from Equation 4.47 as follows:

$$v_p = \frac{v_{15}}{15 \times W_E} \tag{4.47}$$

where

v_p = pedestrian unit flow rate (p/min/m)
v_{15} = peak 15-min flow rate (p/15 min)
W_E = effective walkway width (m)

Table 4.19 summarizes the different criteria for the LOS on walkways. The table allows for using the space, unit flow rate, speed, or the v/c ratio to determine the LOS. For computing the v/c ratio, a value of 76 p/min/m is assumed for the capacity. It should be noted that in the case of significant platooning on the walkway, determination of the LOS should be based on Table 4.20 instead of Table 4.19.

TABLE 4.19
Average Flow LOS Criteria

Source: Adapted from HCM 2000.

LOS	*Space (ft^2/p)*	*Flow Rate (p/min/ft)*	*Speed (ft/s)*	*v/c Ratio*
A	>60	≤5	>4.25	≤0.21
B	>40–60	>5–7	>4.17–4.25	>0.21–0.31
C	>24–40	>7–10	>4.00–4.17	>0.31–0.44
D	>15–24	>10–15	>3.75–4.00	>0.44–0.65
E	>8–15	>15–23	>2.50–3.75	>0.65–1.0
F	≤8	variable	≤2.50	variable

Note: 1 ft^2/p = 0.09 m^2/p; 1 p/min/ft = 3.3 p/min/m; 1ft/s = 0.3 m/s

TABLE 4.20
Platoon-Adjusted LOS Criteria

Source: Adapted from HCM 2000.

LOS	*Space (ft^2/p)*	*Flow Rate[a] (p/min/ft)*
A	>530	≤0.5
B	>90–530	>0.5–3
C	>40–90	>3–6
D	>23–40	>6–11
E	>11–23	>11–18
F	≤11	>18

Note: 1 ft^2/p = 0.09 m^2/p; 1 p/min/ft = 3.3 p/min/m

EXAMPLE 4.17

Calculating LOS for a Sidewalk

Consider a 3.5 m sidewalk segment, bordered by the curb on one side and stores with window displays on the other. The 15-min peak pedestrian flow on the sidewalk is 1200 p/15 min. The effective width of the sidewalk, after accounting for the curb and preempted width for the stores' window displays, is 2.5 m. Determine the LOS during the peak 15 min on the average and within the platoons.

Solution

The first step is to determine the pedestrian unit flow rate, which can be determined from Equation 4.47 as follows:

$$v_p = \frac{v_{15}}{15 \times W_E} = \frac{1200}{15 \times 2.5} = 32 \text{ p/min/m}$$

The LOS can then be determined from Table 4.19 (average conditions) and Table 4.20 (within platoons).

Therefore, for average conditions, the LOS is C from Table 4.19. While within platoons, the LOS is D from Table 4.20.

Shared Pedestrian–Bicycle Facilities

A wide range of users can be found on shared pedestrian–bicycle facilities, including pedestrians, bicyclists, and skateboarders. Bicycles, because of their higher speeds, tend to have a negative impact on pedestrian capacity and LOS. While a number of capacity analysis procedures use equivalency factors to account for negative impacts of one vehicle type on facilities' capacity, for pedestrian–bicycle facilities, researchers found that it was difficult to establish equivalency factors for bicycles relative to pedestrians, and an alternative analysis procedure was needed for evaluating the pedestrian LOS on shared pedestrian–bicycle facilities.

The idea was to base the LOS on the concept of hindrance. The LOS for a pedestrian on a shared path is based upon the frequency of passing (in the same direction) and meeting (from the opposite direction) other users. Since pedestrians rarely pass other users, the LOS is really dependent upon the frequency that the pedestrian is overtaken by bicyclists (either passing or meeting). Equation 4.48 can be used to determine the total number of bicycle passing and meeting events/hour.

$$F_p = Q_{sb}\left(1 - \frac{S_p}{S_b}\right)$$

$$F_m = Q_{ob}\left(1 + \frac{S_p}{S_b}\right) \tag{4.48}$$

where

F_p = number of passing events/h
F_m = number of opposing or meeting events/h
Q_{sb} = bicycle flow rate in the same direction (bicycles/h)
Q_{ob} = bicycle flow rate in the opposing direction (bicycles/h)
S_p = mean pedestrian speed on the path (m/s)
S_b = mean bicycle speed on the path (m/s)

The total number of events is then calculated as follows:

$$F = F_p + 0.5F_m \tag{4.49}$$

The number of meeting events is multiplied by 0.5 because meeting events allow for direct visual contact and therefore the opposing bicycles tend to cause less hindrance to pedestrians. With the total number of events determined, the pedestrian LOS can be determined from Table 4.21.

TABLE 4.21
Pedestrian LOS Criteria for Shared Two-Way Paths

Source: Adapted from HCM 2000.

Pedestrian LOS	*Number of Events/h*
A	≤38
B	>38–60
C	>60–103
D	>103–144
E	>144–180
F	≥180

Example 4.18

Determining the LOS for a Shared Pedestrian–Bicycle Facility

A shared two-way, pedestrian–bicycle facility has a width of 2.5 m. The peak pedestrian flow on the facility is 150 p/15 min. The bicycle flow rate is 100 bicycles/h in the same direction as the pedestrians, and 150 bicycles/h in the opposite direction. Determine the pedestrians' LOS. What would the LOS be if the facility is converted into an exclusive pedestrian facility (i.e., no bicycles are allowed) with an effective width of 1.5 m? Assume a pedestrian speed of 1.2 m/s and a bicycle speed of 4.8 m/s.

Solution

The first step is to determine the number of bicycle passing (F_p) and meeting (F_m) events on the facility/h using Equation 4.48 as follows:

$$F_p = Q_{sb}\left(1 - \frac{S_p}{S_b}\right) = 100\left(1 - \frac{1.2}{4.8}\right) = 75 \text{ events/h}$$

$$F_m = Q_{ob}\left(1 + \frac{S_p}{S_b}\right) = 150\left(1 + \frac{1.2}{4.8}\right) = 187.5 \text{ events/h}$$

The total number of events (F) can then be calculated from Equation 4.49 as follows:

$$F = F_p + 0.5F_m = 75 + 0.5(187.5) = 169 \text{ events/h}$$

From Table 4.21, the corresponding LOS is E.

If the facility is converted into an exclusive pedestrian facility with an effective width of 1.5 m, the pedestrian unit flow rate can be determined from Equation 4.47 as follows:

$$v_p = \frac{v_{15}}{15 \times W_E} = \frac{150}{15 \times 1.5} = 6.6 \text{ p/min/m}$$

From Tables 4.19 and 4.20, this corresponds to a LOS A for average conditions and a LOS B for platooning conditions.

Pedestrian Facilities at Signalized Intersections

On sidewalks and walkways, signalized and unsignalized intersections tend to interrupt pedestrian traffic flow. In this section, we describe the procedures for determining the LOS of pedestrian facilities in the vicinity of signalized intersections having a pedestrian crossing on at least one approach, as an example of interrupted flow pedestrian facilities. Analyzing signalized intersection crossings is complicated by the fact that it involves intersecting sidewalk flows, pedestrians crossing the street, and other pedestrians queued waiting for the signal to change.

Determining the LOS for pedestrians at signalized intersections is typically based upon the average delay experienced by a pedestrian. This average delay, d_p, can be computed from Equation 4.50 as follows:

$$d_p = \frac{0.5(C - g)^2}{C} \tag{4.50}$$

where

d_p = average pedestrian delay in seconds
C = cycle length
g = effective green time *for pedestrians* in seconds

It should be noted that the effective green time for a pedestrian phase would typically be equal to the displayed parallel vehicle green. It should also be noted that, according to Equation 4.50, the average pedestrian delay does not depend upon the pedestrian flow level. This is actually true up to flow levels approaching 5000 p/h.

With the average pedestrian delay determined, Table 4.22 can be used to determine the corresponding LOS. This table also shows the likelihood of pedestrian noncompliance (i.e., their disregard for signal indications) as a function of the average delay. These likelihood values apply to intersections with low to moderate conflicting vehicular volumes. At intersections with heavy vehicular volumes, pedestrians have no choice than to wait for their walk signal.

It should be noted that the HCM includes procedures for determining the pedestrian LOS at street corners and along the crosswalk. The interested reader is directed to the *Highway Capacity Manual* for more details.

TABLE 4.22
LOS Criteria for Pedestrians at Signalized Intersections

Source: Adapted from HCM 2000.

LOS	*Average Delay/Pedestrian (s)*	*Likelihood of Risk-Taking Behavior*[a]
A	<5	Low
B	≥5–10	
C	>10–20	Moderate
D	>20–30	
E	>30–45	High
F	>45	Very high

Note:
a. Likelihood of acceptance of short gaps.

EXAMPLE 4.19

Determining the LOS for Pedestrian Facilities at Signalized Intersections

Determine the LOS for pedestrians at a two-phase signalized intersection with a cycle length of 100 s. The phase serving the major street vehicular traffic gets 60 s of green whereas the phase serving the minor street vehicular traffic gets 30 s of green.

Solution

In order to determine the LOS, we first need to determine the average delay for pedestrians crossing the major and minor streets using Equation 4.50. It should be noted, however, that the green time for pedestrians crossing the major street is equal to the green time displayed for the minor street vehicles, since pedestrians would cross the major street when the minor street vehicles are moving. Similarly, the green time for pedestrians crossing the minor street is equal to the green time displayed for the major street vehicles. Therefore, from Equation 4.50,

d_p (for pedestrians crossing the major street)

$$= \frac{0.5(C - g)^2}{C} = \frac{0.5(100 - 30)^2}{100} = 24.50 \text{ s/p}$$

d_p (for pedestrians crossing the minor street)

$$= \frac{0.5(C - g)^2}{C} = \frac{0.5(100 - 60)^2}{100} = 8 \text{ s/p}$$

Therefore, from Table 4.22,

The LOS for pedestrians crossing the major street is D, and

The LOS for pedestrians crossing the minor street is B.

Pedestrian Facilities on Urban Streets

For extended pedestrian facilities along urban streets, both uninterrupted as well as interrupted flow facilities exist along the length of the facility. For these facilities, the average pedestrian travel speed (which takes into account both uninterrupted and interrupted flow conditions) is the performance measure used for determining the LOS. The urban street under consideration is first segmented, with each segment consisting of a signalized intersection and an upstream segment of pedestrian sidewalk, beginning immediately after the

TABLE 4.23
LOS Criteria for Pedestrian Sidewalks on Urban Streets

Source: Adapted from HCM 2000.

LOS	*Travel Speed (ft/s)*
A	>4.36
B	>3.84–4.36
C	>3.28–3.84
D	>2.72–3.28
E	≥1.90–2.72
F	<1.90

Note: 1 ft/s = 0.3 m/s

nearest upstream intersection. The average travel speed over the entire section can then be computed from Equation 4.51 as follows:

$$S_A = \frac{L_T}{\sum \frac{L_i}{S_i} + \sum d_j} \tag{4.51}$$

where

S_A = average pedestrian speed in m/s
L_T = total length of the urban street being analyzed (m)
L_i = length of segment *I* in m
S_i = pedestrian walking speed over segment *i* in m/s
d_j = intersection delay at intersection *j* in seconds computed from Equation 4.50

The LOS can then be determined using Table 4.23.

EXAMPLE 4.20

Determining the LOS for Pedestrian Facilities on Urban Streets

Determine the LOS for a proposed 2.5 km pedestrian sidewalk on an urban street with three signalized intersections. The sidewalk is segmented into three segments having the following lengths: 975 m, 610 m, and 915 m, respectively. The signals have a 90-second cycle length, and the length of the pedestrian's green phase is equal to 35 s. Assume a pedestrian speed of 1.2 m/s.

Solution

The first step is to determine the average pedestrian delay at the three signalized intersections using Equation 4.50 as follows:

$$d_p = \frac{0.5(C - g)^2}{C} = \frac{0.5(90 - 35)^2}{90} = 16.80 \text{ s/p at each intersection}$$

The average speed can then be computed from Equation 4.51 as follows:

$$S_A = \frac{L_T}{\sum \frac{L_i}{S_i} + \sum d_j} = \frac{2500}{\left(\frac{2500}{1.2}\right) + 3 \times 16.8} = 1.17 \text{ m/s}$$

From Table 4.23, this corresponds to a LOS B.

Bicycle Facilities

Different types of bicycle facilities exist, including exclusive off-street bicycle paths, shared off-street paths, and on-street bicycle lanes. Similar to highways, bicycle facilities can be divided into interrupted and uninterrupted facilities. Off-street paths typically belong to the uninterrupted facilities group. Bicycle traffic on on-street paths is typically interrupted by traffic signals and stop signs.

Bicycle Traffic Flow Characteristics

Although different from vehicles, bicycles still tend to operate in distinct lanes, and therefore the capacity of the bicycle facility depends upon the number of effective lanes in use by bicycles. The best way to determine the number of effective lanes is through a field evaluation. However, if one is planning future facilities, the standard width of a bicycle lane can be assumed to be equal to 1.2 m. Studies have shown that three-lane bicycle facilities operate much more efficiently compared to two-lane facilities. This is because three-lane facilities provide more opportunities for passing and maneuvering.

Capacity and LOS Concepts

As opposed to other types of facilities, bicycle facilities experience severe deterioration in LOS at flow levels well below capacity. Given this, the concept of capacity is not that important for the design and analysis of bicycle facilities. For facilities in the United States, under uninterrupted flow conditions, a value of 2000 bicycles/h/lane can be assumed for the saturation flow rate.

For LOS determination, the performance measures used in conjunction with vehicle traffic are not very appropriate for bicycle facilities. For example, studies have shown that bicycle speeds are not that sensitive to flow rate. Moreover, it is difficult to determine the density for bicycle facilities, especially when they allow for shared use. As an alternative measure, the LOS of bicycle facilities is based on the concept of hindrance (you may recall that we used this same concept to evaluate the LOS for shared pedestrian–bicycle paths). In most cases, the number of events (i.e., passing and meeting events) is used as a surrogate for hindrance. With hindrance used as the performance measure, the LOS E/F is reached at a flow level that is well below the capacity of the facility, as shown in Figure 4.18.

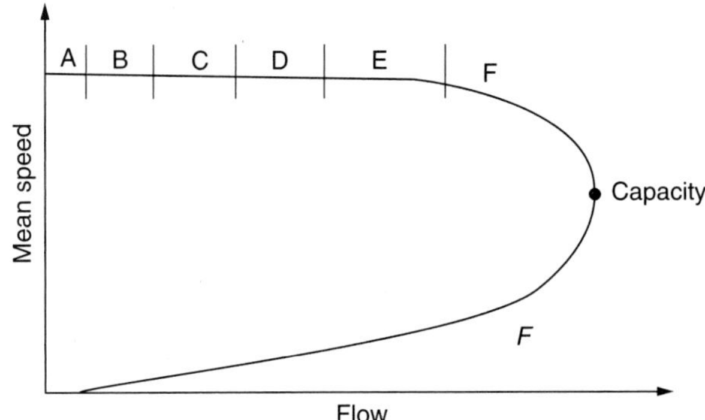

FIGURE 4.18
Bicycle LOS and capacity.

Source: Adapted from HCM 2000.

Analysis Methodology

Uninterrupted flow facilities include exclusive off-street and shared off-street bicycle paths. On-street bicycle paths that are interrupted by signalized intersections represent one example of interrupted bicycle flow facilities. The analysis methodologies for these three types are discussed below.

Exclusive Off-Street Bicycle Paths

Exclusive off-street facilities provide for separation from vehicle traffic and do not allow use by users other than bicyclists. For bicyclists, they provide the best LOS compared to other types of bicycle facilities. Determining the LOS for exclusive off-street bicycle paths is based on determining the number of events encountered by bicyclists using Equations 4.52 through 4.54:

$$F_p = 0.188v_s \tag{4.52}$$

$$F_m = 2v_o \tag{4.53}$$

$$F = 0.5F_m + F_p \tag{4.54}$$

where

F_p = number of passing events with bicyclists in the same direction (events/h)
F_m = number of opposing events with bicyclists in opposing direction (events/h)
F = total number of events on path
v_s = flow rate of bicyclists in direction being analyzed (bicycles/h)
v_o = flow rate of bicycles in opposing direction (bicycles/h)

With the number of events encountered determined, Table 4.24 can be used to determine the corresponding LOS.

TABLE 4.24
LOS for Exclusive Bicycle Paths

Source: Adapted from HCM 2000.

LOS	*Frequency of Events, Two-Way, Two-Lane Paths[a] (events/h)*	*Frequency of Events, Two-Way, Three-Lane Paths[b] (events/h)*
A	≤40	≤90
B	>40–60	>90–140
C	>60–100	>140–210
D	>100–150	>210–300
E	>150–195	>300–375
F	>195	>375

Note:
a. 8.0-ft-wide paths. Also used for on-street bicycle lanes.
b. 10-ft-wide paths.

Note: 1 ft = 0.3 m

EXAMPLE 4.21

Determining LOS for an Exclusive Bicycle Path

Determine the LOS for a north–south exclusive bicycle path which carries a volume of 160 bicycles/h during the peak period. Field observations have determined that 65% of the bicycles move in the peak direction, which is the NB direction, during the peak period. The path is 2.4 m wide and can be assumed to have two effective lanes.

Solution

The first step is to find the directional flows in both the NB and SB directions as follows:

$$v_b(\text{NB}) = 160 \cdot 0.65 = 104 \text{ bicycles/h}$$

$$\text{v}_\text{b}\,(\text{SB}) = 160 \cdot 0.35 = 56 \text{ bicycles/h}$$

We then compute the number of opposing and meeting events for each direction, and the total number of events from Equations 4.52 to 4.54, as follows:

For NB:

$$F_p = 0.188v_s = 0.188 \cdot 104 = 20 \text{ events/h}$$

$$F_m = 2v_o = 2 \cdot 56 = 112 \text{ events/h}$$

$$F = 0.5F_m + F_p = 0.5 \cdot 112 + 20 = 76 \text{ events/h}$$

From Table 4.24, for two-lane paths, this number of events/hr corresponds to LOS C.

For SB:

$$F_p = 0.188v_s = 0.188 \cdot 56 = 11 \text{ events/h}$$

$$F_m = 2v_o = 2 \cdot 104 = 208 \text{ events/h}$$

$$F = 0.5F_m + F_p = 0.5 \cdot 208 + 11 = 115 \text{ events/h}$$

From Table 4.24, for two-lane paths, this number of events/h corresponds to LOS D.

Shared Off-Street Paths

The difference between shared and exclusive off-street paths is that shared paths are open to other nonmotorized modes such as pedestrians, skateboarders, roller skaters, and so on. The analysis methodology for bicycle–pedestrian facilities focuses on the hindrance concept and its surrogate measure, namely the passing and meeting maneuvers. Equations 4.55, 4.56, and 4.57 can be used to determine the number of passing events (F_p) and of meeting events (F_m), as well as the total number of events (F) for shared bicycle and pedestrian situations:

$$F_p = 3v_{ps} + 0.188v_{bs} \tag{4.55}$$

$$F_m = 5v_{po} + 2v_{bo} \tag{4.56}$$

$$F = 0.5F_m + F_p \tag{4.57}$$

where

v_{ps} = flow rate of pedestrians in subject direction (p/h)
v_{bs} = flow rate of bicycles in subject direction (bicycles/h)
v_{po} = flow rate of pedestrians in opposing direction (p/h)
v_{bo} = flow rate of bicycles in opposing direction (bicycles/h)

The LOS can then be determined from Table 4.25.

TABLE 4.25
LOS for Shared Off-Street Bicycle Paths

Source: Adapted from HCM 2000.

LOS	*Frequency of Events, Two-Way, Two-Lane Paths[a] (events/hr)*	*Frequency of Events, Two-Way, Three-Lane Paths[b] (events/hr)*
A	≤40	≤90
B	>40–60	>90–140
C	>60–100	>140–210
D	>100–150	>210–300
E	>150–195	>300–375
F	>195	>375

Note:
a. 8.0-ft-wide paths.
b. 10-ft-wide paths.

1 ft = 0.3 m

EXAMPLE 4.22

Determining LOS for a Shared Bicycle–Pedestrian Facility

Consider a shared bicycle–pedestrian facility that runs in the east–west direction. The facility is 3 m wide and can be assumed to have effectively three lanes. The peak flow rate for the bicycles is 180 bicycles/h, with a directional split of 60/40 (EB/WB). For pedestrians, the peak flow rate is 70 p/h with a directional split of 50/50.

Solution

The first step is to calculate the directional flows for both bicycles and pedestrians, as follows:
EB:

$$\text{Bicycles} = 180 \cdot 0.60 = 108 \text{ bicycles/h}$$

$$\text{Pedestrians} = 70 \cdot 0.50 = 35 \text{ pedestrians/h}$$

WB:

$$\text{Bicycles} = 180 \cdot 0.40 = 72 \text{ bicycles/h}$$

$$\text{Pedestrians} = 70 \cdot 0.50 = 35 \text{ pedestrians/h}$$

We then calculate the number of opposing and meeting events for each direction and the total number of events from Equations 4.55 to 4.57 as follows:
For EB:

$$F_p = 3v_{ps} + 0.188v_{bs} = 3 \cdot 35 + 0.188 \cdot 108 = 126 \text{ events/h}$$

$$F_m = 5v_{po} + 2v_{bo} = 5 \cdot 35 + 2 \cdot 72 = 319 \text{ events/h}$$

$$F = 0.5F_m + F_p = 0.5 \cdot 319 + 126 = 286 \text{ events/h}$$

From Table 4.25, for three-lane paths, this corresponds to LOS D.
For WB:

$$F_p = 3v_{ps} + 0.188v_{bs} = 3 \cdot 35 + 0.188 \cdot 72 = 126 \text{ events/h}$$

$$F_m = 5v_{po} + 2v_{bo} = 5 \cdot 35 + 2 \cdot 108 = 391 \text{ events/h}$$

$$F = 0.5F_m + \mathrm{F_p} = 0.5 \cdot 391 + 126 = 322 \text{ events/h}$$

From Table 4.25, for three-lane paths, this corresponds to LOS E.

TABLE 4.26
LOS for Bicycles at Signalized Intersections

Source: Adapted from HCM 2000.

LOS	*Control Delay (s/bicycle)*
A	<10
B	≥10–20
C	>20–30
D	>30–40
E	>40–60
F	<60

Bicycle Paths at Signalized Intersections

These have a designated on-street bicycle lane on at least one approach. The HCM recommends using a saturation flow rate value of 2000 bicycles/h, assuming that right-turning vehicles will yield the right-of-way to through bicyclists. With the value of the saturation flow rate determined, the bicycle capacity of a signalized intersection approach can be determined in a fashion similar to what was done with regard to the vehicular capacity of signalized intersection approaches using Equation 4.13, as follows:

$$c_i = s_i \frac{g_i}{C} = 2000 \frac{g}{C} \tag{4.58}$$

where g is the effective green time for bicycles and C is the cycle length.

The LOS is then based upon the average control delay. This delay can also be determined in a fashion similar to that previously discussed with respect to vehicles, using Equation 4.21. Table 4.26 gives the delay-based criteria for the bicycles' LOS at signalized intersections.

EXAMPLE 4.23

Determining the LOS for a Bicycle Facility at a Signalized Intersection

Determine the LOS for a 1.2-m-wide bicycle lane at a signalized intersection with a cycle length of 110 s. The bicycle lane gets 50 s of green and carries a peak flow rate of 120 bicycles/h.

Solution

The first step is to compute the bicycle capacity of the approach from Equation 4.58 as follows:

$$c_i = 2000 = 2000 \cdot (50/110) = 909 \text{ bicycles/h}$$

The average control delay is then computed from Equation 4.21 as follows:

$$\text{Average delay} = \frac{1}{2}C\frac{\left[1 - \frac{g}{C}\right]^2}{\left[1 - \left(\frac{g}{C}\right)\left(\frac{v}{c}\right)\right]}$$

$$= \frac{1}{2}(110)\frac{\left[1 - \frac{50}{110}\right]^2}{\left[1 - \left(\frac{50}{110}\right)\left(\frac{120}{909}\right)\right]} = 17.40 \text{ s/bicycle}$$

From Table 4.26 this corresponds to LOS B.

On-Street Bicycle Lanes

Similar to extended pedestrian facilities along urban streets, on-street bicycle lanes experience both uninterrupted and interrupted flow conditions. The LOS for on-street bicycle lanes is therefore based upon the average bicycle travel speed (which takes into account both uninterrupted and interrupted flow conditions). The bicycle lane is divided into segments in the same way as was done with pedestrian facilities along urban streets, and the average speed is computed from the following equation:

$$S_{ats} = \frac{L_T}{\sum \frac{L_i}{S_i} + \sum \frac{d_j}{3600}} \tag{4.59}$$

where

S_{ats} = bicycle travel speed in km/h
L_T = total length of the urban street under analysis in km
L_i = length of segment i in km
S_i = bicycle running speed in km/h (a value of 25 km/h can be assumed)
d_j = average bicycle delay at intersection j computed using Equation 4.21

The LOS can then be determined from Table 4.27.

TABLE 4.27
LOS Criteria for Bicycle Lanes on Urban Streets

Source: Adapted from HCM 2000.

LOS	*Bicycle Travel Speed (mph)*
A	>14
B	>9–14
C	>7–9
D	>5–7
E	≥4–5
F	<4

Note: 1 mph = 1.61 km/h

EXAMPLE 4.24

Determining LOS for an On-Street Bicycle Lane

Determine the LOS of a 2.5-km-long bicycle lane with three signalized intersections and four segments. All three intersections have a cycle length of 90 seconds. The g/C ratios for the intersections are 0.40, 0.30, and 0.50, respectively, and the lengths of the four segments are 0.7, 0.5, 1.0, and 0.3 kilometers. The peak flow rate is 300 bicycles/h.

Solution

The average control delay is first computed for each intersection using Equations 4.58 and 4.21 as follows:

Intersection 1:

$$c_1 = 2000 \cdot (g/C) = 2000 \cdot 0.40 = 800 \text{ bicycles/h}$$

$$\text{Average delay} = \frac{1}{2} C \frac{\left[1 - \frac{g}{C}\right]^2}{\left[1 - \left(\frac{g}{C}\right)\left(\frac{v}{c}\right)\right]}$$

$$= \frac{1}{2}(90) \frac{[1 - 0.40]^2}{\left[1 - (0.40)\left(\frac{300}{800}\right)\right]} = 19.1 \text{ s/bicycle}$$

Intersection 2:

$$c_1 = 2000 \cdot (g/C) = 2000 \cdot 0.30 = 600 \text{ bicycles/h}$$

$$\text{Average delay} = \frac{1}{2} C \frac{\left[1 - \frac{g}{C}\right]^2}{\left[1 - \left(\frac{g}{C}\right)\left(\frac{v}{c}\right)\right]}$$

$$= \frac{1}{2}(90) \frac{[1 - 0.30]^2}{\left[1 - (0.30)\left(\frac{300}{600}\right)\right]} = 27.6 \text{ s/bicycle}$$

Intersection 3:

$$c_1 = 2000 \cdot (g/C) = 2000 \cdot 0.50 = 1000 \text{ bicycles/h}$$

$$\text{Average delay} = \frac{1}{2} C \frac{\left[1 - \frac{g}{C}\right]^2}{\left[1 - \left(\frac{g}{C}\right)\left(\frac{v}{c}\right)\right]}$$

$$= \frac{1}{2}(90) \frac{[1 - 0.50]^2}{\left[1 - (0.50)\left(\frac{300}{1000}\right)\right]} = 12.8 \text{ s/bicycle}$$

The average speed is then computed from Equation 4.59 assuming an average bicycle running speed of 25 km/h as follows:

$$S_{ats} = \frac{L_T}{\sum \frac{L_i}{S_i} + \sum \frac{d_j}{3600}}$$

$$= \frac{2.5}{\frac{0.7 + 0.5 + 1.0 + 0.3}{25} + \left(\frac{19.1 + 27.6 + 12.8}{3600}\right)} = 21.5 \text{ km/h}$$

From Table 4.27, this corresponds to LOS B.

Airport Runway Capacity

Airport systems are made up of several components, including runways, taxiways, aprons, airport passenger terminals, airport passageways, and baggage-handling systems. In any given airport, each of the afore mentioned components would have its own capacity and LOS. In this section, our focus will primarily be on airport runway capacity, which is of critical importance to airport planning and design. From a capacity standpoint, runway capacity can be regarded as the main bottleneck of the air traffic management system and one that typically determines the ultimate capacity of the whole airport. In addition, significantly increasing the capacity of runways at an airport is extremely difficult. The construction of new runways requires a substantial amount of land and could have significant environmental impacts that require lengthy review-and-approval processes that could take years to complete.

This section is divided into three parts. First, we survey the various definitions of runway capacity and illustrate the differences among them. Then we discuss the different factors that impact the capacity of a runway and, finally, we turn our attention to the models and computational procedures that could be used to calculate airport runway capacity.

Measures of Runway Capacity

While there are several measures for expressing the capacity of a runway in common use, they all strive to determine the number of aircraft movements (i.e., arrivals and/or departures) that can be accommodated by a runway system at an airport during a given period of time. Before discussing these different measures, it should be clear to the reader that the capacity of a runway is in fact a probabilistic quantity or a random variable, as discussed in Chapter 2. One should expect the capacity of a runway to vary over time, depending upon wind conditions, visibility, the skill of air traffic controllers, and the mix of aircraft using the runway. Therefore, the number given for the runway capacity of a particular airport should always be regarded as an average number, or, more specifically, the expected number of movements performed during the specified time period. Among the most common measures of runway capacity are (1) maximum throughput capacity, (2) practical hourly capacity (PHCAP), (3) sustained capacity, and (4) declared capacity. Each of these measures is discussed in some detail below.

The maximum throughput capacity (or saturation capacity) is the expected number of aircraft movements that can be performed in an hour on a runway system without violating the Air Traffic Management (ATM) rules and assuming continuous aircraft demand. The definition of the maximum throughput capacity does not make any reference to the associated LOS. The only issue, as far as this definition is concerned, is determining the maximum number of movements that would not violate the ATM rules (such as separation requirements between aircraft, the allocation of movements among runways, etc.), but we are not concerned about the delay associated with accommodating that number of aircraft on the runway.

As opposed to the maximum hourly capacity, the practical hourly capacity, which was originally proposed by the Federal Aviation Administration (FAA) in the 1960s, does concern itself with the resulting LOS. PHCAP is defined as the expected number of movements that can be performed in an hour on a runway system with an average delay/movement of 4 min. A threshold value for the acceptable LOS is specified. A runway would reach its practical capacity once this threshold is exceeded. Typically, PHCAP is equal to 80–90% of the maximum throughput capacity. Currently, runway systems at most airports around the country actually experience delay values in excess of 4 min/movement. This indicates that most airports are currently operating under a LOS that would have been considered unacceptable in the 1960s when the PHCAP was first defined.

The sustained capacity of a runway is another measure that attempts to incorporate LOS in the capacity definition, although the measure is defined in a rather ambiguous way. The sustained capacity is defined as the number of movements/hour that can be reasonably sustained over a period of several hours. This notion of "reasonably sustained" is rather ambiguous. It primarily refers to the workload of the ATM system and air traffic controllers. In general, the sustained capacity is equal to 90% of the maximum throughput capacity for runway configurations with high maximum throughput capacity and can approach 100% of the maximum throughput capacity for configurations with low maximum throughput capacity.

The declared capacity is yet a fourth measure of airport runway capacity that is closely related to the concept of sustained capacity. The declared capacity is defined as the number of aircraft movements/hour that an airport can handle at a reasonable LOS. For airport runways, delay is typically the main performance measure used to describe LOS. Declared capacity is often used at congested airports whereby an airport with congestion problems would "declare" a capacity that sets a limit on the number of movements that would be allowed at that airport. Unfortunately, there is no standard methodology for setting the declared capacity, and therefore setting it is mainly left up to the airport and civil aviation organizations.

Factors Affecting the Capacity of a Runway System

There are a host of factors that impact the capacity of a runway system, including the following:

1. The number and layout of the runways
2. Separation requirements between aircraft imposed by the ATM system
3. Weather conditions, including visibility, cloud ceiling, and precipitation
4. Wind direction and strength
5. Mix of aircraft using the airport (i.e., heavy, large, or small)
6. Mix of movements on the runway (i.e., arrivals, departures, or mixed)
7. Type and location of the taxiway exits
8. The state and performance of the ATM system
9. Noise-related and other environmental considerations

These factors are briefly discussed.

Number and Geometric Layout of the Runways

The number and geometric layout of the runways is perhaps the most important factor affecting the airfield capacity at an airport. For airports, we need to distinguish between the actual number of physical runways at the airport and the number that are active at any given time. An airport, for example, could have five or six runways, but only three or four might be available at any given time. The reason for this difference could be attributed to the fact that some runways might not be useable under certain weather, wind, or environmental conditions. The number of simultaneously active runways is the primary factor in determining airfield capacity. Moreover, the precise geometric layout of the runway also affects airfield capacity since it determines the interdependence among the different runways at an airport.

ATM Separation Requirements

In order to ensure safety of operations, any ATM system must establish a set of required minimum separations between the different types of aircraft. These separation requirements have a direct impact on the number of movements that can be accommodated in a given period of time and hence have an impact on the capacity of the runway. For the purposes of setting these requirements, aircraft are typically categorized into three or four classes according to their size and weight (i.e., Heavy (H), Large (L) and Small (S)). The separation requirements are then specified, either in units of distance or time, for every possible pair of classes and for every possible sequence of movements (by sequence of movements we mean whether we have "arrival followed by arrival (A-A)" or "departure followed by arrival (D-A)", and so on). Table 4.28 lists the separation requirements for a single runway in the United States. There is a separate class for the B-757 because of the strong wake-vortex effects of that type of aircraft.

TABLE 4.28

Separation Requirements for a Single Runway in the U.S.

Source: De Neufville, R. and Odoni, O. R., *Airport Systems: Planning, Design and Management.* McGraw-Hill New York, NY, 2003.

Section I—Arrival Followed by Arrival (A-A)

(a) Throughout final approach, aircraft, at a minimum, must be separated by the distances, in nautical miles, listed in the tables below.

		Trailing Aircraft		
		H	*L + B757*	*S*
Leading Aircraft	**H**	4	5	5/6*
	B757	4	4	5
	L	2.5–3	2.5–3	3/4*
	S	2.5–3	2.5–3	2.5–3

* Distances required at the time when leading aircraft is at the threshold of runway

(b) The trailing aircraft must not touch down before the leading aircraft is totally clear of the runway.

Section II—Arrival Followed by Departure (A-D)

The trailing aircraft is only granted clearance for takeoff after the preceding landing is totally clear of the runway.

Section III—Departure Followed by Departure (D-D)

Aircraft must at a minimum be separated by the amount of time in seconds indicated below.

		Trailing Aircraft		
		H	*L + B757*	*S*
Leading Aircraft	**H**	90	120	120
	B757	90	90	120
	L	60	60	60
	S	45	45	45

Section IV—Departure Followed by Arrival (D-A)

The trailing arriving aircraft must be at least 2 nautical miles away from the runway when the departing aircraft begins its takeoff. The departing aircraft must also be totally clear of the runway before the trailing arriving aircraft can touch down.

Note: 1 nautical mile = 1.85 km

In addition to the longitudinal separation requirements specified above, there are typically additional separation requirements for aircraft landing or departing from a pair of parallel runways, as well as for aircraft operating on intersecting, converging, or diverging runways.

Weather Conditions

Weather conditions could have a significant impact on the capacity of an airport runway. Specifically, cloud ceiling and visibility are the two parameters that determine the weather category in which an airport operates at any given time. For each weather category, different approach, spacing, and sequencing procedures would be used by the ATM. This means that airport capacity is significantly impacted by weather conditions.

Besides ceiling and visibility, precipitation and icing typically adversely impact runway capacity. This is because precipitation and icing often result in poor visibility along with lower surface friction, which leads to poor braking action. Moreover, aircraft would need to be deiced, which requires additional time before take-off. In an extreme case, snowstorms and thunderstorms could result in the temporary closure of the airport.

Wind Direction and Strength

These play a major role in determining the airfield capacity of an airport. As specified by the International Civil Aviation Organization (ICAO), a runway cannot be used if the cross-wind component (which is the component of the surface wind velocity that is perpendicular to the runway centerline) exceeds a certain limit. Given this, the wind direction and strength actually determine which runways can be active at any given time. For airports experiencing strong winds from different directions, therefore, there could be a wide variation in the range of available capacity of the runway system.

Mix of Aircraft

The type of aircraft using a runway can also have a significant impact on the runway capacity. This is because the ATM separation requirements depend upon the type of aircraft (i.e., Heavy, Large, or Small). From Table 4.28, it can be seen easily that a runway that mainly serves large aircraft would have a greater capacity than one where there is a large percentage of heavy aircraft followed by small ones. This is because the separation distance for this second case is about 6 nautical miles (one nautical mile is equal to 1.85 kilometers), which is significantly larger than the separation distance required to separate large aircraft from one another (i.e., 2.5 nautical miles). Generally speaking, a homogeneous mix of aircraft is preferable over a non-homogeneous mix from a capacity standpoint as well as from an air traffic management standpoint.

Mix and Sequencing of Movements

In addition to the type of aircraft, another factor that impacts runway capacity is the mix of the aircraft movements themselves (i.e., arrivals versus departures). Generally speaking, a runway can handle more departures/h than arrivals, assuming that they both have the same mix of aircraft. Air traffic controllers typically prefer to use separate runways for arrivals and for departures, if given the opportunity. This is common at airports with two parallel runways. However, it is not optimal from a capacity standpoint, since runways might be underutilized when the numbers of arrivals and departures are not balanced.

The sequencing of movements on a runway also has a significant impact on runway capacity, particularly when a runway is used for both arrivals and departures. While aircraft are often served on a first-come-first-served basis, air traffic controllers have the flexibility to change that sequence so as to optimize operations.

Type and Location of Runway Exit

The runway occupancy time (O) is the time between the moment an aircraft touches down on the runway and the moment all its parts are clear of the runway. The location of the runway exit plays a major role in determining this occupancy time, and therefore also impacts the runway capacity, since, as was shown in Table 4.28, for certain combinations of aircraft movements (e.g., A-D), clearance for take-off is only granted after the preceding landing aircraft is clear of the runway.

State and Performance of the ATM System

The quality of the ATM system and the skill level of the air traffic controllers also play a major role in determining runway capacity. Tight separations between aircraft, for example, are only possible if the ATM system is highly accurate in displaying information about the positions of the aircraft and if the air traffic controllers are highly skilled in spacing aircraft accurately. Also, if air traffic controllers sense that a certain pilot is inexperienced or is having difficulty understanding air traffic instructions, they would slow down operations to allow for additional margins of safety, which would also impact capacity.

Noise Considerations

Finally, environmental considerations, and in particular noise impacts, can affect runway capacity. For example, if more than one runway can be used at a given time from weather and wind direction standpoints, noise considerations would be among the other factors to be considered in the final decision as to which runway would be used and therefore act as an additional constraint on airport capacity.

Models for Calculating Runway Capacity

A number of mathematical and simulation models have been developed over the years to allow for estimating the capacity of runways under a given set of specified conditions. In this section, we describe one example of these models. Although this is a simple model, experience has shown that it gives reasonably accurate approximations to the capacities actually observed in the real world.

The model described, which was originally proposed by Blumstein in 1959, is based upon a time–space diagram similar to that described in Chapter 2. The model can be used to estimate the capacity of a runway that is to be used only for arrivals. However, the same principles can be applied to runways used for departures and for those used for mixed operations. Consider the single runway shown in Figure 4.19.

Aircraft descend in a single file along the final approach until they hit the runway, as shown in Figure 4.19 (typically, the length of the final approach (r) ranges between 5 to 8 nautical miles, measured from the runway threshold). During their descent, aircraft maintain the longitudinal separation requirements specified by the ATM system. In addition, each aircraft must safely clear the runway before the next landing is allowed. These two rules determine the maximum throughput capacity of the runway, as shown below.
First, let us define the following terms for an aircraft type i:

r = length of the final approach path
v_i = speed of aircraft i on final approach, which is assumed to remain constant throughout the approach
o_i = runway occupancy time, which defines the time from the moment an aircraft touches down to the moment it clears the runway

Now let us consider an aircraft of type i that is landing, followed by another aircraft of type j, and let us denote the minimum separation distance required

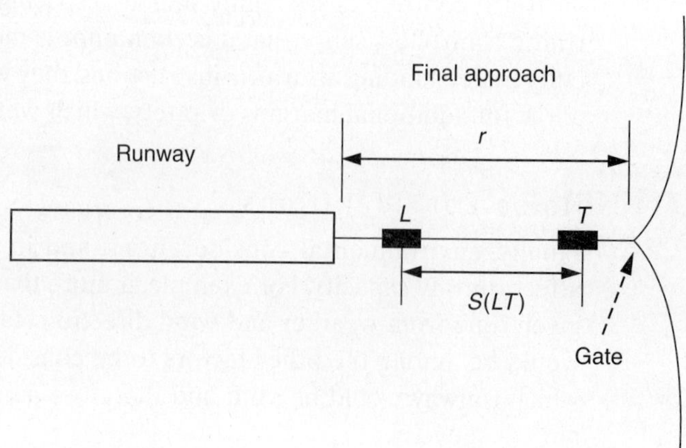

FIGURE 4.19
A single runway used only for arrivals.

between the two aircraft by s_{ij}. Let us also denote the minimum time headway or interval between the successive arrivals of aircraft i and j at the runway by T_{ij}. The value of T_{ij} is governed by the following two equations:

$$T_{ij} = \max\left[\frac{r + s_{ij}}{v_j} - \frac{r}{v_i}, o_i\right] \quad \text{when } v_i > v_j \tag{4.60}$$

or

$$T_{ij} = \max\left[\frac{s_{ij}}{v_j}, o_i\right] \quad \text{when } v_i < v_j \tag{4.61}$$

The rationale behind these two equations is explained as follows. For the case when the speed of the leading aircraft is larger than that of the trailing one (Equation 4.60), the critical case (i.e., the case when the two aircraft are closest) occurs when the leading aircraft is at the gate of the approach (see Figure 4.19) at a distance r from the runway threshold. This is because the distance between the two aircraft would keep increasing as they proceed along the final approach, since the leading aircraft is moving faster than the trailing aircraft (that is the reason behind referring to this case as the "opening case"). If, at that moment, the two aircraft are separated by the required distance, s_{ij}, then aircraft j would at that moment be at a distance $(r + s_{ij})$ from the runway threshold. Therefore, the difference between the times when the leading aircraft i and the trailing aircraft j would touch down on the runway in this case would be equal to $\frac{r + s_{ij}}{v_j} - \frac{r}{v_i}$. However, the interval between successive arrivals must also be at least equal to the occupancy time, o_i, which explains Equation 4.60.

For the case of Equation 4.61 (the "closing case"), when the speed of the leading aircraft i is less than that of the trailing aircraft j, the critical case when the two aircraft are closest is when aircraft i has just landed. In that case, aircraft j would be at distance s_{ij} from the runway threshold, which explains Equation 4.61.

If we denote the probability of the event of "a type i aircraft followed by a type j" by p_{ij}, then the expected value of T_{ij} can be expressed as

$$E[T_{ij}] = \sum_{i=1}^{K}\sum_{j=1}^{K} p_{ij} T_{ij} \tag{4.62}$$

where

$E[T_{ij}]$ = expected value of T_{ij}
K = number of distinct aircraft classes

With the expected value of T_{ij} determined, the maximum throughput capacity is determined since the maximum capacity is equal to the reciprocal of the minimum separation time headway, T_{ij}.

EXAMPLE 4.25

Determining the Capacity of a Runway

For the purposes of setting minimum longitudinal separation requirements, aircraft were classified into the following four groups: (1) Heavy (H); (2) Large (L); (3) Small1 (S1); and (4) Small2 (S2). A runway at a given airport is used for extended periods of time for arrivals only. The runway serves an aircraft population with the characteristics shown in Table 4.29. The longitudinal separation requirements are shown in Table 4.30. If the length of the final approach path, r, can be assumed to be equal to 5 nautical miles, determine the maximum throughput capacity for the runway.

Solution

Step 1: Calculate the minimum separation time, T_{ij}, between each pair of aircraft types. The first step in the solution procedure is to calculate the minimum separation time, T_{ij}, between each pair of aircraft types using Equations 4.60 and 4.61. The results can best be presented in the form of a 4 × 4 matrix, where each cell would give the separation time between the aircraft types specified in the matrix row and column that intersect at that particular cell. So, for example, we start by calculating the minimum separation time between an aircraft of Type (H) followed by another of Type (H). In this case, both the leading and lagging aircraft have the same velocity, and therefore both

TABLE 4.29 Aircraft Population Characteristics

Aircraft Type (i)	*% of Total Population*	*Velocity (v_i) (nautical miles/h or knots)*	*Occupancy Time (o_i) (seconds)*
1 (H)	20%	160	80
2 (L)	30%	140	60
3 (S1)	30%	120	50
4 (S2)	20%	100	40

TABLE 4.30 Longitudinal Separation Requirements

	Trailing Aircraft		
Leading Aircraft	*H*	*L*	*S1 or S2*
H	4	5	6*
L	3	3	4*
S1 or S2	3	3	3

* Indicates that the separation applies when the leading aircraft is at the runway threshold

TABLE 4.31
The Matrix of Minimum Separation Time, T_{ij}

Leading Aircraft	*Trailing Aircraft*			
	H	*L*	*S1*	*S2*
H	90	147	180	216
L	68	77	120	144
S1	68	77	90	138
S2	68	77	90	108

Equations 4.60 and 4.61 would give the same answer. Since it is simpler, we use Equation 4.61. The separation distance for an H-H combination, as shown in Table 4.30, is 4 nautical mile. Therefore, the calculations are as follows:

$$T_{11} = \max\left[\frac{s_{11}}{v_1}, o_1\right] = \max\left[\left(\frac{4}{160}\right) \times 3600, 80\right]$$

$$= \max\,[90, 80] = 90 \text{ s}$$

Note the ratio of $\frac{s_{11}}{v_1}$ has been multiplied by 3600 to convert it from hours to seconds. This value, T_{11}, is then recorded in the first cell in the 4×4 matrix shown in Table 4.31.

We then move to the case where a heavy aircraft is followed by a large aircraft (H-L). For this case, the required separation distance is 5 nautical miles, and Equation 4.60 must be used, since the velocity of the leading aircraft (v_1) is larger than the trailing aircraft (v_2). The calculations proceed as follows:

$$T_{12} = \max\left[\frac{r + s_{12}}{v_2} - \frac{r}{v_1}, o_1\right] = \max\left[\left(\frac{5.0 + 5.0}{140} - \frac{5.0}{160}\right) \times 3600, 80\right]$$

$$= \max\,[147, 80] = 147 \text{ s}$$

Calculations then proceed in the same fashion to fill in the other cells of the matrix shown in Table 4.31. In doing the calculations, we have used Equation 4.60 only for calculating T_{12} and T_{34}. For the diagonal elements, Equation 4.61 was used since it was simpler. Equation 4.61 was also used for computing T_{13}, T_{14}, T_{23}, and T_{24}, since, according to Table 4.31, the separation requirements for these cases apply when the leading aircraft is at the threshold of the runway (i.e., the case of Equation 4.61).

Step 2: Calculate the probabilities of the different *i-j* aircraft type combinations. Since most air traffic controllers service incoming aircraft on a first-come-first-served basis, it can be assumed that the probability of having an aircraft of type *i* as the leading aircraft is simply equal to the percentage of aircraft of type *i* in the traffic mix, and that the

TABLE 4.32
Matrix of Aircraft Pairs Probabilities

Leading Aircraft	*Trailing Aircraft* H	*L*	*S1*	*S2*
H	0.04	0.06	0.06	0.04
L	0.06	0.09	0.09	0.06
S1	0.06	0.09	0.09	0.06
S2	0.04	0.06	0.06	0.04

probability of having a trailing aircraft of type j is similarly equal to the percentage of aircraft of type j in the aircraft mix. Therefore, the probability of an aircraft of type i followed by an aircraft of type j is given by $p_{ij} = p_i \cdot p_j$. This simple equation can therefore be used to develop another 4 × 4 matrix that gives the probability of having each aircraft-pair combinations. The developed matrix is shown in Table 4.32.

Finally, the expected value $E[T_{ij}]$ can be calculated by finding the sum product of the corresponding elements from the two matrices shown in Tables 4.31 and 4.32. This gives a value of $E[T_{ij}] = 106.67$ s. The maximum throughput capacity can then be calculated as follows: 3600/106.67 = 33.72 or approximately 34 aircraft/h.

SUMMARY

In this chapter, the basic concepts of capacity and level of service analyses were introduced. Procedures for conducting such analyses for several transportation facilities were reviewed. This included the capacity analysis procedures for (1) highways, (2) transit, (3) bike paths, (4) pedestrian facilities, and (5) airport runways. Several examples were provided to aid in understanding how these procedures might be applied. Capacity analysis represents a crucial step in almost all transportation analysis, planning, and design exercises. The following chapters will discuss transportation infrastructure planning and design in more detail.

PROBLEMS

4.1 Explain the implication of the use of the word *reasonably* in the *Highway Capacity Manual* (HCM) definition of facility capacity.

4.2 List the performance measures used to define the level of service for the following types of transportation facilities and modes: (1) freeway section; (2) signalized intersections; (3) transit; (4) bike paths; (5) sidewalks; and (6) airport runways.

4.3 Explain the difference between the capacity of a facility and its service flow rate. When is the capacity equal to the service flow rate?

4.4 Give examples for both *uninterrupted* and *interrupted* flow facilities.

4.5 Distinguish between *macroscopic* and *microscopic* parameters of traffic streams. List the three most important *macroscopic* parameters, and the two most important *microscopic* parameters. How are *macroscopic* traffic flow parameters related to *microscopic* parameters?

4.6 Data obtained from an aerial photograph show 12 vehicles on a 275-m-long section of road. For that same section, an observer counts a total of 7 vehicles during a 15-s interval. Determine (a) the density on the road; (b) the flow; and (c) the space mean speed.

4.7 A given traffic stream has an average time headway of 2.7 s and an average spacing of 52 m. Determine the space mean speed for the traffic stream.

4.8 The relationship between the space mean speed, u, and the density, k, on a given transportation facility can be described as $u = 100 - 0.85k$. Determine the free-flow speed and the jam density for the facility.

4.9 For Problem 4.8, develop a relationship between the flow, q, and the density, k. Also determine the maximum flow or capacity for the facility.

4.10 A section of freeway has a speed–flow relationship of the form $q = au^2 + bu$. The section has a maximum flow value or capacity equal to 2000 veh/h, which occurs when the space–mean speed of traffic is 52 km/h. Determine (1) the free-flow speed; (2) the jam density; and (3) the speed when the flow is equal to 900 veh/h.

4.11 A certain road carries an average volume of 1600 veh/h. The closure of some lanes on the road results in reducing its normal capacity down to only 1200 veh/h within the work zone. Observations indicate that traffic flow along the road can be described by a Greenshields model having a free-flow speed of 80 km/h and a jam density of 100 veh/km. Determine the percent reduction in the space mean speed at the vicinity work zone problem.

4.12 The following table shows 5-min vehicle counts that have been recorded for a given transportation facility during the A.M. peak hour:

Time	*Count*
8:00–8:05 a.m.	212
8:05–8:10 a.m.	208
8:10–8:15 a.m.	223
8:15–8:20 a.m.	232
8:20–8:25 a.m.	241
8:25–8:30 a.m.	220
8:30–8:35 a.m.	205
8:35–8:40 a.m.	201
8:40–8:45 a.m.	185
8:45–8:50 p.m.	230
8:50–8:55 p.m.	197
8:55–9:00 p.m.	185

Determine

(a) The maximum flow rate that accounts for the peak 5-min interval within the hour;
(b) The maximum flow rate that accounts for the peak 15-min interval; and
(c) The peak hour factor (PHF), based on peak 15-min counts.

4.13 With reference to signalized intersections, briefly define the following terms: (1) signal cycle length; (2) signal phase; (3) signal interval; and (4) signal offset.

4.14 Determine the capacity for a two-lane approach to a signalized intersection that gets 45 s of green out of a cycle length of 100 s. Studies show that the saturation discharge headway at the intersection is equal to 2.1 s. Assume that the start-up lost time is equal to 2 s, the clearance lost time is equal to 1.2 s, and that the duration of the yellow interval is equal to 3.5 s.

4.15 Determine the maximum sum of critical volumes that a signalized intersection, with four phases and a cycle length of 120 s, can accommodate. Assume that the lost time per phase is equal to 3.5 s, and that the saturation discharge headway is equal to 1.9 s.

4.16 Determine the minimum cycle length for the intersection shown below. The intersection is designed to have the following three phases:

Phase A for the left-turn movements from the EB and WB approaches;
Phase B for the through and right movements from the EB and WB approaches; and
Phase C serving all movements from the NB and SB approaches.

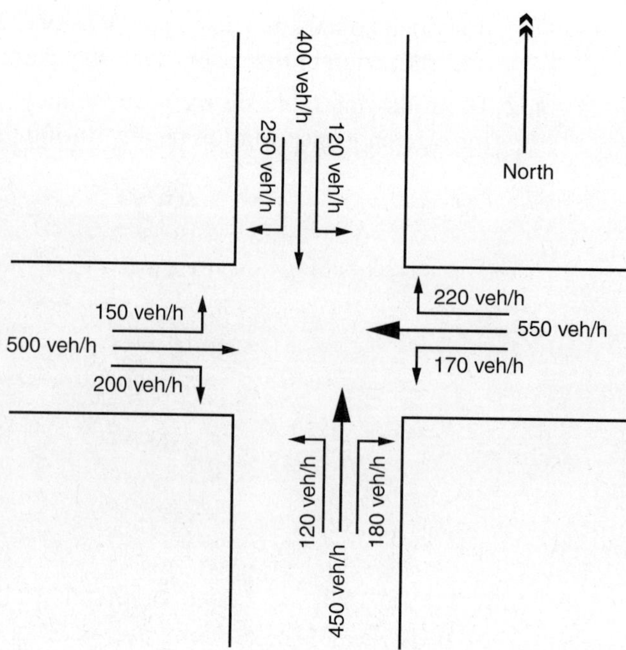

Assume that each intersection approach has three lanes, one for the left-turning vehicles, one for the through vehicles, and one for the right-turning vehicles. The saturation discharge headway is equal to 2.10 seconds.

4.17 For Problem 4.16, calculate the desirable cycle length if it is desired that the volume:capacity ratio does not exceed 0.85. Assume a PHF of 0.92.

4.18 Determine the level of service for an intersection approach that receives 55 seconds of green out of a cycle length of 120 seconds. The approach carries an hourly volume of 720 veh/h and has a saturation flow rate of 1850 veh/h.

4.19 The intersection shown below has two phases and a cycle length of 70 s. Given the critical volumes shown, determine the number of lanes required for each critical movement. Assume that the saturation discharge headway is 2.1 s and that the lost time per phase is equal to 4 s.

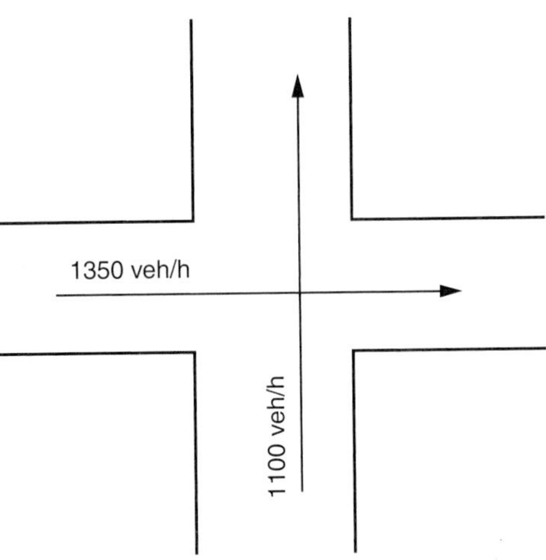

4.20 The intersection shown below has three phases, as follows:

Phase A serves east–west left turns only;
Phase B serves east–west through and right traffic; and
Phase C serves north–south left, through, and right movements.

The hourly flows at the intersection are as shown below:

Eastbound			*Westbound*			*Northbound*			*Southbound*		
L	*T*	*R*	*L*	*T*	*R*	*L*	*T*	*R*	*L*	*T*	*R*
300	900	200	250	1000	150	90	340	50	70	310	60

Assume that

(1) Lost time per phase is equal to 3.5 s/phase
(2) Saturation flow rates are as follows:

$s(TH + \text{RT}) = 1800$ veh/h/lane
$s(TH) = 1900$ veh/h/lane
$s(LT) = 1700$ veh/h/lane

Determine the optimal cycle length, C_o, using the Webster model, as well as the effective green time for phase A.

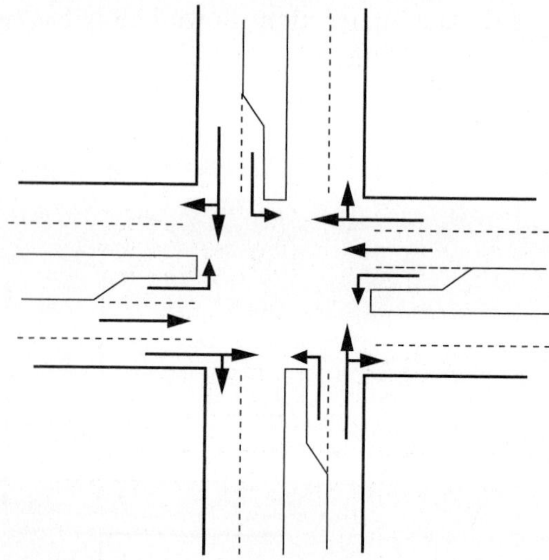

4.21 The intersection shown below has the following phasing scheme:

Phase A is for the EB and WB left-turning movements;
Phase B is for the EB and WB through and right-turning movements;
Phase C is for the NB and SB left, through, and right movements.
The equivalent hourly flows at the intersection are as shown below:

Eastbound			*Westbound*			*Northbound*			*Southbound*		
L	*T*	*R*	*L*	*T*	*R*	*L*	*T*	*R*	*L*	*T*	*R*
280	850	80	320	700	120	50	280	40	35	360	10

Using the Webster model, determine the optimal cycle length for the intersection. Assume lost times equal to 3.5 s/phase, a yellow interval equal to 3 s, and a saturation flow rate of 1800 pc/h/lane for all lane types.

4.22 Discuss the difference between passenger and vehicle capacity.

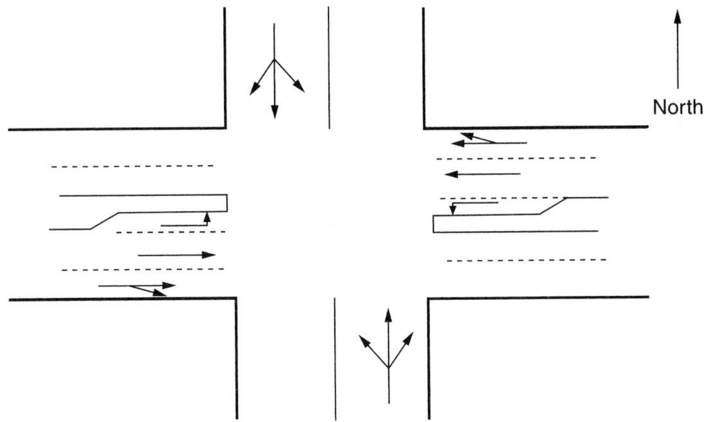

4.23 Identify the locations for which transit capacity is defined.

4.24 Briefly discuss the three factors affecting the capacity of loading areas.

4.25 What are some of the factors affecting dwell time of transit vehicles at transit stops?

4.26 In the context of bus loading areas, what is the clearance time equal to for on-line and off-line facilities?

4.27 A transit system can be regarded as consisting of the following three basic elements: (1) transit stops; (2) route segments; and (3) systems. Briefly explain how the quality of service measures vary depending upon which element is being evaluated.

4.28 A bus route using 42-seat buses has a total of 8 stops. The door opening and closing times can be assumed to be equal to 5 s. All passengers are required to board the bus through the front door and alight through the backdoor. Ridership for the route is given below:

Stop Number	*1*	*2*	*3*	*4*	*5*	*6*	*7*	*8*
Alighting passengers	0	10	14	25	10	15	5	40
Boarding passengers	35	12	18	20	14	20	15	30

Studies have shown that the boarding time is 3 s/passenger when no standees are present, and that the presence of standees increases the boarding time to 3.5 s/passenger. Alighting time is estimated to be equal to 2 s/passenger. Determine the dwell time at the critical bus stop.

4.29 A bus route using 35-seat buses has a total of 7 stops. The door opening and closing times can be assumed to be equal to 4 s. The bus has only one door, which is used for both boarding and alighting. The boarding time can be assumed to be equal to 3 s/passenger, whereas the alighting time is equal to 2 seconds. The presence of standees increases both the boarding as well as the alighting time by 0.5 seconds. Determine the dwell time at the critical stop if the ridership is as follows:

Stop Number	*1*	*2*	*3*	*4*	*5*	*6*	*7*
Alighting passengers	0	20	5	12	20	8	23
Boarding passengers	28	14	21	24	17	14	9

4.30 For the critical bus stop of Problem 4.28, determine the loading area capacity given that

The bus stop is an off-line stop;
The traffic signal where the bus stop is located has a cycle length of 100 s and the bus approach gets 55 s of green; and
The probability of a queue forming behind the bus stop is limited to not more than 5%.

4.31 A transit route with vehicles operating in mixed traffic has a total of 10 bus stops. Based upon ridership observations, it was determined that bus stop 7 is the critical bus stop. This is an on-line stop, located at the far side of a signalized intersection. The following information is provided:

Dwell time at bus stop 7 = 35 seconds;
Curb lane volume = 600 pc/h;
Capacity of right curb lane = 800 pc/h;
The signal has a cycle length of 80 secs, and the bus approach gets the green for 50 s;
Bus stop 7 has 2 loading areas.
Determine the bus lane capacity, given that it is desired to limit the probability of a queue forming behind the bus to less than 10%.

4.32 Redo Problem 4.31 for a nearside stop.

4.33 For Problem 4.31, in an attempt to increase the bus lane capacity, the transit agency is considering implementing skip-stop operations. Observations show that the adjacent lane carries a volume equal to 750 pc/h and has a capacity of 1200 pc/h. Assuming that transit vehicles arrive in a random fashion at the bus stop, determine the increase in the bus lane capacity.

4.34 For Problem 4.33, determine the person capacity of the transit route given that all buses have a capacity of 43 passengers. The transit agency's policy allows for up to 50% standees on all buses except for a total of 5 express buses on that route where standees are not allowed. Assume a PHF equal to 0.75.

4.35 A light rail has two types of right-of-way. The first right-of-way is a single track rail line, which is 0.65 km long, with two intermediate stops. The second right-of-way lies within the median of an arterial street with a speed of 65 km/h. The green cycle length ratio at the critical intersection along the arterial street is 0.45, and the maximum cycle length of the signals is 100 s. The dwell time can be assumed to be equal to 30 seconds for all stations. The train is 27 m long and has an initial service acceleration of 1 m/s^2. The city blocks are 135 m long. Determine both the vehicle capacity as well as the person capacity of the light rail system.

4.36 Determine the *person* capacity of a light rail system that operates within the median of an arterial street with a speed of 55 km/h. The system uses trains that are 35 m long with an initial service acceleration of 1 m/s^2. The green to cycle length of the traffic signal at the critical stop is 0.55, and the maximum cycle length is 110 seconds. The city blocks are only 60 m long. Assume the dwell time at the light rail stations is equal to 35 seconds.

4.37 A heavy rail transit line with a cab signaling system employs trains that have a maximum speed of 105 km/h. The longest train using the line is about 180 m, and the distance from the front of a stopped train to the station exit block is 11 m. Assuming that there are no speed restrictions limiting approach speeds to suboptimal levels, determine the vehicle and person capacity of the transit line.

4.38 Redo Problem 4.37 for a moving block signal system.

4.39 A 3 m sidewalk segment with store window displays on one side carries a 15-min peak flow of 1100 p/15 min. The effective width of the sidewalk after deducting the preempted width for the stores' window displays is 8 ft. Determine the LOS during the peak 15 min for both average conditions as well as for conditions within platoons.

4.40 A 2.7 m sidewalk carries a 15-min peak flow of 1400 p/15 min. Determine the LOS during the peak 15 min for average conditions.

4.41 A shared two-way pedestrian–bicycle facility has a width of 3 m and carries a peak pedestrian flow of 200 p/15 min. The bicycle flow rate is 120 bicycles/h in the same direction as the pedestrians, and 170 bicycles/h in the opposite direction. The average pedestrians' speed is 1.2 m/s, and that for bicycles is 1.8 m/s. Determine the pedestrians' LOS.

4.42 For Problem 4.41, how would the pedestrians' LOS change if the facility is converted into an exclusive pedestrian facility with an effective width of 1.8 m?

4.43 A shared two-way pedestrian-bicycle facility has a width of 2.4 m and carries a peak pedestrian flow of 140 p/15 min. The bicycle flow rate is 160 bicycles/h in the same direction as the pedestrians and 130 bicycles/h in the opposite direction. Assuming a pedestrian average speed of 1.2 m/s and a bicycle speed of 4.5 m/s, determine the pedestrians' LOS.

4.44 Determine the LOS for pedestrians at a two-phase intersection with a cycle length of 120 seconds. The phase serving the major street vehicular movement gets 70 seconds of green and the phase serving the minor movement gets 40 seconds.

4.45 Determine the LOS for a 2.5 km pedestrian sidewalk on an urban street with four signalized intersections. The sidewalk is segmented into four segments having the following lengths: 730 m, 850 m, 425 m, and 495 m, respectively. All the signals have a cycle length of 80 seconds, and the length of the pedestrian green phase is equal to 35 s. Assume a pedestrian speed of 1.2 m/s.

4.46 Determine the LOS for an east–west exclusive bicycle lane that carries a volume of 200 bicycles/h during the peak period. Observations of the bike lane indicate that 70% of the bicycles move in the EB direction during the peak period. The path is 3 m wide and can therefore be assumed to have three effective lanes.

4.47 Determine the LOS for a north–south exclusive bicycle lane that carries a volume of 250 bicycles/h, with around 65% of the bicycles moving in the SB direction. The path is 2.4 m wide and can therefore be assumed to have two effective lanes.

4.48 An 2.4-m-wide, shared bicycle–pedestrian facility runs in the east–west direction. The peak flow rate for bicycles is 150 bicycles/h, with a directional split in the EB/WB direction of 65/35. The facility also carries a peak pedestrian flow rate of 80 p/h, with a directional split of 55/45 in the EB/WB directions. Determine the LOS for the bicycles on the facility for both the EB and WB directions.

4.49 A shared bicycle–pedestrian facility runs in the north–south direction. The facility is 3 m wide and carries a peak bicycle flow rate of 200 bicycles/h, with a 60/40 directional split in the NB/SB direction. For pedestrians, the peak flow rate is 100 p/h with a 45/55 split in the NB/SB direction. Determine the LOS for bicycles on the facility for both the NB and SB directions.

4.50 Determine the LOS for a 1.2-m-wide bicycle lane at a signalized intersection with a cycle length of 100 s. The bicycle lane gets 45 s of green and carries a peak flow rate of 180 bicycles/h.

4.51 Determine the LOS for a 4.2-km-long bicycle lane with four signalized intersections and five segments, with a peak flow rate of 350 bicycles/h. All intersections have a cycle length of 80 s. The green to cycle length ratios for the bicycle lane direction at the four intersections are 0.47, 0.38, 0.50, and 0.35, and the lengths of the five segments are 0.65, 0.70, 0.95, 1.1, and 0.80 km.

4.52 Determine the LOS for a 3-km-long bicycle lane with three signalized intersection and four segments, with a peak flow rate of 400 bicycles/h. The four segments have a length of 1 km, 0.7 km, 0.5 km, and 0.8 km, respectively. The three intersections have a common cycle length of 90 seconds, and the bicycle lane direction at the three intersections receives the following green times: 40 s, 35 s, and 50 s.

4.53 Discuss the differences among the following measures of airport runway capacity: (1) maximum throughput capacity; (2) practical hourly capacity; (3) sustained capacity; and (4) declared capacity.

4.54 Briefly discuss the different factors affecting the capacity of a runway.

4.55 An airport runway, used for arrivals only, serves an aircraft population with the following characteristics. If the length of the final approach path, r, is 4.5 nautical miles, determine the maximum throughput capacity for the runway.

Aircraft Population Characteristics

Aircraft Type (i)	*% of Total Population*	*Velocity (v_i) (nautical miles/h)*	*Occupancy time (o_i) (s)*
1 (Heavy)	25%	170	90
2 (Large)	40%	150	65
3 (Small)	35%	110	45

Longitudinal Separation Requirements

	Trailing Aircraft		
Leading Aircraft	*H*	*L*	*S*
H	4	6	7
L	3.5	4.5	5
S	3	3	3

4.56 An airport runway, used for arrivals only, serves an aircraft population with the following characteristics. If the length of the final approach path, r, is 6 nautical miles, determine the maximum throughput capacity for the runway.

Aircraft Population Characteristics

Aircraft Type (i)	*% of Total Population*	*Velocity (v_i) (nautical miles/h)*	*Occupancy time (o_i) (s)*
1 (Heavy)	15%	170	90
2 (Large)	30%	150	65
3 (Small1)	35%	110	55
4 (Small2)	20%	90	45

Longitudinal Separation Requirements

	Trailing Aircraft			
Leading Aircraft	*H*	*L*	*S1*	*S2*
H	4	6	6.5	7
L	3.5	4.5	5	5.5
S1	3.5	3.5	3.5	3.5
S2	3	3	3	3

References

1. Blumstein, A., "The Landing Capacity of a Runway." *Operations Research*, *7*, pp. 752–763, 1959.
2. Botma, H., "Method to Determine Levels of Service for Bicycle Paths and Pedestrian-Bicycle Paths." In *Transportation Research Record 1502*, TRB, National Research Council, Washington, D.C., pp. 38–44, 1995.
3. De Neufville, R., and Odoni, A. R., *Airport Systems: Planning, Design and Management*, McGraw-Hill, New York, 2003.
4. Garber, N. J., and Hoel, L. A., *Traffic and Highway Engineering.* 3rd ed. Brooks/Cole, Pacific Grove, CA, 2002.
5. *Guide for Development of Bicycle Facilities,* American Association of State Highway and Transportation Officials, Washington, D.C., 1999.
6. *Highway Capacity Manual,* Special Report 209, Fourth Edition, Transportation Research Board, National Research Council, Washington, D.C., 2000.
7. Levinson, H. S., and St. Jacques, K. R., "Bus Capacity Revisited." In *Transportation Research Record 1618*, TRB, National Research Council, Washington, D.C., pp. 189–199, 1998.
8. Roess, R. P., Prassas, E. S., and McShane, W. R., *Traffic Engineering.* 3rd ed. Pearson Education, Upper Saddle River, NJ, 2004.
9. *Transit Capacity and Level of Service Manual*, TCRP Report 100, Second Edition, Transportation Research Board, National Research Council, Washington, D.C., 2003.
10. Webster, F., "Traffic Signal Settings." *Road Research Paper No. 39,* Road Research Laboratory, Her Majesty's Stationery Office, London, UK, 1958.
11. Greenshields, B. D., "A Study in Highway Capacity." Highway Research Board Proceedings, Vol. 14, 1935.

CHAPTER 5

Transportation Planning and Evaluation

This chapter discusses the process transportation planners use to improve the transportation infrastructure of a state, region, or city. Transportation planning and programming involve anticipating, budgeting, and scheduling the acquisition and installation of infrastructure assets such as vehicles, networks, terminals, and control systems. The role of the transportation planning process is to forecast travel demand and to evaluate alternative systems, technologies, and services. Each transportation mode has its own unique characteristics regarding how these assets interact, and thus the planning process will reflect the mode considered and its unique abilities to meet future needs.

Since planning is about a vision of the future, a fundamental element of the planning process is the forecasting of travel demand. Knowledge of the number of passengers or vehicles that can be expected to use an airport, railroad terminal, parking garage, waterway, or highway helps to determine the type and size of the facility that will be required.

In most cases, multiple options exist to meet the travel demand, some more costly than others. Thus the selection of a preferred option from among the list of potential alternatives is another necessary planning task. When the transportation infrastructure plan has been completed, key questions must be addressed: How much will the plan cost, where will the money come from, and in what priority should individual projects be budgeted?

Local, state, and federal laws also govern transportation infrastructure planning. At the local level, zoning ordinances specify the type of development that is allowed and the standards for residential streets. At the state level, there are laws that establish how transportation funds are allocated and suggest design criteria for bridges, bike and pedestrian paths, railroad crossings, and airports. At the federal level, funding for transportation is primarily by mode.

There are laws that require citizen participation in the planning process, stipulate environmental requirements for air and water pollution, regulate the provision of public transit for people with disabilities, and influence land development through access controls.

There are both common elements and unique differences for each transportation technology, including physical, operational, ownership, and legislative histories. The unique characteristics of each mode must be known and understood if they are to be incorporated into the planning process. Key characteristics that are important to the transportation infrastructure planning process for highways, rail, and air transportation are described in the following section.

A Context for Multimodal Transportation Planning

Highway transportation networks consist of streets and arterial roads that are provided by the government. Vehicles are self-propelled and under the ownership and control of a driver, who makes voluntary decisions based on previous experience, current information, and visual cues provided along the roadside and at intersections. The terminal facility contains parking spaces that may be located in a multistory garage, on a street, in a residence or workplace.

The interstate highway system was a bold and far-reaching undertaking and a unique example of a transportation facility plan at the national level. This system of approximately 75,000 km of limited access highway was paid for primarily by a tax on gasoline and was funded 90% by the federal government and 10% by the state (or less if there were substantial amounts of federally owned land).

Uniform standards were adopted for geometric features such as lane widths, shoulders, and medians as well as traffic signs and markings. Research, planning, and construction became a joint effort between the states and the federal government. Later, legislation was established that required a continuing, comprehensive, and cooperative planning process. However, in the view of some critics, planning for the interstate system failed to recognize or consider the impact that this vast highway network would have on other modes, such as public transit. Further, little attention was given to the effects on land use and the environment.

Railroads are rail networks that operate on a fixed track with spacing that matches the dimension between the rail wheels. Control is in the direction of the fixed track and may be visual or by electronic command. Terminals provide for loading and unloading of freight and passengers and other customer services. A terminal can be as simple as an open platform or as complex as a switching yard or multipurpose station. Privately owned railroads ship freight, whereas passenger rail is operated and subsidized by the government through Amtrak. Light rail (or trolleys) and rail rapid transit are rail passenger systems in cities.

In past years, Congress has attempted to influence regional transportation planning through regulating routes and rate structures and, by so doing, assuring competition among modes. Transportation regulation had its beginnings in the nineteenth century, when railroads had a virtual monopoly on the interstate transport of freight and passengers. The railroad's owners exploited the situation and would often vary freight rates and service to their advantage. For example, farmers who were dependent on railroads to transport their crops to market in a timely fashion were captives of the railroads and were discriminated by this injustice. They protested to state legislators and to the federal government for relief from this monopolistic burden. The Interstate Commerce Commission (ICC) was created in 1887 to regulate the railroads.

As new modes emerged, ICC authority was expanded to regulate trucking and water transportation, and the Civil Aviation Board (CAB) was created to regulate air transportation. The purpose of regulation was to preserve the inherent advantages of each mode and promote safe, economic, and efficient service. The intent of Congress, in using regulatory powers to control the transportation marketplace, was to develop, coordinate, and preserve a national transportation system. The result was not as desired because regulatory agencies were unable to implement vague and often contradictory policy guidelines, which in many instances required interpretation and adjudication by the courts.

In the last decades of the twentieth century, regulatory reform occurred and transportation carriers were given the opportunity to develop new and innovative ways of providing services that would utilize the best attributes of each mode. The result has led to positive efforts toward a multimodal integrated transportation system. Changes in the regulatory environment have created a dynamic situation that has impacted transportation planning in both the private and public sector.

Air transport is not confined to a fixed network, and when airplanes are moving in the sky, they can travel in any desired direction. Control systems provided by the government are essential to ensure safety. Pilots are usually required to file a specified flight plan prior to take-off. Air terminal facilities and runways may be publicly or privately owned and operated. Airports are planned and designed for specific aircraft configurations in order to accommodate landing, take-off, and parking requirements and to provide ticketing, security, concessions, and baggage-handling facilities for passengers.

Airport planning at the national level is the responsibility of the Federal Aviation Administration. The FAA prepares a plan called the National Plan of Integrated Airport Systems (NPIAS). This plan is actually a compilation of data furnished on a state-by-state basis of proposed airport improvements expected to occur within a 10-year period. The plan provides data in four airport categories—commercial service, primary, general aviation, and reliever. The categories are based on the type of service provided and annual aircraft arrival and departure volumes. Elements in the NPIAS include airport activity level expected in the next 5- and 10-year planning period and the estimated cost for airport needs such as terminals, runways, lighting, and land. The NPIAS is less of a plan than a "wish list" since the information provided is not based on a planning process that includes goals, objectives, alternatives, and action items. The list contains only those items in which there is potential federal interest and for which funding is available.

The creation of the U.S. Department of Transportation (DOT) in 1967 focused national transportation activities and policies into one cabinet-level agency. Most states followed the lead of the federal government by creating a Department of Transportation. Although they are similar to the U.S. DOT, highways represent the major focus of activity for state-level DOTs. The organization of the U.S. DOT is along modal lines (air, maritime, rail, highway, and transit), which sharpens the modal distinctions rather than emphasizing their interactions. After '9/11', some functions of the U.S. DOT, such as the Coast Guard and the Transportation Security Administration, were transferred to the newly created Department of Homeland Security.

One attempt to establish a national direction for transportation occurred during the Bush administration of 1988–1992, with the publication of the report titled *A Statement of National Transportation Policy: Strategies for Action.* The report identified six policy areas for the nation's transportation system: (1) Maintain and expand the system; (2) foster a sound financial base; (3) keep the industry strong and competitive; (4) ensure public safety and national security; (5) protect the environment and quality of life; and (6) advance U.S. transportation technology and expertise. Similar policy objectives have been espoused by succeeding administrations.

Factors in Choosing a Freight or Passenger Mode

Freight and passenger trips typically rely on more than one mode to complete a journey. The modes complement each other in terms of their functional attributes. A journey may consist of three elements: *collection, delivery,* and *distribution. Collection* refers to the beginning of the trip and involves travel from the point of origin to the nearest line-haul mode. *Delivery* is the portion of the trip that transports the collected passengers or freight on a line-haul mode. *Distribution* involves travel between the line-haul terminal and a final destination.

Collection and distribution modes operate under low speed and low capacity, and may make many stops or travel in mixed traffic. Vehicles include taxis, vans, and personal rapid transit; moving walkways; and pick-up and delivery vehicles. Delivery modes are typically higher in speed and capacity (except on congested freeways) because vehicles travel on exclusive travelways and are designed to carry large numbers of people or huge quantities of freight. Vehicles include jet aircraft, tractor-semitrailers, and railroad trains.

A transfer is required among collection, delivery, and distribution trip elements, usually at terminal facilities. For example, a business trip across the country may involve several elements: a taxi from home to the airport, a cross-country flight on a jet aircraft, and a rental car to a business location. Transporting a freight shipment may require a truck to pick up packages in one city, transfer them to a railroad for shipment to a terminal in another city, and deliver the packages by truck to the final destination.

Every mode has *inherent attributes* that are reflected in service variables such as cost, travel time, convenience, and flexibility. To plan which mode is most appropriate for a given set of circumstances requires that each mode be compared. The mode with the best combination of attributes for the particular requirements of the trip is most likely to be selected. Although modes are usually compared based on measures such as travel time, cost, and frequency of service, these are not the sole measures that explain why a given mode is preferred.

Certain characteristics that are difficult to measure consistently across modes, such as union rules and the ability to use the telephone or sip a beverage while traveling, may also influence mode choice. The marginal cost of a mode may also be an inherent advantage. That is, modes in which the cost of adding a ton of freight or another passenger diminishes with each unit added have an advantage over those in which the unit price remains constant. The marginal cost of rail freight is lower than for trucking. The marginal cost of adding another passenger is lower for auto or taxi (up to a point) than for air or rail. The appearance of cost also is an influencing factor. For example, the direct cost of auto travel is always evident at the gas pump or toll booth, but most drivers do not consider the car insurance bill and other indirect costs to be as relevant when selecting a mode. The relevance of cost can also be a deciding factor in selecting a mode and may depend on who makes the decision. For private passenger travel, it usually is the vehicle owner. For school pupil travel, it is administrators, and for freight travel the decision maker may be a logistics manager. Loyalty to a particular mode or brand is strongest for individuals (repeat customers) and weakest when an administrator or logistics manager makes the decision.

Options available for *passenger travel* are auto, air, rail, bus, or ferry. The automobile is considered to be a reliable, comfortable, flexible, and ubiquitous form of transportation. Accordingly, it remains the mode of choice for many people. Since it can serve all elements of a trip, including collection, delivery, and distribution, it is one mode for which a transfer is unnecessary. When distances are great and time is at a premium, air transportation is preferred, supplemented by auto, bus, or rail for local travel. If cost is important and time is not at a premium or an automobile is not available, then intercity bus or rail is a preferred choice. Ferries are used where sections of a city are separated by water and when there are direct ferry routes to destination points and few if any alternate road or railway options.

Options for *freight travel* are truck, rail, water, air, and pipeline. Selecting a mode combination to haul freight will be based primarily on time and cost factors. Trucks have flexibility and can provide door-to-door service. They have the capability to carry a variety of parcel sizes and usually can pick up and deliver according to customer needs.

Waterways can ship high-density commodities at low cost but at slow speeds and only between two points on ocean, lake, river, or canal. Pipelines are used primarily to ship petroleum products. For international travel with fixed water routes, shipping is the usual mode for freight and is supplemented by rail or truck for collection or distribution.

Railroads can haul a wide variety of commodities between any two points where rail lines exist but typically require trucks to collect and distribute the goods to a freight terminal or to the final destination. Thus, the domestic shipper must examine the time and cost for each mode and select whether goods should be shipped by truck alone or by a combination of truck and rail.

A key factor in the selection of truck or rail modes is the industry's desire to limit inventories by arranging for delivery to the factory when needed rather than having the goods stockpiled in a warehouse. This practice, called just-in-time delivery, has favored the use of trucks since they are capable of making deliveries in smaller than carload lots and on a daily basis, depending on demand. In this case, lower freight rates charged by rail or water modes are not sufficient to compete with truck flexibility.

A shipper or a passenger may have an option among various modes when planning a trip. For the transport of freight, the available modes are typically rail, water, or trucking. For passenger travel the modes are auto, air, bus, or rail. When options are available, the shipper or passenger has a *mode choice*. When options are not available, the trip maker is considered to be a *captive* of the mode. As noted earlier, in the 1880s farmers in the Midwest were captives of the railroads and were charged exorbitant rates. In cities, people who do not have access to an automobile are captive to mass transit.

The selection of a mode is a complex process in which shippers or passengers consider their needs by evaluating the factors that influence this choice. By comparing the ability of each mode to meet a given travel need, the shipper or passenger selects the mode that is perceived to have the greatest or highest *utility*. Passengers place value on factors such as cost, travel time, convenience, flexibility of schedules, and perceived safety. Shippers also consider similar factors but are most sensitive to travel time and cost.

Not all shippers or passengers will value each factor in the same way. A shipper of coal seeks a low-cost mode and is less concerned with travel time than is a shipper of electronic equipment. A shipper of automobile parts must be assured just-in-time delivery to meet production schedules. A shipper of sensitive documents or medical supplies will often require overnight delivery. Similarly, a business traveler may require the use of a private jet to attend a one-day meeting, while a retired couple seeking a leisurely, safe, low-cost, and comfortable trip will select rail for a five-day cross-country journey that would take a total of eight hours by air.

To express the variations in a user's value system, the concept of a utility function can be applied to mode choice. The utility function provides a relative preference value for each mode and estimates the percentage of the total user population that will select each of the modes.

To illustrate, Equation 5.1 is a utility function in which the relevant variables are time and cost expressed as a linear relationship:

$$U_i = K - \beta C_i - \delta T_i \tag{5.1}$$

where

U_i = utility of mode i
C_i = trip cost for mode i
T_i = trip time for mode i
K = constant
β and δ = relative weights of each service variable

The probability that a passenger or shipper will choose one mode over the other may be based on one of the following decision rules:

1. Select the mode with the highest utility;
2. Select modes in proportion to the utility of each mode;
3. Select modes in proportion to an exponential function of utility values.

Not all passengers or shippers will select the same mode even though they have the same time–cost utility value. This is so because other factors, such as safety, security, frequency, reliability of service, and comfort (which are not included in Equation 5.1 of the utility function), are usually relevant to the modal choice decision. A commonly used exponential function is the *logit model,* whose mathematical expression is shown in Equation 5.2.

$$P_i = \frac{e^{U_i}}{\sum_{j=1}^{n} e^{U_j}} \tag{5.2}$$

where

P_i = probability that users with utility values U_i will select mode i
U_i = utility of mode i
n = number of modes being considered

EXAMPLE 5.1

Selecting a Freight Mode Based on Use of the Logit Model

The utility function for selecting a freight mode for transport between a manufacturing facility and a seaport is $U = -(0.05C + 0.10T)$, where C is cost ($/ton) and T is total door-to-door travel time (hours).

The weekly volume of goods transported between the plant and a major port is 1000 containers. There are three possible modes available to the shipper: truck, rail, and water.

The cost and travel time for each mode is as follows:

Truck	$30/metric ton	16 hours
Rail	$17/metric ton	25 hours
Water	$12/metric ton	30 hours

How many containers will be shipped by each mode if

(a) All traffic uses the mode of highest utility,

(b) Traffic is proportional to the utility value, and

(c) Traffic is proportioned based on the logit model?

Solution

Compute the utility value for each mode.

$$U_T = -\{(0.05 \times 30) + (0.10 \times 16)\} = -3.10$$

$$U_R = -\{(0.05 \times 17) + (0.10 \times 25)\} = -3.35$$

$$U_W = -\{(0.05 \times 12) + (0.10 \times 30)\} = -3.60$$

(a) All traffic uses the mode of highest utility.
Under this assumption, since the utility for the truck is highest, all 1000 containers will be shipped by truck.

(b) Traffic is proportional to the utility value. (Use the reciprocal of the utility value in the ratio.)

$$P_{\text{Truck}} = \frac{1/3.10}{1/3.10 + 1/3.35 + 1/3.60} = 0.359$$

Similarly,

Rail = 0.332

Water = 0.309

Thus

359 containers are shipped by truck
332 containers are shipped by rail
309 containers are shipped by water

(c) Traffic is proportioned based on the logit model.

$$P_i = \frac{e^{U_i}}{\sum_{j=1}^{n} e^{U_j}}$$

$$P_{\text{Truck}} = \frac{e^{-3.10}}{e^{-3.10} + e^{-3.35} + e^{-3.60}} = 0.419$$

Similar calculations yield the following values:

Rail = 0.326
Water = 0.255

Thus

419 containers are shipped by truck
326 containers are shipped by rail
255 containers are shipped by water

The quantity of transportation infrastructure available in a region, referred to as the *supply* of transportation, must be in equilibrium with the volume of traffic, which is called the *demand*. The economy produces the demand for transportation such that when times are prosperous, travel increases, and when there is a downturn in the economy, less travel occurs. The state of the transport system (supply) at any point in time refers to the facilities that exist and the quality of service that is provided. More travel will occur when the costs to the user in terms of travel time and out-of-pocket expenses are reduced.

The expression, "If wishes were horses, then beggars would ride" reflects the fact that as transportation costs decrease, travel demand will increase. For example, vehicle volumes usually increase when a road is widened. Airline passenger volumes increase following a fare reduction. If a new mode of transportation is introduced that is significantly less costly in time and money when compared to an existing mode, the new mode will capture market share and possibly replace it. The past century experienced this phenomenon for international passenger travel when air replaced ships and for domestic passenger travel when auto and air travel replaced railroads.

Demand reflects a relationship that describes the willingness of a group of passengers or shippers to pay for a particular transportation service. For example, airlines charge higher fares for business travel or on holidays than they do for travel on weekends. The number of transit riders tends to decline if fares are raised. On the other hand, when the price of gasoline increases there is little effect on traffic volumes, at least in the short run.

Supply is the term used to describe transportation facilities and services available to a user. For example, in planning an airplane trip from Atlanta to Chicago on a weekday, the supply is the number of flights available, the ticket price, travel times, and whether the flights are nonstop, direct, or require a connection. Another example of supply is a tunnel between New York and New Jersey. Included in the cost to use the tunnel are tolls and travel time. During congested hours, traffic may be bumper-to-bumper and the toll might be higher during the week than on weekends to discourage commuter travel by auto.

EXAMPLE 5.2

Computing Travel Cost Due to Congestion

A truck makes a delivery to a warehouse located in the downtown of a large city. The truck costs \$30/h to operate and the driver costs are \$35/h, including labor and fringe benefits. During midday periods when traffic is light, the trip takes 25 minutes. During rush-hour periods, travel time is increased to 55 minutes, including time spent on congested streets in the city. Compute the added cost to the trucking company to deliver goods in peak periods.

Solution

The cost/h to deliver goods is

$$(30 + 35) = \$65/\text{h}$$

The additional time spent in making a delivery during congested periods is

$$(55 - 25)/60 = 0.50 \text{ h/trip}$$

Additional transportation cost during peak periods is thus

$$(65 \text{ \$/h})(0.50 \text{ h/trip}) = 32.5 \text{ \$/trip}$$

At any point in time, the nation's transportation system is in a state of equilibrium. Traffic volumes carried on each mode, whether passenger or freight, are based on the willingness to pay (demand) and the price of travel (supply), expressed as travel attributes of time, cost, frequency, reliability, and comfort. The equilibrium is the result of

- **Market forces**, such as the state of the economy, competition, and cost of service;
- **Government actions**, such as regulation, subsidies, and promotion;
- **Technology**, such as greater increases in speed, range, reliability, and safety.

As these forces shift over time, the transportation system will change, thus altering demand (traffic volume) and supply (transportation infrastructure). Thus, the nation's transportation system is never static. Short-term changes will occur due to revisions in levels of service, such as raising the toll on a bridge, increasing gasoline prices, or raising airline fares. Long-term changes will occur in lifestyles and land use patterns, such as moving to the suburbs from the central city when highways are built or converting auto production from large to small cars. These external forces are illustrated as follows:

Market forces. If gasoline prices were to increase substantially, it is likely that some freight would shift from truck to rail. However, if petroleum prices were to remain high, a shift to other energy sources could occur or fuel-efficient cars and trucks could be developed and manufactured.

Government actions. The decision by the federal government and the states to construct transportation facilities affects transportation equilibrium. For example, the interstate highway system affected the truck–rail balance in favor of truck transportation. It also encouraged long-distance travel by auto and thus influenced the demise of the inter-city bus industry and rail passenger travel. Transit authorities can influence land development by providing bus transit routes and stops at strategic locations. Access management policies can change the balance between access and mobility by providing or withholding traffic signals, driveways, and turning lanes.

Technology. New ideas have also contributed to substantial changes in transportation equilibrium. The most dramatic change occurred with the introduction of jet aircraft that essentially eliminated passenger train travel in the United States as well as international passenger steamship travel. Communications technology has changed transportation dramatically, too, by providing the user with easy access to the system for travel planning and trip routing.

EXAMPLE 5.3

Computing Traffic Volume Based on Supply and Demand Principles

A tunnel has been constructed that connects two cities separated by a river. The cost to use the tunnel, excluding tolls, is expressed as $C = 50 + 0.5V$, where V is the number of veh/h and C is the out-of-pocket driving cost/vehicle trip. Units are in cents. The traffic demand for travel for a given time period can be expressed as $V_t = 2500 - 10C$.

Determine

(a) The volume of traffic across the bridge without a toll

(b) The volume of traffic across the bridge with a toll of 25 cents

(c) The toll charge that would yield the highest revenue and the resulting travel demand

Solution

(a) To determine the volume of traffic without a toll, substitute the cost function, C, into the demand function, V:

$$V = 2500 - 10C = 2500 - 10(50 + 0.5V)$$

$$V = 2500 - 500 - 5V = 2000 - 5V$$

$$6V = 2000$$

$$V = 333 \text{ veh/h}$$

(b) For the volume of traffic if a toll of 25 cents is added, the supply function is $C = 50 + 0.5V + 25$. Again, substitute the cost function, C, into the demand function, V.

$$V = 2500 - 10(75 + 0.5V)$$

$$6V = 1750$$

$$V = 292 \text{ veh/h}$$

(c) To determine the toll to yield the highest revenue, let T = toll rate in cents. The supply function is: $C = 50 + 0.5V + T$. The demand function is

$$V = 2500 - 10(50 + 0.5V + T)$$

$$V = (2000 - 10T)/6$$

Let R = revenue generated by the toll facility.

$$R = VT$$

Substitute V in the equation for R.

$$R = \{(2000 - 10T)/6\}T = (2000T - 10T^2)/6$$

Maximize R by setting $dR/dT = 0$.

$$dR/dT = 2000 - 20T = 0$$

T = 100 cents. Thus the toll charge to maximize revenues is \$1.00. If this toll is used in the supply function, the equilibrium demand is

$$V = (2000 - 10T)/6 = \{2000 - 10(100)\}/6 = 167 \text{ veh/h}$$

Elasticity of demand. Travel demand can also be determined if the relationship between demand and a service variable, such as cost, is known, where V_t is the volume of traffic at a given service level C_s. Thus, the elasticity of demand, $E(V_t)$, with respect to C_s is the percentage change that will occur in volume V_t divided by the percentage change in service level C_s. In other words, demand elasticity is the change in demand/unit change in cost. The relationship is expressed in Equation 5.3:

$$ED = \frac{(V_2 - V_1)/V_1}{(C_2 - C_1)/C_1} = \left[\left(\frac{C_1}{V_1}\right)\left(\frac{\Delta V}{\Delta C}\right)\right] \tag{5.3}$$

where

ED = elasticity of demand
V_1 = initial traffic volume
V_2 = traffic volume after cost change
C_1 = initial cost
C_2 = new cost
$\Delta V = V_2 - V_1$
$\Delta C = C_2 - C_1$

EXAMPLE 5.4

Computing the Decrease in Rail Passenger Demand Due to a Fare Increase

Studies have shown that for certain trip purposes, an increase in railroad fares of 1% will result in a 0.3% reduction in rail passengers. The current passenger volume on a scheduled run between two cities is 1000 passengers when the fare is $10.

(a) What is the new volume if fares are increased to $15?

(b) What is the net change in revenue?

Solution

(a) Use Equation 5.3 for elasticity of demand and solve for ΔV as follows:

$$ED = \left(\frac{C_1}{V_1}\right)\left(\frac{\Delta V}{\Delta C}\right) \qquad \Delta V = \left(\frac{ED * V_1 * \Delta C}{C_1}\right) = \frac{0.3 * 1000 * 5}{10} = 150$$

Thus the new volume will be $1000 - 150 = 850$ passengers.

(b) Revenue earned with the increased fare $= 850 \times 15 = \$12{,}750$.
Revenue earned with current fare $= 1000 \times 10 = \$10{,}000$.
Increased revenue $= (12{,}750) - (10{,}000) = \$2750$.

THE TRANSPORTATION PLANNING PROCESS

Transportation planning is a process that forecasts future travel demand and evaluates alternative systems, technologies, and services. The transportation planning process is also applied to individual modes, including railroads, highways, seaports, and airports. In each case the planning process is intended to address needs related to operations, maintenance and equipment, and the expansion of existing facilities.

Airport planning illustrates the application of processes that are used in planning transportation infrastructure. Airports are continually being expanded and, in some instances, replaced to accommodate growth in air travel. Further, the airport facility contains a variety of elements, including runways and other airside elements, the airport, terminal, and roadways and other access elements that must be included in a plan. The process differs from other modes primarily in the vehicle technology and the nature of the facilities to accommodate its characteristics.

The airport planning process includes evaluating current conditions and forecasting future travel demands for the airport; identifying specific improvements to airside, terminal, and landside facilities that will serve future travel demand; and assuring sources of revenue to finance the construction of proposed improvements.

The results of the planning process are used to develop detailed site plans. These include (1) airport runways, taxiways, and gate locations; (2) passenger and cargo terminal facilities; (3) land uses in the vicinity of the airport, including commercial areas, buffer zones, and hotels; and (4) access facilities to the airport, including parking and terminal area circulation. Typically, airport planning is required because of expected growth in travel demand. Thus, the goals, objectives, and problem definitions reflect a need to reduce travel congestion and accommodate growth.

Evaluation of current conditions is the first activity of the planning process. The results will provide information about the status of existing facilities and equipment at the airport, current air traffic in the service area, and a review of past events that created the current development of air transportation facilities in the region.

Travel demand forecasts for airport planning are intended to provide the following information: (1) the number and type of aircraft that will serve future passenger and freight demand at the airport; (2) the numbers of passengers that will arrive at, depart from, or transfer at the airport; (3) the number of visitors and employees who will arrive each day; and (4) the number and type of vehicles that arrive at and depart from the airport. These results are used in the planning for airside facilities, terminals, and access roads and parking.

Forecasting models that are used for airport planning are in many ways similar to those used for other modes. The models selected will depend on the situation and the availability of historic data. Demand forecasts may be as simple as an extrapolation of time series trend data, and for a 5- to 10-year planning horizon, this approach is often sufficient. Other models, such as a multivariate regression analysis, are more complex and rely on a broad database from which to calibrate constants. Types of dependent variables that may be considered in a multivariate regression model are Gross Domestic Product, consumer spending, income, population, and employment. Variables such as the U.S. population are relatively simple to forecast, whereas a behavioral variable such as future air travel demand is more complex. As with all forecasts, the accuracy of the results will reflect the extent to which the past is a guide to the future.

In the air industry, since technological and economic change is very rapid, the *demand for air transportation is difficult to predict*. While the long-term trends of air traffic demand can be expected to follow a constant rate of growth, the short-term fluctuations in demand can be significant due to factors such as fuel price increases, economic downturns, holiday travel, weather, accidents, competition from other airlines, security issues, and labor conflicts. Since forecasting of air travel demand is speculative at best, the judgment and experience of experts who have considerable experience in observing air transportation market trends may be an equally valid procedure. Individuals with many years of experience in the air industry may provide a more valid and realistic forecast than one produced by a mathematical model. Thus, both approaches should be used if for no other reason than that they provide an independent reality check on the result.

When the process is completed, the results can be used to *compare the capacity with the forecast demand*. Where present capacity is inadequate to serve future travel needs, additional expansion of airport facilities will be determined. Four elements of the demand–capacity analysis are required in developing an airport plan; airfield, terminal, airspace, and roadways. For example, if runway capacity is exceeded by aircraft demand, additional runways may be required. If the terminal space is inadequate, additional waiting areas, corridors, and escalators will be needed. Similarly, if the access roadways serving the airport will be congested, then improvements such as roadway widening, bus or rail transit, and additional terminal curb parking may be provided.

All transportation systems, including airports, create a variety of environmental problems. Thus, as part of the planning process, and in accordance with federal laws, a detailed *environmental impact statement* is required for all major airports. The principal environmental impact created by airports is noise caused by aircraft that land or take off in the vicinity of residential areas. There are several methods used to reduce noise to a tolerable level. These include improving aircraft technology to produce quieter engines, creating land buffer zones such as parks and golf courses, using building materials designed to absorb noise, and restricting the hours that an airport runway may be open or the maximum length of a nonstop flight.

Another environmental problem is air pollution created by aircraft fuel emissions or contamination of water supplies from fuel spills or inadequate sewage treatment facilities. In general, these latter impacts are easier to mitigate or control than airport noise. Accordingly, when new or expanded airports are being planned, participation of the residents in nearby communities is essential.

There are several *types of transportation planning studies*. Among the most common are comprehensive long-range transportation studies; major investment studies; corridor studies; major activity center studies; traffic access and impact studies; and transportation system management studies. These differ in purpose and objective, but the planning process is similar.

Comprehensive long-range transportation studies have been required in order for cities to qualify for federal highway, transit, and airport funds and are intended to produce long-range plans for regional transportation infrastructure needs over a 20-year period.

Major investment studies have been required since the 1991 Intermodal Surface Transportation Efficiency Act (ISTEA), the 1990 Clean Air Act Amendments, and the National Environmental Policy Act of 1969. Major investment studies are conducted at a corridor or subarea level using a 5- to 20-year planning period. Elements of the study include purpose and need, alternatives considered, evaluation criteria, public involvement, technical analysis, demand forecasting, and environmental impact. While these studies remain useful, they are no longer a federal requirement.

Corridor studies focus on a linear segment of an area where high volumes of traffic occur, such as between a suburban area and the central city, a rail corridor for a seaport connection to an inland destination, or an access road or rail line to a major airport. Corridor studies have a planning horizon of 5 to 20 years and are designed to determine the most appropriate mix of transportation infrastructure, including high-occupancy vehicle (HOV) lanes, high-occupancy toll (HOT) lanes, bus rapid transit (BRT), and high-speed rail links. Elements may include advisory committee participation, feasibility analysis, consideration of a wide range of alternatives, and incorporation of economic and environmental concerns. Traffic operations and management may also be included in a corridor study. Short-term improvements such as traffic signal coordination, lane widening, access management, and land use control may be considered.

Major activity center studies cover a concentration of commercial or industrial land uses. Examples of major activity centers include airports, central business districts, shopping center–office complexes, railroad terminals, and container ports. These studies have a planning horizon of 3 to 10 years. The study purpose is to investigate traffic flow, including pedestrian access within the major activity center, and to evaluate options for access and circulation, parking, public transportation, streets, highways, freight delivery patterns, and loading facilities.

Traffic access and impact studies evaluate the potential impact of proposed new developments on the transportation system. A typical time horizon is 3–5 years. For example, if a developer wishes to build a new shopping mall or an airport is to be expanded, the traffic impact study will forecast the traffic generated by the proposed project, assess the impact this traffic will have on current roads and transit facilities, and suggest ways to accommodate new traffic patterns. The study may also be used to determine if additional transportation infrastructure is required and to provide an estimate of the cost. Approval to construct the proposed project may be contingent on a financial assessment to the owners or developers and requirements that the project include specified transportation improvements.

Transportation system management studies are short range (3–5 years) and are intended to complement long-range studies. Typically, improvements are less capital-intensive than those considered in comprehensive long-range or corridor studies. Emphasis is on improving the efficiency of the existing system by *management of supply and demand*. Supply options include intelligent transportation systems, freeway management, priority lanes, and traffic engineering improvements. Demand options include pricing, ride sharing, implementing employee incentive programs, carpools, staggering work hours, and substituting communications for transportation.

The steps in the transportation planning process include the following elements: problem definition, identification of alternatives, analysis of the performance and comparison of each alternative with other candidates, and selection of the alternative that will be implemented.

Problem definition involves two aspects. First is an understanding of the environment within which the transportation facility will function based on knowledge of the present transportation system, current travel characteristics, and prior planning studies. Second, an understanding of the nature of the problems is translated into objectives and criteria. Objectives are statements that identify what is to be achieved by the project, such as to improve safety or decrease travel delays. Criteria are measures of effectiveness that quantify the objectives, such as number of crashes/million vehicle kilometers or delay time.

Among the objectives of transportation planning studies are to preserve the environment, stimulate economic development, improve access to employment, reduce congestion, and reduce air and noise pollution. In the problem definition stage it will be necessary to complete a variety of data-gathering studies that help set the stage for the steps to follow.

Identification of alternatives involves specifying options that could enhance present conditions at an acceptable cost to the transportation agency without damage to the environment. Ideas to solve a transportation problem can come from sources such as citizens, public officials, and technical staff. There are usually many alternatives in any given situation that will be identified in this idea-generating phase. Depending on the situation, options may include various technologies, networks, operation procedures, and pricing strategies.

Analysis of the performance of each alternative is intended to determine how well each of the options will serve present and future conditions. To carry out this step, each option is added to the existing transportation network and the changes in traffic flow are determined. Depending on the time horizon of the project, it may be necessary to forecast future travel demand before determining the changes in the system. For projects that can be completed within a short time period of 1–3 years (such as modifications in operations), a long-range travel demand forecast is not required. For long-term projects of 5 to 15 years, it is necessary to forecast future land use and travel. The result of this step is three sets of information for each alternate: (1) cost, including capital, operation, and maintenance; (2) traffic flow, including hourly peak volumes; and (3) impacts, including environmental effects from noise, air pollution, and displacement of existing homes and businesses.

Comparison of each alternative with other candidates is intended to furnish performance measures for various alternatives to determine how well they achieve the objectives as defined by the criteria. The performance data that are produced in the preceding step are used to compute the benefits and costs that would result if a given option were selected. Usually, these are calculated in monetary terms. Thus, if the benefits created by a given alternative are greater than the costs to build and maintain it, the alternative is considered a candidate for selection.

An economic comparison may involve the computation of the net present worth of all costs to indicate the extent to which an alternative is a sound investment. In some cases, the results used for comparison cannot be reduced to a single monetary value and a ranking method is used that gives a numerical value for each outcome based on the relative value placed on each criterion. In situations where there are multiple criteria, the results may be displayed in a cost-effectiveness matrix that depicts the change of each criterion when compared to the project cost.

Selection of the alternative that will be implemented involves a decision to proceed with one of the alternatives. At this point, a decision maker has a great deal of information available. The problem definition has articulated the issues of concern. The identification of alternatives has shown ways that the problem might be addressed. The analysis of performance has provided measures for each alternative and a comparison of the benefits and costs. For some types of transportation projects the decision is made simply by selecting the alternative with the lowest total cost, a situation that exists only when all other factors are equal. For example, to select a pavement structure, consider designs that perform satisfactorily and select the one with the lowest total life-cycle cost.

For a more complex project, there may be other intangible factors to consider and the selection may be a compromise among viewpoints expressed by the community, the transportation agency, and the user. Often public hearings may be needed. In some cases, an alternative will perform better on one criterion but less well on others and tradeoffs will be required.

The responsibility of the engineer in the selection process is to remain fair and unbiased and to assure that potentially promising alternatives are not discarded. The engineer's role in the planning process is to assist decision makers to make an informed choice and to assure that every feasible alternative is considered.

EXAMPLE 5.5

An Application of the Transportation Planning Process

Traffic conditions have become heavily congested along a major arterial roadway leading to the downtown area of a city. Bicycles, autos, trucks, buses, and pedestrians use the road. There are numerous signalized intersections along a 5-kilometer stretch, which create backups during the morning and afternoon peak hour. The road is 12 m wide. The Department of Transportation is investigating options that could help to relieve the problem. Determine how this situation could be improved by using the steps in the planning process.

Solution

Step 1: Problem definition: Determine current travel characteristics, including traffic volumes for autos, trucks, and buses; turning movements at intersections; speed and delay; bus routes; and bicycle traffic. Prepare a physical inventory of the road, including width, number of lanes, location of bus stops, signal timing, signs and markings, parking locations, turning lanes, and lane striping.

Establish the objectives to be achieved by interviewing public officials, users, business owners, members of community organizations, and property owners. Determine their perception of the problem and establish a set of agreed-upon objectives. These could include the following: (1) Improve travel time for bus riders; (2) increase roadway capacity; (3) improve safety for bikers; (4) minimize disruption to the neighborhood; and (5) maintain costs as low as possible.

Step 2: Identification of alternatives: Prepare a list of possible changes that could improve the present situation. Suggestions could come from the same group that provided input on the objectives as well as from the staff or consultants doing the study. Successful experiences of others who have dealt with a similar problem should be considered by reviewing the literature. It is possible that the solutions proposed by one interest group are ones that provide relief at the expense of others. Possible alternatives, some of which can be combined, are as follows:

- **Widen the roadway**. If right-of-way exists, it may be possible to add one or more lanes. If it were feasible to widen the road an additional 2.5 m, then it would be possible to have four 3-m lanes and a 2.5 m median turning lane.
- **Restripe the existing roadway to permit four 3-m lanes**. This option assumes that widening is not feasible. However, without turning lanes, backups could occur.
- **Use exclusive lanes for buses in peak hours**. This option may encourage transit riding and reduce the number of cars.
- **Add bike lanes on each side of the road**. These would promote safety and encourage greater use of bikes for commuting.
- **Provide left-turn lanes at intersections and consolidate commercial driveways**. If left-turning traffic can be stored in a separate lane, then through traffic will not be delayed when the signal is green. Consolidating commercial driveways will reduce conflicts between through traffic and entering–exiting vehicles.
- **Restrict the road for exclusive use of autos, buses, and bicycles**. Trucks are slow, large, and noisy and have poor acceleration characteristics. If alternate routes exist for trucks, the capacity, safety, and environmental quality of the road will be enhanced.

Step 3: Analysis of the performance of each alternative: Determine the future traffic that is expected to use this road. Since this is a short-term planning study for which actions can be implemented relatively quickly, it will suffice to use existing traffic data modified to reflect growth that may occur in the next five years. Traffic forecasts for the region may be available from the Metropolitan Planning Organization (MPO) or the Department of Transportation (DOT) Planning Division.

Determine the impact of each alternative on cost, travel time, and level of service (LOS). Prepare a matrix that lists the effects.

The results for the six alternatives are shown in the following table. Level of service provides a measure of the ease of travel, where LOS A is the best and LOS D is the worst. For all other criteria, relative values are used (high, medium, and low) to depict performance:

Alternative	*Cost*	*Travel Time Saved*	*Level of Service*
Widen roadway	High	High	B: High
Restripe	Low	Low/medium	C: Medium
Exclusive bus lanes	Medium	Low/medium	D: Low
Bike lanes	Medium	Medium	C: Medium
Left-turn lane—commercial driveways	Low	High	C: Medium
Restrict trucks	Low	Medium	C: Medium

Step 4: Comparison of alternatives:

Alternative 1: **Widen roadway**. The results have indicated that the most costly alternative is to widen the roadway. However, an objective of the plan is to keep costs as low as possible; thus if funds are unavailable or right-of-way is difficult to obtain and will require the taking of adjacent property, this alternative is rejected. However, if funds were available, this could be a viable solution.

Alternative 2: **Restripe**. Delineating travel lanes by restriping is not expensive and could improve the level of service by directing traffic, thus eliminating uncertainties and assuring that traffic will flow more smoothly. This option should be selected regardless of other action taken.

Alternative 3: **Bus lanes**. This alternative could help to speed up transit travel but at the expense of auto congestion. Unless it can be shown that a large number of motorists would switch to transit as the result of this action, this alternative would not be selected.

Alternative 4: **Bike lanes**. A bike lane is 1.5 m wide. Thus if bike lanes were added, the amount of roadway space available to autos would be reduced to 9 m. If the roadway were striped so that there were two 3.25-m lanes and an 2.5-m median for left turns, the level of service might be acceptable while accommodating bicycles.

Alternative 5: **Left-turn lane—commercial driveways**. This alternative provides an improvement in the overall level of service at a relatively low cost. If median turning lanes were provided, as proposed for alternatives 1 and 4, this option would not be required.

Alternative 6: **Restrict trucks**. If the planning study can identify alternative truck routes, additional capacity would be provided. The level of noise and air pollution would also be reduced.

Step 5: Selection of the alternative that will be implemented: Upon consideration of all of the factors involved and in consultation with the affected interests, the Department of Transportation selected an alternative that combined several of the options considered. The highway will remain 12 m wide, due to the high cost of construction and difficulty in acquiring additional right-of-way. The roadway will be restriped to accommodate two moving lanes of traffic each 3.25 m wide, an 2.5-m median for left-turning vehicles, and 1.5-m bike lanes on either side of the road. Through trucks will be prohibited and turnouts provided for bus stops.

Transportation inventories are the starting point for most transportation planning studies because it is essential to assemble data about the characteristics of the system to be studied. In most cases, transportation facilities and services already exist. Prior to deciding how they can be improved, it is essential that the existing conditions be understood. Only in rare instances would a transportation planning study begin with a "clean slate," meaning that the study area is devoid of any transportation infrastructure. Examples of this rarity are new communities constructed on virgin land at the edge of large cities—communities of Reston, Virginia, Columbia, Maryland, and Irvine, California, are examples. The Denver Airport, opened in 1996, was located in a rural area about 30 kilometers from the city. In these instances, an inventory would consist primarily of information related to the geography, topography, land use, utility location, wind patterns, and access roads in the area.

For **highway planning**, an inventory would consist of a classification of the roads to reflect their principal use as major freeways, arterial roads, collectors, and local service streets. Elements of the highway network could include lane width, pavement condition, type and location of traffic control devices, and safety features such as guard rails, medians, and lighting.

For **railroad or urban transit planning**, the inventory would include a map showing all routes, transfer points, schedules, location of bus routes, and parking facilities. Physical assets such as rolling stock and maintenance facilities would also be identified. In addition, the status of administrative, organization, and revenue sources would be established.

Inventories for **airport planning** would include maps of the region showing all existing airports, navigation aids and aviation communications facilities, topography and constructed facilities within potential new air space, restricted air space for instrument flight requirements, current and planned land uses and zoning ordinances, surveys of automobile truck and transit travel to and from the airport, trends in air travel for passengers and cargo, and historical trends in population, employment, and income.

A convenient method for characterizing the elements of an existing transportation system is to produce a computerized network consisting of a series of links and nodes. A link consists of an element in the transportation network for which characteristics such as speed, capacity, and travelway dimensions are constant. The inventory information for a highway link might include its length and width, surface condition, number of travel lanes, capacity, travel time, and accident history. A node represents the endpoint of a link and is the location in the network where the characteristics of a link are changed.

Example 5.6

Selecting Inventory Data for a Transportation Planning Study

A railroad line bisects a two-lane highway in a city of 150,000 people. Traffic has increased on the railroad, as well as on the roadway. Several crashes have occurred within the past year, and a planning study is needed to decide how to improve the situation.

Provide a list of the inventory information needed to carry out the study.

Solution

This is a site-specific project. The study area would include the approaches to the crossing by road and rail as well as the intersection itself. Inventory items that could be required are suggestive and would depend on specific conditions at the site. They include the following:

- Map showing the layout of the road and railroad tracks
- Type and location of warning signs and gates
- Profile of the roadway and railroad tracks
- Location of utilities
- Location of intersecting roads
- Approaching sight distances
- Intersection lighting
- Number of trains/day
- Hourly traffic volumes
- Accident history: fatalities and injuries
- Type of land use in the vicinity

Travel surveys are used to develop a complete understanding of the current travel patterns that will be affected by the transportation plan. The survey gathers data about the *purpose of the trip, the trip origin and destination, and the mode of travel.*

To illustrate, consider a trip by truck to transport computers between San Jose, California, and Chicago, Illinois, or a trip by air to transport a business executive living in San Francisco, California, to attend a meeting in Atlanta, Georgia. A travel survey would record the origin and destination of the trip, the mode of travel, and the trip purpose. In order to organize the travel data, it is convenient to subdivide the planning study area into zones. The number and size of each zone depends on the extent of the study area itself. For example, if the planning study is regional or national in scope, the zones could represent an entire city. If the planning study were for a smaller geographic area, such as an airport, then the zones would represent areas where trip segments would begin or end—for example, the baggage claim area or the parking lot.

A travel survey can elicit the rationale for preference in two ways. One is to ask, "Why did you choose this mode?" and thus rely on respondents to cite explanatory factors. The second way is to examine specific decisions and to relate them to characteristics of the respondent. For example, raw survey data might show that respondents who earn $60,000 or more annually take twice as many air trips for vacation than do respondents in the $30,000 to $45,000 range.

Travel surveys can be conducted in a variety of ways. The most accurate method is to interview at the residence or place of business. Another technique is to question travelers while en route using an on-board or roadside survey. Questions are asked directly and recorded immediately, or the traveler may be requested to complete a survey questionnaire and return it before the trip is over.

A less expensive, but less reliable, method is to mail the survey questionnaire. Recipients are requested to respond in a timely manner, but many recipients choose not to respond, thus potentially biasing the results. Telephone surveys may not be a reliable source of information as it is difficult to obtain a random sample and to convince people to answer questions given the proliferation of telephone solicitations, particularly during the evening hours, now further blocked by the federal Do Not Call registry.

The travel survey produces a sample of all trips made between zones. The results are then expanded to represent the entire population by applying factors that reflect the sample size. Further adjustments may be made when the results are compared with actual "ground" counts of existing travel. The final result is tabulated into a trip matrix that shows the number of trips between each zone.

It is also possible to produce matrices for various trip purposes, modes, and time periods. Furthermore, when the links and modes of the transportation system are incorporated within the matrix cells, it is possible to list land use and economic characteristics for each cell and to develop a travel time matrix for trips from one cell to another.

The information requested in a travel survey will depend on the survey's purpose. If the planning study prospectus includes an estimation of the number of people who will select from a choice of modes, such as among railroad, bus, air, and auto, then the survey will seek to uncover those variables that relate to mode choice. Thus, an origin–destination survey can provide data that are used to explain why people travel as they do. Information gathered could include purpose of the trip, location of beginning and end of trip, time of day, mode used, transfers, age, sex, income, and vehicle ownership. To assure that the information provided is consistent, a count is made of passengers or vehicles that pass through a constricted portion of the system, such as a bridge, tunnel, or a road between two cities. The number obtained from the sample survey can be checked with the actual volumes observed and, if needed, adjustments can be made.

EXAMPLE 5.7

Tabulating and Interpreting Travel Data

Data were collected for intercity bus travel between four cities. The results are shown in the following table in thousands of person trips/weekday.

(a) Which city produces the largest travel demand to or from the other cities?

(b) Which city-pair has the highest travel demand?

	Travel Demand (1000s person trips/day)			
From/To	*A*	*B*	*C*	*D*
A	0	10	20	15
B	40	0	10	50
C	20	10	0	15
D	25	15	30	0

Solution

(a) Compute travel demand to or from the other cities.

Total travel generated by each city is the sum of the number of trips that begin and end in that city and is computed using the data provided:

$$A: \quad (10 + 20 + 15) + (40 + 20 + 25) = 45 + 85 = 130$$

$$B: \quad (40 + 10 + 50) + (10 + 10 + 15) = 100 + 35 = 135$$

$$C: \quad (20 + 10 + 15) + (20 + 10 + 30) = 45 + 60 = 105$$

$$D: \quad (25 + 15 + 30) + (15 + 50 + 15) = 70 + 80 = 150$$

Community D produces the highest demand between cities of 150,000 person trips/day.

(b) Compute city-pairs with the highest travel demand.
If there are four cities, there will be six possible city-pairs, as follows:

City-Pair	*Trips between Each City-Pair*
A–B	10 + 40 = 50
A–C	20 + 20 = 40
A–D	15 + 25 = 40
B–C	10 + 10 = 20
B–D	50 + 15 = 65
C–D	15 + 30 = 45

The highest travel demand is 65,000-person trips/day between communities B and D.

ESTIMATING FUTURE TRAVEL DEMAND

In order to determine what transportation infrastructure will be required in the future, it is necessary to know the travel demand that the facility will serve during its design life. Accordingly, the transportation planning process includes estimation of future travel demand.

There are many methods used to forecast travel demand. A simple yet useful method, particularly for planning studies with a short (3- to 5-year) time horizon, is to *assume a constant growth rate of existing traffic*. In this instance, a growth rate is assumed that will continue for the life of the project. Using a simple compound interest formula, the growth rate is expressed as

$$F = P(1+ i)^n \tag{5.4}$$

where

P = present traffic volume
F = future traffic volume
i = growth rate, expressed as a decimal
n = number of years

A more complex, costly, and time-consuming method is to *develop a set of mathematical models* that incorporate variables such as land use, trip purpose, time of day, travel time, cost, and socioeconomic characteristics of the traveler.

Selection of a forecasting method will depend on factors such as the time horizon for the project, data availability, and financial resources. For example, if a highway-grade crossing or widening project is being considered, then the growth rate method is usually sufficient. If a regional highway, airport, or rail transit network is being planned with a construction schedule spanning 20 or more years, then a large-scale forecasting effort is required. Whatever method is selected, it should assure that the forecast is reliable and that it accurately reflects changing demographics, economic expectations, and current transportation system performance. Unreliable forecasts can lead to outcomes such as underutilization of the new facility or premature overcapacity.

Traffic changes can occur in a variety of ways. Normal traffic growth (or decline) occurs as the result of changes in the economy. Traffic may be diverted from transportation infrastructure when improvements are made to one facility while the other deteriorates. Infrastructure improvements may affect destinations. For example, shifts may occur from downtown to suburban shopping because of better parking availability. Traffic changes also occur when users shift from one mode to another. For example, freight shipments diverted from rail to truck will add traffic to interstate highways. Finally, when transportation infrastructure is improved, new traffic is created that did not previously exist.

EXAMPLE 5.8

Use of Traffic Growth Factors to Forecast Future Travel Volumes

Traffic on a two-lane road has been increasing at the rate of 4%/year. The criterion applied for widening a road to four lanes is that the average daily traffic on the two-lane road should not exceed 13,000 cars/day. Current traffic on the two-lane road is 9500 cars/day. The time period needed to design the roadways, acquire rights-of-way, and build the roadway is estimated to be two years.

(a) How many years will it take for traffic to increase from its current value to the volume that justifies a four-lane roadway?

(b) What year should the design and construction process for the road begin?

(c) After the new road is opened to traffic, how many years will it be before the new four-lane road reaches an average daily traffic (ADT) volume of 1500/veh/lane/h? Because of higher speed limits, better geometric design, and a new traffic signal system, traffic growth on the widened road is 5% annually and peak hour traffic is 15% of ADT.

Solution

(a) Traffic growth on the roadway will increase at the rate of 4%/year until the road is widened to four lanes. The number of years to reach a volume of 13,000/day from its present value of 9500 is

$$F = P(1 + i)^n$$

$$13{,}000 = 9500(1 + 0.04)^n$$

$$13{,}000/9500 = (1 + 0.04)^n$$

$$n = 8 \text{ years}$$

(b) Since traffic will reach capacity for a two-lane road in approximately eight years, and two years are required for design and construction, this project should commence in $(8 - 2) = 6$ years.

(c) When the new road is complete, the volume/lane can be determined by knowing the percentage of traffic that occurs in the peak hour. After a traffic count has been conducted, it is determined that for this area the peak hour represents 15% of annual daily traffic. Thus, when the roadway is opened for traffic, the peak hour volume is

$$V = (ADT)(PH)/N$$

where

V = peak hour lane volume
ADT = average daily traffic
PH = percentage of ADT in the peak hour
N = number of lanes

Thus

$$V = (13{,}000)(0.15)/2 = 975 \text{ vehicles/lane}$$

Since traffic growth is 5%/year, the number of years to reach 1500 veh/lane/h is

$$1500 = 975\,(1 + 0.05)^n$$

$$n = 19$$

A travel forecasting method used in long-range regional transportation studies is called the *four-step process*. The term reflects the fact that travel demand is segmented into four distinct aspects, each with its own set of mathematical models and procedures. In this process, a trip is first *generated* by a given land use for example, a residence or a place of work. Then the generated trip is *distributed* to another land use, for example, a trip between a residence and a shopping center. The distributed trip is then allocated to a *mode of travel,* for example, the trip maker may select trucking, rail, water, auto, transit, walking, or air. (Factors in choosing a freight or passenger mode have been described in a previous section of this chapter.) Finally, for some modes, travelers can *select a route*. For example, an auto trip can use a freeway or a parallel road, a traveler from the East Coast to the West Coast can fly directly between two cities or change planes in Denver or Chicago, and for an auto trip across country by car the options can be I-10, a southern route or I-80, a northern route.

Thus the four-step forecasting process consists of the following elements: (1) **trip generation**, (2) **trip distribution**, (3) **mode choice**, and (4) **route selection**.

Trip generation refers to the number of trips that are produced by an activity unit, such as a shopping center, airport, housing development, or industrial park. Trip generation values are determined from special studies of individual land uses by counting the number of persons or vehicles that enter or leave the facility or by using values from published sources. The data are correlated with land use variables such as acres or dwelling units for residential uses or employees for commercial and industrial units.

EXAMPLE 5.9

Computing Trip Generation Values for Various Land Use Types

A planned retail shopping center expects to feature the following establishments:

- Two supermarkets: gross floor areas of 2,000 m^2 and 2,500 m^2
- A department store: 30 employees
- Two fast food restaurants: 300 m^2 each
- A bank: 20 employees
- A medical office: 15 employees

How many vehicle trips into and out of the center will be generated on a typical day?

Solution

Consult the Institute of Transportation Engineers (ITE) *Trip Generation* guide to obtain the appropriate value of trips/day/employee or per unit area. The calculations to determine the number of trips/day are as follows:

Two supermarkets	(135.3) trips/100 m^2 (45)	= 6089 trips/day
Department store	(32.8) trips/employee (30)	= 984 trips/day
Fast food restaurants	(533) trips/100 m^2 (6)	= 3198 trips/day
Bank	(75) trips/employee (20)	= 1500 trips/day
Medical office	(25) trips/employee (15)	= 375 trips/day

Total trip generation = 6089 + 984 + 3198 + 1500 + 375 = 12,146 trips/day

Other trip generation methods, including regression analysis and cross classification, are described in greater detail in the references provided at the end of this chapter. Regression analysis is similar to the concept of trip rates used in Example 5.9 in that the variable trips/day are related to one or more dependent variables, such as acres, employment, population, dwelling units, and auto ownership. A regression equation is a better approach if a statistical relationship can be identified that demonstrates a strong correlation between the dependent and independent variables.

Cross classification is a method in which trip rates are derived from survey data and cross classified with variables such as income, auto ownership, and trip purpose. Applying the values observed for specific residential areas will produce an estimate of trip generation.

Trip distribution is the process of allocating trips that have been generated by one land use or zone (trip origin), such as a retail shopping center, an airport, or a residential neighborhood, to another land use or zone (trip destination). The process seeks to determine, for trips generated in each zone, where these trips terminate.

The most widely used distribution model is called the *gravity model*. The name comes from the formula that uses an analogy of physical gravitation in which the attraction force of a single body acting on another is directly proportional to the mass of the attracting body and inversely proportional to the square of the distance between them. Thus, if there was more than one body acting on another, the relative force of each attracting body would be its mass divided by the square of the distance between it and the body being attracted divided by the sum of all forces acting on the body.

The number of trips generated by a land use represents the "attractive force" of that land use and the travel time between the "generating" land use or zone and the "attracting" land use or zone represents the "distance" between them. Thus, the gravity model can be used to compute the number of trips between all attraction zones in the study area. To calculate the number of trips between zones i and j, the gravity model can be expressed as follows:

$$T_{ij} = T_i \left[\frac{\dfrac{A_j}{t_{ij}^2}}{\displaystyle\sum_{j=1}^{n} \frac{A_j}{t_{ij}^2}} \right] \tag{5.5}$$

where

T_{ij} = number of trips generated in origin zone i that terminate in attraction zone j
T_i = number of trips generated in zone i
A_j = number of trips generated in zone j
t_{ij} = travel time between zone i and zone j
n = number of zones

There are many variations to this formula. For example, the value of travel time t_{ij} can be replaced by a friction factor, which is a reciprocal of some function of travel time. Further, a correction factor, K, can be used to modify the effect of the attraction factor A_j based on social or economic effects. These and other refinements in the model attempt to replicate actual conditions as closely as possible.

EXAMPLE 5.10

Using the Gravity Model to Forecast Distribution of Truck Trips

A seaport that serves four cities within a 550-km radius generates 25,000 truck trips each day. The population and travel time to each city from the port is shown in the following table. Use the gravity model to estimate the number of trucks that can be expected to arrive at each city/day.

City	*Population (1000s)*	*Travel Time (hours)*
A	40	6
B	75	4
C	120	3
D	150	7

Solution

Use Equation 5.5 to compute, T_{pa}, the number of truck trips between the seaport and city A,

$$T_p = \text{number of trips generated at the port} = 35{,}000$$

$$A_a = \text{population in city A} = 40$$

$$t_{pa} = \text{travel time from port to city A} = 6$$

$$T_{pa} = 25{,}000\left[\frac{1.111}{1.111 + 4.688 + 13.33 + 3.06}\right]$$

$$= 25{,}000\left[\frac{1.111}{22.189}\right] = 1252 \text{ trips/day}$$

Similarly:

$$T_{pb} = 25{,}000\left[\frac{4.688}{22.189}\right] = 5282 \text{ trips/day}$$

$$T_{pc} = 25{,}000\left[\frac{13.33}{22.189}\right] = 15{,}018 \text{ trips/day}$$

$$T_{pd} = 25{,}000\left[\frac{3.06}{22.189}\right] = 3448 \text{ trips/day}$$

Thus, the number of truck trips/day from the seaport to each city is estimated to be as follows:

City A	1252
City B	5282
City C	15,018
City D	3448
Total	25,000

Route selection implies that for some modes there may be more than one route that can be used to travel between two locations. In some cases, however, route selection is limited. For example, waterways are usually constrained by the limitations of the network (i.e., a single river or canal) or by operating conditions such as maximum water depths. Air routes are usually constrained by hub-and-spoke systems used by major carriers such as United Airlines, with hubs in San Francisco, Denver, Chicago, and Washington–Dulles. Route selection for highway, air, or rail travel is made by the user and, in the absence of other considerations, that choice is based on the shortest travel time between two points. When travelways become congested as traffic volumes increase, when accidents occur or weather conditions deteriorate, travelers will seek alternative routes that will take less time than the congested or impassable routes. Information technology is now available to inform travelers of delays and suggested alternatives.

EXAMPLE 5.11

Computing Traffic Route Travel Times and Travel Demand for Noncongested Conditions

The travel time by railroad in hours, between 16 city-pairs, A–P, is depicted in the following diagram. The number of daily trips between city A and all other cities (in thousands) is as follows:

A–B = 10, A–C = 15, A–D = 16, A–E = 20, A–F = 25, A–G = 12,
A–H = 6, A–I = 18, A–J = 8, A–K = 16, A–L = 5, A–M = 14, A–N = 12,
A–O = 4, A–P = 17

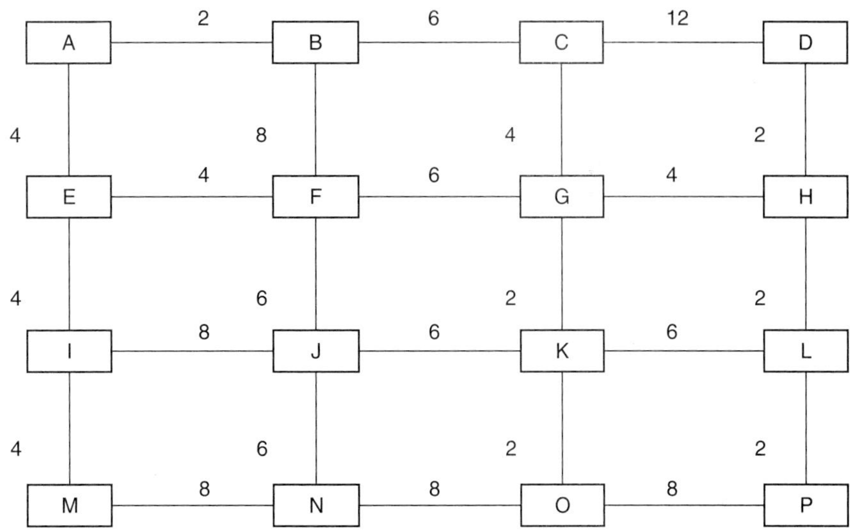

Travel time by rail between city-pairs.

Determine the following:

(a) The shortest time path from city A to all other cities

(b) The shortest time path from city A to all other cities in a straight-line diagram

(c) The number of trips from city A to all other cities (B–P)

(d) The travel demand on each rail line connecting the cities

Solution

(a) The shortest time path from city A to all other cities. Determine the travel time from city A to all other cities B through P. The values are depicted in the following figure. Note that in some cases there may be two or more travel times listed by the city since two or more routes from city A to that city may be possible.

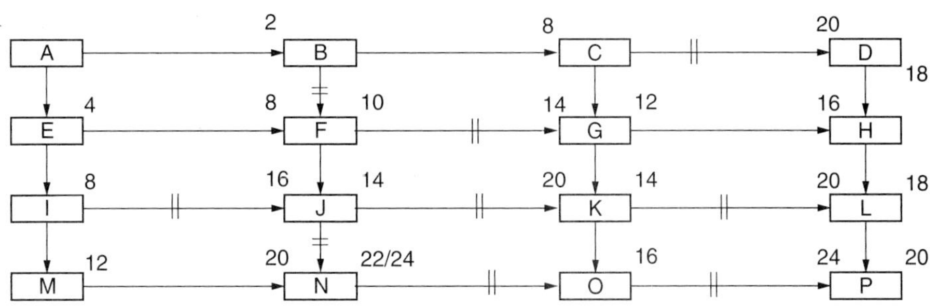

Travel time from city A to the other cities.

For example, there are two possible routes from city A to city F. The first goes through city E and takes 8 hours; the second goes through city B and takes 10 hours. Since city-pair route A–E–F is shorter, city pair B–F is eliminated as noted by the hatch marks through the arrow connecting B–F.

Select city E to determine the travel times from city E to the closest cities, city F and city I. Travel times from city A to city F and city A to J are both 8 hours. The travel time from city A to C is also 8 hours.

Select city C to determine travel times to the next two cities, D and G. The travel times are 20 hours from A to D (via city C) and 18 hours from A to D (via C–G–H) and 12 hours to G. Thus C–D is eliminated.

City F is the next closest to city A. The travel time to city G using city F is 14 hours compared with 12 hours via city C. Thus city-pair F–G is eliminated as noted by the hatch marks.

Next consider city I. Travel time to city J through city I is 16 hours and to city M, 12 hours. Since travel time from city A to city J via city F is 14 hours, city-pair I–J is eliminated.

Continue from city G to city H and K and from city M to N.

Next, go from city J to cities K and N. Eliminate city–pair J–K since a shorter path exists from city A to K via city G.

Proceed from city K to city L and O, and from city H to city L. Eliminate city-pair K–L. Continue from city O to city P and eliminate city pair O–P. Continue from city L to P and eliminate city-pair O–P.

(b) The shortest time path from city A to all other cities in a straight-line diagram. The shortest time path from city A is depicted as follows:

Minimum travel routes between city A and all other cities.

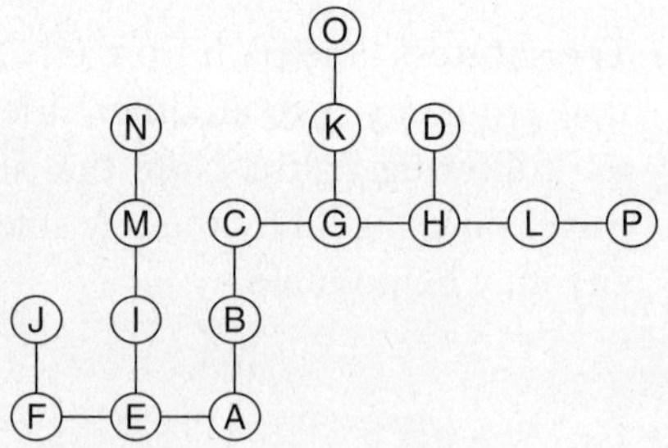

(c) The number of trips between city A and all other cities (B–P). This is the sum of trips AB, AC, . . . AP. Based on the data provided in the problem statement, the value is 198,000 trips/day.

(d) The travel demand on each rail line connecting the cities. First identify the links involved in each trip combination for city A to cities B–P. They are shown in the following table:

From	*To*	*Trips*	*City-Pairs on the Minimum Time Path*
A	B	10	A–B
A	C	15	A–B, B–C
A	D	16	A–B, B–C, C–G, G–H, H–D
A	E	20	A–E
A	F	25	A–E, E–F
A	G	12	A–B, B–C, C–G
A	H	6	A–B, B–C, C–G, G–H
A	I	18	A–E, E–I
A	J	8	A–E, E–F, F–J
A	K	16	A–B, B–C, C–G, G–K
A	L	5	A–B, B–C, C–G, G–H, H–L
A	M	14	A–E, E–I, I–M
A	N	12	A–E, E–I, I–M, M–N
A	O	4	A–B, B–C, C–G, G–K, K–O
A	P	17	A–B, B–C, C–G, G–H, H–L, L–P

Then list the number of trips from city A to all other cities (A–B = 10, A–C = 15, A–D = 16, A–E = 20, A–F = 25, A–G = 12, A–H = 6, A–I = 18, A–J = 8, A–K = 16, A–L = 5, A–M = 14, A–N = 12, A–O = 4, A–P = 17). The following table lists the values for all city combinations:

City-Pair		*Number of Trips between Each City-Pair*
A–B	10, 15, 16, 12, 6, 16, 5, 4, 17	101
A–E	20, 25, 18, 8, 14, 12	97
B–C	15, 16, 12, 6, 16, 5, 4, 17	91
C–G	16, 12, 6, 16, 5, 4, 17	76
E–F	25, 8	33
E–I	18, 14, 12	44
F–J	8	8
G–H	16, 6, 5, 17	44
G–K	16, 4	20
H–D	16	16
H–L	5, 17	22
I–M	14, 12	26
K–O	4	4
L–P	17	17
M–N	12	12

The number of trips between each city-pair is the sum of the number of trips from city A to all other cities that travel between the city-pair. For example, trips between city A and cities B, C, D, G, H, K, L, O, and P all travel between cities A and B. Thus the number of trips between city A and B is $\Sigma(10, 15, 16, 12, 6, 16, 5, 4, 17) = 101{,}000$ trips.

There are two methods for estimating the probable route for a trip: (1) allocating all trips to the shortest time path, assuming that all infrastructure travelways are not so congested that they will alter route choices; and (2) allocating trips to the shortest time path but taking account of congestion that will occur as traffic volumes increase. It is not possible to predict when an accident or delay-causing event will occur; thus the importance of real-time driver information systems.

To calculate the increase in travel time caused by increased traffic, Equation 5.6 may be used:

$$t_1 = t_0\left[1 + 0.15\left(\frac{V}{C}\right)^4\right] \tag{5.6}$$

where

t_1 = travel time in the section where there is congestion
t_0 = travel time in the roadway section under free-flow conditions
V = volume on the section where there is congestion
C = capacity of the line

Transportation planning software packages are available to carry out the multitude of computations required to use the four-step process. An iteration process within the software package can be used that allows trips between cities to alter routes as delay time increases. The process continues until equilibrium is reached such that a minimum total travel time is achieved. When the four-step process is complete, an estimate of traffic demand and travel times is produced for each link on the network.

EVALUATING TRANSPORTATION ALTERNATIVES

The previous sections have described the planning process, including data requirements and methods to forecast travel demand. The result of these activities is a set of possible options that address the problem, which is to select transportation facilities and services that are responsive to present and future travel needs. This section describes how the various options or alternatives that have been proposed can be evaluated and thus provides decision makers with a rational basis for selecting a course of action. The evaluation process involves several concepts that can influence the final project selection:

You get what you pay for. For example, due to the increase in railroad freight traffic, the volume of freight trains moving through a small midwestern town has increased considerably. Citizens have complained about the safety and delays at this unmarked crossing. A planning study considers three alternatives to improve the situation: (1) warning signs and pavement markings; (2) flashing lights and movable gates; and (3) a grade-separated overpass. The proposed solutions will produce different results in terms of safety and delay. Alternative 3, a grade-separated overpass, will eliminate the safety and delay problem but at a much higher cost than alternatives 1 and 2. Thus, the town leaders must decide which alternative provides the best result at an affordable cost.

The information used in the evaluation process must be relevant to the decision. Prior to preparing an evaluation, it is essential to know what information will be important in making a project selection. In some instances, the only relevant criterion is the cost of the project and the cost to users. In other cases the decision may be based on multiple criteria, particularly when nonusers are affected. These criteria can include items such as amount of land and property required, air pollution, and noise effects.

Evaluations can describe the "bottom line" or provide "full disclosure." A single-value evaluation is one in which the final result is reported as either a dollar amount or in alphanumeric terms. This is a "bottom-line" approach in which the decision maker is provided with the total project cost or, if there are other criteria to consider, the "final grade" for each alternative. A more useful evaluation is one that provides all the relevant information regarding each alternative and for which all results are provided separately for every criterion considered.

The evaluation should consider the viewpoint of relevant stakeholders. An alternative can affect a wide range of interest groups. When groups are affected, they become "stakeholders" in the outcome and they will often seek to influence the decision maker to select an alternative in their favor. There are many examples of special interest groups, including shippers and passengers, organized labor, environmentalists, property owners, the business community, and government. Typically, the number of stakeholders will increase as the project scope expands. Further, the project may influence business development, employment, construction activity, and land use. For these reasons, it is important to recognize whose views are being considered in the evaluation.

The criteria selected for an evaluation must be relevant and easy to measure. Measures of effectiveness are stated in terms of a numerical or relative value for each criterion. In the railroad grade crossing example cited earlier, if the goal is to reduce accidents, a criterion could be the number of fatalities/year. If the goal is to reduce delay, a criterion could be the number of vehicles stopped/h or the average time that each stopped vehicle must wait.

Criteria should be closely linked to the stated objective. For example, if the goal of a transportation company is to provide improved on-time delivery, a relevant criterion is the percentage of arrivals within 15 minutes of the stated arrival time.

Measures of effectiveness can be depicted in several ways. One approach is to convert all measures of effectiveness into a common unit and then add them to produce a single result. A common unit is money. If every measure could have an equivalent monetary cost, such as the human and property damage cost of an auto crash and the cost of travel time to a truck driver, then the number of crashes and the total time delay experienced by truckers could be converted into dollar amounts and added.

Another common unit is a numerical score. If every measure can be given a numerical range, say, 1–10, where 1 is poor and 10 is excellent, then the measures can be summed to provide a final rating. This latter system is how grades are determined. Measures such as attendance, homework, midterms, and final exams are scored with a weighting factor attached. The result is the numerical grade achieved in the course.

Finally, each measure of effectiveness can be reported for each alternative in matrix form, providing the decision maker with complete information and a better understanding of what tradeoffs will be required in selecting one alternative over another.

An *economic evaluation* is conducted to determine the actual cost that will be incurred if a given transportation alternative is selected. With similar cost information regarding each of the alternatives, it is possible to compare alternatives to determine which provides the greatest value for the money invested.

Economic evaluations are based on the concept of *time value of money*. To illustrate, if \$1000 is deposited in a savings account at a rate of 6% interest/year, then the amount in the account at the end of the first year will be $(1 + 0.06)(1000) = \$1060$. If the interest amount is not withdrawn, the amount at the end of the second year will be $(1.06)(1060) = \$1123.60$. The amount \$1123.60 is referred to as the future worth F of \$1000 in two years at an interest rate of 6%. The amount \$1000 is referred to as the present worth P of \$1123.60 in two years at an interest rate of 6%. The general expression for this computation is $F = P(1 + i)^n$. Thus, \$1000 in year zero is equivalent to \$1123.60 in year two at 6% interest.

The concept of *present worth* provides a mechanism for converting future costs to their present worth and thus serves as a common basis for comparing costs that occur at different times during the life of the transportation project. The general expression for computing present worth when given a future amount is

$$PW = \sum_{n=1}^{N} \frac{C_n}{(1 + i)^n} \tag{5.7}$$

where

C_n = costs incurred by the project in year n. These may be either infrastructure related, such as construction, maintenance, and operating costs, or user related, such as travel time or accident costs

N = service life of the facility (years)

i = interest rate expressed as a decimal

EXAMPLE 5.12

Evaluating Truck Corridor Alternatives

The Port Authority in a large urban region is considering methods to improve access to its harbor and terminal facility. One element of the project includes the separation of truck traffic from pedestrian and local traffic in the corridor to assure reductions in travel time and accidents. Three alternatives have been proposed:

I: Widen the roadway. Since the current roadway is two lanes wide, increasing the width to four lanes will add capacity and reduce travel time.

II: Provide a new signal control system. If right-of-way is unavailable, it may be possible to improve traffic flow by adding traffic signals and a computerized traffic control system.

III: Add turning lanes and pedestrian overpasses. Options include separate left- and right-turning lanes, medians for pedestrian crossings, and pedestrian overpasses in several high-demand locations.

The total construction cost of each alternative and the annual costs of travel time and maintenance is shown in the following table. Determine which alternative is lowest in total cost. Use an interest rate of 6% and a project life of five years.

Alternative	*Construction Cost*	*Annual Travel*	*Annual Maintenance*
I	1,430,000	42,000	54,000
II	928,000	59,000	74,000
III	765,000	57,000	43,000

Solution

Compute the present worth of each alternative and select the one with the lowest cost. Equation 5.7 is used to compute present worth:

$$PW = \sum_{n=1}^{N} \frac{C_n}{(1 + i)^n}$$

$$PW_{\text{I}} = 1{,}430{,}000 + \sum_{n=1}^{5} \frac{42{,}000 + 54{,}000}{(1 + 0.06)^n}$$

Alternative I

$$PW_{\text{I}} = 1{,}430{,}000 + 404{,}390 = \$1{,}834{,}390$$

Alternative II

$$PW_{\text{II}} = 928{,}000 + \sum_{n=1}^{5} \frac{59{,}000 + 74{,}000}{(1 + 0.06)^n}$$

$$PW_{\text{II}} = 928{,}000 + 560{,}245 = \$1{,}488{,}245$$

Alternative III

$$PW_{\text{III}} = 765{,}000 + \sum_{n=1}^{5} \frac{57{,}000 + 43{,}000}{(1 + 0.06)^n}$$

$$PW_{\text{III}} = 765{,}000 + 421{,}240 = \$1{,}186{,}240$$

Summary of Net Present Worth

Alternative I	\$1,834,390
Alternative II	\$1,488,245
Alternative III	\$1,186,240

Alternative III has the lowest present worth and thus is preferred on the basis of cost.

Multiple criteria evaluation methods are used in many transportation planning studies because not all measures of effectiveness that are relevant in the decision process can be reduced to monetary values. When this occurs, there are two approaches to the evaluation: rating and ranking and cost-effectiveness.

Rating and ranking. Every alternative is assigned a numerical score for each measure of effectiveness. The results are added. The preferred alternative is the one with the highest score.

Cost effectiveness. Rather than assigning a numerical value to each measure of effectiveness, the actual results may be measured in different units. Instead of converting these to an equivalent numerical value, each measure is shown in a cost-effectiveness matrix or chart to illustrate how the value of each measure of effectiveness changes as a function of project cost.

EXAMPLE 5.13

Evaluating Airport Access Alternatives Using Ranking and Rating

The quality of access to a regional airport has become a major concern. The Department of Transportation (DOT) in cooperation with the Regional

Airport Authority (RAA) is considering alternatives to improve service. Use ranking-rating and cost-effectiveness to evaluate the following alternatives:

I: A high-speed rail line from the downtown to the airport

II: Express bus service from the downtown and several suburban office complexes supplemented with super shuttle service

III: Expanded parking facilities and increased roadway capacity

The DOT and RAA have established four major criteria for evaluation:

C-1 Average travel time between the airport boundary and the terminal (minutes)

C-2 Air quality (metric tons of carbon monoxide produced)

C-3 One-way trip cost ($)

C-4 Total project cost ($ millions)

A planning study was performed and the results for each alternative determined as follows:

	C-1	*C-2*	*C-3*	*C-4*
Alternative/Criteria	*Time*	*CO_2*	*Trip$*	*Project ($ millions)*
I	12	230	10	26
II	17	360	11	11
III	22	420	22	14

The following values have been assigned to each criterion:

Criteria	*Value*
1	30
2	10
3	20
4	40

The criteria point values are computed for each alternative by assigning the maximum point value to the best performer and a proportionate amount to the lower-performing ones. Thus for criterion 1, Alternative I is assigned 30 points, Alternative II: 12/17(30) = 21.1 points, and Alternative III: 12/22(30) = 16.4 points. The results for each criterion are shown in the following table.

Alternative/Criteria	*C-1*	*C-2*	*C-3*	*C-4*	*Total Score*
I	30.0	10.0	20.0	16.9	76.9
II	21.1	6.4	18.2	40.0	85.7
III	16.4	5.5	9.1	31.4	62.4

Based on this evaluation, improved express bus and van shuttle service (Alternative II) is preferred.

When a cost-effective matrix is used, the actual values for each alternative/criteria combination are provided to the decision maker. In this example the data are provided in the statement of the problem.

The lower-cost alternative (II) would be compared with higher-cost alternatives (I, III) to determine the benefit that results from selecting a higher-cost plan. Clearly, Alternative III is unacceptable since it is more costly than Alternative II and results in greater time and trip costs as well as more pollution. Alternative I is more costly than II but reduces travel time and trip cost as well as CO_2 emissions.

SUMMARY

This chapter has described how the transportation planning process is utilized to develop a strategy to meet the needs of future travel. The process is applicable to all modes because it follows a systematic and rational approach that includes problem definition, identification of alternatives, analysis of performance, comparison of alternatives, and selection.

To carry out the process, it is necessary to acquire appropriate data that provide a baseline for the study, assist in problem definition, and suggest appropriate methods to forecast future demand. Data requirements include facility inventories, travel patterns, and traffic studies such as traffic volumes and parking.

Forecasting of future travel demand may be as simple as following a trend line or as complex as the four-step process of trip generation, trip distribution, mode choice, and route selection. Embedded within each step are a variety of mathematical models whose accuracy depends on the quality of data collected.

Many alternatives will be considered in the planning process. The desirability of each alternative will be determined in the evaluation phase. Evaluation is intended to provide decision makers with information for selecting a project. The evaluation can be based on an economic criterion, a numerical ranking of economic and noneconomic factors, or a set of cost-effectiveness relationships for each criterion. Thus, the planning process is a rational approach to transportation decision making and a useful tool to assist in choosing among alternatives. The success of the process is directly related to the extent to which it provides useful and relevant information that results in an informed decision.

PROBLEMS

5.1 What are the elements of transportation planning and programming and how does the process vary among modes?

5.2 What is one key element of the transportation planning process that is required to validate a vision of the future?

5.3 Explain how laws and ordinances influence the transportation planning process.

5.4 The Interstate Highway System, which was authorized in 1956, was perhaps one of the most comprehensive systems planned in U.S. history. Provide three examples of successful and three unsuccessful planning outcomes of this project.

5.5 How does planning for highways differ from planning for railroads?

5.6 What agencies are responsible for airport planning? How does air transport differ from other modes such as railroads or motor vehicles?

5.7 What were the six policy areas as priorities for the nation's transportation system in the early 1990s? Refer to the U.S. DOT Web site www.dot.gov to determine how priorities have changed in the twenty-first century.

5.8 Define the following terms: *collection*, *delivery*, *distribution*. Explain how the terms are used by describing the modes taken in a trip between downtown Washington, D.C., and suburban Los Angeles.

5.9 List five service variables that can be measured when evaluating the competitiveness of a transportation mode. What are three examples of characteristics that are difficult to measure but could affect a decision regarding which mode to use?

5.10 Under what circumstances would a passenger select the following mode: automobile, air, rail, bus, or ferry?

5.11 Under what circumstances would a shipper select the following mode: truck, rail, water, air, or pipeline?

5.12 What is meant by a "captive" trip? Illustrate your answer with reference to passenger and freight travel.

5.13 What is meant by the utility of a transportation mode? What factors will affect a user's perception of the utility for a given mode?

5.14 The utility function for selecting a freight mode for shipping computers is

$$U = -(0.03C + 0.15T)$$

Where C is cost (\$/computer unit) and T is total door-to-door travel time (hours). The weekly volume of goods transported between the plant and a major distribution center is 25,000 units. There are three possible modes available to the shipper: truck, rail, and air.

The cost and travel time for shipping by each mode is as follows:

Truck	$10/unit	8 hours
Rail	$6/unit	17 hours
Air	$18/unit	5 hours

How many computer units will be shipped by each mode based on the following assumptions?

(a) All traffic uses the mode of highest utility.
(b) Traffic is proportional to the utility value.
(c) Traffic is proportioned based on the logit mode.

5.15 What is meant by supply and demand as applied to transportation?

5.16 A taxi cab operates in both peak and off-peak hours in a large city. The vehicle operating costs are $20/h and the operator costs are $40/h, including labor and fringe benefits. During midday periods when traffic is light, a typical trip takes 10 min, whereas during rush-hour periods, travel time is increased to 30 min. What fare should be charged during the peak and off-peak hours if the company earns a 10% profit on each trip?

5.17 What is meant by *equilibrium*, and what are the influencing factors in the context of transportation?

5.18 A toll bridge carries 5000 vehicles per day when the toll is $1.50/vehicle. It is estimated that when the toll is increased by 25 cents, the traffic on the bridge will decline by 10% of the current volume. What is the toll that should be charged if it is desired to maximize the amount of money collected from motorists? How much revenue will be generated, and what is the traffic volume? How much additional revenue would be generated with this toll policy?

5.19 Elasticity of demand for bus travel on an express line in a major city is 0.33. What will be the effect on travel demand and revenue if fares are increased from $1.25 to $1.50 if the present demand is 6000 passengers/day?

5.20 Describe the purpose and function of the following transportation planning studies: comprehensive long-range studies, major investment studies, corridor studies, major activity center studies, traffic access and impact studies, and transportation management studies.

5.21 What are the most important environmental problems facing airport planners? Describe approaches to these problems.

5.22 What are the steps in the transportation planning process?

5.23 List five inventory items that would be included in an airport planning study.

5.24 The origin–destination data collected for truck traffic between four regional terminals are shown in the following table, in thousands trucks/week. Determine (a) the number of truck trips generated at each terminal, and (b) the traffic volume between terminals.

From/To	*A*	*B*	*C*	*D*
A	0	52	75	41
B	25	0	64	26
C	126	79	0	95
D	65	31	47	0

5.25 If traffic on a two-lane road is 6500 veh/day and increases at the rate of 4%/year, using compound growth, in how many years will traffic volumes reach 10,000 veh/day?

5.26 Describe the four-step forecasting process used in transportation planning.

5.27 A regional airport generates 8000 passenger arrivals and departures/day and serves four employment centers within a 300-km radius. The employment and travel time to each location is shown in the following table. Use the gravity model (Equation 5.5) to determine how many passengers travel to each employment center/day.

Center	*Employment*	*Travel Time (minutes)*
A	2500	150
B	1500	75
C	1000	45
D	1750	90

5.28 What are the six concepts that relate to the outcome of a transportation evaluation?

5.29 Three transportation alternatives have been proposed for a railroad safety improvement program:

I grade separations
II signal control
III advance warning systems

The cost for each alternative is shown in the following table. Determine which alternative is lowest in total cost. Use an interest rate of 8% and a project life of 5 years.

Alternative	*Initial Cost*	*Annual Operating Cost*
I	$550,500	$39,000
II	$454,000	$43,000
III	$440,850	$57,000

5.30 Another stakeholder group is evaluating the airport access problem described in Example 5.12. The stakeholders examined the criteria for project selection and agreed upon the following weighting values. Using this revised information, determine the weighted score for each alternative.

Criteria	*Value*
1	20
2	20
3	30
4	30

Alternative/Criteria	*C-1*	*C-2*	*C-3*	*C-4*
	Time	CO_2	Trip$	Project$
I	12	230	10	26
II	17	360	11	11
III	22	420	22	14

References

1. de Neufville, Richard, and Odoni, Amedeo, *Airport Systems: Planning, Design and Management*, McGraw-Hill Professional, 2002.
2. Institute of Transportation Engineers, *Transportation Planning Handbook*, 2nd ed., Washington, D.C., 1999.
3. Institute of Transportation Engineers, *Trip Generation,* 7th ed., Washington, D.C., 2003.
4. Meyer, Michael B., and Miller, Eric J., *Urban Transportation Planning*, McGraw-Hill, 2000.
5. Ortuzar, Juan de Dios, and Willunsen, Louis G., *Modelling Transport*, John Wiley and Sons, 2001.
6. Wells, Alexander T., and Young, Seth, *Airport Planning and Management*, McGraw-Hill Professional, 2003.

CHAPTER 6

Geometric Design of Travelways

This chapter discusses the geometric design of the travelways of the highway, air, and rail modes. The material covered includes the geometric design of the roadway for the highway mode, the rail track for the rail mode, and runways and taxiways for the air mode. In each case, the human, vehicle, and travelway characteristics discussed in Chapter 3 are used to proportion the different elements of the travelway. For example, the minimum sight distance required for the highway is used to determine the minimum length of a vertical curve. Similarly, for the air mode, the airplane design group that the airport is being designed to serve is used to determine the dimensional standards for runways and taxiways; and for the rail mode, the lengths of vertical curves depend on the type of service the track will be expected to carry.

CLASSIFICATION OF TRANSPORTATION TRAVELWAYS

The design of any transportation facility is based on the classification of that facility. The basis for classification differs from one mode to the other, but the basic principle used is that transportation facilities should be grouped according to their respective functions in terms of the character of the service they are providing. For example, the classification system used for the highway mode facilitates the systematic development of highways and the logical assignment of highway responsibilities among different jurisdictions.

Classification System of Highways and Streets

The American Association of State Highway and Transportation Officials (AASHTO) developed the classification system for highways. These classifications are given in *A Policy on Geometric Design of Highways and Streets* published by AASHTO and are referred to as the *functional classification of highways*. First, highway facilities are classified as urban or rural, depending on the area in which they are located. Urban roads are those located in areas designated as such by local officials with populations of 5000 or more, although some states use other values. Rural roads are those located outside the urban areas. Second, highways are then classified separately for urban and rural areas under the following categories:

- Principal arterials
- Minor arterials
- Major collectors
- Minor collectors
- Local roads and streets

Freeways such as interstate highways are not classified separately, as they are usually considered as principal arterials. It should be noted, however, that freeways and interstate highways have unique geometric criteria that should be considered during their design. Figures 6.1 and 6.2 give schematics of the functional classes of suburban and rural highways.

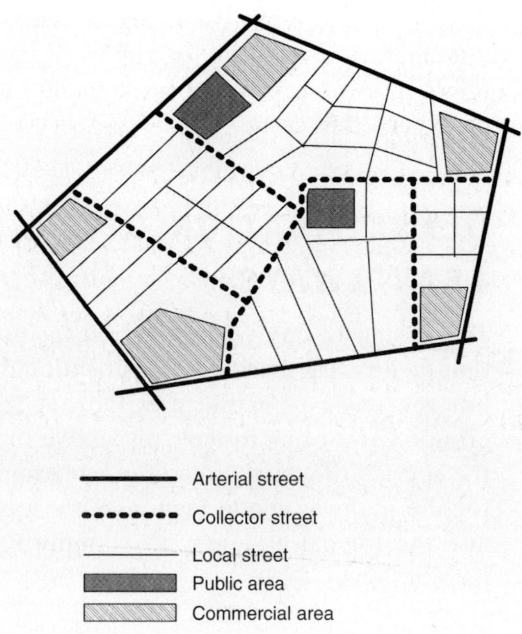

FIGURE 6.1
Schematic of the functional classes of suburban roads.

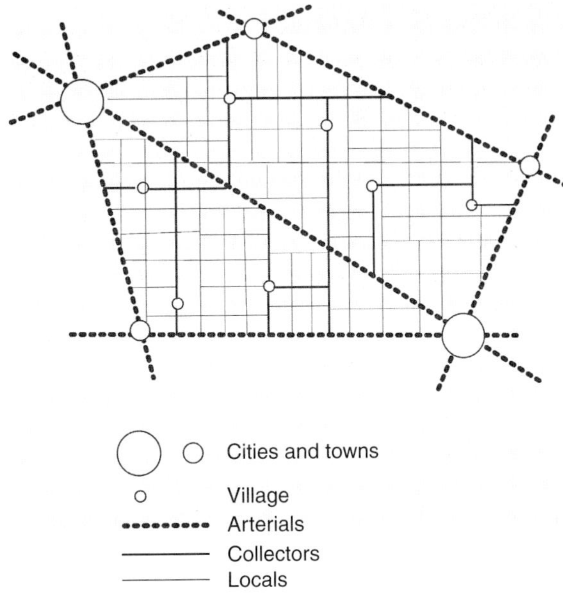

FIGURE 6.2
Schematic of the functional classes of rural roads.

Urban Principal Arterial Roads

Urban principal arterials serve the major activities centers of the urban area and carry the highest traffic volumes, including most trips that start and end within the urban area and all trips that bypass the central business district (CBD) of the urban area. As a result, they carry the highest proportion of vehicle kilometers traveled in the urban area. This category of roads includes all controlled access facilities, although controlled access is not necessarily a requirement for a highway to be included. Highways within this category are further divided into the following subcategories based on the type of access to the facility: (i) interstate highways, with fully controlled access and grade-separated interchanges; (ii) other freeways; and (iii) other principal arterials, which may have partial or no controlled access.

Urban Minor Arterial Roads

Roads within this category interconnect with and augment the urban principal arterials. This category includes all arterial highways that are not classified as principal arterials. Such roads also serve trips of moderate length and provide more access to land use than the principal arterials. These roads do not usually go through identifiable neighborhoods, but they can be used as bus routes and may connect communities within urban areas. Urban minor arterials are usually spaced at distances of not less than 1.5 km in fully developed urban areas, but can also be spaced at distances of 3–5 km in suburban fringes and as low as 0.15 km in central business districts.

Urban Collector Streets

Streets within this category mainly collect traffic from the local streets and convey it to the arterial system. Therefore, these streets usually go through residential areas and support the circulation of traffic within residential, commercial, and industrial areas.

Urban Local Streets

Streets within the urban area that are not included in any of the systems described previously are considered to be within this category. These streets provide access to abutting land and the collector streets, but through traffic is deliberately discouraged on them.

Rural Principal Arterial Roads

Roads within this category serve most of the interstate trips and a substantial amount of intrastate trips. They serve trips between most urban areas with populations higher than 50,000 and a substantial proportion of those between urban areas with populations of over 25,000. They are also further classified as (i) freeways or interstate highways (which are divided highways with fully controlled access and no at-grade intersections; and (ii) other principal arterials, consisting of all principal arterials not classified as freeways.

Rural Minor Arterial Roads

This category of roads supports the rural principal arterials to form a system of roads that connects cities, large towns, and other traffic generators such as large resorts. The spacing of these usually depends on the population density such that reasonable access to the arterial system is provided from all developed areas. Speeds on these roads are usually similar to those on the principal arterials and should be designed to avoid significant interference to through traffic.

Rural Major Collector Roads

Rural major collector roads usually carry trips that have shorter lengths than the arterials, as they primarily carry traffic that originates or ends in county seats or large cities that are not on arterial routes. They also serve other traffic generators such as consolidated schools, shipping points, county parks, and important agricultural and mining areas. In general, major collector roads tend to link the locations they serve with adjacent larger towns, cities, and higher-classification routes.

Rural Minor Collector Roads

The rural minor collector roads system consists of roads that collect traffic from local roads and convey it to other facilities to provide reasonable access to collector roads from all developed areas. An important function of these roads is that they provide linkage between rural hinterland and locally important traffic generators such as small communities.

Rural Local Roads

All rural roads that are not classified within any of the previous classifications form this road system. Roads within this category usually connect adjacent land with the collector streets and carry trip lengths that are relatively shorter than those on the rural collector streets.

Classification of Airport Runways

Airport runways can be generally classified into three main groups:

(i) Primary runways
(ii) Crosswind runways
(iii) Parallel runways

Figures 6.3a and b give a schematic layout of the relative orientation of the different runways.

Primary Runways

Primary runways serve as the primary take-off and landing facility for airports. The lengths of runways are based on either the family of airplanes with similar performance characteristics expected to use the airport or a specific airplane that needs the longest runway. The length is based on a family of airplanes when the maximum gross weight of each airplane that is expected to use the airport is 272000 N or less. It is usually based on a specific airplane when the maximum gross weight of the airplane expected to use the airport is greater than 272000 N. The most desirable orientation for these runways is the one with the largest wind coverage and minimum crosswind component—this is the component of the wind speed perpendicular to the runway direction. Maximum values for the crosswind components are 10.5 knots for Airport Reference Codes (ARCs) of A-1 and B-1; 13 knots for ARCs of A-II and B-II; 16 knots for ARCs of A-III, B-III, and C-I through D-III; and 20 knots for ARCs of A-IV through D-VI (see Chapter 3 for the definition of ARC). The wind coverage is the percentage of time the crosswind components are below the acceptable level. The desirable wind coverage for an airport is 95%.

Crosswind Runways

A crosswind runway is oriented at an angle to the primary runway. Crosswind runways are provided to augment the primary runway and to obtain the desired wind coverage at the airport.

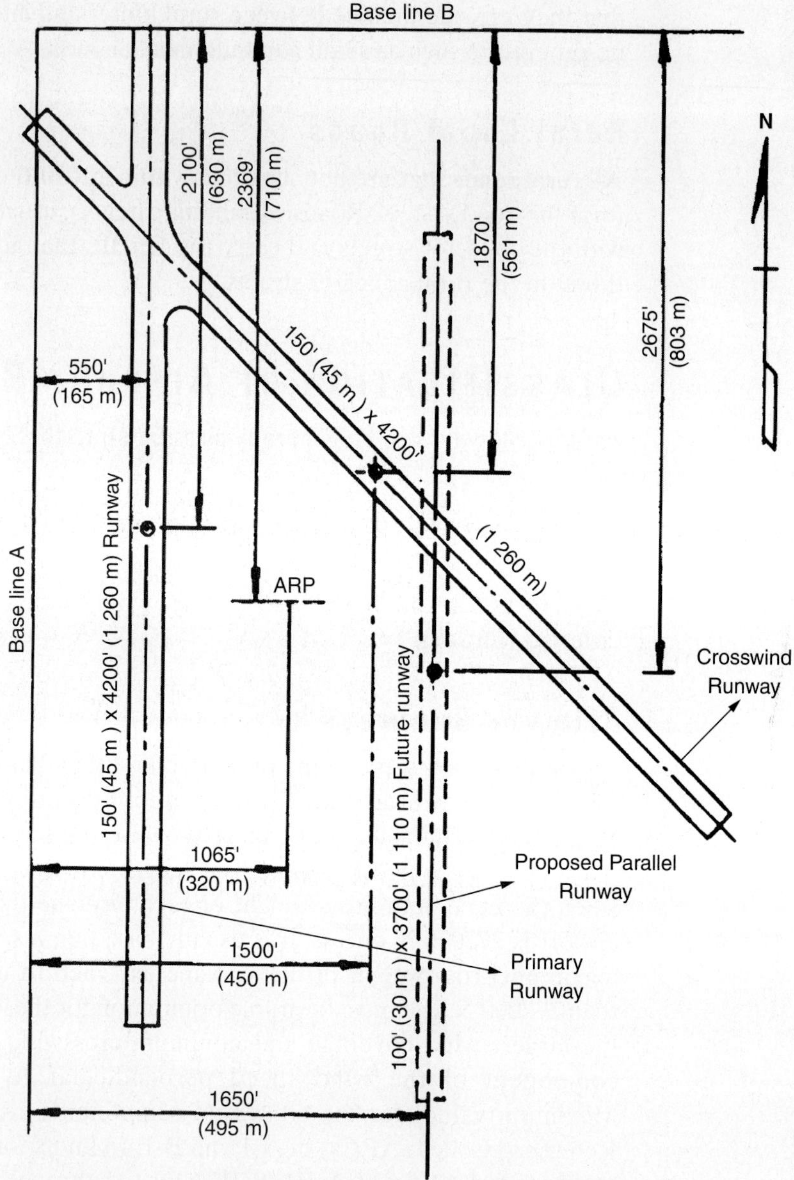

FIGURE 6.3a

Schematic layout of different classes of runways.

Source: Advisory Circular AC 150/5300-13, Federal Aviation Administration, Department of Transportation, Washington, D.C. (Incorporating Changes 1 through 8), September 2004.

Parallel Runways

Parallel runways are constructed parallel to the primary runway to augment it if the traffic volume exceeds the primary runway's operational capability. When visual flight rules (VFRs) are used for simultaneous landings and take-offs at parallel runways, the distance between their centerlines should be not less than 210 m. However, for airplane design groups V and VI, centerlines should be at

FIGURE 6.3b

Layout of Washington–Dulles International Airport—satellite photo.

Source: TerraServer–USA and U.S. Geological Survey.

least 750 m apart (see Chapter 3 for the definition of airplane design groups). Consideration should also be given to increasing these distances to accommodate air traffic control practices, such as holding airplanes between runways. Minimum separation distances between centerlines of primary runways and taxiways and aircraft parking areas are also stipulated as shown in Tables 6.1 and 6.2.

Classification of Airport Taxiways

Taxiways provide access from the runways to the apron, terminal areas, and service hangars. They are located in a way that prevents conflict between an aircraft that has just landed and one that is taxiing to take off. The taxiway system at an airport could become the limiting operational factor as the runway traffic increases. Airport taxiways can be generally classified into the following groups:

(i) Parallel taxiways
(ii) Entrance taxiways
(iii) Bypass taxiways
(iv) Exit taxiways
(v) Apron taxiways and taxilanes

TABLE 6.1

Runway Separation Standards for Aircraft Approach Categories A and B

Source: Adapted from *Airport Design: Advisory Circular AC 150/5300-13,* Federal Aviation Administration, Department of Transportation, Washington, D.C. (Incorporating Changes 1 through 8), September 2004.

ITEM		AIRPLANE DESIGN GROUP				
		I[a]	I	II	III	IV
Visual runways and runways with not lower than ¾-statute mile (1200 m) approach visibility minimums.						
Runway Centerline to:						
Taxiway/Taxilane Centerline[b]		150 ft 45 m	225 ft 67.5 m	240 ft 72 m	300 ft 90 m	400 ft 120 m
Aircraft Parking Area		125 ft 37.5 m	200 ft 60 m	250 ft 75 m	400 ft 120 m	500 ft 150 m
Runways with lower than ¾-statute mile (1200 m) approach visibility minumums.						
Taxiway/Taxilane Centerline[b]		200 ft 60 m	250 ft 75 m	300 ft 90 m	350 ft 105 m	400 ft 120 m
Aircraft Parking Area		400 ft 120 m	400 ft 120 m	400 ft 120 m	400 ft 120 m	500 ft 150 m

[a]These dimensional standards pertain to facilities for small airplanes exclusively.

[b]The taxiway/taxilane centerline separation standards are for sea level. At higher elevations, an increase to these separation distances may be required to keep taxiing and holding airplanes clear of the object-free zone (OFZ).

Parallel Taxiways

Parallel taxiways run parallel to primary runways. They provide access from the primary runways to the terminal areas. Distances between parallel taxiways should comply with the minimum distances given in Table 6.3.

Entrance Taxiways

Entrance taxiways provide direct access to the runway. They are usually L shaped and have a right-angle connection to runways. Entrance taxiways that serve bidirectional runways also serve as the final exit taxiways for these runways.

Bypass Taxiways

Bypass taxiways provide the flexibility that is sometimes required in busy airports to move airplanes that are ready for departure to the desired take-off runways. This often occurs at busy airports when a preceding airplane that is not ready to take off blocks the access taxiway. The bypass taxiway can then be used to bypass the blockage. These taxiways therefore facilitate the maneuvering of steady streams of departing airplanes.

TABLE 6.2

Runway Separation Standards for Aircraft Approach Categories C and D

Source: Adapted from *Airport Design: Advisory Circular AC 150/5300-13*, Federal Aviation Administration, Department of Transportation, Washington, D.C. (Incorporating Changes 1 through 8), September 2004.

ITEM		AIRPLANE DESIGN GROUP					
		I	II	III	IV	V	VI
Visual runways and runways with not lower than ¾-statute mile (1200 m) approach visibility minimums.							
Runway Centerline to:							
Taxiway/Taxilane Centerline[a]		300 ft 90 m	300 ft 90 m	400 ft 120 m	400 ft 120 m	3[b] 3[b]	600 FT 180 M
Aircraft Parking Area	G	400 ft 120 m	400 ft 120 m	500 ft 150 m	500 ft 150 m	500 ft 150 m	500 ft 150 m
Runways with lower than ¾-statute mile (1200 m) approach visibility minumums.							
Taxiway/Taxilane Centerline[a]	D	400 ft 120 m	400 ft 120 m	400 ft 120 m	400 ft 120 m	3[b] 3[b]	600 FT 180 M
Aircraft Parking Area	G	500 ft 150 m	500 ft 150 m	500 ft 150 m	500 ft 150 m	500 ft 150 m	500 ft 150 m

[a]The taxiway-taxilane centerline separation standards are for sea level. At higher elevations, an increase to these separation distances may be required to keep taxiing and holding airplanes clear of the OFZ.

[b]For Airplane Design Group V, the standard runway centerline to parallel taxiway centerline separation distance is 400 ft (120 m) for airports at or below an elevation of 1345 feet (410 m); 450 feet (135 m) for airports between elevations of 1345 feet (410 m) and 6560 feet (2000 m); and 500 feet (150 m) for airports above an elevation of 6560 feet (2000 m).

Exit Taxiways

Exit taxiways are used by airplanes to exit from the runways. They can be right-angled or standard acute-angled taxiways. The acute-angled taxiways, commonly referred to as high-speed exits, allow landing airplanes to exit the runway at higher speeds than right-angled taxiways. The Federal Aviation Administration suggests that when the total landings and take-offs during the peak hour are less than 30, right-angled exit taxiways will achieve an efficient flow of traffic.

Apron Taxiways and Taxilanes

Apron taxiways and taxilanes provide through taxi routes across an apron to gate positions or to other terminal areas. Taxilanes usually provide access from apron taxiways to airplane parking positions and other terminals. Apron taxiways can be located inside or outside the movement area of the apron, but taxilanes should only be provided outside the movement area. The centerline of a taxilane or apron taxiway that is located at the edge of the apron should be inward from the apron edge at one and a half times the width of the taxiway structural pavement.

TABLE 6.3
Taxiway and Taxilane Separation Standards

Source: Adapted from *Airport Design: Advisory Circular AC 150/5300-13,* Federal Aviation Administration, Department of Transportation, Washington, D.C. (Incorporating Changes 1 through 8), September 2004.

ITEM	AIRPLANE DESIGN GROUP					
	I	II	III	IV	V	VI
Taxiway Centerline to:						
Parallel Taxiway/ Taxilane Centerline	69 ft 21 m	105 ft 32 m	152 ft 46.5 m	215 ft 65.5 m	267 ft 81 m	324 ft 99 m
Fixed or Movable Object[a,b]	44.5 ft 13.5 m	65.5 ft 20 m	93 ft 28.5 m	129.5 ft 39.5 m	160 ft 48.5 m	193 ft 59 m
Taxilane Centerline to:						
Parallel Taxilane Centerline	64 ft 195. m	97 ft 29.5 m	140 ft 42.5 m	198 ft 60 m	245 ft 74.5 m	298 ft 91 m
Fixed or Movable Object[a,b]	39.5 ft 12 m	57.5 ft 17.5 m	81 ft 24.5 m	112.5 ft 34 m	138 ft 42 m	167 ft 51 m

[a]This value also applies to the edge of service and maintenance roads.
[b]Consideration of the engine exhaust wake impact from aircraft should be given to objects located near runway/taxiway/taxilane intersections.

Note:
The values obtained from the following equations may be used to show that a modification of standards will provide an acceptable level of safety.

Taxiway centerline to parallel taxiway/taxilane centerline equals 1.2 times airplane wingspan plus 10 feet (3 m).
Taxiway centerline to fixed or movable object equals 0.7 times airplane wingspan plus 10 feet (3 m).
Taxilane centerline to parallel taxilane centerline equals 1.1 times airplane wingspan plus 10 feet (3 m).
Taxilane centerline to fixed or movable object equals 0.6 times airplane wingspan plus 10 feet (3 m).

CLASSIFICATION OF RAILROAD TRACKS

Rail tracks are grouped under the following primary general categories:

(i) Light rail transit tracks
(ii) Urban rail transit tracks
(iii) Freight and intercity passenger tracks
(iv) High-speed railway tracks

In addition to the four primary groups, rail tracks are also grouped into the following secondary categories:

(i) Main line tracks
(ii) Secondary tracks
(iii) Yard and nonrevenue tracks

Railroad tracks can be first classified using the primary categories and then further classified using the secondary categories.

Light Rail Transit Tracks

Light rail transit tracks carry a system of passenger vehicles that are propelled electrically by power obtained from an overhead distribution system of wires. The propulsion power is transmitted by means of a pantograph and returned to the electrical substations through the rails. Operating speeds of light rail transit systems are usually between 65 and 90 km/h. Although the materials used to construct these tracks are the same as those used for other rail systems, such as urban rail transit and freight and intercity passenger rails, the geometric characteristics of light rail tracks have subtle differences from those of other track systems. For example, light rail tracks often have horizontal curves that are as sharp as 82 ft radius. This is due mainly to the type of vehicles used, the necessity for light rail tracks to be capable of accommodating the interaction of automobile traffic and pedestrians, and the need to traverse city streets.

Urban Rail Transit Tracks

Urban rail transit tracks carry urban rail transit vehicles that are usually propelled by direct current electrification at moderate voltages. Running speeds of trains on these tracks can be as high as 130 km/h. Examples of these tracks include the Washington Metropolitan Area Transit Authority Rail system, the Bay Area Rapid Transit system, and the Port Authority Transit Corporation system connecting Philadelphia and New Jersey. They are typically located in major corridors carrying large volumes of passengers.

Freight and Intercity Passenger Tracks

Freight and intercity passenger tracks connect cities and generally carry line-haul rail traffic consisting of both passenger and freight movements. Operations on these lines generate most of the railroad industry's revenue with potential operating speeds of trains on these tracks higher than 160 km/h. These tracks carry the U.S. national railroad passenger service known as Amtrak and freight services. Southern Pacific, Conrail, and CSX Corporation are examples of freight services using these tracks.

High-Speed Railway Tracks

High-speed railway tracks carry high-speed trains traveling at speeds of 145 km/h to 480 km/h, such as the "TGV" high-speed track between Paris and Lyons in France. Several states are planning high-speed rail systems and are upgrading existing tracks to accommodate high-speed trains. For example, the Port Authority of Allegheny County, the Maryland Department of Transportation, the California–Nevada Super Speed Train Commission, the Greater New Orleans Expressway Commission, and the Georgia/Atlanta Regional Commission have all received Federal Railroad Administration grants for pre-construction planning for magnetic levitation (Maglev) high-speed ground transportation. Two approaches can be used in designing these high-speed tracks. The first approach assumes only passenger trains, run on the tracks, and the second allows for both passenger and freight trains. When these tracks are designed for passenger trains only, relatively higher grades may be allowed because of the low load/axle. The high-speed track between Paris and Lyons is an example of this type of design. However, it is now common for these tracks to be designed for both passenger and freight trains. Design standards for these tracks are not given as their design is outside the scope of this book.

Main Line Tracks

Main line tracks form the primary network of railroads and connect the primary origins and destinations of the system.

Secondary Tracks

Secondary tracks are sometimes referred to as branch lines and include tracks that connect the mainline to a station not on the mainline, and tracks that connect the mainline with railroad yards.

Yard and Nonrevenue Tracks

Yard and nonrevenue tracks enter railroad yards where cars are sorted and maintenance and repairs to cars and locomotive engines are carried out.

DESIGN STANDARDS FOR TRAVELWAYS

In the design of the travelway of any transportation system, the first step is to determine the appropriate standards that should be used for the specific facility being designed. For example, in the highway mode, the classification of the road being designed should first be ascertained and then the specific geometric standards for highways under that classification determined. Similarly, in the design of a runway for an airport, the designer should know the classification of the runway (i.e., primary, parallel, or crosswind) and the airplanes that will be using the runway on a regular basis. These standards are then used as the basis for design. The standards considered in this chapter are those related to the alignment design and not those related to the characteristics of the soil support. The standards related to soil characteristics are discussed in Chapter 7.

Highway Design Standards

In addition to the specified design volume, standards are usually given for the design speed and cross-sectional elements such as lane width, shoulder width, medians, and grades. Standards are also given for roadside appurtenances, including median and roadside barriers, curbs and gutters, and guardrails.

Specified Design Volume

This is the projected volume used for design and is given either as a daily (24 hr) volume or as a design hourly volume (DHV). When given as a daily volume, it is either given as the average annual daily traffic (AADT) or as the average daily traffic (ADT). The AADT is the average of 24-hr counts collected every day of the year, while the ADT is the average of 24-hr counts collected over a number of days but less than a year. When the specified design volume is given as an hourly volume, it is usually taken as a percentage of either the expected AADT or the expected ADT. The relationship between hourly traffic volumes as a percentage of ADT on rural highways and the number of hours in one year with higher volumes is shown in Figure 6.4. Data collected from traffic counts on highways with a wide range of volumes and geographic locations were used to develop this relationship. Note that these curves have a unique characteristic—between 0 and about 25 highest hours, a significant increase in the percentage of ADT is observed for a small increase in the number of hours. However, only a

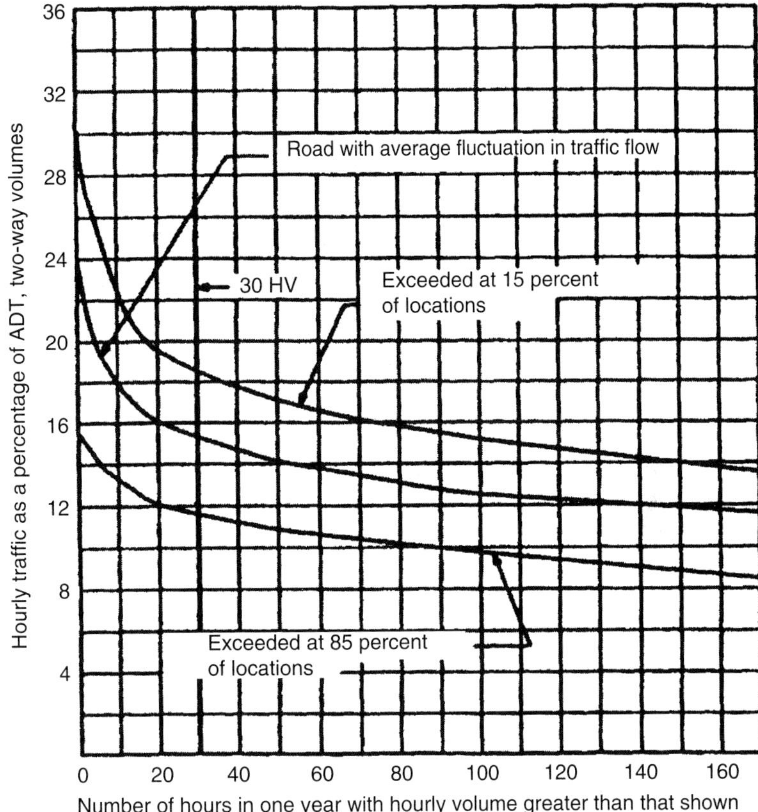

FIGURE 6.4

Relationship between hourly volume (two-way) and annual average daily traffic on rural roads.

Source: A Policy on Geometric Design of Highways and Streets, American Association of State Highway and Transportation Officials, Washington, D.C., 2004. Used by permission.

slight reduction in the percentage of ADT is observed for changes in the number of hours at the right of the 30th highest hour. It was therefore concluded that it is uneconomical to select a DHV that will be exceeded during less than 30 hours in the year. The 30th hourly volume is therefore usually selected as the DHV. Experience has also shown that there is very little variation from year to year in the percentage of the ADT represented by the 30th highest hour volume, even when significant changes in ADT are observed.

The 30th highest hourly volume for rural highways is usually between 12 and 18% of the ADT, with the average being 15%. Care should, however, be taken in using the 30th highest hour volume on highways with unusual or high seasonal fluctuation in traffic flow. It is likely that on these roads a high percentage of high-volume hours and a low percentage of low-volume hours may occur as a result of the seasonal fluctuation. This fluctuation may occur even though the percentage of annual average daily traffic represented by the 30th highest hour may not be significantly different from those on most rural roads. Under such conditions, the 30th highest hourly volume may be so high that it may not be economical to use it as the DHV. At the same time, however, the DHV selected should not be so low as to result in severe congestion during the peak hour. A compromise is usually made in such cases by selecting a DHV that will not result in severe congestion during the peak hour, but will result in traffic operating at a lower level of service than that which normally exists on rural roads with normal fluctuations. One suggested alternative is to use 50% of the average of a few expected highest hour volumes of the design year.

The 30th highest hourly volume may also be used as the DHV for urban highways. However, there is only a slight variation between the 30th and the 200th highest volume, as morning and after-peak flows are similar in urban areas during the year. On these roads, the 30th highest hourly volume is usually between 8 and 12% of the ADT. An alternative method used to determine an appropriate DHV for urban highways is to compute the average of the afternoon highest volumes for each of the 52 weeks in a year. When there are high seasonal variations on an urban highway, it may be necessary to evaluate specific traffic volumes appropriate for the highway to determine the DHV.

Design Speed

The design speed of a highway is the speed on which the different features of the highway are based. The factors commonly used to guide the selection of an appropriate design speed of a highway are the functional classification, the topography of the area in which the highway is located, and the land use of the adjacent area. For this purpose, the topography of a highway is usually classified as one of three groups: level, rolling, or mountainous terrain.

Level terrain is used to describe a topography that has grades that are 2 degrees or less. Safe sight distances can easily be achieved without extensive earthwork. Trucks and passenger cars can achieve similar speeds on grades.

Rolling terrain describes topography in which the natural slopes generally fall below and rise above the grade of the highway. In areas with this topography, steep grades are sometimes encountered. Truck speeds are reduced below that of passenger cars on some grades, although they do not reduce to a crawling speed.

Mountainous terrain describes topography with steep grades and abrupt changes in the longitudinal and transverse elevations relative to the road. Extensive earthwork is usually required to obtain minimum sight distances, and speeds of trucks on grades are significantly reduced in comparison to those of passenger cars. Trucks may also operate at crawl speeds on some grades.

It is important that an appropriate design speed be selected for a given highway. Highways should not be built with standards that conform to high design speeds when maximum operating speeds, as indicated by posted speed limits, are expected to be much lower. Drivers will generally ignore the posted speed limit and drive at speeds that are near the design speed. The design speed should also be selected to achieve the desired level of operation while ensuring a high level of safety on the highway. Many of the design factors of a road depend on the design speed, which makes it one of the first parameters selected in the design process.

TABLE 6.4
Recommended Design Speeds for Arterials

Source: A Policy on Geometric Design of Highways and Streets, American Association of State Highway and Transportation Officials, Washington, D.C., 2004. Used by permission.

Arterial Classification	*Type of Terrain*		
	Level	*Rolling*	*Mountainous*
Rural	60–75 mi/hr	50–60 mi/hr	40–50 mi/hr
Urban	20–45 mph depending on location (e.g., crowded CBDs)		

Note: 1 mph = 1 mi/hr = 1.61 km/h

Design speeds range from 30 km/h to 130 km/h, with intermediate values at 5 km/h intervals. Freeways are usually designed for speeds of 80–130 km/h. It is recommended that when design speeds as low as 80 km/h are used on freeways, the speed limit should be properly posted and enforced, particularly during off-peak hours, to ensure maximum compliance of the posted speed limit. Experience has also shown that a design speed of 95 km/h or higher can be used on many freeways in developing urban areas without significant increase to the cost of the freeway. A design speed of 110 km/h should be used for urban freeways when the alignment is straight and the locations of the interchanges permit this. A design speed of 80–95 km/h that is consistent with driver expectancy may be used for urban freeways in mountainous terrain. A design speed of 110 km/h is recommended for rural freeways. Design speeds for other arterials, collectors, and local roads could be as low as 30 km/h. Tables 6.4, 6.5, and 6.6 show recommended design speeds for different classes of roads.

TABLE 6.5
Minimum Design Speeds for Collectors

Source: A Policy on Geometric Design of Highways and Streets, American Association of State Highway and Transportation Officials, Washington, D.C., 2004. Used by permission.

Collector Classification	*Type of Terrain*	*Design Speed (mph) for Specified Design Volume (veh/day)*		
		Under 50	*400–2000*	*Over 2000*
Rural	Level	40	50	60
	Rolling	30	40	50
	Mountainous	20	30	40
Urban	All	30 mph*		

*May be higher depending on right-of-way availability, terrain, pedestrian presence, and so on.
Note: 1 mph = 1 mi/hr = 1.61 km/h

TABLE 6.6
Minimum Design Speeds for Local Roads

Source: A Policy on Geometric Design of Highways and Streets, American Association of State Highway and Transportation Officials, Washington, D.C., 2004. Used by permission.

Local Road Classification	*Type of Terrain*	*Design Speed (mph) for Specified Design Volume (veh/day)*					
		Under 50	*50–250*	*250–400*	*400–1500*	*1500–2000*	*2000 and over*
Rural	Level	30	30	40	50	50	50
	Rolling	20	30	30	40	40	40
	Mountainous	20	20	20	30	30	30
Urban	All	20–30-mi/hr mph*					

*Depending on available right-of-way adjacent development and likely pedestrian presence.
Note: 1 mph = 1 mi/hr = 1.61 km/h

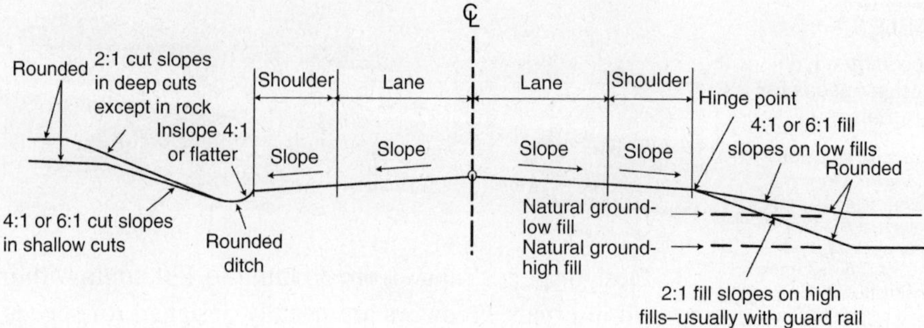

FIGURE 6.5

Typical cross section for two-lane highways.

Source: Redrawn from *A Policy on Geometric Design of Highways and Streets,* American Association of State Highway and Transportation Officials, Washington, D.C., 2004. Used by permission.

Cross-Sectional Elements

The main elements of an undivided highway cross-section are the travel lanes and shoulders. For divided highways, they are the travel lanes, shoulders, and medians. The importance of other elements, such as roadside barriers, curbs, gutters, guardrails, and sidewalks, depends on the type of road being designed. For example, in the design of rural principal arterial roads, it is not important to provide sidewalks, while it may be important to do so in the design of an urban minor arterial road. Figures 6.5 and 6.6 show cross-sectional elements for two-lane highways and divided arterial roads, respectively.

Width of Travel Lanes

This has a significant impact on the operation and safety of the highway. It has been shown that lane widths less than 3.6 m may reduce the capacity of the highway. Research has also shown that the crash rates of large trucks are higher on two-lane highways with lane widths less than 3.3 m than on those with lane widths in excess of 3.3 m. Lane widths generally vary between 3 and 3.9 m, with 3.6 m being the predominant width. Lane widths of 3 and 3.3 m are sometimes used on two-lane, two-way rural roads. Under extreme right-of-way conditions in urban areas, lanes of 2.7 m wide are sometimes used when the expected traffic volume is low.

Width of Shoulders

Figures 6.5 and 6.6 also show the elements of a highway that are designated as shoulders. This is a section contiguous with the traveled lane and has two primary functions. First, it provides a stopping facility for vehicles, particularly during an emergency, and second, it provides lateral support for the pavement structure. The shoulder is sometimes used as a temporary travel lane to avoid excessive congestion, particularly when one of the travel lanes is closed. When shoulders are used as travel lanes, adequate signing must be provided to avoid the shoulder being used as a stopping facility. The shoulder width is known as either graded or useable, depending on the section of the shoulder being considered. The whole width of the shoulder, measured from the edge of the pavement to the intersection of the shoulder slope and the side slope, is the graded shoulder width. The useable shoulder width is the section of the graded shoulder that can be used by vehicles

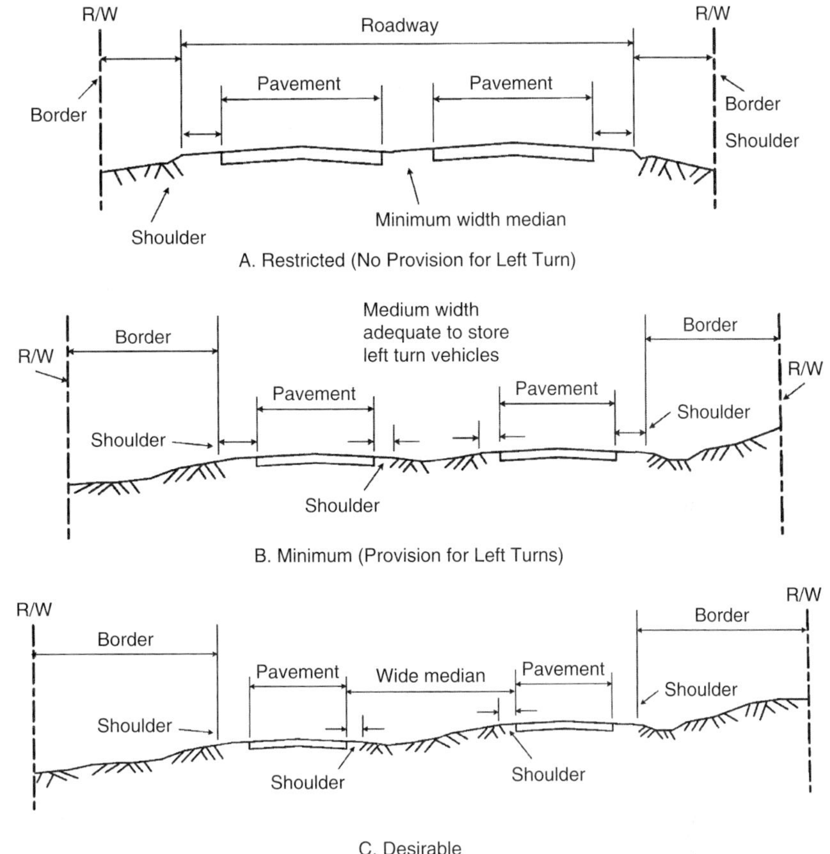

FIGURE 6.6

Typical cross section and right-of-way on divided arterial roads.

Source: A Policy on Geometric Design of Highways and Streets, American Association of State Highway and Transportation Officials, Washington, D.C., 2004. Used by permission.

stopped along the road. When the side slope is equal or flatter than 4:1 (horizontal:vertical) the useable shoulder width is the same as the graded shoulder width, because the break between the shoulder and the side slope is usually rounded to a width between 1.2 and 1.8 m. This, in turn, increases the useable width.

AASHTO recommends that useable shoulder widths should be at least 3 m and preferably 3.6 m on roads that carry a significant number of trucks and on highways with high traffic volumes and high speeds. This is based on the desirability of providing at least a distance of 0.3 m, and preferably 0.6 m, between the edge of the travelway and a stopped vehicle on the shoulder. When it is not feasible to provide this minimum width, 1.8–2.4 m may be used. On low-volume roads, a minimum of 0.6 m ft may be used. The useable width of the inside (median) shoulder may be reduced to 0.9 m on divided four-lane (both directions) arterial roads, as these shoulders are very seldom used by drivers as a stopping facility. However, on six- or eight-lane arterial roads, the useable median shoulder should be at least 2.4 m in order to provide ample space for a driver in difficulty in the lane next to the median to stop.

Shoulders should facilitate the drainage of surface water on the travel lane. They should therefore be flush with the edge of the travelway and be sloped away from the travel lane. Recommended slopes are 2 to 6% for concrete-surfaced shoulders, and 4 to 6% for crushed-rock shoulders.

Medians

The section of a divided highway that separates traffic moving in opposite directions is the median. Its width is measured from the edge of the inside lane in one direction to the edge of the inside lane in the opposite direction. The main functions of medians include

- Providing a recovery area for out-of-control vehicles
- Separating opposing traffic
- Providing stopping areas during emergencies
- Providing storage areas for left-turning and U-turning vehicles
- Providing refuge for pedestrians
- Reducing the effect of headlight glare
- Providing temporary lanes and crossovers during maintenance operations

Medians can be raised, flush, or depressed. In urban areas where it is necessary to control left-turn traffic at intersections of arterial streets, raised medians are often deployed so that part of the median width can be used as a left-turn lane. Flush medians are more commonly used on urban arterial streets but are also used on freeways if a median barrier is provided. Depressed medians are mainly deployed on freeways where they are often used as a means to facilitate drainage of surface water from the travel lanes. AASHTO recommends a slope of 6:1 for depressed medians, although a slope of 4:1 is adequate.

To facilitate safety, medians should be as wide as possible. Median width should, however, be balanced with other elements of the cross-section and the cost involved. The widths of medians generally range from 0.6 m to 24 m or more. AASHTO, however, recommends that a divided arterial should have a minimum median width of 1.2 m. A minimum width of 3 m for four-lane urban freeways is recommended. This includes two 1.2 m shoulders and a 0.6 m median barrier. For freeways of six or more lanes, a minimum of 6.6 m is recommended but 7.8 m is preferable. The widths of medians on urban collector streets vary from 0.6 to 12 m depending on the median treatment. The narrowest widths (0.6 to 12 m) are usually for medians that are separated by paint stripes and the widest (4.8 to 12 m) are for medians in curbed areas. Intermediate widths (0.6 to 1.8 m) are usually for narrow raised curbed areas. Figure 6.6 also shows varying widths of medians on divided highways.

Median and Roadside Barriers

Barriers are generally used to prevent errant vehicles from encroaching into areas where they will be at risk of crashing with other vehicles or will be roadside hazards. Median barriers provide protection for errant vehicles encroaching onto the travelway carrying traffic in the opposite direction, while roadside barriers protect errant vehicles from crashing onto hazards along the side of the road. Consideration should be given to the provision of median barriers when the road is expected to carry high traffic volumes and when access to multilane highways and other highways is only partially controlled. Consideration should also be given to the use of a median barrier when the median on a divided arterial road creates unsafe conditions, such as sudden lateral drop-off or obstacles, even though the expected traffic volume is not high. Conditions that warrant roadside barriers include a high embankment slope and the existence of a roadside object.

Curbs and Gutters

Curbs are used mainly to delineate pavement edges and pedestrian walkways in urban areas, but they can also be used to control drainage. They are constructed of Portland cement concrete or bituminous concrete (rolled asphalt curbs) and are classified as either barrier or mountable curbs. Barrier curbs are designed to prevent vehicles from leaving the highway and are therefore higher (15–20 cm), while mountable curbs are designed to permit vehicles to cross over them if necessary and have heights varying from 10 to 15 cm. Barrier curbs should not be used at the same location with traffic barriers as they may contribute to vehicles rolling over the traffic barriers.

Gutters or ditches provide the principal drainage facility for the roadway. They are usually located at the pavement side of the curb and, in conjunction with storm sewer systems, are used mainly in urban areas to control street runoff. They are usually 0.3 to 1.8 m wide. In order to prevent any hazard to traffic, gutters are usually constructed with cross slopes of 5 to 8% on a width of 0.6 to 0.9 m adjacent to the curb.

Guardrails

Guardrails are used to prevent errant vehicles from leaving the roadbed at sharp horizontal curves and at high embankments. They are normally placed at embankments with heights greater than 2.4 m and when side slopes are greater than 4:1. Guardrails should be properly designed to avoid creating a hazardous situation when they are placed at a location. The proper design of guardrails has therefore become a topic of extensive research, which has resulted in significant improvement to the treatment of the end section of guardrails and barriers.

Sidewalks

Sidewalks are provided mainly on urban streets to facilitate the safe movement of pedestrians. For example, on urban collector streets, sidewalks are usually located on both sides of the street to provide access for pedestrians, particularly at areas adjacent to schools, transit stops, parks, and shopping centers. Although they are not normally provided in rural areas, consideration should be given to providing them at locations with high pedestrian concentrations, such as areas adjacent to schools, industrial plants, and local businesses. Sidewalks should also be provided along arterials with no shoulders, even though pedestrian traffic is low. Sidewalks should have a minimum clear width of 1.2 m in residential areas with a range of 1.2 to 2.4 m in both residential and commercial areas.

Cross Slopes

To facilitate the drainage of surface water, pavements on straight sections of two-lane and multilane highways without medians are sloped in the transverse direction from the middle of the road toward the edges on both sides of the road, and their cross-section can be plain, curved, or a combination of the two. When cross slopes are plain, uniform transverse slopes are provided at both sides of the crown. The curved cross-section usually takes the shape of a parabola, with the highest point (the crown) of the pavement slightly rounded and the cross slope increasing toward the edge of the pavement. The increasing transverse slope of the curved cross-section enhances the flow of water from the pavement surface, which gives the curved surface an advantage. However, the curved sections are more difficult to construct.

The cross slopes on divided highways are achieved by sloping the pavement of each roadway section in two directions, by providing a crown, or by sloping the entire pavement of each section in one direction. When the cross slopes are provided in two directions on divided highways, the drainage of surface water is enhanced as the surface water is quickly removed from the travelway. The disadvantage, however, is that this type of construction requires additional drainage facilities such as inlets and underground drains.

Although a high rate of cross slope is better for drainage purposes, this requirement should be balanced with the necessity to provide a rate that will not cause vehicles to drift to the edge of the pavement, particularly during icy conditions. AASHTO recommends that rates of cross slopes for high-type pavements should be 1.5–2% and for intermediate-type pavements 1.5–3%. High-type pavements are defined as those that have wearing surfaces that can adequately support the expected traffic load without visible distress due to fatigue and are not susceptible to weather conditions. Intermediate-type pavements have surfaces that vary from qualities just below that of high type pavements to surface treated pavements.

Grades

It is well known that the operating speed of a heavy vehicle can be significantly reduced on a steep and/or long grade and that performance of passenger cars can be affected by steep grades. It is therefore necessary to select maximum grades judiciously. Maximum grades are based on the design speed and the design vehicle.

When the design vehicle is a passenger car, grades of up to 4 or 5% can be used without any significant impact on the car's performance, except for those with high weight/horsepower ratios such as compact and subcompact cars. When grades are higher than 5%, speeds of passenger cars increase on downgrades and decrease on upgrades. When the design vehicle is a truck, particular attention should be given to the maximum grade of the highway, as grade has a greater impact on trucks than on passenger cars. For example, the speed of a 120 kg/kw or 90 kg/hp truck will reduce from 88 km/h to about 61 km/h after traveling a distance of about 60 m on a grade of 4%.

Although the impact of grades on recreational vehicles is not as severe as that for trucks, it is more significant than that for passenger cars. The problem, though, is that it is not easy to establish maximum grades for recreational routes. When the percentage of recreational vehicles is high on a road, it may be necessary to provide climbing lanes on steep grades.

Table 6.7 gives recommended values for maximum grades for different road classifications. Every effort should be made to use these maximum values

TABLE 6.7
Recommended Maximum Grades

Source: Adapted from *A Policy on Geometric Design of Highways and Streets*, American Association of State Highways and Transportation Officials (AASHTO), Washington, D.C., 2004. Used by permission.

Rural Collectors[a]

Type of Terrain	*20*	*30*	*40*	*50*	*60*	*70*
			Grades (%)			
Level	7	7	7	6	5	4
Rolling	10	9	8	7	6	5
Mountainous	12	10	10	9	8	6

Design Speed (mi/hr)

Urban Collectors[a]

Type of Terrain	*20*	*30*	*40*	*50*	*60*	*70*
			Grades (%)			
Level	9	9	9	7	6	5
Rolling	12	11	10	8	7	6
Mountainous	14	12	12	10	9	7

Design Speed (mi/hr)

Rural Arterials

Type of Terrain	*40*	*50*	*60*	*70*
		Grades (%)		
Level	5	4	3	3
Rolling	6	5	4	4
Mountainous	8	7	6	5

Design Speed (mi/hr)

Freeways[b]

Type of Terrain	*50*	*60*	*70*
	Grades (%)		
Level	4	3	3
Rolling	5	4	4
Mountainous	6	6	5

Design Speed (mi/hr)

[a]Maximum grades shown for rural and urban conditions of short lengths (less than 500 ft) and on one-way downgrades may be 1% steeper.

[b]Grades that are 1% steeper than the value shown may be used for extreme cases in urban areas where development precludes the use of flatter grades and for one-way downgrades, except in mountainous terrain.

Note: 1 mi/hr = 1.61 km/h

only when necessary, particularly when grades are long and the percentage of trucks in the traffic stream is high. However, when grades are less than 150 m long and are one way in the downgrade direction, maximum grades may be increased by 1 or 2%, particularly on low-volume rural roads.

It is also necessary to stipulate a minimum grade as this facilitates drainage along the highway. The minimum grade on an uncurbed pavement can be as low as 0% if the pavement has adequate cross slopes to drain the surface water away from the travelway. A longitudinal grade should, however, be provided on curbed pavements to facilitate the longitudinal flow of the surface water. This grade is usually 0.5%, although 0.3% can be used on high-type pavement constructed on suitable crowned, firm ground.

EXAMPLE 6.1

Determining Appropriate Design Standards for a Rural Collector Road

A four-lane divided rural collector road with a design volume of 500 veh/day and expected to carry very low truck volume is to be designed for a level area. Determine

(a) An appropriate design speed
(b) The maximum grade based on the selected design speed
(c) An appropriate useable shoulder width
(d) An appropriate median width

Solution

(a) Use Table 6. 5 to select design speed.

Appropriate design speed is 80 km/h.

(b) Use Table 6.7 to determine maximum grade.

Maximum grade is 6%. Note that grades should be made flatter if possible.

(c) For useable shoulder width, AASHTO recommends a minimum of 3 m, preferably 3.6 m on high-speed roads with high truck volumes. In this case, although truck volume is low, no restriction on right-of-way is indicated; therefore, use 3.6 m. However, if right-of-way restriction is encountered, 1.8–2.4 m may be used.

(d) The median width should be as wide as possible. A minimum width of 3 m is desirable.

Runway and Taxiway Design Standards

The standards presented in this section are those recommended by the Federal Aviation Administration as given in its advisory circulars. The recommended standards have been developed to ensure the safety, economy, and longevity of an airport. It should be noted that in using these standards, a designer should be aware of the significant interrelationship among the various features of the airport. Therefore, it is necessary to ensure that the requirements for other airport facilities that are related to the runways and taxiways are also met.

Runway and Taxiway Location and Orientation

The safety, efficiency, economics, and environmental impacts of an airport depend on the runway location and orientation. The extent to which any of these impacts is considered depends on the airport reference code (ARC), the topography of the airport, and the expected air traffic volume. For example, as noted earlier, the maximum crosswind component for a given airport depends on the ARC of the airport, which significantly influences the orientation of the runway. Specific guidelines are given for wind, obstructions to air navigation, topography, airport traffic control, tower visibility, and wildlife hazards.

Wind

The best alignment for a runway is in the direction of the prevailing wind, as this gives the maximum wind coverage and minimum crosswind component. However, when this is not feasible, the runway should be oriented so as to achieve a minimum wind coverage of 95%. When a single runway cannot achieve this wind coverage, an additional runway should be considered. A wind analysis is presented later in this chapter.

Obstruction to Air Navigation

The orientation of runways should ensure that the airport areas associated with the ultimate development of the airport are clear of hazards to air navigation.

Topography

Considering that the topography affects the extent to which extensive grading is required, the runway should be oriented to minimize the amount of grading. Also, grades should not exceed the recommended maximum, nor should lengths of vertical curves be less than the recommended minimum.

The Federal Aviation Administration recommended maximum grades and minimum lengths of vertical curves for runways are as follows:
For Aircraft Approach Categories A and B

- Maximum longitudinal grade: ±2%.
- Maximum allowable longitudinal grade change: ±2%.

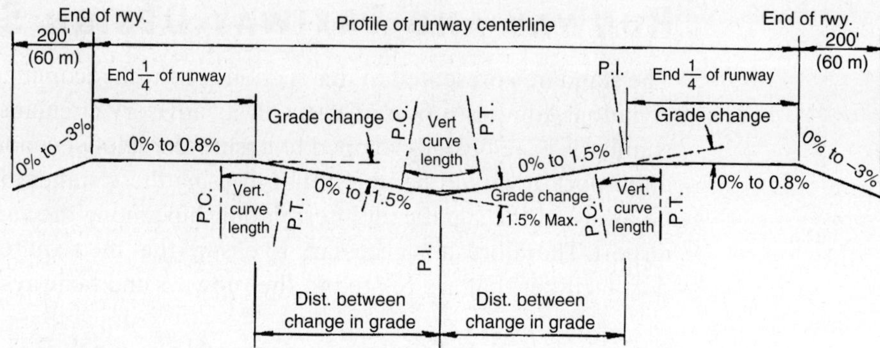

FIGURE 6.7 Longitudinal grade requirements for aircraft approach categories C and D.

Source: Advisory Circular AC 150/5300-13, Federal Aviation Administration, Department of Transportation, Washington, D.C. (Incorporating Changes 1 through 8), September 2004.

- Minimum length of vertical curves: 90 m for each 1% of grade change. A vertical curve is not necessary if the grade change is lower than 0.4%.
- Minimum distance between the points of intersections of consecutive vertical curves is 75 m multiplied by the sum of the grade changes, in percent, associated with the two vertical curves.

For Aircraft Approach Categories C and D

- Maximum longitudinal grade: ±1.5% but not to exceed ±0.8% in the first and last quarter of the runway.
- Maximum allowable grade change: ±1.5%.
- Minimum length of vertical curves: 300 m for each 1% of grade change (algebraic difference) in grades (A).
- Minimum distance between the points of intersections of consecutive vertical curves is 300 m multiplied by the sum of the grade changes, in percent, associated with the two vertical curves.

Figure 6.7 illustrates these requirements for aircraft approach categories C and D. Note that these grades are much lower than the maximum grades for highways. Although the longitudinal maximum grades just given are permissible, it is recommended that longitudinal grades be kept to a minimum. Also, longitudinal grade changes should be used only when absolutely necessary.

Maximum longitudinal grades on taxiways are similar to those for runways. The maximum grades are as follows:

Aircraft Approach Categories A and B: ±2%

Aircraft Approach Categories C and D: ±1.5%

It is also recommended that changes in longitudinal grades of taxiways be avoided unless absolutely necessary. When longitudinal grade change is necessary, it should not be greater than 3%. The vertical curves connecting different grades on taxiways are also parabolic, and their lengths should be not less than 30 m for each 1% of grade change. Also, the distance between consecutive vertical curves should not be less than 30 m multiplied by the sum of the grade changes, in percent, associated with the two vertical curves.

Airport Traffic Control Tower Visibility

It is essential that all runways and taxiways be oriented in a manner that provides a clear line of sight for all traffic patterns and all operational surfaces under airport traffic control. These include the final approaches to all runways and all runway structural pavements. It is also desirable to have a clear line of sight to taxiway centerlines.

Runway Safety Area (RSA)

An area symmetrically located along the centerline of the runway should be provided to enhance the safety of airplanes that may undershoot, overrun, or veer off the runway. This area should be structurally capable of supporting the loads applied by aircraft without causing any structural damage to the aircraft or injury to occupants of the aircraft. Objects should not be located within this area, except those that are needed because of their function. When these objects are over 0.9 m high, they should be on low-impact resistance support so that they can be broken on impact, with the height of the frangible point not greater than 7.5 cm. The dimensional standards for the runway safety area are given in Tables 6.8, 6.9, and 6.10.

Taxiway Safety Area (TSA)

This is an area located along the taxiway similar to the runway safety area. Its functions are similar and its dimensional standards are given in Table 6.11.

Runway Object-Free Area (OFA)

All objects higher than the elevation of the runway safety area edge should be precluded from this area except those that are needed for air navigation or the ground maneuvering of aircraft. However, these objects should not be placed within the OFA if they are precluded by other clearing standards. The OFA is also located symmetrically along the centerline of the runway with dimensions of Q and R shown in Figure 6.8 for which standards are given in Tables 6.8, 6.9, and 6.10.

Taxiway and Taxilane Object-Free Area

This area is similar to the OFA for runways in that service vehicle roads, parked airplanes, and above-ground objects are prohibited. Exceptions to this requirement include objects for air navigation and those required for aircraft ground maneuvering. The operation of motor vehicles within this area may be permitted, but these vehicles must give the right-of-way to oncoming aircraft by keeping a safe distance. Alternatively, exiting facilities could be provided along the outside of the OFA to facilitate vehicles leaving the area to allow the aircraft through.

TABLE 6.8

Runway Design Standards for Aircraft Approach Categories A and B Visual Runways and Runways with Approach Visibility of ¾-statute Mile (1200 m) Minimum*

Source: Adapted from *Airport Design: Advisory Circular AC 150/5300-13,* Federal Aviation Administration, Department of Transportation, Washington, D.C. (Incorporating Changes 1 through 8), September 2004.

ITEM	*DIM*[a]	*AIRPLANE DESIGN GROUP*				
		I[b]	*I*	*II*	*III*	*IV*
Runway Length		Refer to section on runway length on page 376				
Runway Width		60 ft 18 m	60 ft 18 m	75 ft 23 m	100 ft 30 m	150 ft 45 m
Runway Shoulder Width		10 ft 3 m	10 ft 3 m	10 ft 3 m	20 ft 6 m	25 ft 7.5 m
Runway Blast Pad Width		80 ft 24 m	80 ft 24 m	95 ft 29 m	140 ft 42 m	200 ft 60 m
Runway Blast Pad Length		60 ft 18 m	100 ft 30 m	150 ft 45 m	200 ft 60 m	200 ft 60 m
Runway Safety Area Width		120 ft 36 m	120 ft 36 m	150 ft 45 m	300 ft 180 m	500 ft 150 m
Runway Safety Area Length Prior to Landing Threshold		240 ft 72 m	240 ft 72 m	300 ft 90 m	600 ft 180 m	600 ft 180 m
Runway Safety Area Length beyond RW End[c]		240 ft 72 m	240 ft 72 m	300 ft 90 m	600 ft 180 m	1000 ft 300 m
Runway Object-Free Area Width	Q	250 ft 75 m	400 ft 120 m	500 ft 150 m	800 ft 240 m	800 ft 240
Runway Object-Free Area Length beyond RW End[c]	R	240 ft 72 m	240 ft 72 m	300 ft 90 m	600 ft 180 m	1000 ft 300 m

*Visual runways and runways with not lower than ¾-statute mile (1200 m) approach visibility minimums.

[a]Letters correspond to the dimensions in Figure 6.8.

[b]These dimensional standards pertain to facilities for small airplanes exclusively.

[c]The runway safety area and runway object free area lengths begin at each runway end when stopway is not provided. When stopway is provided, these lengths begin at the stopway end. The runway safety area length and the object-free area length are the same for each runway end. Use Table 6.8 or 6.9 to determine the longer dimension. RSA length beyond the runway end standards may be met by provision of an Engineered Materials Arresting System or other FAA-approved arresting system providing the ability to stop the critical aircraft using the runway exiting the end of the runway at 70 knots. See AC 150/5220-22.

TABLE 6.9

Runway Design Standards for Aircraft Approach Categories A and B Visual Runways and Runways with Approach Visibility Less Than ¾-statute Mile (1200 m) Minimum*

Source: Adapted from *Airport Design: Advisory Circular AC 150/5300-13,* Federal Aviation Administration, Department of Transportation, Washington, D.C. (Incorporating Changes 1 through 8), September 2004.

ITEM	*DIM*[a]	*AIRPLANE DESIGN GROUP*				
		I[b]	*I*	*II*	*III*	*IV*
Runway Length		Refer to section on runway length on page 360				
Runway Width		75 ft 23 m	100 ft 30 m	100 ft 30 m	100 ft 30 m	150 ft 45 m
Runway Shoulder Width		10 ft 3 m	10 ft 3 m	10 ft 3 m	20 ft 6 m	25 ft 7.5 m
Runway Blast Pad Width		95 ft 29 m	120 ft 36 m	120 ft 36 m	140 ft 42 m	200 ft 60 m
Runway Blast Pad Length		60 ft 18 m	100 ft 30 m	150 ft 45 m	200 ft 60 m	200 ft 60 m
Runway Safety Area Width		300 ft 90 m	300 ft 90 m	300 ft 90 m	400 ft 120 m	500 ft 150 m
Runway Safety Area Length Prior to Landing Threshold		600 ft 180 m	600 ft 180 m	600 ft 180 m	600 ft 180 m	600 ft 180 m
Runway Safety Area Length beyond RW End[c]		600 ft 180 m	600 ft 180 m	600 ft 180 m	800 ft 240 m	1000 ft 300 m

ITEM	DIM[a]	AIRPLANE DESIGN GROUP				
		I[b]	I	II	III	IV
Runway Object-Free Area Width	Q	800 ft 240 m	800 ft 240 m	800 ft 240 m	800 ft 240 m	800 ft 240 m
Runway Object-Free Area Length beyond RW End[c]	R	600 ft 180 m	600 ft 180 m	600 ft 180 m	800 ft 240 m	1000 ft 300 m

*runways with lower than ¾-statute mile (1200 m) approach visibility minimums.

[a]Letters correspond to the dimensions in Figure 6.8.

[b]These dimensional standards pertain to facilities for small airplanes exclusively.

[c]The runway safety area and runway object-free area lengths begin at each runway end when stopway is not provided. When stopway is provided, these lengths begin at the stopway end. The runway safety area length and the object-free area length are the same for each runway end. Use Table 6.8 or 6.9 to determine the longer dimension. RSA length beyond the runway end standards may be met by provision of an Engineered Materials Arresting System or other FAA-approved arresting system providing the ability to stop the critical aircraft using the runway exiting the end of the runway at 70 knots. See AC 150/5220-22.

TABLE 6.10

Runway Design Standards for Aircraft Approach Categories C and D

Source: Adapted from *Airport Design: Advisory Circular AC 150/5300-13*, Federal Aviation Administration, Department of Transportation, Washington, D.C. (Incorporating Changes 1 through 8), September 2004.

ITEM	DIM[a]	AIRPLANE DESIGN GROUP					
		I	II	III	IV	V	VI
Runway Length		*Refer to text*					
Runway Width		100 ft 30 m	100 ft 30 m	100 ft[b] 30 m[b]	150 ft 45 m	150 ft 45 m	200 ft 60 m
Runway Shoulder Width[c]		10 ft 3 m	10 ft 3 m	20 ft[b] 6 m[b]	25 ft 7.5 m	35 ft 10.5 m	40 ft 12 m
Runway Blast Pad Width		120 ft 36 m	120 ft 36 m	140 ft[b] 42 m[b]	200 ft 60 m	220 ft 66 m	280 ft 84 m
Runway Blast Pad Length		100 ft 30 m	150 ft 45 m	200 ft 60 m	200 ft 60 m	400 ft 120 m	400 ft 120 m
Runway Safety Area Width[d]		500 ft 150 m	500 ft 150 m	500 ft 150 m	500 ft 150 m	500 ft 150 m	500 ft 150 m
Runway Safety Area Length Prior to Landing Threshold		600 ft 180 m	600 ft 180 m	600 ft 180 m	600 ft 180 m	600 ft 180 m	600 ft 180 m
Runway Safety Area Length beyond RW End[e]		1000 ft 300 m	1000 ft 300 m	1000 ft 300 m	1000 ft 300 m	1000 ft 300 m	1000 ft 300 m
Runway Object-Free Area Width	Q	800 ft 240 m	800 ft 240 m	800 ft 240 m	800 ft 240 m	800 ft 240	800 ft 240
Runway Object-Free Area Length beyond RW End[e]	R	1000 ft 300 m	1000 ft 300 m	1000 ft 300 m	1000 ft 300 m	1000 ft 300 m	1000 ft 300

[a]Letters correspond to the dimensions in Figure 6.8.

[b]For Airplane Design Group III serving airplanes with maximum certificated take-off weight greater than 150,000 pounds (68,100 kg), the standard runway width is 150 feet (45 m), the shoulder width is 25 feet (7.5 m), and the runway blast pad width is 200 feet (60 m).

[c]Design Groups V and VI normally require stabilized or paved shoulder surfaces.

[d]For Airport Reference Code C-I and C-II, a runway safety area width of 400 feet (120 m) is permissible. For runways designed after 2/28/83 to serve Aircraft Approach Category D, the runway safety area width increases 20 feet (6 m) for each 1000 feet (300 m) of airport elevation above MSL.

[e]The runway safety area and runway object-free area lengths begin at each runway end when stopway is not provided. When stopway is provided, these lengths begin at the stopway end. (Use Table 6.8 and 6.9 to determine the longer dimension.) Use Tables 6.8 or 6.9 that results in the dimension. RSA length beyond the runway end standards may be met by provision of an Engineered Materials Arresting System or other FAA-approved arresting system providing the ability to stop the critical aircraft using the runway exiting the end of the runway at 70 knots. See AC 150/5220-22.

TABLE 6.11
Taxiway Dimensional Standards

Source: Adapted from *Airport Design: Advisory Circular AC 150/5300-13,* Federal Aviation Administration, Department of Transportation, Washington, D.C. (Incorporating Changes 1 through 8), September 2004.

ITEM	*AIRPLANE DESIGN GROUP*					
	I	*II*	*III*	*IV*	*V*	*VI*
Taxiway Width	25 ft 7.5 m	35 ft 10.5 m	50 ft[a] 15 m[a]	75 ft 23 m	75 ft 23 m	100 ft 30 m
Taxiway Edge Safety Margin[b]	5 ft 1.5 m	7.5 ft 2.25 m	10 ft[c] 3 m[c]	15 ft 4.5 m	15 ft 4.5 m	20 ft 6 m
Taxiway Shoulder Width	10 ft 3 m	10 ft 3 m	20 ft 6 m	25 ft 7.5 m	35 ft[d] 10.5 m[d]	40 ft[d] 12 m[d]
Taxiway Safety Area Width	49 ft 15 m	79 ft 24 m	118 ft 36 m	171 ft 52 m	214 ft 65 m	262 ft 80 m
Taxiway Object-Free Area Width	89 ft 27 m	131 ft 40 m	186 ft 57 m	259 ft 79 m	320 ft 97 m	386 ft 118 m
Taxilane Object-Free Area Width	79 ft 24 m	115 ft 35 m	162 ft 49 m	225 ft 68 m	276 ft 84 m	334 ft 102 m

[a]For airplanes in Airplane Design Group III with a wheelbase equal to or greater than 60 feet (18 m), the standard taxiway width is 60 feet (18 m).

[b]The taxiway edge safety margin is the minimum acceptable distance between the outside of the airplane wheels and the pavement edge.

[c]For airplanes in Airplane Design Group III with a wheelbase equal to or greater than 60 feet (18 m), the taxiway edge safety width is 15 feet (4.5 m).

[d]Airplanes in Airplane Design Group V and VI normally require stabilized or paved taxiway shoulder surface.

Consideration should be given to objects near runway/taxiway/taxilane intersections that can be impacted by exhaust wake from a turning aircraft.

The values obtained from the following equations may be used to show that a modification of standards will provide an acceptable level of safety.

Taxiway safety area width = airplane wingspan;
Taxiway OFA width = 1.4 × airplane wingspan + 20 feet (6m); and
Taxilane OFA width = 1.2 × airplane wingspan + 20 feet (6m).

Wildlife Hazards

The presence of a large number of birds or other wildlife may create a hazardous situation. The relative locations of bird sanctuaries, sanitary landfills, or other land uses that may attract a large number of wildlife should therefore be considered in designing the location and orientation of the runway.

Runway and Taxiway Lengths

As indicated earlier, the length of an airport runway depends on many factors, including the performance characteristics of the particular type of airplane the airport will be serving, the altitude and temperature of the airport, and the length of the trip. In general, runway lengths vary from 600 m to over 3,000 m, and the design procedure for runway lengths is given later in this chapter.

Runway Widths and Runway Shoulder Widths

Standards for the widths of runways and runway shoulders are provided for different airplane design groups and aircraft approach categories and are given in Tables 6.8, 6.9, and 6.10. Shoulders are provided along the edges of the runway

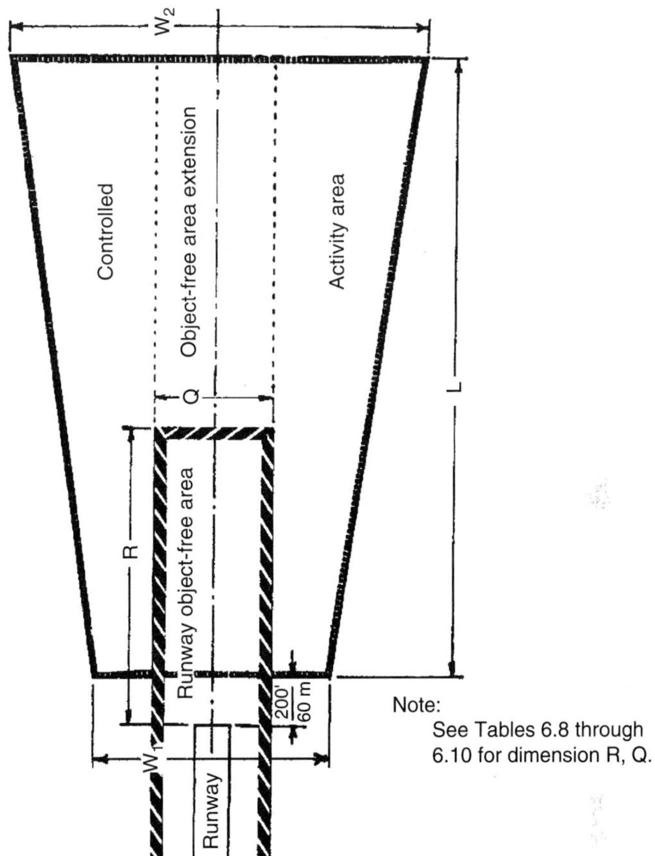

FIGURE 6.8
Runway protection zone.

Source: Advisory Circular AC 150/5300-13, Federal Aviation Administration, Department of Transportation, Washington, D.C. (Incorporating Changes 1 through 8), September 2004.

similar to those provided for highways. Their functions are similar to those of highway shoulders in that they provide an area along the runway for the occasional airplane that veers from the runway and for the movement of emergency and maintenance equipment.

Taxiway Widths and Taxiway Shoulder Widths

Standards for the widths of taxiways and taxiway shoulders are also provided and are given in Table 6.11.

Runway and Taxiway Transverse Slopes

Figures 6.9 and 6.10 give the transverse grade limitations for airport approach categories A and B, and C and D, respectively. The figures also show the main sections of the runway and taxiway. These include the side slopes, the shoulders, the runway safety area, a ditch or gutter, and the taxiway safety area. Note that the transverse cross-sections are similar to those for a highway. The safety area

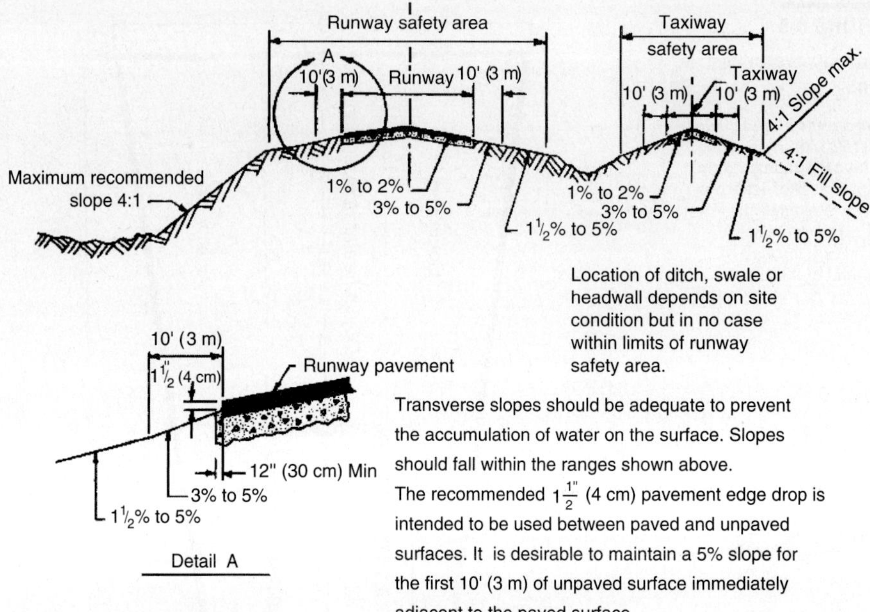

FIGURE 6.9

Transverse grade limitations for aircraft approach categories A and B.

Source: Advisory Circular AC 150/5300-13, Federal Aviation Administration, Department of Transportation, Washington, D.C. (Incorporating Changes 1 through 8), September 2004.

is an area surrounding the runway or taxiway that provides a reduced risk of damage to airplanes that unintentionally leave the runway or taxiway.

Figure 6.9 shows that side slopes for aircraft approach categories A and B cannot be greater than 4:1 for both fills and cuts. The unpaved surface immediately adjacent to the paved surface should have a slope of 3–5%

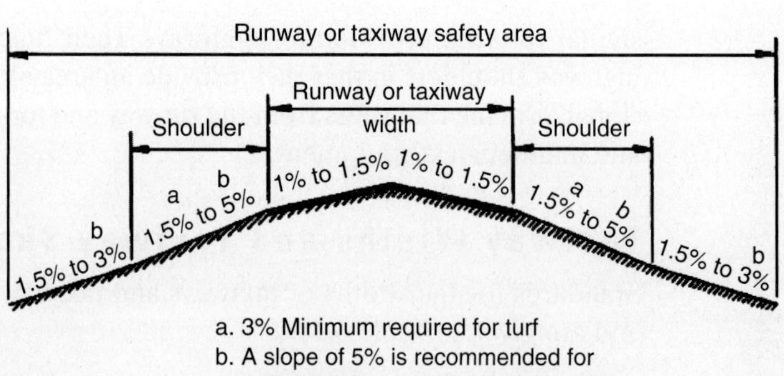

FIGURE 6.10

Transverse grade limitations for aircraft approach categories C and D.

Source: Advisory Circular AC 150/5300-13, Federal Aviation Administration, Department of Transportation, Washington, D.C. (Incorporating Changes 1 through 8), September 2004.

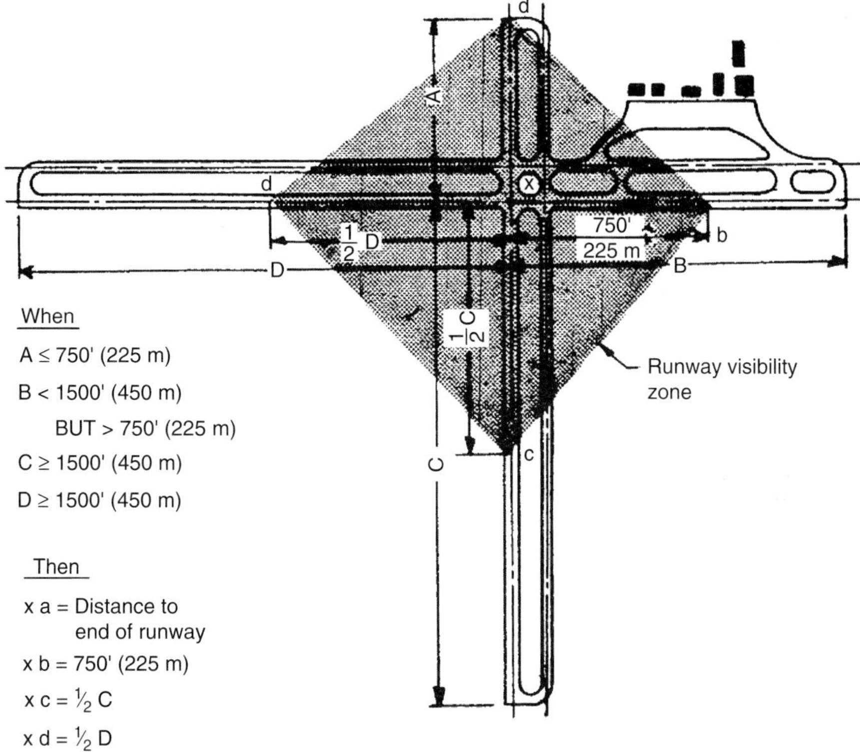

FIGURE 6.11

Runway visibility zone.

Source: Advisory Circular AC 150/5300-13, Federal Aviation Administration, Department of Transportation, Washington, D.C. (Incorporating Changes 1 through 8), September 2004.

(preferably 5%) for a distance of 3 m ft from the paved area. This facilitates drainage of surface water away from the paved surface. In order to facilitate the drainage of surface water from the travelway, the paved surfaces of both the runway and the taxiway are crowned similar to that of a two-lane highway.

Line of Sight along Runway

It is required that a clear line of sight be available along the runway that allows any two points 1.5 m above the runway to be mutually visible for the entire length of the runway.

Line of Sight between Intersecting Runways

It is also recommended that a clear line of sight between intersecting runways be available. This should provide an unobstructed line of sight from any point with a height of 1.5 m above the centerline of one runway, to any similar point above the centerline of an intersecting runway. A runway visibility zone shown in Figure 6.11 should also be provided.

EXAMPLE 6.2

Determining Appropriate Design Standards for an Airport

An airport is being designed for aircraft approach category C. Determine the following:

(a) The maximum longitudinal grade of the primary runway

(b) The minimum length of a vertical curve connecting tangents of +0.5% and −1% grades

(c) The minimum distance between points of intersections of consecutive vertical curves, for the maximum allowable change of grades

Solution

(a) The maximum longitudinal grade for aircraft approach C and D is 1.5%

(b) Minimum length of vertical curve on a primary runway is 300 m for each 1% of grade change (i.e., 300*A* where *A* is an algebraic difference in grades)

$$\text{Grade change} = 0.5 - (-1\%) = 1.5\%$$

$$\text{Minimum length} = (300)(1.5) = 450 \text{ m}$$

(c) Minimum distance between the points of intersection of two curves = $300(A_1 + A_2)$, where A_1 and A_2 are the grade changes in the two curves. Minimum distance is therefore $300(1.5 + 1.5) = 900$ m

RAILROAD TRACK DESIGN STANDARDS

Geometric standards for light rail, urban rail, and freight and passenger intercity tracks are given, including those for longitudinal gradient, circular and vertical curves, and superelevation. Figures 6.12 and 6.13 show transverse cross-sections of a single track and a multiple track respectively. The standards for high-speed rail tracks are beyond the scope of this book and are not discussed in this section.

Longitudinal Gradient

Recommended maximum longitudinal grades for light rail transit and other commuter rails carrying only passenger traffic are similar to those for highways. Those for tracks that also carry freight traffic are much lower than those for highways but are similar to those for airport runways. Maximum grades are specified for different categories of tracks.

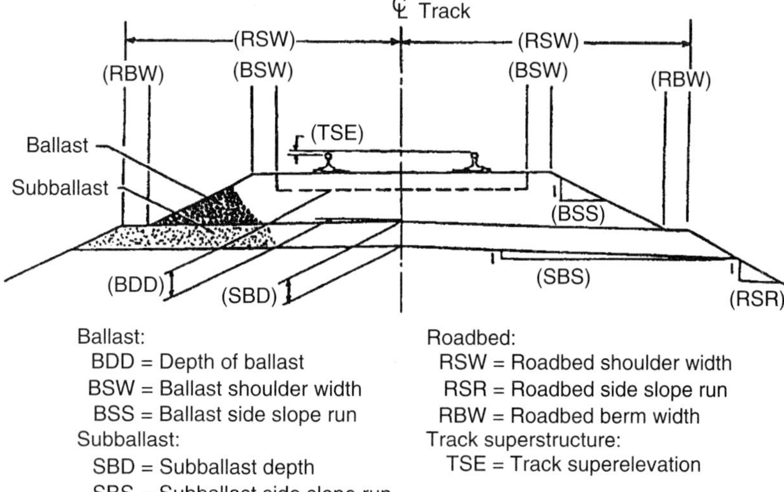

FIGURE 6.12

Cross-section of a single superelevated track.

Source: American Railway Engineering and Maintenance-of-Way Association (AREMA) *Manual for Railway Engineering,* Volume 1, 2005.

Light Rail Main Line Tracks

Maximum grades for these tracks as given in the Transportation Research Board's report *Track Design Handbook for Light Rail Transit* are as follows:

Maximum sustained grade (unlimited length), 4%.

Maximum sustained grade (up to 750 m between points of vertical intersections (PVIs) of vertical curves), 6%.

Maximum short sustained grade (up to 150 m between PVIs of vertical curves), 7%.

Minimum grade for drainage, 0.2%.

No minimum grade is specified for passenger stations on light rail main line tracks. Adequate drainage of the track is, however, required.

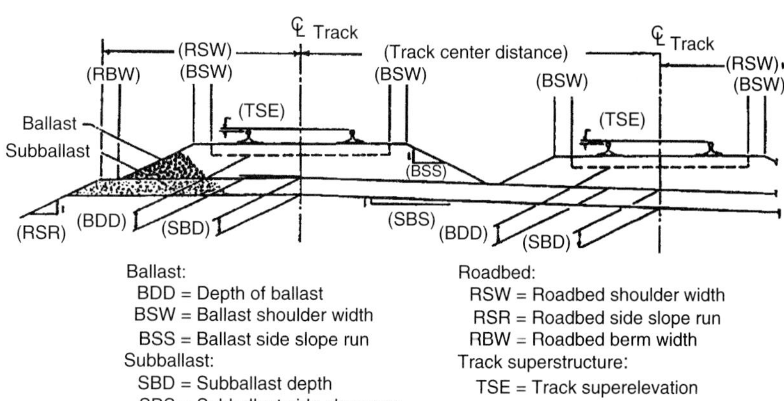

FIGURE 6.13

Cross-section of a dual superelevated rail track.

Source: American Railway Engineering and Maintenance-of-Way Association (AREMA) *Manual for Railway Engineering,* Volume 1, 2005.

Urban Rail Transit Main Line Tracks

Maximum grades of up to 4% have been used, although lower grades are preferred.

Freight and Intercity Main Line Tracks

Grades on these tracks are usually not greater than 1.5%, although grades of up to 3% have been used. An important standard is the requirement for the rate of change in grade. It is recommended by the American Railway Engineering and Maintenance-of-Way Association (AREMA) that the rate of change in grade on high-speed main tracks should not be greater than 3 cm/station of 30 m on crest vertical curves and not greater than 1.5 cm/station of 30 m on sag vertical curves (see Figure 6.14 for crest and sag vertical curves).

Secondary Tracks

Light rail secondary tracks connecting the main line and the rail yard should be designed to prevent rail vehicles from rolling out of the yard to the main line. This is achieved by sloping the secondary line downward away from the main line or by providing a dish on the track between the main line and the rail yard.

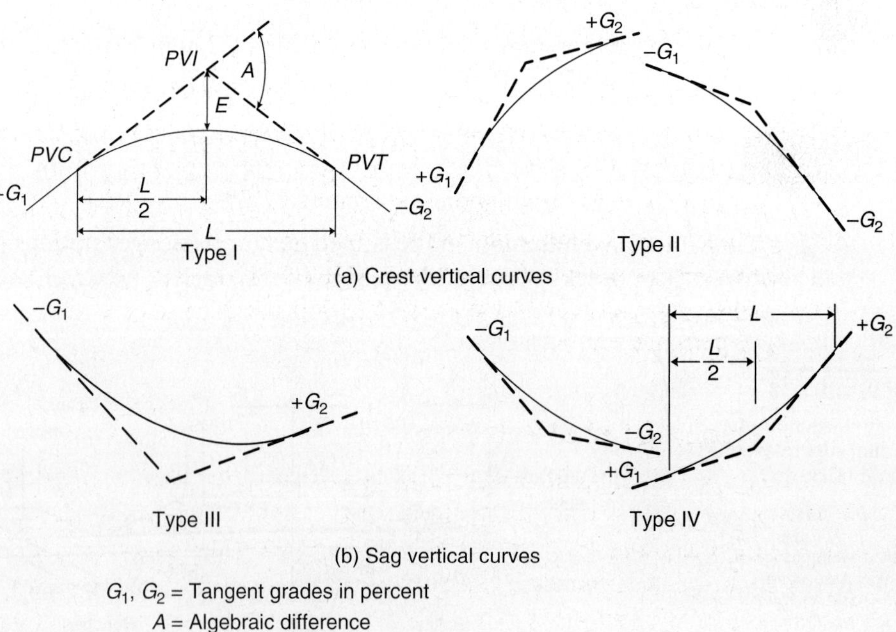

FIGURE 6.14 Types of vertical curves.

It is also recommended that, in order to achieve adequate drainage, grades on these tracks be between 0.35 and 1%.

An additional requirement for freight and intercity secondary tracks is that the rate of change in grade should not be greater than 6 cm/station of 30 m on crest vertical curves and 3 cm/station of 30 m on sag vertical curves.

Light Rail Yard and Nonrevenue Tracks

The desirable grade for these tracks is 0.00%. However, maximum grades of 1% on yard tracks and 0.2% on yard storage and pocket tracks are acceptable.

Design Speed

The most important factors considered in setting the maximum operating speed on any railway track are passenger comfort and safety. In order to achieve passenger comfort, the maximum speed is set so as not to exceed the maximum rate of lateral acceleration that can be comfortably tolerated by passengers. This rate is usually taken as 0.1g. In addition, curved tracks are designed for speeds that do not result in excessive forces on the track work and vehicles. In general, design speeds for light rail transit tracks are usually between 65 and 90 km/h, while design speeds for passenger intercity tracks can be higher than 210 km/h.

Minimum Tangent Length between Horizontal Curves

An important design standard for railroads is the minimum tangent length between horizontal curves. The basic requirement stipulated by AREMA is that the length of a tangent (L_T) between curves should at least be equal to the length of the longest car that the track is expected to carry. This requirement is usually satisfied if the distance of a tangent between curves is at least 30 m. In addition, light rail transit and other passenger tracks must also take passenger comfort into consideration. This has resulted in the following additional requirements for light rail transit and other tracks carrying passenger trains:

Main Line Preferred: The greater of

(i) $L_T = 60$ m

(ii) $L_T = 0.57u$ (u = maximum operating speed km/h)

Main Line Desired: The greater of

(i) L_T = length of light rail transit vehicle over couplers

(ii) $L_T = 0.57u$ (u = maximum operating speed km/h)

Main Line Absolute Minimum: The greater of

(i) $L_T = 9.5$ m

(ii) L_T = (vehicle truck center distance) + (axle spacing)

Main Line Embedded Tracks:

(i) $L_T = 0$; if speed is less than 30 km/h, no track superelevation is used and the vehicle coupler angles are not exceeded, or

(ii) L_T = main line absolute minimum

Note that the minimum tangent length between horizontal curves on main line tracks that are used by both light rail transit and freight trains should be 30 m, although the desired length is 90 m.

It is not practical to achieve the minimum prescribed for main line yard and nonrevenue tracks, as speeds in these facilities are much lower than on the main lines. Also, superelevations are not commonly used on these tracks. AREMA has suggested minimum lengths that are based on the smaller radii of the curves being connected:

$L_T = 9.1$ m for $R > 175$ m

$L_T = 7.6$ m for $R > 195$ m

$L_T = 6.1$ m for $R > 220$ m

$L_T = 3.0$ m for $R > 250$ m

$L_T = 0.0$ m for $R > 290$ m

EXAMPLE 6.3

Designing a Freight and Passenger Intercity Main Line Track

Determine the following on an intercity main line track:

(a) The maximum grade

(b) The maximum rate of change in grade for a crest vertical curve, 180 m long

(c) The maximum rate of change in grade for a sag vertical curve, 180 m long

Solution

(a) Maximum grade of 1.5% is preferred.

(b) Maximum rate of change in grade for a crest vertical curve is 3 cm/station of 30 m

where L = length of curve (m).

Maximum change in grade for a crest:

$= 3 \times 180/30$

$= 18$ cm/station of 30 m

(c) Maximum rate of change in grade for sag vertical curve is 1.5 cm/station of 30 m

where

L = length of curve m

= 1.5 × 180/30

= 9 cm/station of 100 ft

Design of the Vertical Alignment

The vertical alignment of a roadway, runway, or railroad consists of straight sections known as grades or tangents connected by vertical curves. The basic tasks involved in the design of the vertical alignment therefore consist of the selection of appropriate grades for the tangents and the design of appropriate vertical curves to connect these grades. The selection of the appropriate grade depends on the topography on which the travelway is to be located and the standards given for the specific mode. The shape of the vertical curve for each of these modes is the parabola. There are two types of vertical curves: crest vertical curves and sag vertical curves. The different types of vertical curves are shown in Figure 6.14.

It should be noted that points on the centerline of any travelway are identified by their distances from a fixed reference point which is usually the start of the project. These distances are usually given in 30-m stations. For example, if the station of a point is (350 + 8.20), it means that this point is 350 whole stations plus 8.20 m or 10508.2 m from the fixed reference point.

Selection of Appropriate Grades for Highway Vertical Curves

Recommended maximum grades for different types of roads given in Table 6.7 are used to select an appropriate grade for the highway. It should be emphasized that, whenever possible, every effort should be made to select lower grades than those given in the table. Example 6.1 illustrates the use of the table to determine the maximum grade allowable for a given road.

Design of Highway Vertical Curves

After selecting the grades, the next stage is to design an appropriate vertical curve to connect the two intersecting tangents or grades. The main criterion used for designing highway vertical curves is the provision of the minimum stopping sight distance (see Chapter 3). Two conditions exist for the minimum length of highway vertical curves: (1) when the sight distance is greater than the length of the curve; and (2) when the sight distance is less than the length of the curve. Let us first consider the crest vertical curve with the sight distance greater than

FIGURE 6.15
Sight distance on crest vertical curve ($S > L$).

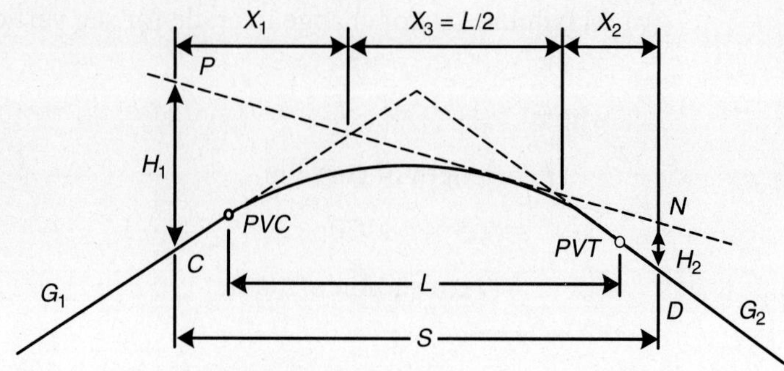

L = Length of vertical curve (ft)
S = Sight distance (ft)
H_1 = Height of eye above roadway surface (ft)
H_2 = Height of object above roadway surface (ft)
G_1 = Slope of first tangent
G_2 = Slope of second tangent
PVC = Point of vertical curve
PVT = Point of vertical tangent

the length of the curve, as shown in Figure 6.15. This figure schematically presents a vehicle on the tangent at C with the driver's eye at height H_1. An object of height H_2 is also located at D. The line of sight, which allows the driver to see the object, is PN. The sight distance is S. Note that the length of the vertical curve (L) and the sight distance (S) are the horizontal projections and not the distances along the curve. The reason is that in laying out these vertical curves the horizontal distance is used. Considering the properties of the parabola

$$X_3 = \frac{L}{2} \tag{6.1}$$

The sight distance S is then given as

$$S = X_1 + \frac{L}{2} + X_2 \tag{6.2}$$

Since the algebraic difference between grades G_1 and G_2 is given as $(G_1 - G_2) = A$, the minimum length (L_{min}) of the crest vertical curve for the required sight distance is obtained as

$$L_{min} = 2S - \frac{200(\sqrt{H_1} + (\sqrt{H_2})^2}{A} \quad \text{(for } S > L) \tag{6.3}$$

AASHTO recommends that the height of the driver's height H_1 above the road surface be taken as 1.1 m and the height of the object H_2 be 0.61 m. Substituting these values for H_1 and H_2 in Equation 6.3, we obtain

$$L_{min} = 2S - \frac{670}{A} \quad \text{(for } S > L) \tag{6.4}$$

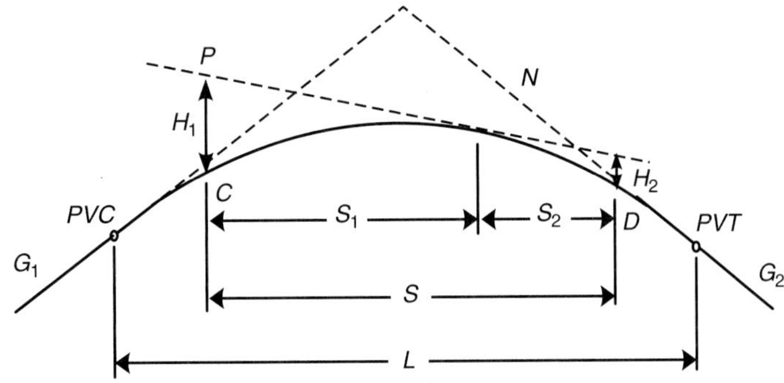

L = Length of vertical curve (ft)
S = Sight distance (ft)
H_1 = Height of eye above roadway surface (ft)
H_2 = Height of object above roadway surface (ft)
G_1 = Slope of first tangent
G_2 = Slope of second tangent
PVC = Point of vertical curve
PVT = Point of vertical tangent

FIGURE 6.16
Sight distance on crest vertical curve ($S < L$).

Now let us consider the length of a highway crest vertical curve where the sight distance is less than the length of the curve, as shown in Figure 6.16. It can be shown that for this case the minimum length (L_{min}) of the vertical curve is given as

$$L_{min} = \frac{AS^2}{200(\sqrt{H_1} + H_2)^2} \quad \text{(for } S < L\text{)} \tag{6.5}$$

Substituting 1.1 m for H_1 and 0.61 m for H_2, we obtain

$$L_{min} = \frac{AS^2}{670} \quad \text{(for } S < L\text{)} \tag{6.6}$$

EXAMPLE 6.4

Determining the Minimum Length of a Highway Crest Vertical Curve

Determine the minimum length of a highway crest vertical curve connecting a +3.5% grade to a −3.5% grade on a rural interstate highway if the design speed is 110 km/h. Assume $S < L$.

Solution

Determine the minimum sight distance. The minimum stopping sight distance (SSD) for a design speed of 110 km/h is first determined. Use Equation 3.21

of Chapter 3:

$$SSD = 0.28ut + \frac{u^2}{254.3(\frac{a}{g} - G)}$$

Use $a = 3.41$ m/s^2 and $g = 9.81$ m/s^2 (see Chapter 3):

$$SSD = 0.28 \times 110 \times 2.5 + \frac{110^2}{254.3(0.35 - 0.035)}$$

$$= 77 + 151.05 \text{ m} = 228.05 \text{ m}$$

Use Equation 6.6 to determine the minimum length of the vertical curve:

$$L_{\text{min}} = \frac{AS^2}{670}, \quad A = (+3.5) - (-3.5) = 7$$

$$L_{\text{min}} = \frac{7 \times 228.05^2}{670}$$

$$= 543.35 \text{ m}$$

EXAMPLE 6.5

Determining Maximum Safe Speed on a Highway Crest Vertical Curve

In designing a vertical curve joining a +2% grade and a −2% grade on a rural arterial highway, the length of the curve must be limited to 210 m because of topographical and right-of-way restrictions. Determine the maximum safe speed on this section of the highway.

Solution

Determine the stopping sight distance (SSD) using the length of the curve. In this case it is not known whether $S < L$ or $S > L$. Let us first assume that $S < L$. Use Equation 6.6:

$$L_{\text{min}} = \frac{AS^2}{670}, \quad A = (+2.0) - (-2.0) = 4$$

$$L_{\text{min}} = 4 \times S^2/670$$

$$210 = 4 \times S^2/670$$

$$S = 187.55 \text{ m}$$

$S < L$; assumption is correct

Use Equation 3.21 to determine the maximum speed for the SSD of 187.55 m.

$$SSD = 0.28ut + \frac{u^2}{254.3(0.35 - G)}$$

$$187.55 = 0.28 \times 2.5 \times u + u^2/254.3(0.35 - 0.02)$$

which gives:

$$u^2 + 58.74u - 14{,}785.13 = 0$$

Solve the quadratic equation to determine the maximum speed u:

$$u = 95.72 \text{ km/h}$$

The maximum safe speed is therefore 95 km/h.

Let us now consider the highway sag vertical curve. The minimum length of a highway sag vertical curve is usually based on the following criteria:

(i) Sight distance provided by the headlight;
(ii) Rider comfort;
(iii) Control of drainage; and
(iv) General appearance.

The requirement for the headlight distance takes into consideration the distance that is lighted by the headlight beam and can therefore be seen at night as a vehicle is driven on a sag vertical curve. This distance depends on the position of the headlight and the inclination of the headlight beam. Figure 6.17 gives a schematic representation of the situation when the sight distance is greater than the length of the curve ($S > L$). Let the headlight be located at a height H above the ground, and the inclination of the headlight beam upward to the horizontal is β. The headlight beam will intersect the roadway at D at a distance S. The available sight distance will then be limited to S. Values recommended by

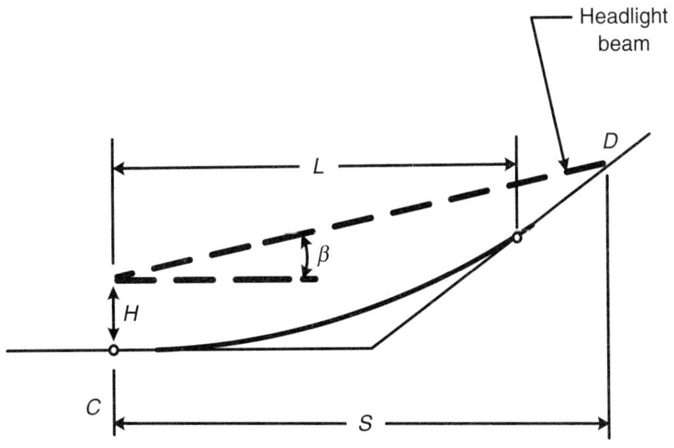

FIGURE 6.17
Headlight sight distance on sag vertical curves ($S > L$).

AASHTO for H and β are 0.6 m and 1°, respectively. Using the properties of a parabola, it can be shown that

$$L = 2S - \frac{200(H + S\tan\beta)}{A} \qquad (\text{for } S > L) \tag{6.7}$$

Substituting 0.6 m for H and 1° for β makes Equation 6.7 become

$$L = 2S - \frac{(120 + 3.5S)}{A} \qquad (\text{for } S > L) \tag{6.8}$$

Similarly, it can be shown that for the condition when $S < L$,

$$L = \frac{AS^2}{200(H + S\tan\beta)} \qquad (\text{for } S < L) \tag{6.9}$$

and substituting 0.6 m for H and 1° for β gives

$$L = \frac{AS^2}{120 + 3.5S} \qquad (\text{for } S < L) \tag{6.10}$$

In using Equations 6.8 and 6.10 to compute the minimum length of a sag vertical curve, S is taken as the stopping sight distance for the design speed at the location of the sag vertical curve. This will provide a safe condition, as the driver will see a distance that is at least equal to the stopping sight distance.

The comfort criterion considers the fact that both the gravitational and centrifugal forces act in combination on sag vertical curves. This results in a greater effect of these forces on the occupants of the vehicle than on crest vertical curves, where these forces act in opposition. The sag vertical curve is therefore designed so that the radial acceleration observed by occupants of a vehicle on the curve does not exceed an acceptable level. An acceptable value for radial acceleration, (i.e., one that will provide a comfortable ride) is usually taken as 0.3 m/s^2. An expression that has been used for the minimum length of a sag vertical curve to satisfy the comfort criterion is

$$L = \frac{Au^2}{395} \tag{6.11}$$

where u is the design speed in km/h and L and A are the same as previously used. This length is usually less than that required to satisfy the headlight sight distance requirement.

The drainage criterion for highway sag vertical curves is important on curbed roads. In this case, the requirement calls for a maximum length rather than a minimum as is required by the other criteria. In order to satisfy this criterion, it is usually stipulated that a minimum grade of 0.35% should be provided within 15 m of the level point of the curve. Experience has shown that the maximum length satisfying this criterion is usually greater than the minimum required for the other criteria.

The criterion of general appearance is usually satisfied by the use of a rule of thumb expressed as

$$L = 30A \tag{6.12}$$

where L is the minimum length of the sag vertical curve and A is the algebraic difference of the intersecting grades.

EXAMPLE 6.6

Determining the Length of a Sag Vertical Curve on a Highway Segment

Determine the length of a highway sag vertical curve joining a −3% grade with a +3% grade on a rural arterial highway if the design speed of the highway is 95 km/h. Assume $S < L$.

Solution

Determine the stopping sight distance. Use Equation 3.21 to determine the minimum stopping sight distance for 95 km/h:

$$SSD = 0.28ut + \frac{u^2}{254.3(\frac{a}{g} - G)}$$

$$\begin{aligned} \text{SSD} &= 0.28 \times 2.5 \times 95 + 95^2/254.3(0.35 - 0.03) \\ &= 177.4 \text{ m} \end{aligned}$$

Determine the length of the sag vertical curve to provide a minimum sight distance of 177.4 m (i.e., to satisfy sight distance criterion). Use Equation 6.10 to determine the length of the sag vertical curve:

$$\begin{aligned} L &= \frac{AS^2}{400 + 3.5S} \\ &= 6 \times (177.4^2/(120 + 3.5 \times 177.4) \\ &= 254.86 \text{ m} \end{aligned}$$

Determine the minimum length required to satisfy comfort criterion. Use Equation 6.11:

$$\begin{aligned} L &= \frac{Au^2}{395} \\ &= 6 \times 95^2/395 \\ &= 137.1 \text{ m} \end{aligned}$$

Determine the minimum length to satisfy the comfort criterion. Use Equation 6.12:

$$L = 30A$$

$$= 30 \times 6$$

$$= 180 \text{ m}$$

The minimum length that satisfies all criteria is that required to satisfy the sight distance criterion, which is 254.86 m.

Selection of an Appropriate Airport Runway Grade

This process is similar to that used for highway curves in that the grade criteria given earlier in this chapter are used to select appropriate grades for the runway. The main difference is that maximum allowable grades for runways are much lower than those for highways. Example 6.3 illustrates the use of these criteria to select appropriate grades for airport runways.

Design of Airport Runway Vertical Curves

The procedure used for the design of airport runway vertical curves is similar to that used for highways in that the main objective is to determine the length of the vertical curve. The difference is that, rather than the stopping sight distance used in highways, the main criterion used is the minimum length (L_{min}) requirements given earlier in this chapter, in which the minimum length is a constant multiplied by the algebraic difference of the grades. These requirements are repeated here to enhance comprehension of the procedure:

For Aircraft Approach Categories A and B:

$$L_{min} = 90A \tag{6.13}$$

For Aircraft Approach Categories C and D:

$$L_{min} = 300A \tag{6.14}$$

where

A = percent grade change (algebraic difference in percent grade)

Example 6.7

Determining the Length of a Runway Vertical Curve

Determine the length of a crest vertical curve connecting a +0.75% grade with a −0.75% grade on the primary runway of an airport with aircraft approach categories B and D.

Solution

Use Equation 6.13 to determine minimum length for an airport with aircraft approach category B:

$$\begin{aligned} L_{min} &= 90A \\ &= 90 \times 1.5 \\ &= 135 \text{ m} \end{aligned}$$

Use Equation 6.14 to determine minimum length for an airport with aircraft approach category D:

$$\begin{aligned} L_{min} &= 300A \\ &= 300 \times 1.5 \\ &= 450 \text{ m} \end{aligned}$$

Selection of an Appropriate Grade for a Railroad Track

A procedure similar to those used for highway and runway vertical curves is also used for rail track vertical curves in that railroad track grades suitable for the topography of the area in which the track is located are selected to satisfy the criteria given earlier in this chapter for maximum grades. Example 6.3 illustrates the use of these criteria in selecting appropriate grades for railroad tracks. However, it should be noted that maximum grades for railroad tracks are usually less than those for highways although somewhat similar to those for airport runways.

Design of Railroad Vertical Curves

Having selected appropriate grades, the next task in the design of the vertical alignment is to design the vertical curves connecting consecutive grades. The procedure used is similar to those of the highways and railroad tracks in that the main objective is to determine the length of the vertical curve. Recommended minimum lengths have been provided for light rail transit main line tracks and freight and passenger intercity tracks.

For *light rail transit main line tracks,* the absolute minimum length of the vertical curve depends on the design speed of the track and the algebraic difference of the grades connected by the curve. Recommended criteria are given in the *Track Design Handbook for Light Rail Transit* for the desired length, the preferred minimum length, and the absolute minimum length as follows:

$$\text{Desired length } (L_{mindes}) = 60A \text{ (m)} \tag{6.15}$$

$$\text{Preferred minimum length } (L_{minpref}) = 30A \text{ (m)} \tag{6.16}$$

Absolute minimum length (L_{minabs})

$$L_{minabs} = \frac{Au^2}{212} \text{ (for crest vertical curves) (m)} \tag{6.17}$$

$$L_{minabs} = \frac{Au^2}{382} \text{ (for sag vertical curves) (m)} \tag{6.18}$$

where

$A = (G_2 - G_1)$ algebraic difference of the grades connected by the vertical curve

G_1 = percent grade of approaching tangent

G_2 = percent grade of departing tangent

u = design speed in km/h

For *freight and passenger intercity tracks,* the length of the vertical curve depends on the algebraic difference between grades (A), the vertical acceleration and the speed of the train. The minimum length L_{min} is given as:

$$L_{min} = \frac{Au^2K}{100a} \tag{6.19}$$

where

A = algebraic difference between grades in percent

u = speed of the train km/h

K = 0.077, conversion factor to give L_{min} in m

a = vertical acceleration, m/s^2

= 0.03 m/s^2 for freight trains

= 0.18 m/s^2 for passenger trains

However, the length of any vertical curve cannot be less than 30 m.

EXAMPLE 6.8

Determining the Minimum Length of a Crest Vertical Curve on a Light Rail Transit Main Line Track

The distance between the PVIs of two consecutive vertical curves (a crest vertical curve followed by a sag vertical curve) on a light rail transit main line track is 1605 m. The grade of the approaching tangent of the crest vertical curve is 6% and that of the departing tangent of the sag vertical curve is 5%. Determine the desired, preferred minimum and absolute minimum lengths of each of these curves if the design speed of the track is 90 km/h.

Solution

Determine maximum allowable grade of the common tangent. Since the distance between the PVIs is greater than 750 m, the length of the common tangent should be taken as unlimited. The maximum sustained grade is therefore −4% (see page 321).

Determine required lengths for the crest vertical curve. The grades of the approaching and departing grades of the crest vertical curves are +6% and −4%, respectively.
Desired length—use Equation 6.15:

$$\begin{aligned} LVC &= 60A \\ &= 60(6 - (-4)) \\ &= 600 \text{ m} \end{aligned}$$

Preferred minimum—use Equation 6.16:

$$\begin{aligned} LVC &= 30A \\ &= 30(6 - (-4)) \\ &= 300 \text{ m} \end{aligned}$$

Absolute minimum—use Equation 6.17:

$$\begin{aligned} LVC &= \frac{Au^2}{212} \\ &= (6 - (-4))(90^2)/212 \\ &= 382 \text{ m} \end{aligned}$$

(Note that in this case the absolute minimum is longer than the preferred minimum. Therefore, if it is not feasible to use the desired length of 600 m, the absolute minimum length of 382 m should be used.)

Determine the required lengths for the sag vertical curve. The grades of the approaching and departing tangents of the sag vertical curve are −4% and 5%, respectively.
Desired length—use Equation 6.15:

$$\begin{aligned} L_{\text{mindes}} &= 60A \\ &= 60(-4 - (+5)) \\ &= 540 \text{ m} \end{aligned}$$

Preferred length—use Equation 6.16:

$$\begin{aligned} L_{\text{minpref}} &= 30A \\ &= 30(-4 - (5)) \\ &= 270 \text{ m} \end{aligned}$$

Absolute minimum length—use Equation 6.18:

$$\begin{aligned} L_{\text{minabs}} &= \frac{Au^2}{382} \\ &= (-4 - (+5))(90^2)/382 \\ &= 191 \text{ m} \end{aligned}$$

EXAMPLE 6.9

Determining Minimum Length of a Crest Vertical Curve on Freight and Passenger Main Line Track

In designing a vertical curve on a main line railroad track for freight and passenger cars, the engineer used a length of 1500 m for a curve joining a +2% grade with a −2% grade. Determine whether this curve satisfies the minimum length requirement if trains are expected to run at 80 km/h.

Solution

Use Equation 6.19 to determine minimum length of the curve:

$$L_{\text{min}} = \frac{Au^2K}{100a}$$

$$A = 2-(-2) = 4$$

$$L_{\text{min}} = \frac{4 \times 80^2 \times 0.077}{100 \times 0.03} \quad \text{(for freight trains)}$$

$$= 657 \text{ m} \quad \text{(for freight trains)}$$

$$L_{\text{min}} = \frac{4 \times 80^2 \times 0.077}{100 \times 0.18} \quad \text{(for passenger trains)}$$

$$= 110 \text{ m} \quad \text{(for passenger trains)}$$

The minimum length requirement is satisfied.

Layout of Vertical Curves

Having determined the length of a vertical curve, it is necessary to determine the elevation on the curve at regular intervals to facilitate the construction of the curve in the field. The properties of a parabola are again used to carry out this task. Consider a crest vertical curve in the form of a parabola shown in Figure 6.18.

From the properties of a parabola, $Y = ax^2$, where a is a constant. Rate of change of slope is

$$\frac{d^2Y}{dx^2} = 2a$$

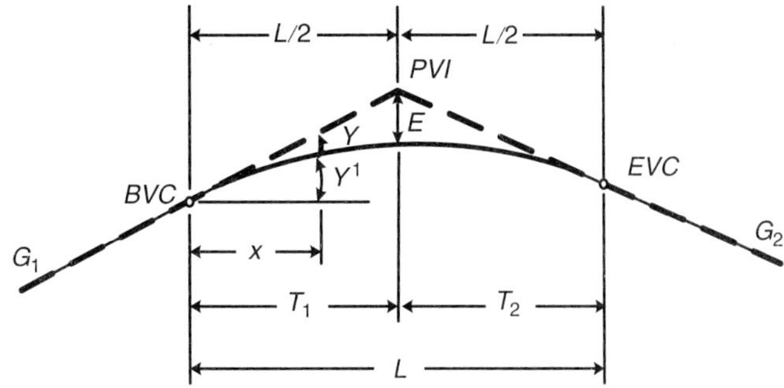

PVI = Point of vertical intersection
BVC = Beginning of vertical curve (same point as PVC)
EVC = End of vertical curve (same point as PVT)
E = External distance
G_1, G_2 = Grade of tangents (%)
L = Length of curve
A = Algebraic difference of grades, $G_1 - G_2$

FIGURE 6.18
Layout of a crest vertical curve for design.

but $T_1 = T_2 = T$, which gives

$$L = 2T$$

If the total change in slope is A, then

$$2a = \frac{A}{100L}$$

The equation of the curve can therefore be written as

$$Y = \left[\frac{A}{200L}\right]x^2 \tag{6.20}$$

When $x = L/2$, the external distance E from the point of vertical intersection (PVI) to the curve is determined by substituting $L/2$ for x in Equation 6.20. This gives

$$E = \frac{A}{200L}\left(\frac{L}{2}\right)^2 = \frac{AL}{800} \tag{6.21}$$

Since stations are given in 30-m intervals, E can be given as

$$E = \frac{AN}{26.67} \tag{6.22}$$

where N is the length of the curve in stations of 30 m and E is in meters. The vertical offset Y from any point of the curve to the tangent can also be determined in terms of E, by substituting $800E/L$ for A in Equation 6.20. This gives

$$Y = \left(\frac{x}{L/2}\right)^2 E \tag{6.23}$$

Because of clearance and drainage requirements, it is sometimes necessary to determine the location and elevation of the highest and lowest points of the crest and sag vertical curves, respectively. The expression for the distance between the beginning of the vertical curve (BVC) and the highest point on a crest vertical curve can be determined in terms of the grades. Note that this distance is T only when the curve is symmetrical. For unsymmetrical curves (i.e., G_1 not equal to G_2), this distance may be shorter or longer than T depending on the values of G_1 and G_2.

Consider the expression for Y^1 (see Figure 6.18):

$$\begin{aligned} Y^1 &= \frac{G_1 x}{100} - Y \\ &= \frac{G_1 x}{100} - \frac{A}{200L}x^2 \\ &= \frac{G_1 x}{100} - \left(\frac{G_1 - G_2}{200L}\right)x^2 \end{aligned} \tag{6.24}$$

Differentiating Equation 6.24 and equating it to zero will give us the value of x_{high} for the maximum value of Y (i.e., the distance the highest point is located from the BVC):

$$\frac{dY^1}{dx} = \frac{G_1}{100} - \left(\frac{G_1 - G_2}{100L}\right)x = 0 \tag{6.25}$$

which gives

$$\begin{aligned} x_{\text{high}} &= \frac{100L}{G_1 - G_2}\,\frac{G_1}{100} \\ &= \frac{LG_1}{G_1 - G_2} \end{aligned} \tag{6.26}$$

Similarly, it can be shown that the difference in elevation between the BVC and the highest point Y^1_{high} can be obtained by substituting the value for x_{high} for x in Equation 6.24. This gives

$$Y^1_{\text{high}} = \frac{LG_1^2}{200(G_1 - G_2)} \tag{6.27}$$

The complete design of the vertical curve for the railroad track, highway, or airport runway will generally use the following steps:

Step 1: Determine the minimum length of the curve to satisfy the requirements for the specific mode and type of vertical curve (crest or sag).

Step 2: Using the profile drawing (drawing of the vertical alignment of the travelway), determine the point of vertical curve (PVI) (point of intersection of the grades).

Step 3: Compute stations and elevations of the beginning of vertical curve (BVC) and end of vertical curve (EVC).

Step 4: Compute offset Y from the tangent to the curve at equal distances usually 30 m apart (i.e., multiple of 30 m) beginning with the first whole station. It is common for lengths of vertical curves on railroad tracks to be in multiples of 30 m. In such cases, offsets could be determined at equal distances of 30 m and are not necessary at whole stations.

Step 5: Compute elevations on the curve. Note that for crest curves the offset is subtracted from the corresponding tangent elevation to obtain the elevation on the curve, while for sag curves the offset is added to the corresponding tangent elevation. The procedure is illustrated in Example 6.10.

EXAMPLE 6.10

Design of a Crest Vertical Curve on a Railway Track

A crest vertical curve joining a +0.75% grade and a −0.75% grade is to be designed for a freight rail track with trains traveling at 108 km/h. If the tangents intersect at station (350 + 22.5) at an elevation of 138 m, determine the stations and elevations of the BVC and EVC. Also calculate the elevations on the curve at 30-m intervals. A sketch of the curve is shown in Figure 6.19.

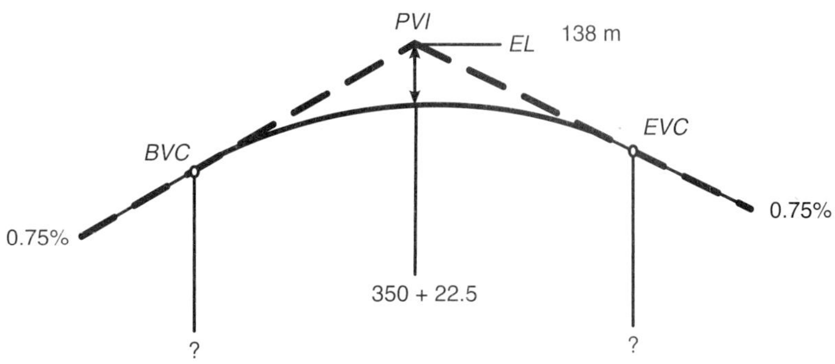

FIGURE 6.19
Layout of a vertical curve for Example 6.10.

Solution

Determine the length of the curve (L) ft. Use Equation 6.19 to determine the minimum length of the curve:

$$L_{min} = \frac{Au^2K}{100a}$$

$$G_1 = +0.75$$

$$G_2 = -0.75$$

$$A = 0.75 - (-0.75) = 1.5$$

$$L_{min} = \frac{1.5 \times 108^2 \times 0.077}{100 \times 0.03}$$

$$= 449.06 \text{ m (say 450 m)}$$

Compute station and elevation of the beginning of vertical curve (BVC):

$$\text{Station of BVC is } (350 + 22.5) - \frac{15 + 00.00}{2} = 343 + 7.5$$

$$\text{Elevation of BVC is } 138 - 0.0075 \times \frac{450}{2} = 136.31 \text{ m}$$

Compute station and elevation of end of vertical curve (EVC):

$$\text{Station of EVC is } (350 + 22.5) + \frac{15 + 00.00}{2} = (358 + 7.5)$$

$$\text{Elevation of EVC is } 138 - 0.0075 \times \frac{1500}{2} = 136.31 \text{ m}$$

The remainder of the solution is shown in Table 6.12.

TABLE 6.12

Elevation Computations for Example 6.10

Station	*Distance from BVC(x) m*	*Tangent Elev.*	*Offset* $y = \frac{Ax^2}{200L}$	*Curve Elevation (Tangent El – offset) m*
343 + 07.50	0	136.31	0	136.31
344 + 00.00	22.5	136.479	0.0084	136.4703
345 + 00.00	52.5	136.704	0.0459	136.6578
346 + 00.00	82.5	136.929	0.1134	136.8153
347 + 00.00	112.5	137.154	0.2109	136.9428
348 + 00.00	142.5	137.379	0.3384	137.0403
349 + 00.00	172.5	137.604	0.4959	137.1078
350 + 00.00	202.5	137.829	0.6834	137.1453
351 + 00.00	232.5	138.054	0.9009	137.1528
352 + 00.00	262.5	138.279	1.1484	137.1303
353 + 00.00	292.5	138.504	1.4259	137.0778
354 + 00.00	322.5	138.729	1.7334	136.9953
355 + 00.00	352.5	138.954	2.0709	136.8828
356 + 00.00	382.5	139.179	2.4384	136.7403
357 + 00.00	412.5	139.404	2.8359	136.5678
358 + 00.00	442.5	139.629	3.2634	136.3653
358 + 07.50	450	139.685	3.3750	136.3100

Design of the Horizontal Alignment

Design of the horizontal alignment is similar to the design of the vertical alignment in that straight horizontal sections of the road known as tangents are connected by horizontal curves. The main difference is that vertical curves are parabolic while horizontal curves are circular. A circular horizontal curve is designed by determining an appropriate radius that will provide a smooth flow around the curve. This radius mainly depends on the maximum speed at which the vehicle transverses the curve and the maximum allowable superelevation.

Types of Horizontal Curves

Horizontal curves can be divided into four general types: simple, compound, reverse, and spiral. Let us now consider how the computation is carried out for each of them.

Simple Curves

This is a single segment of a circular curve of radius R. Figure 6.20 shows the layout of a simple horizontal curve. Several important locations on the curve should be noted as they play an important role in the calculation and layout of the curve. The point at which the curve begins—that is, where the curve meets the tangent—is the point of curve (PC) and is also sometimes referred to as the

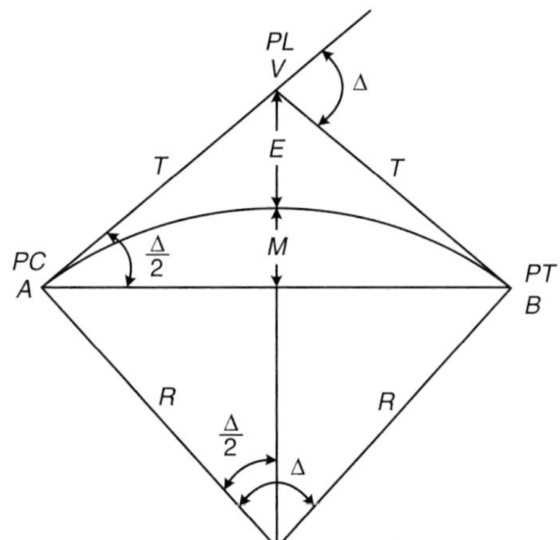

FIGURE 6.20
Layout of a simple horizontal curve.

beginning of curve (BC). The point at which the curve ends is the point of tangent (PT) or end of curve (EC). The intersection point of the two tangents is known as the point of intersection (PI) or the vertex (V). The simple curve is defined either by its radius (for example, 255 m radius curve) or by the degree of curve.

Either the arc definition or the chord definition is used to give the degree of curve. Using the arc definition, the curve is defined by the angle that is subtended at the center by a circular arc, 30 m long. For example, a 3° curve is one in which an arc of 30 m subtends an angle of 3° at the center. Figure 6.21a illustrates the arc definition. Using the chord definition, the curve is defined by the angle that is subtended at the center by a chord 30 m in length. In this case, a 3° curve is one in which a chord of 30 m subtends an angle of 3° at the center. Figure 6.21b illustrates the chord definition. The arc definition is used for highway and runway design while the chord definition is commonly used for railroad design. It is useful to determine the relationship between the radius of the curve and the degree of the curve.

Let us first consider the arc definition. The length of an arc of a circular curve is given as

$$L_{\text{arc}} = R\theta$$

where

L_{arc} = length of the arc
R = radius of the curve
θ = angle in radians subtended at the center by the arc of length L

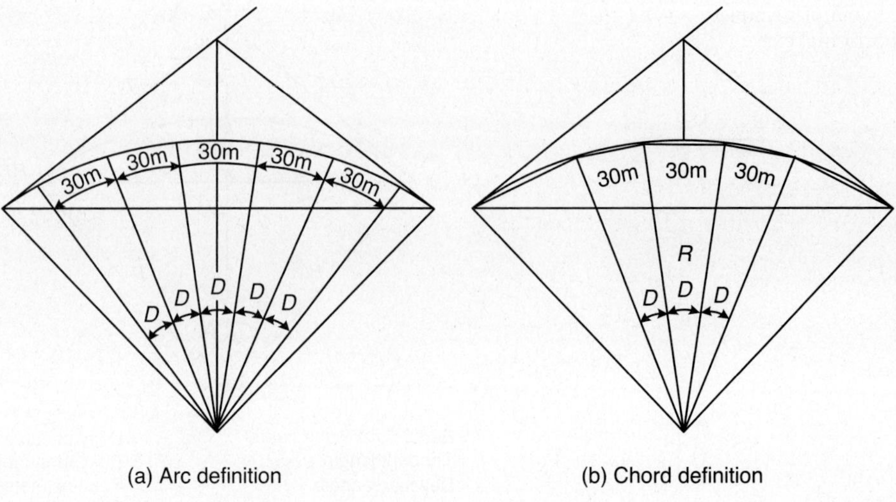

FIGURE 6.21
Arc and chord definitions for a circular curve.

If the angle subtended at the center by an arc of 30 m is D_a° degrees, then

$$\theta = \frac{\pi D_a^\circ}{180} \text{ radians and}$$

$$30 = \frac{R\pi D_a^\circ}{180}$$

which gives

$$R = \frac{1718.18}{D_a^\circ} \tag{6.28}$$

The radius of the curve can then be determined if the degree of curve is known or the degree of curve can be determined if the radius is known.

In the case of the chord definition, since the curve is defined by the angle subtended at the center by a 30 m chord

$$R = \frac{15}{\sin \frac{D_c^\circ}{2}} \tag{6.29}$$

where

R = radius of the curve
D_c° = angle subtended at the center in degrees by a chord of 30 m

For a curve of 1°, $R = 1718.89$ m, which gives

$$R = \frac{1718.89}{D_c^\circ} \tag{6.30}$$

for the range of angles normally used in railroad design.

Several basic relationships can be developed for the simple horizontal curve. Using the properties of the circle, and referring to Figure 6.20, the two tangents AV and BV have equal lengths that are designated as T. The angle formed by the tangents is known as the deflection angle, Δ. The tangent length is given as

$$T = R \tan \frac{\Delta}{2} \tag{6.31}$$

The chord AB is the long chord and its length C is given as

$$C = 2R \sin \frac{\Delta}{2} \tag{6.32}$$

The distance from the point of intersection of the two tangents and the curve is the external distance E and is given as

$$E = R \sec \frac{\Delta}{2} - R$$

$$E = R\left[\frac{1}{\cos\frac{\Delta}{2}} - 1\right] \tag{6.33}$$

The distance M between the midpoint of the long chord and the midpoint of the curve is the middle ordinate and is given as

$$M = R - R\cos\frac{\Delta}{2}$$

$$M = R\left(1 - \cos\frac{\Delta}{2}\right) \tag{6.34}$$

The length of the curve L_c is given as

$$L_c = \frac{R\Delta\pi}{180} \tag{6.35}$$

Compound Curves

Compound curves are formed when any two successive curves of a series of two or more successive simple curves turning in the same direction have a common tangent point. Figure 6.22 shows the layout of a compound curve formed by two simple curves. Compound curves are used mainly to obtain a desirable shape of the alignment at a particular location. Figure 6.22 shows seven different variables, $R_1, R_2, \Delta_1, \Delta_2, \Delta, T_1, T_2$, that are associated with a compound curve. Many equations can be developed relating two or more of these variables. The equations given below are more commonly used in setting out compound curves:

$$\Delta = \Delta_1 + \Delta_2 \tag{6.36}$$

$$t_1 = R\tan\frac{\Delta_1}{2} \tag{6.37}$$

$$t_2 = R\tan\frac{\Delta_2}{2} \tag{6.38}$$

$$\frac{\overline{VG}}{\sin\Delta_2} = \frac{\overline{VH}}{\sin\Delta_1} = \frac{t_1 + t_2}{\sin(180 - \Delta)} = \frac{t_1 + t_2}{\sin\Delta} \tag{6.39}$$

$$T_1 = \overline{VG} + t_1 \tag{6.40}$$

$$T_2 = \overline{VH} + t_2 \tag{6.41}$$

where

R_1 and R_2 = radii of simple curves forming compound curves
Δ_1 and Δ_2 = deflection angles of simple curves
t_1 and t_2 = tangent lengths of simple curves
T_1 and T_2 = tangent lengths of compound curves
Δ = deflection angle of compound curve

FIGURE 6.22
Layout of a compound curve.

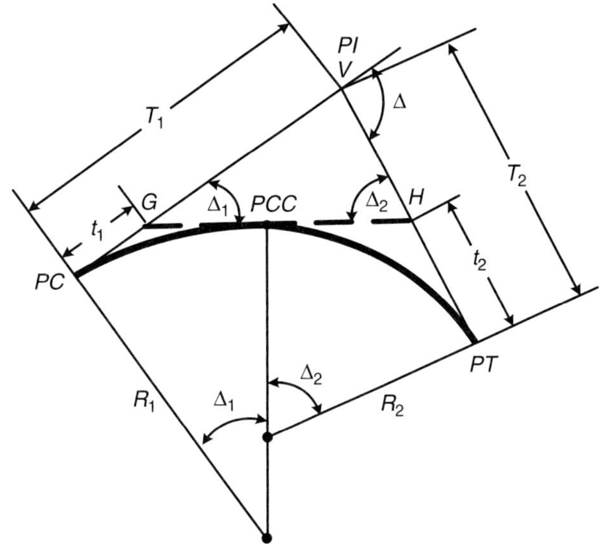

R_1, R_2 = Radii of simple curves forming compound curve
Δ_1, Δ_2 = Deflection angles of simple curves
Δ = Deflection angle of compound curve
t_1, t_2 = Tangent lengths of simple curves
T_1, T_2 = Tangent lengths of compound curves
PCC = Point of compound curve
PI = Point of intersection
PC = Point of curve
PT = Point of tangent

Reverse Curves

Reverse curves are formed when two consecutive simple curves turning in opposite directions have a common tangent as shown in Figure 6.23. These curves are mainly used when the horizontal alignment needs to be changed. It can be seen from Figure 6.23 that

$$\Delta = \Delta_1 = \Delta_2$$

$$\text{angle } OWX = \frac{\Delta_1}{2} = \frac{\Delta}{2}$$

$$\text{angle } OYZ = \frac{\Delta_1}{2} = \frac{\Delta_2}{2}$$

Therefore, line *WOY* is a straight line:

$$\tan\frac{\Delta}{2} = \frac{d}{D}$$

$$d = R - R\cos\Delta_1 + R - R\cos\Delta_2$$

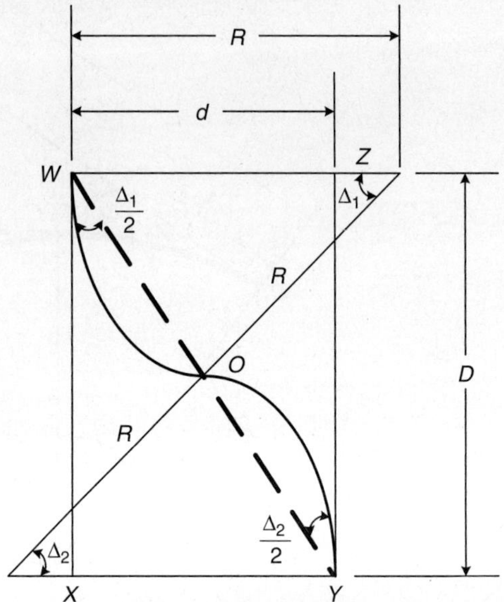

FIGURE 6.23
Layout of reverse curve.

$$d = 2R(1 - \cos \Delta)$$

$$R = \frac{d}{2(1 - \cos\Delta)}$$

Transition or Spiral Curves

Transition or spiral curves are curves with varying radii and are usually placed between a tangent and a horizontal curve or between two horizontal curves with significantly different radii. Transition curves provide a gradual change of degree and easier riding in going from the tangent to a full curvature, or from one circular curve to another having a substantially different radius. When placed between a tangent and a horizontal curve, the degree of a transition curve varies from zero to the degree of the curve, and when placed between two curves, its degree varies from that of the first circular curve to that of the second. Figure 6.24 shows a schematic of a spiral curve between a tangent and a circular curve.

Let us consider a spiral curve between a tangent and a circular curve. Since the degree of the spiral curve varies from zero at the tangent (i.e. at T.S.) to the degree of the circular curve D_a at the start of the circular curve (i.e. at S.C.) (see Figure 6.24), the rate of change in degrees (K) of the spiral is given as

$$K = \frac{30D_a}{L_s} \tag{6.43}$$

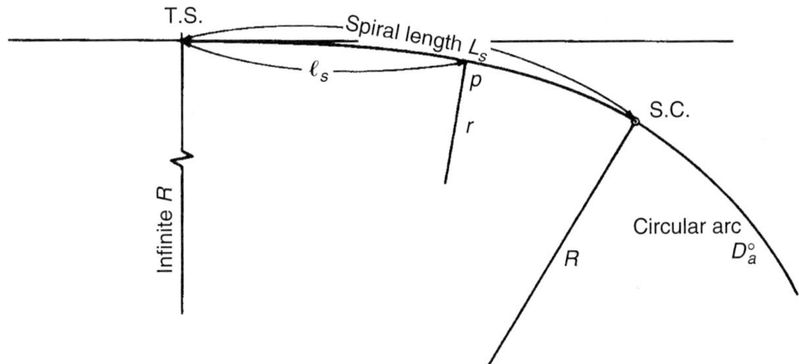

FIGURE 6.24
Schematic of a spiral curve.

Source: Surveying Theory and Practice, Davis, Foote, Anderson, and Mikhail, McGraw-Hill Book Company, 1997.

where

K = rate of change/station of 30 m
D_a = degree of simple curve
L_s = length of the spiral in m

Since K is a constant, it follows that the radius r or the degree of curve D_p of the spiral at any point p at distance ℓ_s from T.S. can be determined as

$$D_p = \frac{\ell_s K}{30} \text{ and}$$

$$r = \frac{1718.2}{D_p} = \frac{1718.2(30)}{\ell_s K} \tag{6.44}$$

Similarly, the radius at the S.C. is

$$R = \frac{1718.2(30)}{L_s K} \tag{6.45}$$

from which we obtain

$$\frac{r}{R} = \frac{L_s}{\ell_s} \tag{6.46}$$

We shall now discuss how the basic relationships given in Equations 6.28 through 6.46 are used in designing the horizontal alignment of the travelway for different modes.

Design of the Highway Horizontal Alignment

Highway Simple Curves

The first task in the design of a simple horizontal curve for a highway is to determine the minimum radius that is required for the curve. This radius is based on the design speed selected for the highway for the following two conditions:

(i) Maximum superelevation rate
(ii) Minimum stopping sight distance on the curve

Maximum Superelevation Rate

The relationship that governs the radius of a highway horizontal curve for this condition was developed in Chapter 3 and is given as

$$R = \frac{u^2}{127(e + f_s)} \tag{6.47}$$

where

R = radius of the circular curve, m
u = vehicle speed, km/h
e = superelevation
f_s = coefficient of side friction

It was noted that several factors control the maximum value that can be used for the superelevation (e). These include the location of the highway (that is, whether it is in an urban or rural area); weather conditions (such as the occurrence of snow); and the distribution of slow-moving traffic within the traffic stream. It was also noted that a maximum value of 0.1 is used in areas with no snow and ice and, for areas with snow and ice, maximum values have ranged from 0.08 to 0.1. For expressways in urban areas, a maximum value of 0.08 is used.

Minimum Stopping Sight Distance on the Curve

This condition applies at locations where an object is located near the inside edge of the road, as shown in Figure 6.25. The object may interfere with the view of the driver, resulting in a reduction of sight distance ahead. It is therefore necessary for the horizontal curve to be designed in a way that provides a sight distance at least equal to the stopping sight distance.

Figure 6.25 shows a schematic representation in which the vehicle is at point A and the object is at point T. Chord AT is the line of sight that will just permit the driver to see the object at T. However, it should be noted that the actual horizontal distance traveled by the vehicle from point A to point T is the arc AT. It is therefore the distance S actually available for the vehicle to stop at T. Let the angle subtended at the center by arc AT be $2\theta°$. Then

$$S = \frac{2R\theta\pi}{180}$$

$$\theta = \frac{28.65}{R}S \tag{6.48}$$

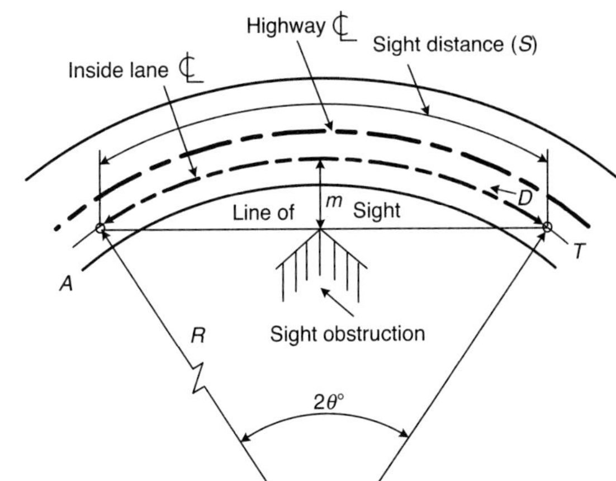

FIGURE 6.25
Sight distance on horizontal curve with an object located near the inside of the curve.

Source: A Policy on Geometric Design of Highways and Streets, American Association of State Highway and Transportation Officials, Washington, D.C., 2004. Used by permission.

$$m = \frac{5730}{D} \text{vers} \frac{SD}{200}\text{; also } m = R\left(\text{vers} \frac{28.65S}{R}\right)\text{, and } S = \frac{R}{28.65} \cos^{-1}\left[\frac{R-m}{R}\right]$$

Note: vers $\theta = (1 - \cos\theta)$

Also note: Here m and R are in ft and 1 ft = 0.3 m

where

R = radius of curve, m
S = sight distance = length of arc AT, m

However,

$$\cos\theta = \frac{R - m}{R}$$

$$\cos\left(\frac{28.65}{R} S\right) = \frac{R - m}{R}$$

which gives

$$m = R\left(1 - \cos\frac{28.65}{R} S\right) \tag{6.49}$$

Equation 6.49 can be used to determine m, R, or S depending on the information known. Note that the minimum stopping sight distance S is obtained from Equation 3.24 and is given as

$$SSD = 0.28ut + \frac{u^2}{254.3(\frac{a}{g} \pm G)} \tag{6.50}$$

where

u = speed, km/h
t = perception reaction time, s
a = deceleration rate, m/s^2
g = acceleration of gravity, m/s^2
G = grade of tangent

EXAMPLE 6.11

Determining the Radius of a Simple Horizontal Curve on a Highway

A horizontal curve is being designed to connect two tangents that intersect on a rural collector road with a specified design volume of 4000 veh/day. It is expected that, because of right-of-way restriction, a large billboard will be placed at a distance of 13.5 m from the centerline of the inside lane. If the terrain can be described as level and only a minimal amount of snow is expected on the road, determine the minimum radius of the curve.

Solution

Use Table 6.5 to determine the design speed. For level terrain and specified design volume of over 2000 veh/day, design speed is 95 km/h.

Determine minimum radius based on maximum superelevation rate. For a rural collector road and minimal amount of snow, a maximum superelevation rate (e) of 0.1 can be used. Use Equation 6.47 to determine minimum radius:

$$R = \frac{u^2}{127(e + f)}$$

Note: For a design speed of 95 km/h, the coefficient of side friction is 0.12 (see Table 3.9):

$$R = \frac{95^2}{127(0.1 + 0.12)}$$

$$= 323 \text{ m}$$

Determine minimum radius based on sight distance requirement. Use Equation 6.50 to determine sight distance for a design speed of 95 km/h:

$$SSD = 0.28ut + \frac{u^2}{254.3(\frac{a}{g} \pm G)}$$

$$= 0.28 \times 95 \times 2.5 + \frac{95^2}{254.3(.35 + 0)} \text{ (assuming } a = 3.5 \text{ m/s}^2\text{)}$$

$$= 66.5 + 101.40$$

$$= 167.9 \text{ m}$$

Use Equation 6.49 to determine minimum radius:

$$m = R\left(1 - \cos\frac{28.65}{R}S\right)$$

$$13.5 = R\left(1 - \cos\frac{28.65}{R}167.9\right)$$

from which we obtain $R \sim 259$ m, which is less than the 323 m for the maximum superelevation requirement. This therefore governs, and the minimum radius of the curve is 323 m.

Highway Compound Curves

As noted earlier, compound curves are used mainly to obtain desirable shapes of the horizontal alignment. On highways, they are particularly used at at-grade intersections, ramps of interchanges, and highway sections located in difficult topographic conditions. When compound curves are used on highways, every effort should be made to avoid abrupt changes in the alignment. AASHTO has therefore recommended that the ratio of the flatter radius to the sharper radius should not be greater than 1.5:1. At locations where drivers can adjust to sudden changes in curvature and speed such as intersections, AASHTO suggests that the ratio of the flatter radius to the sharper radius can be increased to 2:1. However, the maximum desirable ratio recommended by AASHTO is 1.75:1. When it is necessary to use ratios higher than 2:1, a spiral curve should be placed between the two curves.

Other factors that should be considered in the design of highway compound curves are as follows: (i) a smooth transition from a flatter curve to a sharp curve should be provided; and (ii) a reasonable deceleration rate should exist as a driver traverses a series of curves of decreasing radii. These conditions are usually satisfied if the length of each curve is not less than the minimum values specified by AASHTO. Table 6.13 shows these recommended values.

Highway Reverse Curves

Reverse curves are not often used in highway design as they may result in a sudden change in the alignment, which in turn may result in drivers finding it difficult to keep to their lanes. It is suggested that two simple curves separated by a sufficient length of tangent between them or by an equivalent length of spiral is a preferable design.

TABLE 6.13

Lengths of Circular Arc for a Compound Intersection Curve when Followed by a Curve of One-Half Radius or Preceded by a Curve of Double Radius

Source: Adapted from *A Policy on Geometric Design of Highways and Streets,* American Association of State Highway and Transportation Officials, Washington D.C., 2004. Used by permission.

Length of Circular Arc (ft)	*Radius (ft)*						
	100	*150*	*200*	*250*	*300*	*400*	*500 or more*
Minimum	40	50	60	80	100	120	140
Desirable	60	70	90	120	140	180	200

Note: 1 ft = 0.3 m

Highway Spiral Curves

In designing a highway spiral curve, the first task is to determine the length of the curve. The minimum length is given as

$$L = \frac{0.0214u^3}{RC} \tag{6.51}$$

where

L = minimum length of curve (m)
u = speed (km/h)
R = radius of curve (m)
C = rate of increase of radial acceleration (m/s²/s)

C is an empirical factor that indicates the level of comfort and safety. In highway design, the values used for C have ranged from 0.3 to 0.9. A practical alternative way to determine the minimum length of the spiral curve is to use the length required for superelevation runoff. This is the length of the highway that is required to transition from the normal cross slope to the fully superelevated section of the curve.

Table 6.14 gives recommended values of lengths required for superelevation runoff for different design speeds and superelevation rates. The values given in this table are for the rotation of one and two lanes. A close examination will show that the values for rotating two lanes are not double those required for one lane, as would be expected. AASHTO has recommended empirically derived adjustment factors that consider that it is not always practicable to provide superelevation runoffs that are based on the value for one lane multiplied by the number of lanes, as this tends to be excessive in some cases. When two or more lanes are rotated, AASHTO recommends that the multiplication factor obtained from Equation 6.52 be used:

$$b_w = [1 + 0.5(n_1 - 1)]/n_1 \tag{6.52}$$

where

b_w = multiplication factor to be applied to the value for one lane rotation
n_1 = number of lanes to be rotated

EXAMPLE 6.12

Determining the Length of a Spiral Curve on a Highway

Determine the minimum length of a spiral curve connecting a tangent and a circular curve of 240 m radius on a four-lane undivided rural highway, with 3.6 m lanes and a design speed of 105 km/h. Assume $C = 3$. If the transportation agency of the state in which the road is located requires that the length of any

TABLE 6.14

Superelevation Runoff L_r (ft) for Horizontal Curves

e	V_d = 15 mi/hr		V_d = 20 mi/hr		V_d = 25 mi/hr		V_d = 30 mi/hr		V_d = 35 mi/hr		V_d = 40 mi/hr		V_d = 45 mi/hr		V_d = 50 mi/hr		V_d = 55 mi/hr		V_d = 60 mi/hr		V_d = 65 mi/hr		V_d = 70 mi/hr		V_d = 75 mi/hr		V_d = 80 mi/hr	
	Number of Lanes Rotated. Note that 1 lane rotated is typical for a 2-lane highway, 2 lanes rotated is typical for a 4-lane highway, etc.																											
(%)	1 L_r (ft)	2 L_r (ft)	1 L_r (ft)	2 L_r (ft)	1 L_r (ft)	2 L_r (ft)	1 L_r (ft)	2 L_r (ft)	1 L_r (ft)	2 L_r (ft)	1 L_r (ft)	2 L_r (ft)	1 L_r (ft)	2 L_r (ft)	1 L_r (ft)	2 L_r (ft)	1 L_r (ft)	2 L_r (ft)	1 L_r (ft)	2 L_r (ft)	1 L_r (ft)	2 L_r (ft)	1 L_r (ft)	2 L_r (ft)	1 L_r (ft)	2 L_r (ft)	1 L_r (ft)	2 L_r (ft)
1.5	0	0	0	0	0	0	0	0	0	0	0	0	0	0	0	0	0	0	0	0	0	0	0	0	0	0	0	0
2.0	31	48	32	49	34	51	36	55	39	58	41	62	44	67	48	72	51	77	53	80	56	84	60	90	63	95	69	103
2.2	34	51	36	54	38	57	40	60	43	64	46	68	49	73	53	79	56	84	59	88	61	92	66	99	69	104	75	113
2.4	37	55	39	58	41	62	44	65	46	70	50	74	53	80	58	86	61	92	64	96	67	100	72	108	76	114	82	123
2.6	40	60	42	63	45	67	47	71	50	75	54	81	58	87	62	94	66	100	69	104	73	109	78	117	82	123	89	134
2.8	43	65	45	68	48	72	51	76	54	81	58	87	62	93	67	101	71	107	75	112	78	117	84	126	88	133	96	144
3.0	46	69	49	73	51	77	55	82	58	87	62	93	67	100	72	108	77	115	80	120	84	126	90	135	95	142	103	154
3.2	49	74	52	78	55	82	58	87	62	93	66	99	71	107	77	115	82	123	85	128	89	134	96	144	101	152	110	165
3.4	52	78	55	83	58	87	62	93	66	99	70	106	76	113	82	122	87	130	91	136	95	142	102	153	107	161	117	175
3.6	55	83	58	88	62	93	65	98	70	105	74	112	80	120	86	130	92	138	96	144	100	151	108	162	114	171	123	185
3.8	58	88	62	92	65	98	69	104	74	110	79	118	84	127	91	137	97	146	101	152	106	159	114	171	120	180	130	195
4.0	62	92	65	97	69	103	73	109	77	116	83	124	89	133	96	144	102	153	107	160	112	167	120	180	126	189	137	206
4.2	65	97	68	102	72	108	76	115	81	122	87	130	93	140	101	151	107	161	112	168	117	176	126	189	133	199	144	216
4.4	68	102	71	107	75	113	80	120	85	128	91	137	96	147	106	158	112	169	117	176	123	184	132	198	139	208	151	226
4.6	71	106	75	112	79	118	84	125	89	134	95	143	102	153	110	166	117	176	123	184	128	193	138	207	145	218	158	237
4.8	74	111	78	117	82	123	87	131	93	139	99	149	107	160	115	173	123	184	128	192	134	201	144	216	152	227	165	247
5.0	77	115	81	122	86	129	91	136	97	145	103	155	111	167	120	180	128	191	133	200	140	209	150	225	158	237	171	257
5.2	80	120	84	126	89	134	95	142	101	151	108	161	116	173	125	187	133	199	139	208	145	218	156	234	164	246	178	267
5.4	83	125	88	131	93	139	98	147	105	157	112	168	120	180	130	194	138	207	144	216	151	226	162	243	171	256	185	278
5.6	86	129	91	136	96	144	102	153	108	163	116	174	124	187	134	202	143	214	149	224	156	234	168	252	177	265	192	288
5.8	89	134	94	141	99	149	105	158	112	168	120	180	129	193	139	209	148	222	155	232	162	243	174	261	183	275	199	298
6.0	92	138	97	146	103	154	109	164	116	174	124	186	133	200	144	216	153	230	160	240	167	251	180	270	189	284	206	309
6.2	95	143	101	151	106	159	113	169	120	180	128	192	138	207	149	223	158	237	165	248	173	260	186	279	196	294	213	319
6.4	98	148	104	156	110	165	116	175	124	186	132	199	142	213	154	230	163	245	171	256	179	268	192	288	202	303	219	329
6.6	102	152	107	161	113	170	120	180	128	192	137	205	147	220	158	238	169	253	176	264	184	276	198	297	208	313	226	339
6.8	105	157	110	165	117	175	124	185	132	197	141	211	151	227	163	245	174	260	181	272	190	285	204	306	215	322	233	350
7.0	108	162	114	170	120	180	127	191	135	203	145	217	156	233	168	252	179	268	187	280	195	293	210	315	221	332	240	360
7.2	111	166	117	175	123	185	131	196	139	209	149	223	160	240	173	259	184	276	192	288	201	301	216	324	227	341	247	370
7.4	114	171	120	180	127	190	135	202	143	215	153	230	164	247	178	266	189	283	197	296	207	310	222	333	234	351	254	381
7.6	117	175	123	185	130	195	138	207	147	221	157	236	169	253	182	274	194	291	203	304	212	318	228	342	240	360	261	391

Note: 1 mi/hr = 1.61 km/h; 1 ft = 0.3 m

(continued)

TABLE 6.14

Superelevation Runoff L_r (ft) for Horizontal Curves (*continued*)

	V_d = 15 mi/hr		V_d = 20 mi/hr		V_d = 25 mi/hr		V_d = 30 mi/hr		V_d = 35 mi/hr		V_d = 40 mi/hr		V_d = 45 mi/hr		V_d = 50 mi/hr		V_d = 55 mi/hr		V_d = 60 mi/hr		V_d = 65 mi/hr		V_d = 70 mi/hr		V_d = 75 mi/hr		V_d = 80 mi/hr	
	Number of Lanes Rotated. Note that 1 lane rotated is typical for a 2-lane highway, 2 lanes rotated is typical for a 4-lane highway, etc.																											
e (%)	1 L_r (ft)	2 L_r (ft)	1 L_r (ft)	2 L_r (ft)	1 L_r (ft)	2 L_r (ft)	1 L_r (ft)	2 L_r (ft)	1 L_r (ft)	2 L_r (ft)	1 L_r (ft)	2 L_r (ft)	1 L_r (ft)	2 L_r (ft)	1 L_r (ft)	2 L_r (ft)	1 L_r (ft)	2 L_r (ft)	1 L_r (ft)	2 L_r (ft)	1 L_r (ft)	2 L_r (ft)	1 L_r (ft)	2 L_r (ft)	1 L_r (ft)	2 L_r (ft)	1 L_r (ft)	2 L_r (ft)
7.8	120	180	126	190	134	201	142	213	151	226	161	242	173	260	187	281	199	299	208	312	218	327	234	351	246	369	267	401
8.0	123	185	130	195	137	206	145	218	155	232	166	248	178	267	192	288	204	306	213	320	223	335	240	360	253	379	274	411
8.2	126	189	133	199	141	211	149	224	159	238	170	254	182	273	197	295	209	314	219	328	229	343	246	369	259	388	281	422
8.4	129	194	136	204	144	216	153	229	163	244	174	261	187	280	202	302	214	322	224	336	234	352	252	378	265	398	288	432
8.6	132	198	139	209	147	221	156	235	166	250	178	267	191	287	206	310	220	329	229	344	240	360	258	387	272	407	295	442
8.8	135	203	143	214	151	226	160	240	170	255	182	273	196	293	211	317	225	337	235	352	246	368	264	396	278	417	302	453
9.0	138	208	146	219	154	231	164	245	174	261	186	279	200	300	216	324	230	345	240	360	251	377	270	405	284	426	309	463
9.2	142	212	149	224	158	237	167	251	178	267	190	286	204	307	221	331	235	352	245	368	257	385	276	414	291	436	315	473
9.4	145	217	152	229	161	242	171	256	182	273	194	292	209	313	226	338	240	360	251	376	262	393	282	423	297	445	322	483
9.6	148	222	156	234	165	247	175	262	186	279	199	298	213	320	230	346	245	368	256	384	268	402	288	432	303	455	329	494
9.8	151	226	159	238	168	252	178	267	190	285	203	304	218	327	235	353	250	375	261	392	273	410	294	441	309	464	336	504
10.0	154	231	162	243	171	257	182	273	194	290	207	310	222	333	240	360	255	383	267	400	279	419	300	450	316	474	343	514
10.2	157	235	165	248	175	262	185	278	197	296	211	317	227	340	245	367	260	391	272	408	285	427	306	459	322	483	350	525
10.4	160	240	169	253	178	267	189	284	201	302	215	323	231	347	250	374	266	398	277	416	290	435	312	468	328	493	357	535
10.6	163	245	172	258	182	273	193	289	205	308	219	329	236	353	254	382	271	406	283	424	296	444	318	477	335	502	363	545
10.8	166	249	175	263	185	278	196	295	209	314	223	335	240	360	259	389	276	414	288	432	301	452	324	486	341	512	370	555
11.0	169	254	178	268	189	283	200	300	213	319	228	341	244	367	264	396	281	421	293	440	307	460	330	495	347	521	377	566
11.2	172	258	182	272	192	288	204	305	217	325	232	348	249	373	269	403	286	429	299	448	313	469	336	504	354	531	384	576
11.4	175	263	185	277	195	293	207	311	221	331	236	354	253	380	274	410	291	437	304	456	318	477	342	513	360	540	391	586
11.6	178	268	188	282	199	298	211	316	225	337	240	360	258	387	278	418	296	444	309	464	324	486	348	522	366	549	398	597
11.8	182	272	191	287	202	303	215	322	228	343	244	366	262	393	283	425	301	452	315	472	329	494	354	531	373	559	405	607
12.0	185	277	195	292	206	309	218	327	232	348	248	372	267	400	288	432	306	460	320	480	335	502	360	540	379	568	411	617

Source: A Policy on Geometric Design of Highways and Streets, American Association of State Highway and Transportation Officials, Washington D.C., 2004. Used by permission.

Note: 1 mi/hr = 1.61 km/h; 1 ft = 0.3 m

spiral curve should at least be equal to that for superelevation runoff when two lanes are rotated, determine the length that should be used for design.

Solution

Use Equation 6.51 to determine the length of a spiral curve based on the design speed and radius of curve:

$$L = \frac{0.0214u^3}{RC}$$

$$= \frac{0.0214 \times 105^3}{240 \times 0.9}$$

$$= 114.7 \text{ m}$$

Use Table 6.14 to determine the length required for superelevation runoff. Use superelevation of 0.1 (10%) as the road is in a rural area. For a two-lane pavement, $e = 0.1$ and lane width = 3.6 m. Length of superelevation runoff for one lane = 83.7 m (see Table 6.14).

Use Equation 6.52 to determine the adjustment factor for two lanes rotated:

$$b_w = [1 + 0.5(n_1 - 1)]/n_1$$

$$b_w = [1 + 0.5(2 - 1)]/2$$

$$= 0.75$$

For two-lane rotation, length of superelevation runoff = 0.75 × 2 × 83.7 m = 125.55 m = 126 m, which is the same given in Table 6.14.

Assume a value of 0.9 for C gives a spiral length that is less than the length required for superelevation runoff. The length of 126 m should therefore be used.

Design of Railroad Horizontal Curves

Railroad Simple Curves

Although actual and unbalanced superelevations are briefly discussed in Chapter 3, a detailed discussion is given here to facilitate the use of the relevant equations for computing these superelevations.

When a train is moving around a horizontal curve, it is subjected to a centrifugal force acting radially outward, similar to that discussed for highways. It is therefore necessary to raise the elevation of the outer rail of the track by a value E_q, which is the superelevation that provides an equilibrium force similar to that on highways. For any given equilibrium elevation, there is an equilibrium speed. This is the speed at which the resultant weight and the centrifugal force are perpendicular to the plane of the track. When this occurs, the components of the centrifugal force and the weight of the plane of the track are balanced.

If all trains travel around a curve at the equilibrium speed, both smooth riding and minimum wear of the track will be obtained. This is not always the case as some trains may travel at higher speeds than the equilibrium speed while others may travel at lower speeds. Trains traveling at a higher speed will cause more than normal wear on the outside rail, while traveling at lower speeds will cause more than normal wear on the inside track. Also, when the train is traveling faster than the equilibrium speed, the centrifugal force is not completely balanced by the elevation, which results in the car body tilting toward the outside of the curve. Consequently, under normal conditions, the inclination of the car body from the vertical is less than the inclination of the track from the vertical. The difference between the inclination of the car from the vertical (car angle) and that of the track from the vertical (track angle) is known as the roll angle. The higher the roll angle, the less comfort is obtained as the train traverses the curve. Full equilibrium superelevation (e_q) is, however, rarely used in practice for two main reasons. First, the use of a full equilibrium superelevation may require long spiral transition curves. Second, full equilibrium superelevation can result in discomfort for passengers on a train traveling at a speed that is much less than the equilibrium speed or if the train is stopped along a highly superelevated curve. The portion of the equilibrium superelevation used in the design of the curve is known as the actual superelevation (e_a), and the difference between the actual superelevation and the equilibrium superelevation is known as the unbalanced superelevation (e_u).

Equations relating to the different superelevations of the curve, the design speed, and the radius of the curve have been developed for different track classifications.

The *Track Design Handbook for Light Rail Transit* gives the relationship shown in Equation 6.53 for computing desired values of the actual superelevation of horizontal curves on these tracks:

$$e_a = 0.79\left(\frac{u^2}{R}\right) - 1.68 \tag{6.53}$$

and the desired relationship between the actual superelevation and the unbalanced superelevation is given as

$$e_u = \left[1 - \left(\frac{e_a}{2}\right)\right] \tag{6.54}$$

where

e_a = actual superelevation in cm
e_u = unbalanced superelevation in cm
u = curve design speed, km/h
R = radius, m

It is recommended that values obtained for e_a using Equation 6.53 should be rounded up to the nearest 0.5 cm. Also, when the sum of the unbalanced and actual elevations ($e_a + e_u$) is 2.5 cm or less, it is not necessary to provide any actual superelevation. For tracks that are jointly used by light rail transit and freight vehicles, Equation 6.53 should be used until the calculated value reaches 7.5 cm. Higher values of up to 10 cm may be used to achieve operating speed if this is approved by both the transit authority and the railroad authorities.

The equation for the equilibrium superelevation for *light rail transit track* is given as

$$e_q = e_a + e_u = 1.184\left(\frac{u^2}{R}\right) \tag{6.55}$$

or

$$e_q = e_a + e_u = 0.00068u^2D_c \tag{6.56}$$

where

e_q = equilibrium superelevation in cm
e_a = actual track elevation to be constructed in cm
e_u = unbalanced superelevation in cm
u = design speed through the curve, km/h
R = radius of the curve, m
D_c = degree of curve (chord definition)

AREMA gives a similar relationship for computing the equilibrium superelevation for *freight and passenger intercity tracks* as that given in Equation 6.55 for light rail tracks. This relationship is given in Equation 6.57:

$$e_q = 0.00068u^2D_c^\circ \tag{6.57}$$

where

e_q = equilibrium superelevation
u = design speed through the curve, km/h
D_c = degree of curve (chord definition)

However, note that the horizontal curvature on railroad main line tracks should not be greater than 3° for new tracks or the maximum on an existing track that is being realigned, and in no case should it be greater than 9° 30". Experience has also shown that baggage cars, passenger coaches, diner cars, and Pullman cars can ride comfortably with an unbalanced superelevation of up to 7.5 cm. This can be increased to 11.25 cm if the roll angle is less than 1.5°.

EXAMPLE 6.13

Determining the Adequacy of the Actual Superelevation on a Freight and Passenger Intercity Rail Track

An existing freight and passenger railway track has actual superelevation of 15 cm on a curve of 840 m. If the track is being upgraded for a design speed of 120 km/h, determine whether the existing superelevation is adequate.

Determine degree of curve—use Equation 6.30:

$$R = \frac{1718.89}{D_c^\circ}$$

$$840 = \frac{1718.89}{D_c^\circ}$$

$$D_c = 2.046^\circ$$

Determine equilibrium superelevation—use Equation 6.57:

$$e_q = 0.00068u^2D^\circ$$

$$e_q = 0.00068(120)^2(2.046)$$

$$= 20.03 \text{ cm}$$

Determine unbalanced superelevation:

$$e_u = e_q - e_a$$

$$= (20.03 - 15) \text{ cm}$$

$$= 5.03 \text{ cm}$$

Since the unbalanced equilibrium is less than 7.5 cm, the existing actual superelevation is acceptable.

Railroad Compound Curves

The compound curve is seldom used in railroad design. It is recommended that a spiral curve be used to connect the two or more legs of simple curves forming the compound curve.

Railroad Spiral Curves

It is recommended that on railroads a spiral or transition curve be used to connect a tangent and a curve, unless this is not practicable. The design of the spiral curve for railroads is similar to that for highways in that it starts with the determination of the length of the curve. The minimum length of the spiral curve depends on the classification of the track on which it will be constructed.

AREMA gives two conditions that govern the length of a railroad spiral curve on an entirely reconstructed or new main track for *freight and passenger intercity tracks:*

(i) The unbalanced lateral acceleration acting on a passenger in a passenger car of average roll tendency should not exceed 0.03 g/s. To satisfy this condition, it is recommended by AREMA that the length should not be less than that obtained from Equation 6.58:

$$L_{\text{minspiral}} = 0.122(e_u)u \tag{6.58}$$

where

$L_{minspiral}$ = desirable length of spiral, m
e_u = unbalanced elevation, cm (usually taken as 7.5 cm for comfortable speed)
u = maximum train speed, km/h

(ii) In order to limit the possible racking and torsional forces, the longitudinal slope of the outer rail with respect to the inner rail should not be greater than 1/744. This condition is satisfied if the length of the spiral curve is not less than L_{min} given in Equation 6.59, which is based on an 25.5-m-long car:

$$L_{minspiral} = 7.44e_a \tag{6.59}$$

where

$L_{minspiral}$ = desirable length of spiral curve (m)
e_a = actual elevation, cm

When existing tracks are being realigned, the use of Equation 6.59 may give a spiral length for which the construction cost is excessive. In such cases, the unbalanced lateral acceleration acting on a passenger in a passenger car of average roll tendency may be increased to 0.04 g/s. This condition is satisfied if the length of the spiral curve is not less than that obtained from Equation 6.60. Equation 6.60 can then be used instead of Equation 6.58:

$$L_{minspiral} = 0.091e_u u \tag{6.60}$$

where

$L_{minspiral}$ = desirable length of spiral, m
e_u = unbalanced elevation, cm
u = maximum speed of train, km/h

When Equation 6.59 is used to determine the length of the spiral curve, the maximum slope condition should also be satisfied, which means that the longer length obtained from Equation 6.58 and Equation 6.59 should be used.

EXAMPLE 6.14

Determining the Length of a Spiral Curve on a Freight and Passenger Intercity Railway Track

A new freight and passenger intercity railway track is being designed with a design speed of 120 km/h. Determine the minimum length of a spiral curve connecting a tangent to a 2° horizontal curve on this track if the actual superelevation is 15 cm.

Solution

Determine equilibrium superelevation—use Equation 6.57:

$$\begin{aligned} e_q &= 0.00068u^2D \\ &= 0.00068(120)^2(2.00) \\ &= 19.584 \text{ cm} \end{aligned}$$

Determine unbalanced superelevation

$$\begin{aligned} e_u &= e_q - e_a \\ &= 19.584 - 15 \\ &= 4.58 \text{ cm} \end{aligned}$$

Determine the minimum length of the spiral curve that satisfies unbalanced lateral acceleration requirements—use Equation 6.58:

$$\begin{aligned} L_{\text{minspiral}} &= 0.122(e_u)u \\ &= (0.122)(4.58)(120) \\ &= 67.05 \text{ m} \end{aligned}$$

Determine the minimum length of spiral curve to satisfy the racking and torsional forces limitation—use Equation 6.59:

$$\begin{aligned} L_{\text{minspiral}} &= 7.44e_a \\ &= (7.44)15 \\ &= 111.6 \text{ m} \end{aligned}$$

In order to satisfy both requirements, the length of the spiral should be 111.6 m.

Passenger comfort and the rate of change in superelevation are also the two factors that influence the length of the spiral curve on a *light rail transit track*. In order to avoid excessive lateral unbalanced acceleration acting on passengers, the length of a spiral connecting a tangent to a horizontal curve on a light rail transit track should not be less than that obtained from Equation 6.61:

$$L_{\text{minspiral}} = 0.061e_uu \tag{6.61}$$

where

e_u = unbalanced superelevation, cm
u = design speed, km/h

In order to limit the rate of change of superelevation on the spiral curve on a light rail transit track so as to avoid excessive stress on the vehicle frame, the minimum length of the spiral curve is obtained from Equations 6.62 and 6.63:

$$L_{\text{minspiral}} = 0.082e_au \tag{6.62}$$

$$L_{minspiral} = 3.72e_a \tag{6.63}$$

where

$L_{minspiral}$ = minimum length of spiral curve, m
e_a = actual track superelevation, cm
u = design speed, km/h

However, the length of the spiral should not be less than 18 m. It can be reduced to 9.3 m when the geometric conditions are extremely restricted; for example, on an embedded track in a central business district (CBD).

It is also necessary to insert a transition spiral curve between the two curves of a compound curve. Criteria similar to those of the tangent-to-curve spirals are used. In this case, the desired minimum length of the spiral is obtained as the longest computed from Equations 6.64, 6.65, and 6.66:

$$L_{minspiral} = 3.72(e_{a2} - e_{a1}) \tag{6.64}$$

$$L_{minspiral} = 0.061(e_{u2} - e_{u1})u \tag{6.65}$$

$$L_{minspiral} = 1.082(e_{a2} - e_{a1})u \tag{6.66}$$

where

$L_{minspiral}$ = minimum length of spiral curve, m
e_{a1} = actual track superelevation for first circular curve, cm
e_{a2} = actual track superelevation for second circular curve, cm
e_{u1}, e_{u2} = unbalanced superelevation for first and second curves, cm
u = design speed, km/h

However, the absolute minimum length of the spiral curve on main line light rail transit tracks as well as the minimum length for yard and nonrevenue tracks is obtained as the longer computed from Equations 6.64 and 6.67:

$$L_{minspiral} = 0.081e_u u \tag{6.67}$$

where

$L_{minspiral}$ = minimum length of spiral curve, m
e_u = unbalanced superelevation, cm
u = design speed, km/h

EXAMPLE 6.15

Determining the Length of a Spiral Transition Curve Connecting the Two Curves of a Compound Curve on a Light Rail Transit Track

A spiral curve is being designed to connect the two curves of a compound curve on a light rail transit track with a design speed of 75 km/h. The first curve has a radius of 825 m and actual superelevation of 3.8 cm and the second curve has a radius of 600 m and an actual superelevation of 4.37 cm. Determine

(a) Desired length of the spiral curve
(b) Absolute minimum length of the spiral curve
(c) The length that should be used in constructing the curve

Solution

Determine equilibrium superelevations—use Equation 6.55:

$$e_q = 1.184\left(\frac{75^2}{R}\right)$$

For the first curve

$$e_q = 1.184\left(\frac{75^2}{825}\right)$$
$$= 8.07 \text{ cm}$$

For the second curve

$$e_q = 1.184\left(\frac{75^2}{600}\right)$$
$$= 11.1 \text{ cm}$$

Determine unbalanced superelevations:

$$e_u = e_q - e_a$$

For the first curve

$$e_u = 8.07 - 3.8$$
$$= 4.27 \text{ cm}$$

For the second curve

$$e_q = 11.1 - 4.37$$
$$= 6.73 \text{ cm}$$

Determine the desired length of spiral curve—use Equations 6.64, 6.65, and 6.66:

$$L_{\text{minspiral}} = 3.72(e_{a2} - e_{a1})$$
$$= 3.72(4.37 - 3.8)$$
$$= 2.12 \text{ m}$$

$$L_{\text{minspiral}} = 0.061(e_{u2} - e_{u1})u$$
$$= 0.061(6.73 - 4.27)75$$
$$= 11.25 \text{ m}$$

$$L_{\text{minspiral}} = 0.082(e_{a2} - e_{a1})u$$
$$= 0.082(4.37 - 3.8)75$$
$$= 3.50 \text{ m}$$

The computed desired length is therefore 11.25 m.

Determine the absolute minimum length of the spiral—use Equation 6.67:

$$L_{\text{minspiral}} = 0.081(e_u)u$$
$$= 1.09(6.73)(75)$$
$$= 40.88 \text{ m}$$

The results indicate that the computed desired length is 11.25 m, which in this case is less than the absolute minimum length of 40.88 m. The length of the spiral should therefore be 40.88 m.

Layout of Horizontal Curves for Highways and Railroads

Having determined the type and length of a horizontal curve on the travelway of a specific mode, it is necessary to compute certain properties of the curve that are required for laying out the curve in the field. There are several ways of setting out a simple horizontal curve, which include deflection angles, tangent offsets, and middle ordinates. However, the most commonly used method in all modes is the deflection angle method, which is described here.

Deflection Angle Method for Setting Simple Horizontal Curves

This method involves staking out points on the curve using deflection angles measured from the tangent at the point of curve (PC) and the lengths of arcs joining consecutive whole stations. Figure 6.26 is a schematic of the procedure involved. Angle *VAp* is the first deflection angle formed by the tangent *VA* and the chord joining the point of curve (PC) and the first whole station. Note that in many cases of highway design, the length of the arc *Ap* is less than 30 m as the PC may not be at a whole station. Using the geometry of a circle

$$\text{angle } VAp = \frac{\delta_1}{2}$$

we note that the next deflection angle to the next whole station is angle *VAq*, which is formed by the tangent *VA* and the chord joining the PC and *q* (i.e., the next whole station), is

$$\frac{\delta_1}{2} + \frac{D}{2}$$

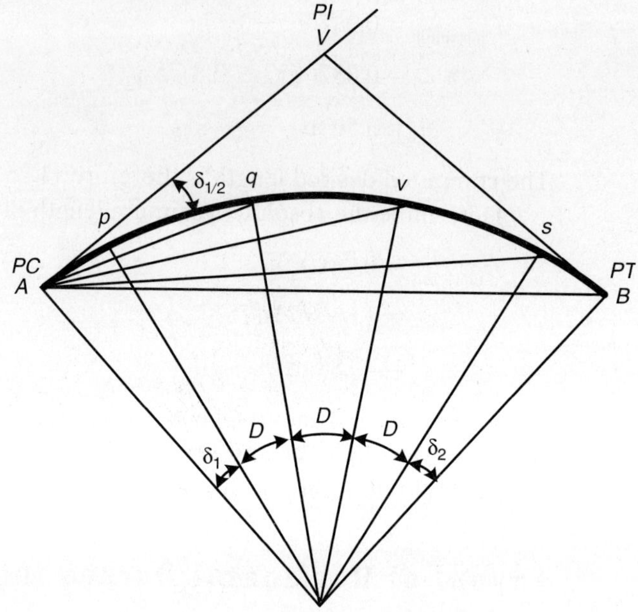

FIGURE 6.26
Deflection angles on a simple circular curve.

where

D = degree of the curve

The next deflection angle VAv is

$$\frac{\delta_1}{2} + \frac{D}{2} + \frac{D}{2} = \frac{\delta_1}{2} + D$$

and the next deflection angle VAs is

$$\frac{\delta_1}{2} + \frac{D}{2} + \frac{D}{2} + \frac{D}{2} = \frac{\delta_1}{2} + \frac{3D}{2}$$

and the last deflection angle is

$$\frac{\delta_1}{2} + \frac{D}{2} + \frac{D}{2} + \frac{D}{2} + \frac{\delta_2}{2} = \frac{\delta_1}{2} + \frac{3D}{2} + \frac{\delta_2}{2} \tag{6.68}$$

Note that the number of deflection angles required depends on the length of the curve. In order to set out the curve, it is necessary for us to determine δ_1 and δ_2 as we already know D. If l_1 is the length of the arc Ap, then

From Eq. 6.35 we know that

$$R = \frac{180L}{\Delta\pi}$$

Therefore

$$\frac{l_1}{\delta_1} = \frac{L}{\Delta} = \frac{l_2}{\delta_2} \tag{6.69}$$

where

L = length of the circular curve

Δ = deflection angle of the curve

In setting out the simple horizontal curve, using the deflection angle method, the following steps are taken:

Step 1: Locate PC and PT.

Step 2: Determine the length (l_1) of the arc of the curve between the PC and the first whole station. Note that if the length of the curve is a number of whole stations (i.e., multiples of 30 m), l_1 is automatically 30 m.

Step 3: Determine the first deflection angle δ_1 using Equation 6.69. Note that if the length of the curve is a number of whole stations, δ_1 is also D.

Step 4: Mount a transit over the PC and sight the PI.

Step 5: Locate the first whole station using l_1 and δ_1.

Step 6: Repeat step 5 for the other stations.

EXAMPLE 6.16

Design of a Simple Horizontal Curve for a Highway

A horizontal curve on a rural collector road in a level terrain is to be designed for a specified DHV of 2500 to connect two tangents that deflect at an angle of 48°. If the intersection of the tangents is located at station (586 + 20.52), determine

(a) Recommended minimum design speed of the highway

(b) Radius of the curve for the recommended minimum design speed

(c) Station of BC

(d) Station of PT

(e) Deflection angles to whole stations for laying out the curve.

Use a superelevation rate of 0.08.

Solution

Determine minimum design speed—use Table 6.5:

Specified DHV = 2500

Recommended design speed = 95 km/h

Determine minimum radius of the curve—use Equation 6.47:

$$R = \frac{u^2}{127(e + f_s)}$$

From Table 3.9, $f_s = 0.12$.

$$R = \frac{95^2}{127(0.08 + 0.12)}$$

$$R = 355 \text{ m}$$

Determine the length of the tangent (T)—use Equation 6.31:

$$T = R \tan \frac{\Delta}{2}$$

$$T = 355 \tan \frac{48}{2}$$

$$T = 158.06 \text{ m}$$

Determine the length of the curve—use Equation 6.35:

$$L = \frac{R\,\Delta\pi}{180}$$

$$L = \frac{355 \times 48 \times \pi}{180}$$

$$L = 297.5 \text{ m}$$

Determine the stations of BC and PT:

Station of BC = (586 + 20.52) − (5 + 8.06) = (581 + 12.46)

Station of PT = (581 + 12.46) + (9 + 27.5) = (591 + 9.96)

Determine the first, intermediate, and final deflection angles (δ_1), D, and δ_2—use Equation 6.69:

$$\frac{l_1}{\delta_1} = \frac{L}{\Delta} = \frac{l_2}{\delta_2}$$

$$l_1 = -(581 + 12.46) + (582 + 00.00) = 17.54 \text{ m}$$

$$l_2 = (591 + 9.96) - (591 + 00.00) = 9.96 \text{ m}$$

$$\delta_1 = \frac{48 \times 17.54}{297.5} = 2.83°$$

$$\delta_2 = \frac{48 \times 9.96}{297.5} = 1.607°$$

$$D = \frac{48 \times 30}{297.5} = 4.84°$$

Table 6.15 gives the computation for the intermediate whole stations.

TABLE 6.15
Deflection Angles and Chord Lengths for Example 6.16

Station	*Deflection Angle*	*Chord Length (m)*
BC 581 + 12.46	0	0
582 + 00.00	1.415	17.524
583 + 00.00	3.835	29.964
584 + 00.00	6.255	29.964
585 + 00.00	8.675	29.964
586 + 00.00	11.095	29.964
587 + 00.00	13.515	29.964
588 + 00.00	15.935	29.964
589 + 00.00	18.355	29.964
590 + 00.00	20.775	29.964
591 + 00.00	23.195	29.964
PT 591 + 09.96	23.999	9.951

Layout of Compound and Reverse Curves

Since compound and reverse curves are based on simple curves, the same procedure used to set out the simple curve is also used to lay out the compound or reverse curve. In each case, the first curve is laid out. The PT of the first curve is then considered as the PC of the second curve to lay out the second curve and so on.

Deflection Angle Method for Laying Out the Spiral Curve

In setting out the spiral curve, the original circular curve is shifted toward the center away from the main tangent, as shown in Figure 6.27. This shift provides room for the spiral curve. The section CC' of the circular curve is then retained and the spirals are placed from A to C and from C' to B. Note that the point where the spiral starts is the point of spiral and is usually designated as T.S. and the point where the spiral ends is usually designated as S.T., as shown in Figure 6.27.

Let Δ = central angle of the spiral

I = central angle of the circular curve

The central angle of a spiral is given as

$$\Delta = \frac{L_s D_a}{60} \tag{6.70}$$

where

L_s = length of the spiral curve, m
D_a – degree of the circular curve

Assuming that the spirals at both ends of the circular curve have the same length and central angle I, the central angle of the remaining simple curve CC' is $I - 2\Delta$. In order to set out the spiral curve, the deflection angles from the

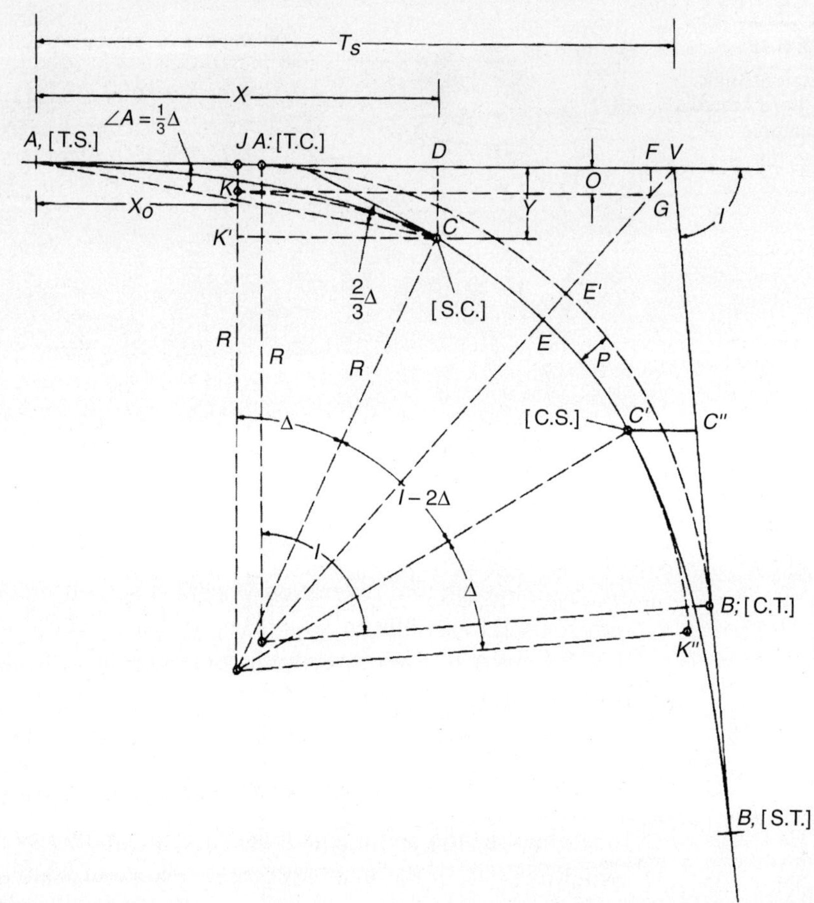

FIGURE 6.27

Field layout of a spiral curve.

Source: Surveying Theory and Practice, Davis, Foote, Anderson, and Mikhail, McGraw-Hill Book Company, 1981.

tangent should be computed. Consider a point p on the spiral located at distance l_s from T.S. (i.e., the point where the spiral joins with the tangent), as shown in Figure 6.28. It can be shown that

$$\partial = \frac{\ell_s^2}{L_s}\Delta \tag{6.71}$$

$$y \approx \frac{\ell_s^3}{6RL_s} \tag{6.72}$$

$$a \approx \frac{\ell_s^2}{L_s^2}A \tag{6.73}$$

$$a \approx \frac{\delta}{3} \tag{6.74}$$

where

δ = angle subtended at the center by length l_s of the spiral curve
a = deflection angle from the tangent at TS to any point p on the spiral curve
A = total deflection angle from the tangent at T.S. to S.C.

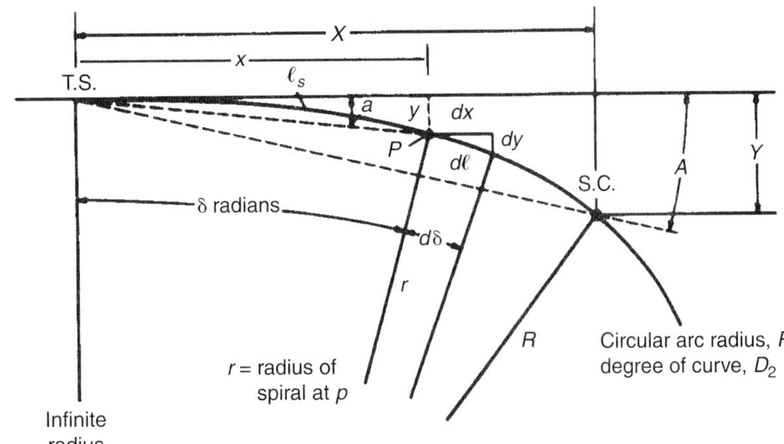

FIGURE 6.28
Mathematical development of a spiral curve.

Source: Surveying Theory and Practice, Davis, Foote, Anderson, and Mikhail, McGraw-Hill Book Company, 1981.

Note that in the development of Equations 6.73 and 6.74, it is assumed that both δ and a are small angles and

$$\sin\delta = \frac{dy}{dl_s} = \delta \tag{6.75}$$

$$\sin a = \frac{y}{l_s} = a \tag{6.76}$$

The equations are therefore approximate but they are sufficiently accurate for fieldwork in most practical situations. The values for X, Y, o, VA, and the external distance EV (see Figure 6.27) are also required for the spiral to be laid out. It can be shown that

$$X = \ell_s\left[1 - \frac{\delta_r^2}{(5)(2!)} + \frac{\delta_r^4}{(9)(4!)} - \frac{\delta_r^6}{(13)(6!)} + \cdots\right] \tag{6.77}$$

$$Y = \ell_s\left[\frac{\delta_r}{3} - \frac{\delta_r^3}{(7)(3!)} + \frac{\delta_r^5}{(11)(5!)} - \frac{\delta_r^7}{(15)(7!)} \cdots\right] \tag{6.78}$$

$$o = Y - KK' = Y - R(1 - \cos\Delta_s^\circ) \tag{6.79}$$

$$T_s = X - R\sin\Delta_s + (R + o)\tan\frac{I}{2} \tag{6.80}$$

$$EV = EG + GV + R\left[\frac{1}{\cos\frac{I}{2}} - 1\right] + \frac{o}{\cos\frac{I}{2}} \tag{6.81}$$

Note that in Equations 6.76 and 6.77, δ_r is the value of δ in radians.

In setting out the spiral curve using the deflection angle method, the following steps are taken:

Step 1: Determine the length of the spiral curve based on the requirement for the specific mode of transportation system under consideration.

Step 2: Set the transit at the vertex V (the intersection of the tangents), back sight along the tangent and locate the point A (T. S.), by measuring the distance T_s from V along the tangent (see Figure 6.27).

Step 3: Locate point D at a distance of $(T_s - X)$ from V.

Step 4: Rotate the transit through an angle of $(180 + I)°$ to sight along the forward tangent. Locate points S.T. and C'' at distances T_s and $(T_s - X)$ from V, respectively. Note that if different spiral lengths are used for the approach and leaving spirals, the appropriate value for T_s should be computed for each spiral.

Step 5: Set the transit at point D. Sight along the tangent to point A. Locate point C perpendicularly from the tangent at a distance of Y from point D.

Step 6: Set the transit at point C'' and locate point C' by repeating step 5.

Step 7: Set the transit at A (T.S.) and set the stations on the approach spiral using deflection angles a and chords. Points on the spiral are usually located at equal distances.

Step 8: Set the transit at C (S.C.), back sight the T.S. with $180 \pm 2A$ $(180 \pm 2/3\Delta_s)$ set on the horizontal circle. Rotate the upper circle of the transit through 180° and set the stations on the circle on whole stations as discussed earlier.

Step 9: Set the transit on S.T. and repeat step 7.

EXAMPLE 6.17

Determining Setting Out Properties of a Railroad Spiral Curve

If the spiral curve in Example 6.14 should connect a tangent with a 4° curve and the intersection angle I is 35°, determine

(a) The station of the point of spiral (TS) if the station of the point of intersection of the tangents is 885 + 9

(b) The stations of S.C., C.S. and S.T.

Solution

Determine the central angle of spiral—use Equation 6.70:

$$\Delta = \frac{L_s D_a}{60}$$

Length of spiral L_s from Example 6.14 = 111.6 m (note that the minimum length is still 111.6 m as the requirement for racking and torsional forces limitation still governs).

$$\Delta = \frac{111.6(4)}{60} = 7.44°$$

Determine the central angle of the spiraled circular curve $(I - 2\Delta) = (35 - 2 \times 7.44) = 20.12°$.
Determine the radius of the circular curve—use Equation 6.30:

$$R = \frac{1718.89}{D_c^{\circ}}$$

$$R = 1718.89/4$$

$$= 429.72 \text{ m}$$

Determine the length of the circular curve—use Equation 6.35:

$$L = \frac{R(I - 2\Delta_s)\pi}{180}$$

$$L = \frac{R(20.12)\pi}{180}$$

$$= 150.96 \text{ m}$$

Determine the distance along the tangent from T.S. to S.C. (i.e., X)—use Equation 6.77:

$$\delta = \Delta \text{ and}$$

$$\ell_s = L_s$$

$$\Delta = 7.44\pi/180 \text{ rad}$$

$$= 0.1299 \text{ rad}$$

$$X = L_s\left(1 - \frac{\delta^2}{10} + \frac{\delta^4}{216}\right) \text{(using only the first three terms in the brackets)}$$

$$= 111.41 \text{ m}$$

Determine the perpendicular distance from the tangent to S.C. (i.e., Y)—use Equation 6.78:

$$Y = L_s\left(\frac{\delta}{3} + \frac{\delta^3}{42} + \frac{\delta^5}{132}\right)$$

$$= 4.95 \text{ m}$$

Determine the throw (o)—use Equation 6.79:

$$o = Y - R(1 - \cos \Delta)$$

$$= 4.95 - 429.72(1 - 0.99158)$$

$$= 4.95 - 3.62 = 1.33 \text{ m}$$

Determine the distance along the tangent from T.S. to the vertex V (PI) (i.e., T_s)—use Equation 6.80:

$$T_s = X - R\sin\Delta + (R + o)\tan\frac{I}{2}$$

$$T_s = 111.41 - 429.72 \sin 7.44 + (429.72 + 1.33) \tan\frac{35}{2}$$

$$= 111.41 - 55.64 + 135.91$$

$$= 191.68$$

Determine the station at T.S.

Station of $PI - T_s$

$(885 + 9.00) - (6 + 11.68) = 878 + 27.32$

Determine the station at S.C.

Station at S.C. = station at T.S. + length of spiral curve

$= (878 + 27.32) + (3 + 21.6) = 882 + 18.92$

Determine the station at C.S.

Station at C.S. = station at S.C. + length of horizontal curve

$= (882 + 18.92) + (5 + 0.96)$

$= 887 + 19.88$

Determine the station at S.T.

Station at S.T. = station at C.S. + length of spiral curve

$= (887 + 19.88) + (3 + 21.6)$

$= 891 + 11.48$

Determination of the Orientation and Length of an Airport Runway

Designing the geometric alignment of an airport runway is significantly different from designing the travelways of other modes in that, in addition to the design of vertical curves, the orientation and minimum length of the runway should be determined. The runway orientation is necessary as the runway should be in the direction of the prevailing wind or be orientated so as to achieve wind coverage of at least 95%. Recall that wind coverage is the percentage of time crosswind components are below an acceptable velocity.

The length of the runway should be adequate to allow safe landings and take-offs by the current and future aircraft expected to use the airport. The runways considered here are fully usable in both directions and have clear approaches and departures to each runway end. The design of runways that are not fully usable for landing and take-off in both directions is outside the scope of this book. Interested readers may refer to the Federal Aviation Administration *Advisory Circular AC 150/5300-13,* which discusses the "declared distance" concept for determining the minimum runway lengths for those runways.

Runway Orientation

The best runway orientation can be determined by using one of the existing computer programs or graphically using a wind rose as described in the FAA *Advisory Circular 150/5300-13*. This consists of concentric circles each representing a different velocity in km/h or knots, and radial lines indicating wind direction. The outermost circle indicates a graduated scale in degrees around its circumference. The division between the speed groupings is indicated by the perimeter of each circle, and the area between any two successive radial lines is centered in the direction of the wind being considered. Figure 6.29 shows a wind rose constructed from the wind data shown in Table 6.16. As it is essential that the most reliable and up-to-date data be used, it is recommended that data based on the most recent 10 consecutive years be used to construct the wind rose. The National Oceanic and Atmospheric Administration (NOAA), National Climatic Data Center (NCDC), is the best source of wind data.

The first step in constructing the wind rose is to use the recorded data to determine the percentage of time wind velocities within a given range can be expected to be in a certain direction. The values obtained are rounded up to the nearest one-tenth of 1%. For example, using the data shown in Table 6.16, the percentage of time a wind with a speed of 11 to 16 knots will be expected in the direction 01 is 212/87864 (i.e., 0.2%). The values obtained are then entered in the appropriate segments of the wind rose, as shown in Figure 6.29. When the value obtained for any segment is less than 0.1%, the plus (+) symbol is used in that segment. The objective of the analysis is to determine the orientation of the runway that will provide the highest wind coverage within allowable crosswind limits.

The wind rose procedure to determine the most appropriate orientation involves the use of a transparent template with three parallel lines drawn to the same scale as that of the wind rose circles. The middle line represents the centerline of the runway. The distance between the middle line and the outer lines is the value of the design (allowable) crosswind. For example, in Figure 6.29, the design crosswind component is 13 knots. To determine the most appropriate

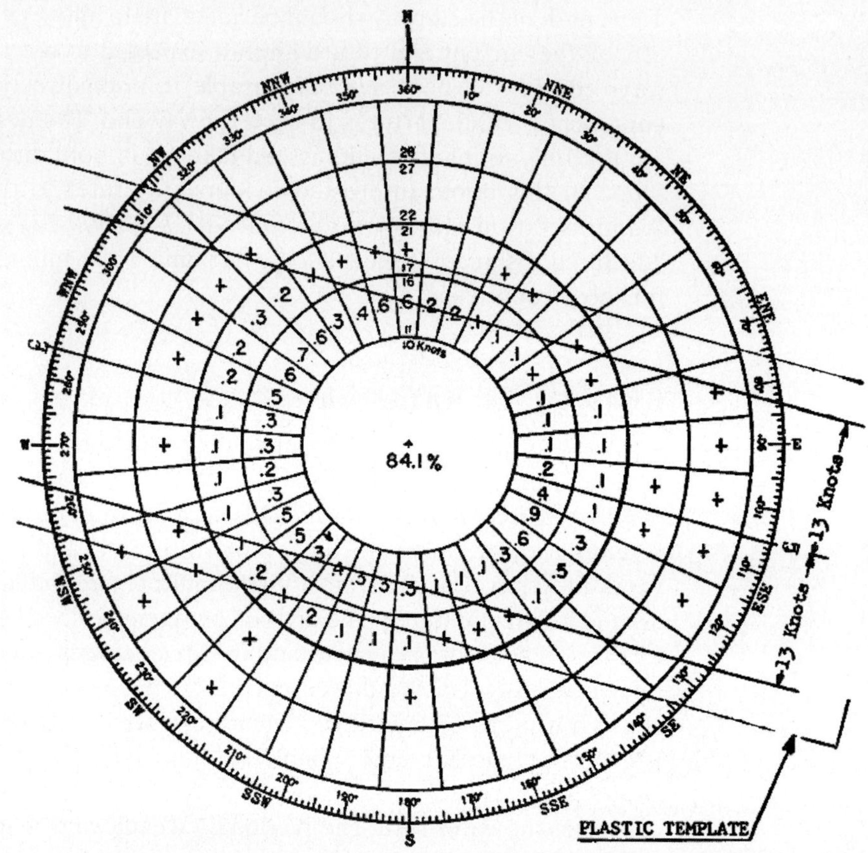

FIGURE 6.29
Completed wind rose.

Source: Advisory Circular AC 150/5300-13, Federal Aviation Administration, Department of Transportation, Washington, D.C. (Incorporating Changes 1 through 8), September 2004.

A runway oriented 105°–285° (true) would have 2.72% of the winds exceeding the design crosswind/crosswind component of 13 knots.

Note: 1 knot = 1.85 kmph

WIND SPEED DIVISIONS		RADIUS OF CIRCLE (KNOTS)
KNOTS	M.P.H.	
0 – 3.5	0 – 3.5	* 3.5 Units
3.5 – 6.5	3.5 – 7.5	* 6.5 "
6.5 – 10.5	7.5 – 12.5	10.5 – "
10.5 – 16.5	12.5 – 18.5	16.5 – "
16.5 – 21.5	18.5 – 24.5	21.5 – "
21.5 – 27.5	24.5 – 31.5	27.5 – "
27.5 – 33.5	31.5 – 38.5	*33.5 – "
33.5 – 40.5	38.5 – 46.5	*40.5 – "
40.5 – over	46.5 – over	

*May not be needed for most windrose analyses.

Note: 1 mph = 1.61 km/h

orientation and the percentage of time the orientation satisfies the crosswind standards, we take the following steps:

1. Locate the middle of the transparent template on the wind rose with the middle line passing through the center, as shown in Figure 6.29.
2. Rotate the template about the center of the wind rose until the sum of the percentages within the outer lines of the template is at a maximum.

TABLE 6.16

Wind Direction versus Wind Speeds

Source: Adapted from *Airport Design: Advisory Circular AC 150/5300-13,* Federal Aviation Administration, Department of Transportation, Washington, D.C. (Incorporating Changes 1 through 8), September 2004.

WIND DIRECTION VERSUS WIND SPEED

STATION: Anywhere, USA HOURS: 24 Observations/Day PERIOD OF RECORD: 1964–1973

DIRECTION	HOURLY OBSERVATIONS OF WIND SPEED									TOTAL	AVERAGE SPEED	
	KNOTS											
	0-3	4-6	7-10	11-16	17-21	22-27	28-33	34-40	41 OVER		KNOTS	mi/hr
	mi/hr											
	0-3	4-7	8-12	13-18	19-24	25-31	32-38	39-46	47 OVER			
01	469	842	568	212						2091	6.2	7.1
02	568	1263	820	169						2820	6.0	6.9
03	294	775	519	73	9					1670	5.7	6.6
04	317	872	509	62	11					1771	5.7	6.6
05	268	861	437	106						1672	5.6	6.4
06	357	534	151	42	8					1092	4.9	5.6
07	369	403	273	84	36	10				1175	6.6	7.6
08	158	261	138	69	73	52	41	22		814	7.6	8.8
09	167	352	176	128	68	59	21			971	7.5	8.6
10	119	303	127	180	98	41	9			877	9.3	10.7
11	323	586	268	312	111	23	28			1651	7.9	9.1
12	618	1397	624	779	271	69	21			3779	8.3	9.6
13	472	1375	674	531	452	67				3571	8.4	9.7
14	647	1377	574	281	129					3008	6.2	7.1
15	338	1093	348	135	27					1941	5.6	6.4
16	560	1399	523	121	19					2622	5.5	6.3
17	587	883	469	128	12					2079	5.4	6.2
18	1046	1984	1068	297	83	18				4496	5.8	6.7
19	499	793	586	241	92					2211	6.2	7.1
20	371	946	615	243	64					2239	6.6	7.6
21	340	732	528	323	147	8				2078	7.6	8.8
22	479	768	603	231	115	38	19			2253	7.7	8.9
23	187	1008	915	413	192					2715	7.9	9.1
24	458	943	800	453	96	11	18			2779	7.2	8.2
25	351	899	752	297	102	21	9			2431	7.2	8.2
26	368	731	379	208	53					1739	6.3	7.2
27	411	748	469	232	118	19				1997	6.7	7.7
28	191	554	276	287	118					1426	7.3	8.4
29	271	642	548	479	143	17				2100	8.0	9.3
30	379	873	526	543	208	34				2563	8.0	9.3
31	299	643	597	618	222	19				2398	8.5	9.8
32	397	852	521	559	158	23				2510	7.9	9.1
33	236	721	324	238	48					1567	6.7	7.7
34	280	916	845	307	24					2372	6.9	7.9
35	252	931	918	487	23					2611	6.9	7.9
36	501	1568	1381	569	27					4046	7.0	8.0
00	7729									7729	0.0	0.0
TOTAL	21676	31828	19849	10437	3357	529	166	22		87864	6.9	7.9

Note: * 1 knot = 1.85 kmph
** 1 mph = 1.61 kmph

3. The bearing of the centerline of the runway is then obtained from the scale on the outermost circle of the wind rose.
4. The sum of the percentages between the outside lines gives the percentage of time a runway orientated along the bearing determined above, which will satisfy the crosswind requirement. For example, as shown in Figure 6.29, the orientation of the runway is 105–285° and the design crosswind/crosswind component of 13 knots is exceeded only 2.72% of the time.

It may be necessary to try several orientations by rotating the centerline of the crosswind template about the wind rose to obtain the maximum coverage. When it is not possible to obtain at least 95% coverage by using a single orientation, the provision of another runway (crosswind) should be considered. The orientation of the crosswind runway should provide adequate coverage such that the combined coverage of the two runways is at least 95%.

TABLE 6.17
FAA Maximum Permissible Crosswind Components

Airport Reference Codes	*Allowable Crosswind Component*
A-I and B-I	10.5 kt
A-II and B-II	13.0 kt
A-III, B-III, and C-I through D-III	16.0 kt
A-IV through D-VI	20.0 kt

Note: 1 knot = 1.85 kmph

Source: Adapted from *Airport Design: Advisory Circular AC 150/5300-13,* Federal Aviation Administration, Department of Transportation, Washington, D.C. (Incorporating Changes 1 through 8), September 2004.

The design crosswind/crosswind component for different Airport Reference Codes was given earlier but is repeated here for easy access. Table 6.17 gives FAA maximum permissible crosswind components based on Airport Reference Codes. It should be noted that in the wind rose procedure, it is assumed that the winds are uniformly distributed over the area of each segment in the wind rose. The accuracy of this assumption decreases with the increasing size of the segment. Also, note that computer programs are now available for determining runway orientation. Interested readers may visit the Web site of the Federal Aviation Administration at http://www.fhwa.dot.gov.

Runway Length

An important task in the design of an airport is the selection of the runway length. The length selected has a significant impact on the overall cost of the airport and governs the type of aircraft that can safely use the airport. In general, factors that influence the selection of a runway length include

- Type of runway (e.g., primary, crosswind, or parallel)
- Landing and take-off gross weights
- Elevation of the airport
- Mean daily maximum temperature at the airport (°F)
- Gradient of the runway

Length of a Primary Runway

The procedure for determining this length is based on that given by the FAA in its *Advisory Circular 150/5325-4A*. The length of a primary runway is determined by considering one of two conditions:

- Length for a family of airplanes that have similar performance characteristics; and
- Length for a specific airplane that needs the longest runway.

Regardless of the conditions used, the selection of the runway length should be based on the airplanes that are expected to use the airport on a regular basis, which is defined as at least 250 operations/year. When the gross weight of the airplanes expected to use the airport does not exceed 272000 N), the family of airplanes condition is used. When the gross weight exceeds 272000 N, the specific airplane condition is used.

Runway Length Based on Airplane Grouping

Design guidelines are given for different airplane groupings based on approach speed and maximum certificated take-off weights.

Approach speeds of less than 30 knots: Airplanes with approach speeds less than 30 knots are categorized as ultralight or short take-off and landing aircraft. The recommended runway minimum length at sea level for this type of aircraft is 90 m. For higher elevations, this minimum length should be increased by 9 m for every 30 m increase in elevation.

Approach speeds of 30 knots or more but less than 59 knots: The minimum length recommended for this type of airplane is 240 m at sea level. For higher elevations, this length should be increased by 24 m for every 300 m increase in elevation.

All Airplanes with Maximum Certificated Takeoff Load of up to 56700 N

The minimum runway length recommended for this group of airplanes can be obtained from Figures 6.30 and 6.31 for planes with less than 10 and 10 or more passenger seats, respectively. Note that Figure 6.30 gives three sets of charts. These give lengths that will be adequate for 75, 95, and 100% of the fleet respectively (percentage of type of planes covered in this category of planes).

In some cases when runways for this class of airplanes are located at elevations higher than 1500 m above sea level, the lengths obtained may be greater than those required for turbo-powered jet airplanes within this class. In such cases, the longer length should be used.

Airplanes with Maximum Certificated Takeoff Weight of More Than 56700 N and Up to and Including 272000 N

Required runway lengths can be obtained from the charts given in Figures 6.32 and 6.33. These charts give runway lengths for 75 and 100% of the fleet and for 60 or 90% of useful load. Examples of the types of planes that make up 75% of the fleet are given in Table 6.18. The useful load is the difference

TABLE 6.18

Examples of Airplanes that Make Up 75% of the Fleet

Manufacturer	*Model*
Gates Lear Jet Corporation	Lear Jet (20, 30, 50 series)
Rockwell International	Sabreliner (40, 60, 75, 80 series)
Cessna Aircraft	Citation (II, III)
Dassault–Breguet	Fan Jet Falcon (10, 20, 50 series)
British Aerospace Aircraft Corporation	HS-125 (400, 600, 700 series)
Israel Aircraft Industries	1124 Westwind

Source: Advisory Circular ACI #150/5325-4A, Federal Aviation Administration, U.S. Department of Transportation, Washington D.C., January 1990.

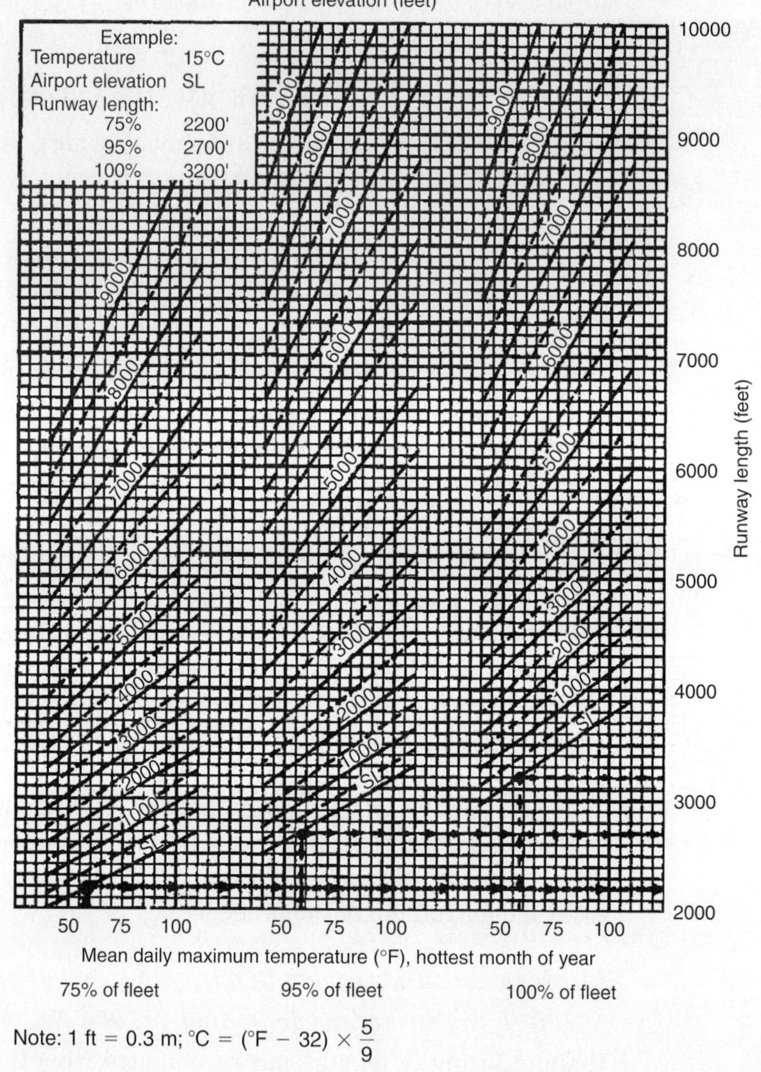

FIGURE 6.30

Runway lengths to serve small airplanes having less than 10 passenger seats.

Source: Advisory Circular AC 150/5325-4A, Federal Aviation Administration, Department of Transportation, Washington, D.C. (Incorporating Changes 1 through 8), September 2004.

between the airplane's maximum certificated load and its operating empty weight. This is considered to include the weights of the empty airplane, crew, crew baggage and supplies, removable passenger service equipment and emergency equipment, engine oil, and unuseable fuel. The useful load is therefore considered to include the weights of passengers and baggage, cargo, and useable fuel.

It is necessary to increase the runway lengths obtained from Figures 6.32 and 6.33 to account for either the maximum difference in center elevations or for wet and slippery runway conditions. The former correction is for take-offs while the latter is for landings. These are therefore mutually exclusive, and when corrections are required for both conditions, only the larger of the two is used.

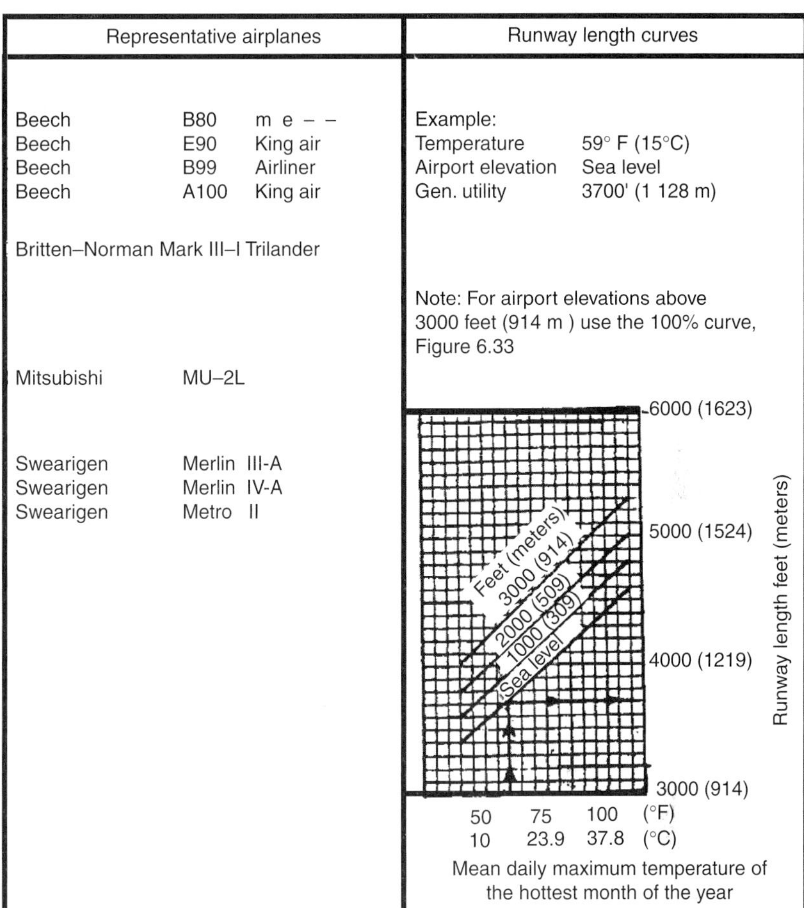

FIGURE 6.31

Runway lengths to serve small airplanes having 10 passenger seats or more.

Source: Advisory Circular AC 150/5325-4A, Federal Aviation Administration, Department of Transportation, Washington, D.C. (Incorporating Changes 1 through 8), September 2004.

In order to allow for the additional length that may be required for take-off on upgrades, the runway lengths obtained from the chart are increased by 10 ft for every foot of elevation difference between the high and low points of the runway centerline. To account for wet and slippery conditions, the runway lengths obtained from the 60% useful load curves should be increased by 15% or up to a maximum increase of 1680 m. Runway lengths obtained from the 90% useful load curves should be increased by 15% or up to a maximum increase of 2130 m.

Airplanes with Maximum Certificated Take-Off Load of More Than 27,0000 N

The Federal Aviation Administration suggests that the minimum runway length for this group of planes may be estimated from Equation 6.82, which

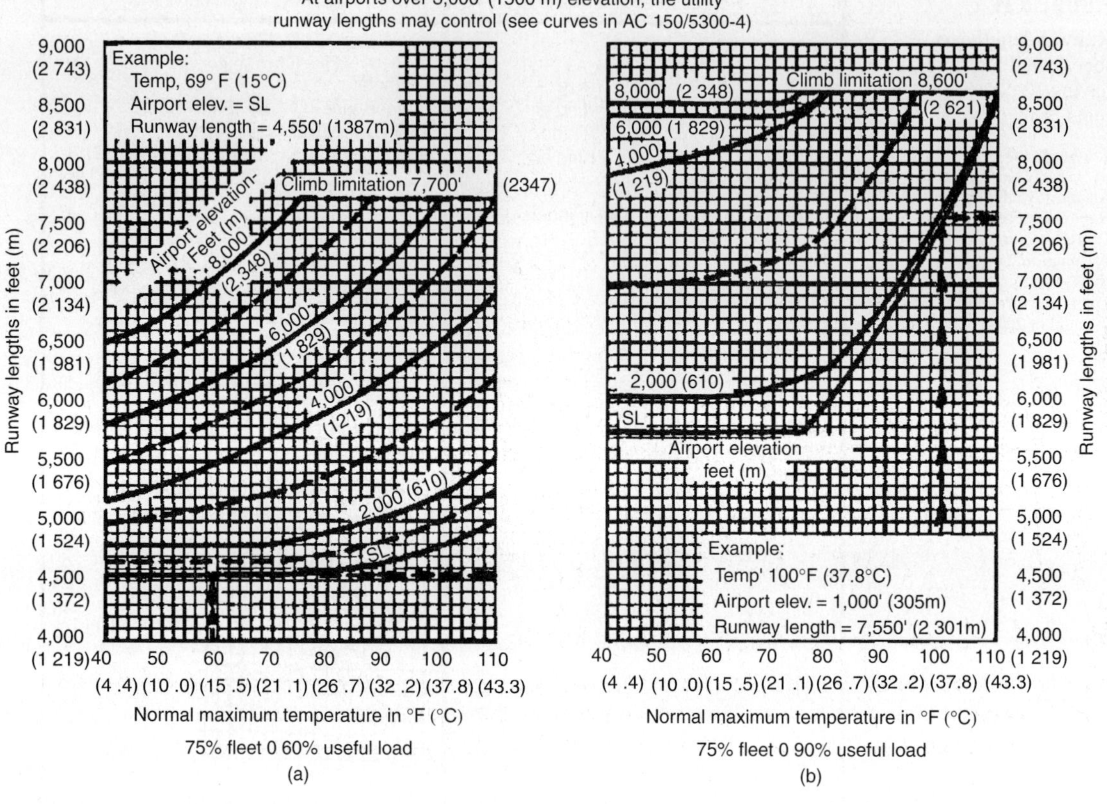

FIGURE 6.32

Runway lengths to serve 75% of large airplanes of 272000 N or less.

Source: Advisory Circular AC 150/5325-4A, Federal Aviation Administration, Department of Transportation, Washington, D.C. (Incorporating Changes 1 through 8), September 2004.

gives a general relationship between the minimum runway length and the length of haul:

$$Rwy = 1200 + 0.3915(Haul) - 0.000017(Haul)^2 \tag{6.82}$$

where

Rwy = minimum runway length in m
$Haul$ = maximum haul for the airplane group (km)

It should be noted that it is not necessary to adjust the minimum lengths obtained from Equation 6.82 for surface conditions as this equation is based on wet or slippery conditions. It is, however, recommended that lengths obtained from Equation 6.82 be increased by 7% for every 300 m of elevation above sea level. It is also suggested that runway lengths of up to 4900 m may be taken as

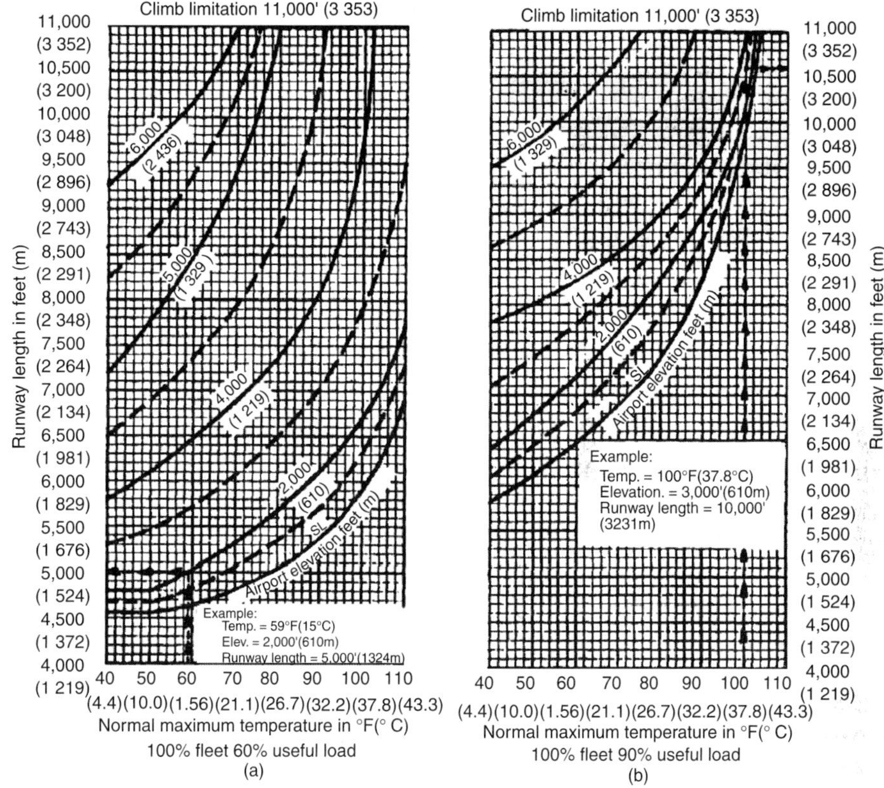

FIGURE 6.33

Runway length to serve 100% of large airplanes of 60,000 lb (27,200 kg or less).

Source: Advisory Circular ACI #150/5325-4A, Federal Aviation Administration, U.S. Department of Transportation, Washington, D.C., January, 1990

a recommended length for this group of airplanes. However, as will be seen later, runways expected to serve airplanes within this category are normally designed for specific airplanes.

Runway Length Based on Specific Airplanes

Runways that are designed to serve airplanes having a gross weight of 272000 N or more are usually designed for specific airplanes. Recommended landing and take-off lengths for a specific type of airplane can be obtained from performance curves that have been prepared by the Federal Aviation Administration. These curves are based on actual flight tests and operational data. The use of these charts to determine the minimum runway length requires the following information:

- Specific airplane to be served
- Mean daily maximum temperature (°C) for the hottest month of the year at the airport
- Longest length of haul flow on a regular basis
- Maximum difference in runway centerline elevation

Note that landing lengths obtained from these charts should be adjusted for wet and slippery conditions by increasing them by 15% for piston and turbo-prop-powered airplanes or by 7% for turbo-jet-powered airplanes. It is necessary to increase the lengths obtained for turbo-prop-powered airplanes by only 7% as these charts take a 5-knot tail wind into consideration. It should be noted that these corrections for wet and slippery conditions are made only for turbo-jet landing lengths and are not required for take-off lengths. Also note that the take-off lengths obtained from these charts should be increased by 10 m for every meter of elevation difference between the high and low points of the runway centerline to account for the additional length that is required during take-off on upgrades. In addition to the charts, tables have also been prepared that can be used to determine the recommended lengths for turbo-jet-powered airplanes. Only the procedures using the curves are presented. Both the landing and take-off lengths should be determined and the longer length selected as the design length. Examples of these charts are given in Figures 6.34 through 6.39. These charts are for the Convair 340/440, Boeing 720-000 series, and Douglas DC-9-10 series aircraft. The procedure for using these charts is described below and illustrated in Figures 6.36 and 6.37 for an airplane of the Boeing 720-000 series, having a maximum landing weight of 697500 N, a maximum take-off weight of 810000 N using a level runway in an airport at an elevation of 900 m, a haul distance of 644 km and a mean daily maximum temperature of 27°C.

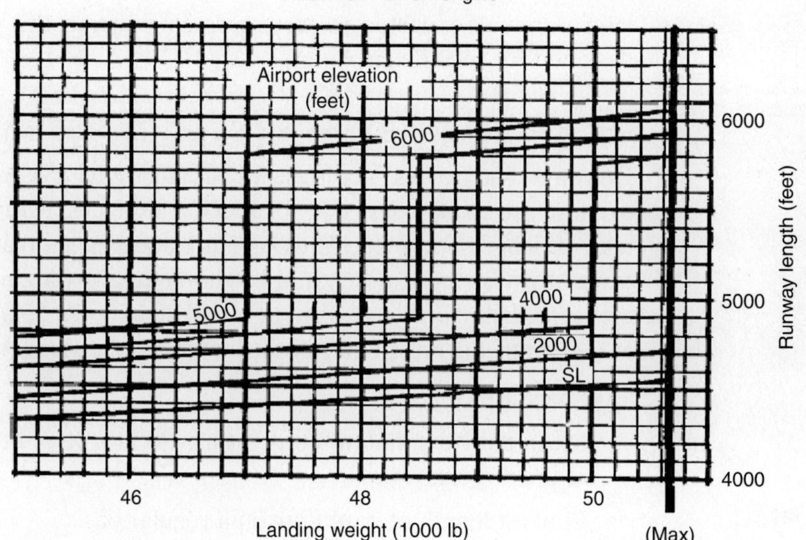

FIGURE 6.34

Aircraft performance curve, landing, Convair 340/440.

Source: Advisory Circular ACI #150/5325-4A, Federal Aviation Administration, U.S. Department of Transportation, Washington, D.C., January, 1990.

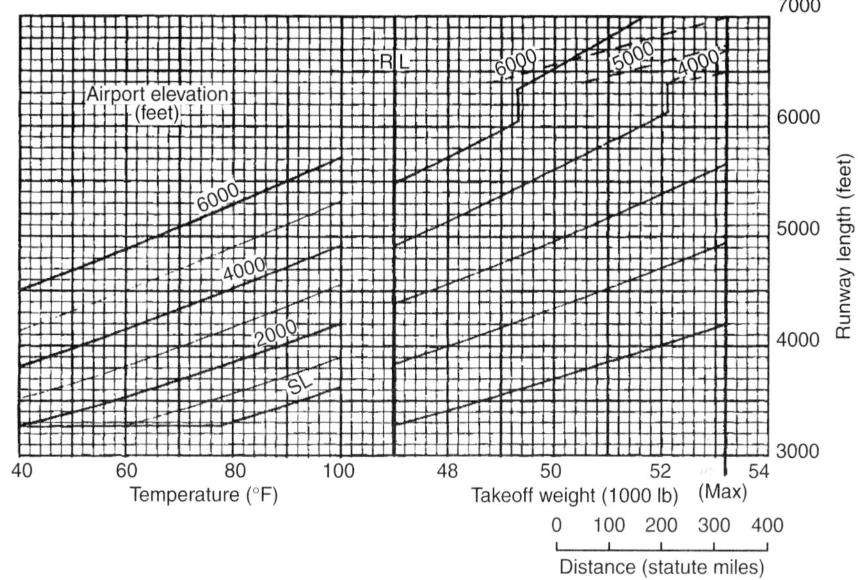

FIGURE 6.35

Aircraft performance curve, takeoff, Convair 340/440.

Source: Advisory Circular ACI #150/5325-4A, Federal Aviation Administration, U.S. Department of Transportation, Washington, D.C., January, 1990.

Boeing 720-000 series

Pratt & Whitney JT3C-12 engine

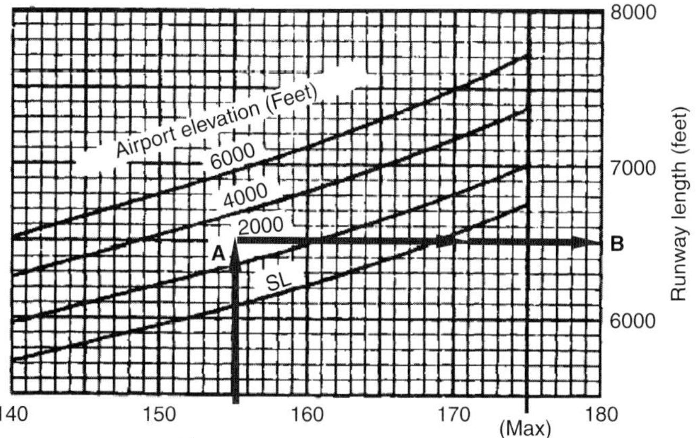

Note: 1 lb = 4.5 N; 1 ft = 0.3 m

FIGURE 6.36

Aircraft performance curve, landing, Boeing 720-000 series.

Source: Advisory Circular ACI #150/5325-4A, Federal Aviation Administration, U.S. Department of Transportation, Washington, D.C., January, 1990.

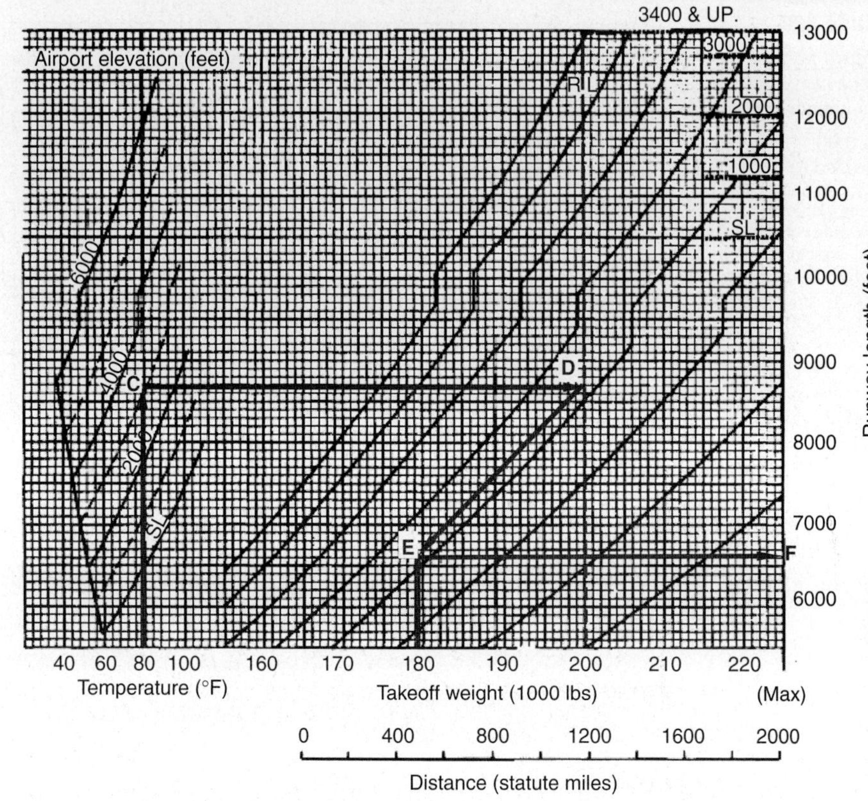

Note: 1 statute mile = 1.61 km; 1 ft = 0.3 m; 1 lb = 4.5 N; $°C = \frac{5}{9}(°F - 32)$

FIGURE 6.37

Aircraft performance curve, takeoff, Boeing 720-000 series.

Source: Advisory Circular ACI #150/5325-4A, Federal Aviation Administration, U.S. Department of Transportation, Washington, D.C., January, 1990.

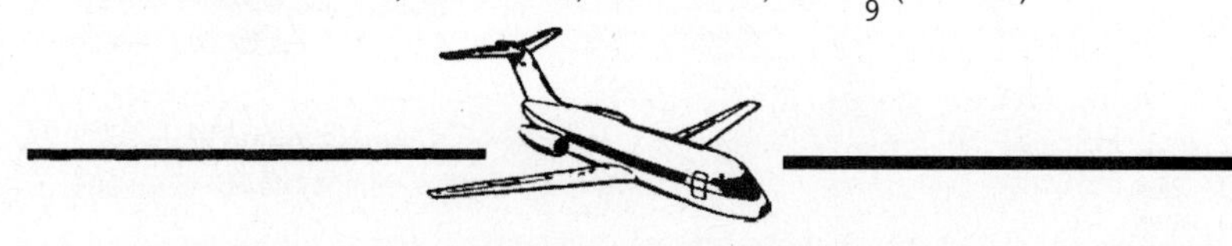

Douglas DC-9-10 series

Pratt & Whitney JT8D-5 engine

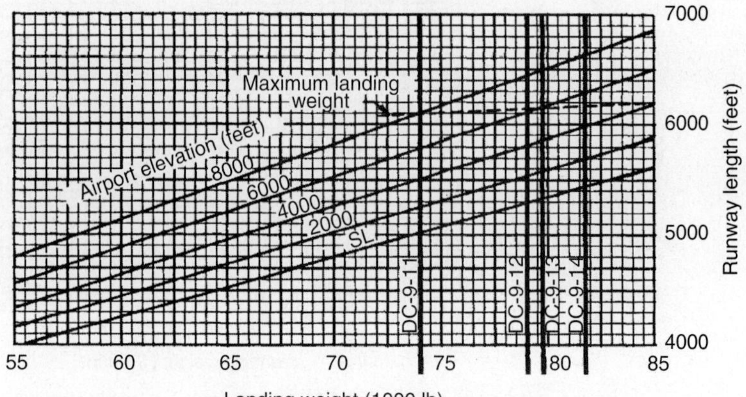

Note: 1 lb = 4.5 N; 1 ft = 0.3 m

FIGURE 6.38

Aircraft performance curve, landing, Douglas DC-9-10 series.

Source: Advisory Circular ACI #150/5325-4A, Federal Aviation Administration, U.S. Department of Transportation, Washington, D.C., January, 1990.

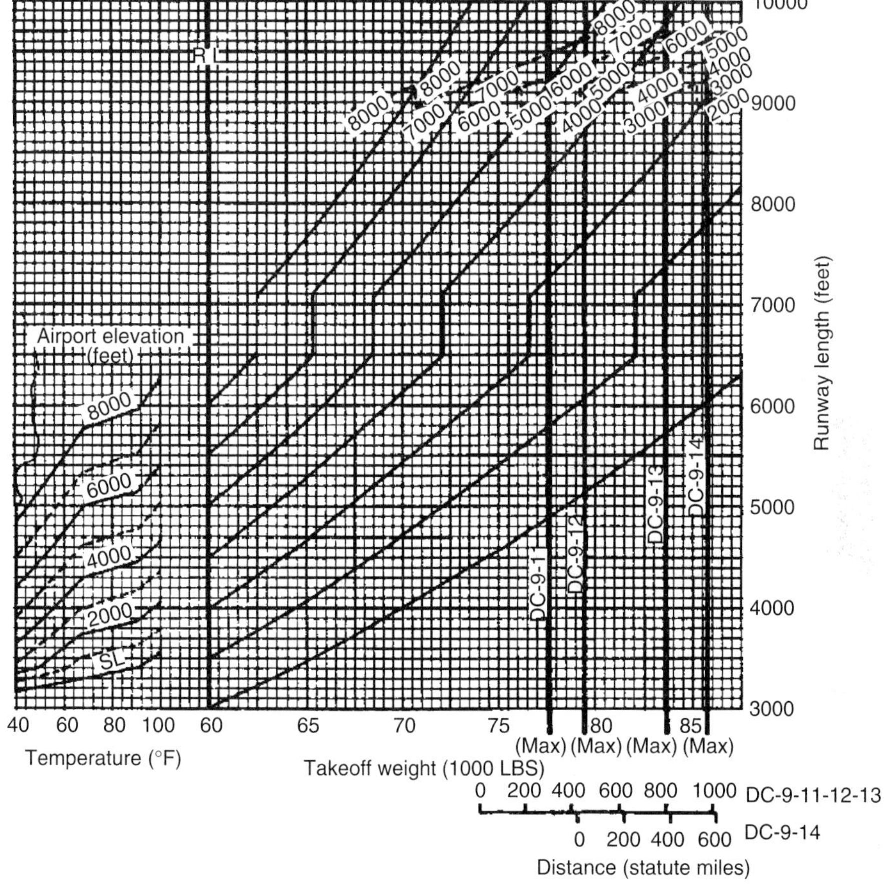

FIGURE 6.39

Aircraft performance curve, takeoff, Douglas DC-9-10 series.

Source: Advisory Circular ACI #150/5325-4A, Federal Aviation Administration, U.S. Department of Transportation, Washington, D.C., January, 1990.

Determination of Landing Runway Length

- Enter the landing runway length chart for the specific airplane (Figure 6.36) on the horizontal scale at the maximum landing weight of 697500 N.
- Project this point vertically to the line representing the elevation of the airport (900 m) at *A*. Interpolate between lines if necessary.
- Draw a horizontal line from *A* to intersect with the runway length at *B*, to give the length of about 1950 m.
- For piston and turbo-prop-powered airplanes, increase this landing length by 7% to account for wet and slippery conditions (this length should be increased by 15% for piston and turbo-prop-powered airplanes and by 7% for turbo-jet-powered airplanes)—in this case the plane is turbo-jet powered. The required landing length is therefore $(1950 \times 1.07) = 2086.5$ m (i.e., 2100 m since runway lengths are usually rounded up to the nearest 30 m).

Determination of Take-Off Length

- Enter the take-off runway length chart for the specific airplane (Figure 6.37) for the mean daily maximum temperature (27°C) on the horizontal scale at the left of the chart.
- Project this point vertically to the line representing the elevation of the airport (900 m) at *C*.
- Draw a horizontal line from *C* to intersect with the reference point at *D*.
- Project upward to the right or downward to the left between the slanted lines as necessary to intersect either the elevation limit at *D*, a line drawn vertically from the maximum take-off weight (810000 N) at *E* or a line drawn vertically upward from the haul distance. The shortest runway length obtained from these is the minimum take-off length.
- Draw a horizontal line from the point that yields the shortest distance to obtain the take-off length. In this case, the shortest take-off length is obtained for the maximum take-off weight (*E*) to obtain 1980 m at *F*.

Since the required minimum length for landing is higher than that for take-off, the minimum length of the airport is 2100 m.

In addition to the factors listed earlier, the runway lengths for specific airplanes also depend on flap settings. The charts are based on flap settings that produce the shortest runway lengths. These settings are incorporated in the tables and should be known if the tables are to be used. Turbo-jet-powered airplanes expected to use airports at elevations higher than 1500 m are normally modified for higher altitudes to reduce their runway length requirement. The minimum runway length for these airplanes may be less than the minimum required for an airport serving a family of airplanes of 56250 N or less maximum certificated take-off weight. In these cases, the longer length should be used.

EXAMPLE 6.18

Determining the Minimum Runway Length Based on Airplane Grouping

Determine the minimum length of the runway for an airport located at an elevation of 600 m above sea level expected to serve ultralight airplanes with approaching speeds of 25 knots.

Solution

Determine minimum runway length at sea level:

Approach speed less than 30 knots, minimum runway length is 90 m

Adjust for elevation:

(600/300) × 9 = 18 m

Determine minimum runway length elevation of 600 m:

Minimum length at 600 m = minimum length at sea level + adjustment for elevation = (90 + 18) = 108 m

EXAMPLE 6.19

Determining the Minimum Runway Length for an Airport Serving Airplanes of 272000 N or Less

An airport is being designed to serve 100% fleet and 90% useful load of a family of airplanes having a maximum certificated load of 272000 N. The airport will be located at an elevation of 1200 m, and the normal maximum temperature is 32°C. Determine the minimum length of the runway if the difference in centerline elevation between the high and low points of the runway is 7.5 m.

Solution

Determine unadjusted runway minimum length—use Figure 6.33b. From Figure 6.33b the unadjusted minimum length is 2880 m. Adjust for wet and slippery conditions:

Increase minimum length by 15%

Adjusted length = 2880 m (1.15) = 3312 m

Adjust for difference in centerline elevation:

Increase minimum length by 10 m per m of the difference in centerline elevations

Adjusted minimum length = 2880 + (10)(7.5)

= 2955 m

Since these adjustments are mutually exclusive, the minimum length of the runway is 3330 m rounded up from 3312 m.

Minimum Length of Crosswind Runways

The Federal Aviation Administration recommends that the minimum length of a crosswind runway should be at least 80% of the primary runway length.

Minimum Length of Parallel Runways

The Federal Aviation Administration recommends that the minimum length of a parallel runway should be based on the airplanes that will use it. Also, all parallel runways at an airport should have approximately equal lengths.

SUMMARY

This chapter presents the fundamental principles used for the design of the geometric alignments of the travelways of the highway, rail, and air modes. It is important to note that in the design of the different components of the travelway, a classification system is used for each of the modes. For example, just as a highway can be classified as a principal arterial, minor arterial, collector road, and so on, so a runway can be classified as a primary, crosswind, or parallel runway. This principle of classification serves as the fundamental basis on which the design of the geometric alignment is carried out for all modes.

The chapter also shows that the basic mathematical principles used for the design of the geometric alignment of the travelway are the same for all the modes considered, but different standards are required for different modes. These standards vary mainly because of the differences in the characteristics of the vehicles used in each mode. For example, because of the unique characteristic of the airplane, it is necessary to determine the minimum length of a runway so as to satisfy the take-off and landing requirements of the airplanes expected to use the airport, which is not necessary for either the highway or the railroad. Similarly, maximum grades allowable on runways are much lower than those allowed on highways and railroads. However, the same mathematical principle is used for the design of a vertical curve regardless of the specific mode being considered.

Several computer programs are available that can be used to carry out the design procedures presented in this chapter. However, emphasis has been placed on understanding the basic principles, which is a necessary requirement for the use of any of the available computer programs.

PROBLEMS

6.1 Compare and contrast the classification systems used as the basis for the design of the geometric alignment of the travel ways of the highway, rail, and air modes.

6.2 A rural collector road located in a rolling terrain is to be designed to carry a specified design volume of 1500 veh/day. Determine the following:

(i) Suitable design speed
(ii) Suitable lane and shoulder widths

(iii) The maximum desirable grade
(iv) The minimum radius of horizontal curves

6.3 For a rural principal arterial road in a mountainous terrain with a design speed of 105 km/h, determine the following:

(i) Suitable lane and shoulder widths
(ii) The maximum desirable grade
(iii) The minimum radius of horizontal curves

6.4 A light rail main line track is to be designed for a major metropolitan area. Determine the following:

(i) The maximum sustained longitudinal grade
(ii) The maximum sustained grade for a distance of 465 m between the PVIs of consecutive vertical curves
(iii) The maximum sustained grade for a distance of 1050 m between the PVIs of consecutive vertical curves
(iv) The minimum grade

6.5 A crest vertical curve is to be designed on a freight and intercity main line track. Determine the following:

(i) Maximum grade
(ii) Maximum change in grade for the two tangents connected by a crest vertical curve of 225 m
(iii) Maximum change in grade for the two tangents connected by a sag vertical curve of 225 m

6.6 Determine the minimum length of a crest vertical curve on a freight and passenger intercity main line track, connecting two tangents of +0.75% and −1.5% traveling at 112.5 km/h.

6.7 Repeat Problem 6.6 for a crest vertical curve on a light rail transit main line with a design speed of 90 km/h.

6.8 Repeat Problem 6.6 for a sag vertical curve on a light rail transit main line track with a design speed of 75 km/h.

6.9 An airport is being designed for aircraft approach category B. Determine the following:

(i) The maximum longitudinal grade of the primary runway
(ii) The maximum longitudinal grade of a taxiway
(iii) The minimum length of a vertical curve joining two tangents of maximum permissible grades
(iv) The minimum distance between the points of intersection of two consecutive vertical curves connecting tangents of maximum grades

6.10 Repeat Problem 6.9 for an aircraft approach category D.

6.11 A 3% grade on a principal arterial road intersects with a –2% grade at station (355 + 16.35) at an elevation of 97 m. If the design speed of the road is 95 km/h, determine the stations and elevations of the BVC and EVC and the elevations on the curve at each 30-m station.

6.12 A vertical curve connects a +2% grade and a –2% grade on the primary runway of an airport. The grades intersect at station (650 + 10.05) and at an elevation of 60 m. Determine the stations and elevations of the BVC and EVC and the elevations on the curve at 30-m stations.

6.13 A sag vertical curve connects a –1.5% grade and a +1.5% grade on a freight and intercity main line railway track. If the grades intersect at station (300 + 7.61) and an elevation of 105.15 m, determine the stations and elevations of the BVC and EVC and the elevations on the curve at 30 m intervals.

6.14 A primary runway is being designed for a new airport to serve airport approach category B. As part of this work it is required to design two consecutive vertical curves (a crest vertical curve followed by a sag vertical curve) to be located midway along the length of the runway. The design should satisfy all the grade and minimum length requirements. If conditions are such that the crest vertical curve connects grades of +0.5% and +1.5%, determine all the properties of both curves that will be necessary to set out the curves. The elevation and station of the point of intersection of the two tangents of the crest vertical curve are 166.95 ft and 595 + 13.5, respectively, and the distance between the points of intersections (PVIs) of the vertical curves is 292.5 m.

6.15 Repeat Problem 6.14 for an airport approach category C, but in this case the distance between the points of intersections (PVIs) of the vertical curves is not given and the common tangent between the two curves is 1.0%.

6.16 A horizontal curve is to be designed to connect two tangents on a rural principal arterial with a design speed of 110 km/h. The station of the BC is 545 + 13.65. It is expected that an existing building will be located at a distance of 15 m from the centerline of the inside lane. Determine the minimum radius that will satisfy the sight distance and superelevation requirements.

6.17 A horizontal curve connects two tangents that deflect at an angle of 45° on an urban arterial highway. The point of intersection of the tangents is located at station (658 + 16.13). If the design speed of the highway is 105 km/h, determine the point of the tangent and the deflection angles to whole stations for setting out the curve from the PC.

6.18 A curved section of a light rail transit track with an actual superelevation of 14 cm is to be upgraded to allow its use by freight and passenger intercity trains. If the existing radius cannot be improved because of land use restrictions, determine whether it is feasible to upgrade the track such that freight and passenger intercity trains traveling at 80 km/h can use it. The existing maximum speed of the light rail transit trains is 72.5 km/h.

6.19 A spiral curve connects two consecutive circular curves on a light rail transit with a design speed of 80 km/h. The first curve has a radius of 900 m and the second curve has a radius of 675 m. Determine the minimum length of the spiral curve for the limiting value of the unbalanced superelevation.

6.20 A spiral curve connects a tangent and a 3° circular curve with an intersection angle of 40° on a light rail transit track. The design speed is 95 km/h and the actual superelevation is 12.5 cm. The tangents of the circular curve intersect at station (586 + 16.91) with an intersection angle of 40°.

Determine:

(i) The station of the point of spiral (T.S.)
(ii) The station of the start of the circular curve (S.C.)
(iii) The station of the end of the circular curve (C.S.)
(iv) The station of the end of spiral (S.T.)

6.21 An airport at an elevation of 750 m above sea level is being designed. Determine its minimum primary runway length for each of the following family of airplanes:

(i) Airplanes with approach speeds of less than 30 knots
(ii) Airplanes with approach speeds of 30 knots or more but less than 59 knots

6.22 An airport is being considered for a location at an elevation of 1350 m. If the topographic and other conditions restrict the maximum length of the primary runway to 330 m, determine the highest approach speed family of airplanes for which the airport will be suitable.

6.23 An airport located at an elevation of 1200 m and a mean daily maximum temperature of the hottest month of the year of 32°C is to be designed to serve airplanes with a maximum certified take-off load of 47250 N and capable of carrying 9 passengers. Determine the minimum primary runway length that is required for

(i) 75% of the fleet
(ii) 95% of the fleet
(iii) 100% of the fleet

6.24 An existing airport located at a site with an elevation of 1200 m and a mean daily maximum temperature of the hottest month of the year of 27°C was designed to serve all airplanes with a maximum certificated take-off load of up to 56250 N carrying less than 10 passengers. If the airport is to be improved so as to serve all airplanes with a maximum certified take-off weight of 272000 N and a 90% useful load, determine by how much the length of the existing airport should be increased.

6.25 An airport located at an elevation of 1200 m is planned to serve the Douglas DC-9-10 series aircraft. The mean daily maximum temperature is 27°C. Determine the minimum length of the primary runway if the maximum take-off weight is 360000 N, a haul distance is 644 km and the maximum landing weight is 270000 N.

References

1. *A Policy on Geometric Design of Highways and Streets,* American Association of State Highway and Transportation Officials, Washington, D.C., 2004.
2. *Airport Design: Advisory Circular AC 150/5300-13,* Federal Aviation Administration, Department of Transportation, Washington, D.C., (Incorporating Changes 1 through 8), October 2002.
3. Garber, Nicholas J., and Black, Kirsten, *Advanced Technologies for Improving Large Truck Safety on Two-Lane Highways,* Report No. FHWA/VA-95-R4, September 1994.
4. *Highway Capacity Manual,* Special Report 209, Fourth Edition, Transportation Research Board, National Research Council, Washington, D.C., 2000.
5. *Manual for Railway Engineering,* Vol. 1, American Railway Engineering and Maintenance-of-Way Association, Landover, MD, 2005.
6. *Roadside Design Guide,* Washington, D.C., 1989.
7. *Runway Length Requirements for Airport Design: Advisory Circular 150/5325-4A,* Federal Aviation Administration, Department of Transportation, Washington, D.C., 1990.
8. *Surveying Theory and Practice,* Davis, Foote, Anderson, and Mikhail, McGraw-Hill Book Company, 1997.
9. *Track Design Handbook for Light Rail Transit,* Transit Cooperative Research Program, TCRP Report 57, Transportation Research Board, National Research Council, National Academy Press, Washington, D.C., 2000.

CHAPTER 7

Structural Design of Travelways

The structural design of the travelway of any transportation mode is carried out to ensure that it is structurally sound and can withstand the loads imposed upon it by the vehicles of that mode over its design life. In this chapter, we present design methodologies for the highway, air, and rail modes. The methods presented are those of the American Association of State Highway and Transportation Officials (AASHTO) for highway pavements, the Federal Aviation Administration (FAA) for airport pavements, and the American Railway Engineering and Maintenance-of-Way Association (AREMA) for railroad tracks. Only a brief description is presented on the soil characteristics that are required for the soil that supports the travelway, as a detailed discussion of soil characteristics is beyond the scope of this book.

The design procedures used for highway and airport pavements are similar, but the loads imposed on an airport pavement by aircraft taking off or landing are much higher than those imposed by automobiles on highway pavements. Also, although the structure of the rail track is in some ways similar to that of a highway pavement, the design of the railroad track is significantly different from that for the highway or airport pavement, mainly because of the way the load is transferred from the railway vehicles to the natural ground. While the loads from an aircraft or automobile are transferred directly from the vehicle to the ground through the pavement (which may consist of layers of different materials), the loads from railway vehicles are first transferred to the rails, which transfer these loads to individual cross ties, which in turn transfer these loads to the ground through layers of different materials. This chapter discusses the similarity and dissimilarity of the design procedures of the different modes, using the design methods presented.

STRUCTURAL COMPONENTS OF TRAVELWAYS

The travelway of the highway, air, or rail mode consists of two or more structural components, through which the load applied by the traveling vehicle is transferred to the ground. The performance of the travelway depends on the satisfactory performance of each component. This requires that each of these be properly designed to ensure that the load applied by the traveling vehicle does not overstress any of these structural components. Figure 7.1 shows typical cross-sections of the travelways for the highway, air, and rail modes. The travelways for the highway and air modes are usually referred to as the pavement while that for the rail mode is referred to as the track. It can be seen from Figure 7.1 that the structural components for the highway and airport pavements are very similar, while that for the rail track is somewhat different. The structural components of the highway and airport pavements consist of the subgrade or prepared roadway, the subbase, the base, and the wearing surface, while those for the rail track are the subgrade, the subballast, the ballast, the cross ties, and the rail. The assembly of rail and cross ties is usually referred to as the *track superstructure,* and the ballast and subballast as the *track substructure*. Each of these components is now briefly described.

Subgrade (Prepared Roadbed)

The subgrade or roadbed is a common component for all three modes. It is usually the natural material located along the horizontal alignment of the pavement or track and serves as the foundation of the pavement or track structure. The subgrade may also consist of a layer of selected material that is obtained from somewhere else and properly compacted to meet certain specifications. In some cases, the subgrade material is treated to achieve certain strength properties required for the type of pavement or track to be constructed. This treatment is usually referred to as stabilization. The textbook *Traffic and Highway Engineering,* by Garber and Hoel, gives a detailed description of the different stabilization processes. The load imposed by the vehicle using the travelway is eventually transmitted to the subgrade through the different structural components of the travelway, such that the load is spread over a greater area than that of the vehicle's contact area. It follows, therefore, that the lower the strength of the subgrade, the greater the required area of load distribution and therefore the greater the required depth.

Subbase Course

This is located immediately above the subgrade of the highway and airport pavement, as shown in Figures 7.1a and 7.1b, and consists of a higher-quality soil material than that for the subgrade. Materials used for subbase construction should meet certain particle size distributions (gradation), strength, and plasticity requirements. When the subgrade material satisfies these requirements, the subbase course is usually omitted. Materials not meeting these

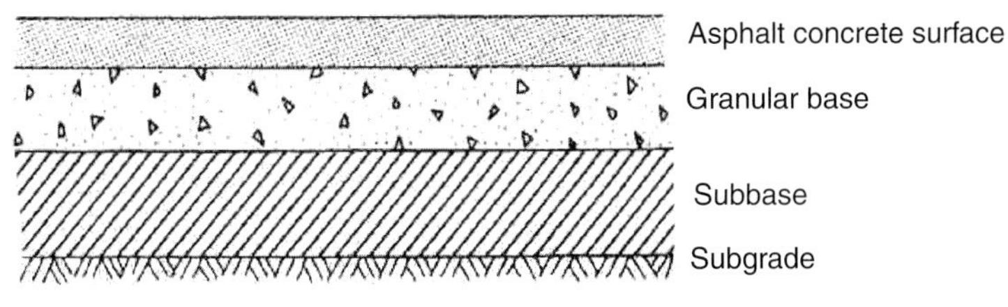

FIGURE 7.1

Schematic of highway, airport runway and rail track pavement
(a) Typical cross-section for highway pavement.

Source: Garber and Hoel, *Traffic and Highway Engineering,* 2002.

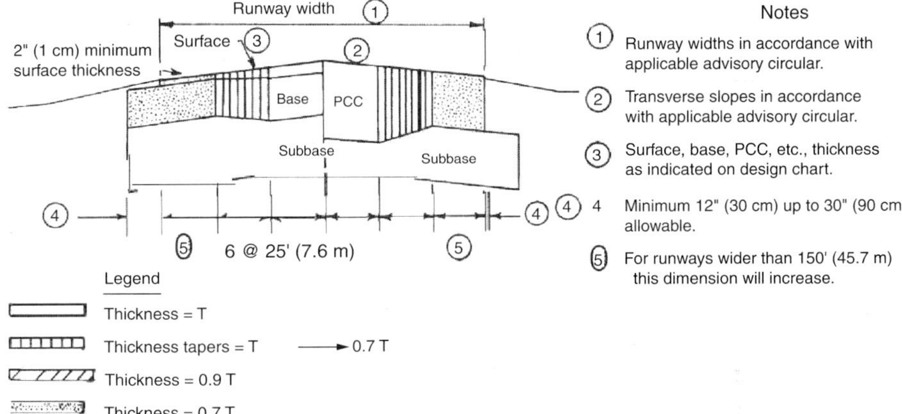

(b) Typical cross-section for airport runway pavement.

Source: Airport Pavement Design and Evaluation, Advisory Circular AC 150/5320-6D (Incorporating Changes 1 through 5), Federal Aviation Administration, Department of Transportation, Washington, D.C., April, 2004.

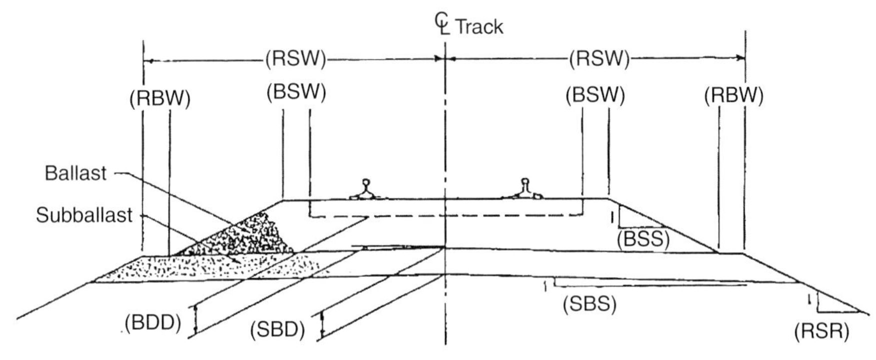

(c) Typical cross-section for rail track.

Source: Manual for Highway Engineering, American Railway Engineering and Maintenance-of-Way Association, Landover, MD, 2005.

requirements can be treated with other materials (stabilization) to achieve the necessary properties.

Subballast Course

This is located immediately above the roadbed (subgrade) of the rail track, as shown in Figure 7.1c. It occupies a similar location within the track structure as the subbase of the highway and airport pavements. The subballast is a graded aggregate material that must also meet specified requirements for gradation, plasticity, and strength. Its purpose is to augment the ballast course in the provision of adequate drainage, stability, flexibility, and uniform support for the rail and ties.

Base Course

This lies immediately above the subbase of the highway or airport pavement, as shown in Figures 7.1a and 7.1b, and is usually constructed with materials that have higher quality than that of the subbase course. These materials should also meet specified requirements for gradation, plastic characteristics, and strength. They are usually of a more granular nature than those for the subbase course, so as to facilitate the drainage of subsurface water.

Ballast Course

This lies immediately above the subballast course, as shown in Figure 7.1c. It occupies a similar location within the track structure as the base course of the highway and airport pavements. It provides drainage, stability, flexibility, uniform support for the rail ties, and distribution of the track loadings to the subgrade through the subballast. Common materials used in constructing ballast courses include granites, traprocks, quartzites, limestones, dolomites, and slags.

Surface Course

This is the upper course of highway and airport pavements and is constructed immediately above the base course. While the base and subbase courses of the highway and airport pavements are comparable to the subballast and ballast courses of the rail track, the surface course has no comparable course on the rail track. It can be either of Portland cement concrete or asphalt concrete. Portland cement surfaces are known as rigid pavements, and asphalt concrete pavements are known as flexible pavements.

Cross Ties

These are used only on rail tracks and are made of treated timber, concrete, or steel. They are transversely placed at regular intervals along the length of the rail track, immediately above the ballast course. Their main purpose is to evenly distribute the load from the rails to the ballast. There is no structural component of the highway or airport pavement that is directly comparable to the ties, as the loads from an automobile or aircraft are transmitted directly from the wheels of the vehicle to the pavement.

Rails

These are usually constructed of high-quality steel and are sometimes referred to as the guideway. Their main purpose is to guide the train and ensure that it travels along the required path. They also transfer the loads from the train wheels to the ties.

There is also no structural component of the highway or airport pavement that is directly related to the rails, as the travel paths of automobiles and airplanes are not as restricted as that of the rail.

General Principles of Structural Design of Travelways

The general principle incorporated in the structural design of any travelway is to ensure the integrity of each structural component to withstand the stress imposed on it by the vehicles using it. For example, the design ensures that the stress imposed on the subgrade or roadbed is less than the maximum allowable on it, while assuming that it is infinite in the horizontal direction. This is achieved through the construction of several structural components above the subgrade that spread the loads that are imposed by the vehicles using the travelway. For example, the wheels of railway vehicles impose their load onto the rails, which transmit it to the ties. The ties then transmit the load to the ballasts, which transfer the loads to the subballast, and finally the subballast transmits the load to the subgrade or road bed. As the load is transmitted from one structural component to the other, a stress distribution is caused within each structural component. An example of the stress distribution for a flexible highway or airport pavement is shown in Figure 7.2. The maximum vertical stresses are compressive and occur directly under the wheel load. These decrease with an increase in depth from the surface. The maximum horizontal stresses also occur directly under the wheel load but can be either tensile or compressive. The structural design of the travelway is therefore generally based on stress/strain characteristics for each structural component that limit both the horizontal and vertical strains/stresses below those that will cause permanent deformation.

Several software packages are now available for the design of different travelways. Commonly used methods include the AASHTO methods for highway pavements, the Federal Aviation Administration *Airport Pavement Design and Evaluation* as given in its advisory circular for airport pavements, and the

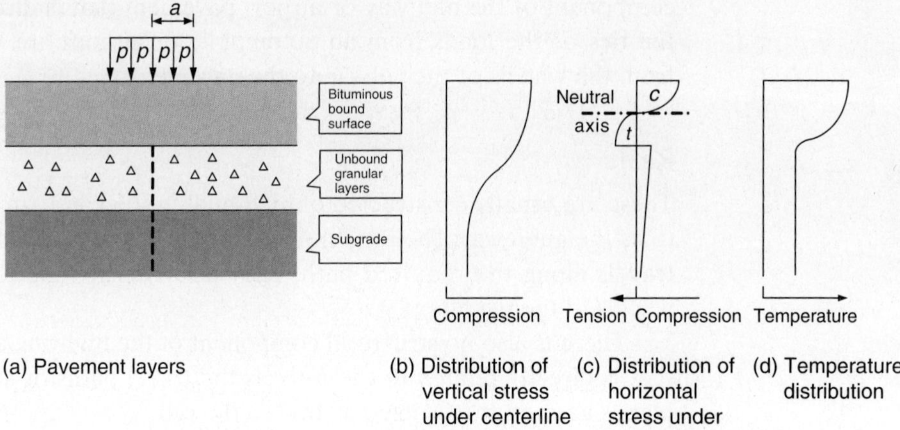

FIGURE 7.2
Typical stress and temperature distribution in a flexible pavement under a wheel load.

p = Wheel pressure applied on pavement surface
a = Radius of circular area over which wheel load is spread
c = Compressive horizontal stress
t = Tensile horizontal stress

AREMA procedure for rail tracks. The procedure generally followed in the structural design of the travelway consists of the following steps:

Step 1. Determine input load.
Step 2. Select material for each structural component.
Step 3. Determine minimum size and/or minimum thickness for each structural component.
Step 4. Carry out economic analysis of alternative designs and select the best design.

Only steps 1 through 3 are presented.

Step 1 Determine Input Load

The input load is usually the cumulative or maximum wheel load applied by the vehicles during the lifetime of the travelway. The input load for the highway pavement depends on the traffic characteristics (i.e., the distribution of the different types of vehicles) expected on the road being designed. The input load for the airport pavement depends on the gross weight of the aircraft for which the runway is being designed, usually referred to as the design aircraft. It should be noted that the design aircraft is not necessarily the heaviest aircraft expected to use the runway. The input load for the rail track depends on the wheel loads applied by the locomotive or loaded car to the rails.

Traffic Characteristics for Highway Pavement Design

In the AASHTO methods, traffic characteristics are determined in terms of the equivalent single-axle load (ESAL), which is the number of repetitions of an

18,000-lb (80-kilonewtons (kN)) single-axle load applied to the pavement on two sets of dual tires. The dual tires are represented as two circular plates, each 0.114 m in radius, spaced at 0.344 m apart, which is equivalent to a contact pressure of 0.495 MPa. The use of an 80 kN axle load is based on the results of experiments that have shown that the effect of any load on the performance of a highway pavement can be represented in terms of the number of single applications of an 80 kN single axle. Table 7.1 shows equivalency factors for converting axle loads to ESAL when using the AASHTO design procedure for flexible pavements. It should be noted that the values given in Table 7.1 are for a final serviceability index of 2.5 and various levels of *Pavement Structural Number* (*SN*). These terms will be defined and discussed in this chapter in the section "AASHTO Method for Design of Flexible Pavements." When axle loads are not readily available, some states have recommended factors for different vehicle types. For example, the Commonwealth of Virginia recommends a factor of 0.0002 for passenger vehicles, 0.37 for single-unit trucks, and 1.28 for tractor trailer trucks.

The design life of the pavement and the traffic growth rate are required for the computation of the total ESAL. The design life is the number of years the highway pavement is expected to carry the traffic load without requiring an overlay. The traffic growth rate is required as it is likely that the traffic will not remain constant throughout the life of the pavement. Transportation planning agencies can usually provide traffic growth rates for their jurisdictions. Growth factors (G_{jt}) for different growth rates j and design lives n can be computed using Equation 7.1:

$$G_{it} = \frac{[(1 + r)^N - 1]}{r} \tag{7.1}$$

where

N = design life of the pavement (yrs)
r = the annual growth rate, (%/100)

Note when $r = 0$ (no growth), the growth factor is the design period (life of pavement). Also, only a percentage of the total ESAL will be imposed on the design lane, as all the traffic will not be traveling on the same lane. The design lane for two-lane highways can be either of the two lanes, whereas the outside lane is usually considered as the design lane for multilane highways. It is essential that the design lane be properly selected, as in some cases more trucks may travel in one direction or trucks may travel empty in one direction and loaded in the other direction. States have stipulated values that should be used for design. For example, the Commonwealth of Virginia stipulates that 100% of the accumulated traffic load in one direction should be used for roads with one lane in each direction, 90% for two lanes in one direction, 70% for three lanes in one direction, and 60% for roads with four or more lanes in one direction.

TABLE 7.1a
Axle Load Equivalency Factors for Flexible Pavements, Single Axles, and P_t of 2.5

Source: Adapted from *AASHTO Guide for Design of Pavement Structures,* American Association of State Highway and Transportation Officials, Washington, D.C., 1993. Used by permission.

Axle Load (kips)	*Pavement Structural Number (SN)*					
	1	*2*	*3*	*4*	*5*	*6*
2	0.0004	0.0004	0.0003	0.0002	0.0002	0.0002
4	0.003	0.004	0.004	0.003	0.002	0.002
6	0.011	0.017	0.017	0.013	0.010	0.009
8	0.032	0.047	0.051	0.041	0.034	0.031
10	0.078	0.102	0.118	0.102	0.088	0.080
12	0.168	0.198	0.229	0.213	0.189	0.176
14	0.328	0.358	0.399	0.388	0.360	0.342
16	0.591	0.613	0.646	0.645	0.623	0.606
18	1.00	1.00	1.00	1.00	1.00	1.00
20	1.61	1.57	1.49	1.47	1.51	1.55
22	2.48	2.38	2.17	2.09	2.18	2.30
24	3.69	3.49	3.09	2.89	3.03	3.27
26	5.33	4.99	4.31	3.91	4.09	4.48
28	7.49	6.98	5.90	5.21	5.39	5.98
30	10.3	9.5	7.9	6.8	7.0	7.8
32	13.9	12.8	10.5	8.8	8.9	10.0
34	18.4	16.9	13.7	11.3	11.2	12.5
36	24.0	22.0	17.7	14.4	13.9	15.5
38	30.9	28.3	22.6	18.1	17.2	19.0
40	39.3	35.9	28.5	22.5	21.1	23.0
42	49.3	45.0	35.6	27.8	25.6	27.7
44	61.3	55.9	44.0	34.0	31.0	33.1
46	75.5	68.8	54.0	41.4	37.2	39.3
48	92.2	83.9	65.7	50.1	44.5	46.5
50	112.0	102.0	79.0	60.0	53.0	55.0

Note: 1 kips = 4.5 kN

It should be noted that only the truck traffic is considered as passenger cars impose insignificant loads on the pavement.

The accumulated ESAL for each category of axle load is given as

$$ESAL_i = (f_d)(G_{jt})(AADT_i)(365)(N_i)(F_{Ei}) \tag{7.2}$$

where

$ESAL_i$ = equivalent accumulated 80 kN single-axle load for axle category i
f_d = design lane factor
G_{jt} = growth factor for a given growth rate j and design life t
$AADT_i$ = first year annual average daily traffic for axle category i
N_i = number of axles on each vehicle in category i
F_{Ei} = load equivalency factor for axle category i

TABLE 7.1b

Axle Load Equivalency Factors for Flexible Pavements, Tandem Axles, and P_t of 2.5

Source: Adapted from *AASHTO Guide for Design of Pavement Structures*, American Association of State Highway and Transportation Officials, Washington, D.C., 1993. Used by permission.

Axle Load (kips)	*Pavement Structural Number (SN)*					
	1	*2*	*3*	*4*	*5*	*6*
2	0.0001	0.0001	0.0001	0.0000	0.0000	0.0000
4	0.0005	0.0005	0.0004	0.0003	0.0003	0.0002
6	0.002	0.002	0.002	0.001	0.001	0.001
8	0.004	0.006	0.005	0.004	0.003	0.003
10	0.008	0.013	0.011	0.009	0.007	0.006
12	0.015	0.024	0.023	0.018	0.014	0.013
14	0.026	0.041	0.042	0.033	0.027	0.024
16	0.044	0.065	0.070	0.057	0.047	0.043
18	0.070	0.097	0.109	0.092	0.077	0.070
20	0.107	0.141	0.162	0.141	0.121	0.110
22	0.160	0.198	0.229	0.207	0.180	0.166
24	0.231	0.273	0.315	0.292	0.260	0.242
26	0.327	0.370	0.420	0.401	0.364	0.342
28	0.451	0.493	0.548	0.534	0.495	0.470
30	0.611	0.648	0.703	0.695	0.658	0.633
32	0.813	0.843	0.889	0.887	0.857	0.834
34	1.06	1.08	1.11	1.11	1.09	1.08
36	1.38	1.38	1.38	1.38	1.38	1.38
38	1.75	1.73	1.69	1.68	1.70	1.73
40	2.21	2.16	2.06	2.03	2.08	2.14
42	2.76	2.67	2.49	2.43	2.51	2.61
44	3.41	3.27	2.99	2.88	3.00	3.16
46	4.18	3.98	3.58	3.40	3.55	3.79
48	5.08	4.80	4.25	3.98	4.17	4.49
50	6.12	5.76	5.03	4.64	4.86	5.28
52	7.33	6.87	5.93	5.38	5.63	6.17
54	8.72	8.14	6.95	6.22	6.47	7.15
56	10.3	9.6	8.1	7.2	7.4	8.2
58	12.1	11.3	9.4	8.2	8.4	9.4
60	14.2	13.1	10.9	9.4	9.6	10.7
62	16.5	15.3	12.6	10.7	10.8	12.1
64	19.1	17.6	14.5	12.2	12.2	13.7
66	22.1	20.3	16.6	13.8	13.7	15.4
68	25.3	23.3	18.9	15.6	15.4	17.2
70	29.0	26.6	21.5	17.6	17.2	19.2
72	33.0	30.3	24.4	19.8	19.2	21.3
74	37.5	34.4	27.6	22.2	21.3	23.6
76	42.5	38.9	31.1	24.8	23.7	26.1
78	48.0	43.9	35.0	27.8	26.2	28.8
80	54.0	49.4	39.2	30.9	29.0	31.7
82	60.6	55.4	43.9	34.4	32.0	34.8
84	67.8	61.9	49.0	38.2	35.3	38.1
86	75.7	69.1	54.5	42.3	38.8	41.7
88	84.3	76.9	60.6	46.8	42.6	45.6
90	93.7	85.4	67.1	51.7	46.8	49.7

Note: 1 kips = 4.5 kN

EXAMPLE 7.1

Computing Accumulated Equivalent Single-Axle Load for a Proposed Six-Lane Highway, Using Load Equivalency Factors

A six-lane divided highway is to be constructed on a new alignment in the Commonwealth of Virginia. Traffic volume forecasts indicate that the average annual daily traffic (AADT) in one direction during the first year of operation is 5000 with the following vehicle mix and axle load:

Percentage cars (50 kN/axle) = 65%

2-axle single-unit trucks (300 kN/axle) = 25%

3-axle single-unit trucks (600 kN/axle) = 10%

Assumed percentage of trucks in design lane = 70%

The vehicle mix is expected to remain the same throughout the design life of the pavement. If the expected annual traffic growth rate is 5% for all vehicles, determine the design ESAL for a design life of 20 years. Assume a terminal serviceability index of 2.5 and a structural number (SN) of 2.

Solution

The following data apply:

Percent of truck volume on design lane = 70

Load equivalency factors (from Table 7.1a)

2-axle single-unit trucks (300 kN/axle) = 0.02621

3-axle single-unit trucks (600 kN/axle) = 0.29616

$f_d = 0.7$ for four or more lanes in Virginia

Determine the growth factor—use Equation 7.1:

$$G_{it} = \frac{[(1 + r)^N - 1]}{r} = G_{it} = \frac{[(1 + .05)^{20} - 1]}{0.05} = 33.07$$

Determine the ESAL for each class of vehicle using Equation 7.1:

$$ESAL_i = (f_d)(G_{jt})(AADT_i)(365)(N_i)(F_{Ei})$$

2-axle single-unit trucks $= 0.7 \times 33.07 \times 5{,}000 \times 365 \times 0.25 \times 2 \times 0.02621$

$= 0.55 \times 10^6$

3-axle single-unit trucks $= 0.7 \times 33.07 \times 5{,}000 \times 0.1 \times 365 \times 3 \times 0.29616$
$= 3.75 \times 10^6$

Determine the accumulated ESAL $= (0.55 + 3.75) \times 10^6$:

$$= 4.3 \times 10^6$$

Note that the ESAL contributed by the passenger cars is negligible. Therefore, passenger cars are usually omitted in the calculation of ESAL values.

EXAMPLE 7.2

Computing Accumulated Equivalent Single-Axle Load for a Proposed Four-Lane Highway with Varying Growth Rates for Different Types of Vehicles Using Load Equivalency Factors

Determine the accumulated equivalent axle load for a four-lane (two lanes in each direction) divided highway in Virginia, with a design life of 20 years if traffic forecast indicates the following:

AADT in first year of operation in one direction= 9500 vehicles

Percentage of passenger cars in first year of operation = 70%

Percentage of 2-axle single-unit trucks (300 kN/axle) in first year of operation = 15%

Percentage of 3-axle single-unit trucks (500 kN/axle) in first year of operation = 15%

Growth rate of passenger cars = 5%

Growth rate of 2-axle single-unit trucks = 4%

Growth rate of 3-axle single-unit trucks = 0% (no growth)

Assume a terminal serviceability index of 2.5 and a structural number (SN) of 2.

Solution

Ignore passenger vehicles load, as it is negligible.

Determine the growth factor—use Equation 7.1:

$$G_{it} = \frac{[(1 + r)^N - 1]}{r}$$

2-axle single-unit trucks

$$G_{it} = \frac{\left[(1 + r)^N - 1\right]}{r} = \frac{\left[(1 + 0.05)^{20} - 1\right]}{0.05} = 29.78$$

3-axle single-unit trucks

Growth rate = 0; therefore growth factor = 20

Determine the load equivalency factors based on Table 7.1a.

2-axle single-unit trucks = 0.02621

3-axle single-unit trucks = 0.1511

Determine the ESAL for each type of truck (passenger cars are omitted) using Equation 7.2:

$$ESAL_i = (f_d)(G_{jt})(AADT_i)(365)(N_i)(F_{Ei})$$

2-axle single-unit trucks $= 0.90 \times 29.78 \times 9500 \times 0.15 \times 365 \times 2 \times 0.02621$
$= 0.731 \times 10^6$

3-axle single-unit trucks $= 0.90 \times 20.0 \times 9500 \times 0.15 \times 365 \times 3 \times 0.1511$
$= 4.243 \times 10^6$

Determine the accumulated equivalent axle load

$= (0.731 + 4.243) \times 10^6 := 4.97 \times 10^6$

Note: 90% is used for design lane factor as stipulated by the Virginia Department of Transportation.

Design Gross Weight for Airport Pavements

The design gross weight for an airport pavement depends on the design aircraft, which is the aircraft requiring the greatest pavement thickness. This is determined from the gross take-off weight and the number of annual departures of the aircraft. The respective pavement thickness for each aircraft (landing gear) type is determined using the design method presented later in this chapter. However, since the distribution of the load imposed on the pavement depends on the gear type and configuration of the aircraft, it is first necessary to convert all aircraft to the design aircraft, using the appropriate multiplication factors that have been established. These factors, given in Table 7.2, account for the relative fatigue effect of different gear types and are the same for rigid and flexible pavements. These factors are comparable to the equivalency factors for converting axle load to equivalent axle loads (ESAL). It should be noted, however, that while these aircraft (landing gear) conversion factors are the same for rigid and flexible pavements, the axle equivalency factors for flexible highway pavements are different from those for rigid pavements. Next, the number of annual departures for each aircraft (landing gear) type expected to use the runway is converted to

TABLE 7.2

Conversion Factors for Converting from One Aircraft (Landing Gear) Type to Another

Source: Airport Pavement Design and Evaluation, Advisory Circular AC No 150/5320 – 6D, Federal Aviation Administration, U.S. Department of Transportation, (Incorporating Changes 1 through 5), Washington D.C., April 2004.

To Convert From	*To*	*Multiply Departures By*
single wheel	dual wheel	0.8
single wheel	dual tandem	0.5
dual wheel	single wheel	1.3
dual wheel	dual tandem	0.6
dual tandem	single wheel	2.0
dual tandem	dual wheel	1.7
double dual tandem	dual tandem	1.0
double dual tandem	dual wheel	1.7

an equivalent annual departure of the designed aircraft using Equation 7.3:

$$\log R_1 = \log R_2 \left(\frac{W_2}{W_1} \right)^{\frac{1}{2}} \tag{7.3}$$

where

R_1 = equivalent annual departures by design aircraft
R_2 = annual departures expressed in design aircraft landing gear
W_1 = wheel load of design aircraft
W_2 = wheel load of aircraft in question

This computation assumes that the main landing gears carry 95% of the gross weight with the remaining 5% carried by the nose gear. It should be noted, however, that because wide-body aircraft have landing gear assembly spacings that are significantly different from those of other aircraft, special considerations should be given to this type of aircraft in order to maintain the relative effects. This is accounted for by taking each wide-body aircraft as a 1360 kN (300,000-lb) dual tandem aircraft when computing equivalent annual departures. This is done in every case, even when the design aircraft is a wide body. The total of the equivalent annual departures is then used to determine the required thickness of the pavement using the appropriate design curve. Note that the information on the expected number of departures for each aircraft type can be obtained from publications such as *Airport Master Plans* and *FAA Aerospace Forecasts*.

EXAMPLE 7.3

Computing the Design Load for an Airport Pavement

The following table gives the average annual departures and maximum take-off weight of each aircraft type expected to use an airport pavement. Determine the equivalent annual departures and design load for the pavement.

TABLE 7.3
Data for Example 7.3

Aircraft	*Gear Type*	*Average Annual Departures*	*Max Take-Off Weight (N)*
727-100	Dual	3500	680380 N
727-200	Dual	9100	864090 N
707-320B	Dual tandem	3000	1483240 N
DC-10-30	Dual	5800	489880 N
737-200	Dual	2650	523900 N
747-100	Double dual tandem	80	3175130 N

Solution

Determine design aircraft: The pavement thickness required for each of the aircraft types is determined and the aircraft type requiring the largest thickness is the design aircraft. This procedure will be described later. Assume that the 727-200 is the design aircraft. Convert each annual departure to an annual departure expressed in the design aircraft landing gear by multiplying the average annual departures by the appropriate factor given in Table 7.2.

For the 727-100, conversion factor = 1, since it has the same gear type as the design aircraft:

$$\text{Equivalent dual-gear departures} = 3500 \times 1 = 3500$$

For the 707-320B, conversion factor = 1.7 (from Table 7.2) (i.e., converting from dual tandem to dual):

$$\text{Equivalent dual-gear departures} = 3000 \times 1.7 = 5100$$

For the DC-10-30, conversion factor = 1:

$$\text{Equivalent dual-gear departures} = 5800 \times 1 = 5800$$

For the 737-200, conversion factor = 1:

$$\text{Equivalent dual-gear departures} = 2650 \times 1 = 2650$$

For the 747-100, conversion factor = 1.7:

$$\text{Equivalent dual-gear departures} = 80 \times 1.7 = 136$$

Determine wheel load for each airport type:

This is given as 0.95 (maximum take-off weight)/number of wheels on landing gears.

For the 727-100:

$$0.95 \times 680380/4 = 161590 \text{ N}$$

For the 727-200:

$$0.95 \times 864090/4 = 205220 \text{ N}$$

For the 707-320B:

$$0.95 \times 1483240/8 = 176135 \text{ N}$$

For the DC-10:

$$0.95 \times 489880/4 = 116346 \text{ N}$$

For the 737-200:

$$0.95 \times 523900/4 = 124426 \text{ N}$$

For the 747-100:

$0.95 \times 1360000/8 = 161500$ N (1360000 N is used as the maximum take-off weight for this purpose as the 747 is a wide-body aircraft)

Convert each aircraft annual average departure to equivalent annual departures of the design aircraft using Equation 7.3:

$$\log R_1 = \left(\frac{W_2}{W_1}\right)^{\frac{1}{2}} \log R_2$$

where

R_1 = equivalent annual departures based on the design aircraft (in this case the 727-200)
R_2 = annual departures expressed in design aircraft landing gear
W_1 = wheel load of design aircraft
W_2 = wheel load of aircraft in question

For the 727-100:

$$\log R_1 = \left(\frac{16159}{20522}\right)^{\frac{1}{2}} \log 3500 = 3.1448$$

$$R = 1396$$

The remaining results obtained are shown in Table 7.4.

TABLE 7.4 Computation Results for Example 7.3

Aircraft	*Equivalent Dual-Gear Departures (R_2)*	*Wheel Load (N)(W_2)*	*Wheel Load of Design Aircraft (N)(W_1)*	*Equivalent Annual Departures for Design Aircraft (R_1)*
727-100	3500	161590	205220	1396
727-200	9100	205220	205220	9100
707-320B	5100	176135	205220	2721
DC-10-30	5800	116346	205220	682
737-200	2650	124426	205220	462
747-100	136	161500	205220	78

Compute total equivalent annual departures based on design aircraft:

$$\text{Total annual departures based on design aircraft} = (1396 + 9100 + 2721 + 682 + 462 + 78) = 14{,}439$$

For this example, the pavement will be designed for 14,500 annual departures of a dual-wheel aircraft with a maximum take-off weight of 864090 N. It should be noted, however, that the requirements of the heaviest aircraft in the traffic mixture should be considered in determining the depth of compaction, thickness of asphalt, and drainage structures.

Imposed Wheel Loads from Locomotives

The dynamic load is considered in determining the maximum stresses within each structural component. It depends on the train speed, transfer of load due to rolling, traction augment (torque reaction), and track irregularities and is obtained from the static wheel load as recommended by AREMA in Equation 7.4:

$$P^d = (1 + \theta)P \tag{7.4}$$

where

P^d = dynamic wheel load
P = static wheel load
θ = coefficient of impact = $33u/(100D) \times 0.057$
u = dominant train speed, m/s
D = diameter of vehicle wheels, m

It should be noted that in some cases when the maximum deflection and maximum moment in the rail are being determined, it may be necessary to use more than one wheel load if the axles of a truck are closely spaced. This issue is discussed later.

EXAMPLE 7.4

Determining the Dynamic Wheel Load for the Design of a Railway Track

Given the static wheel load of a locomotive is 113400 N, determine the associated dynamic wheel load for design purposes if the dominant speed on the track is 36 m/s and the diameter of the locomotive wheel is 0.914 m.

Solution

Determine the coefficient of impact θ:

$$\theta = 33u/(100D) \times 88$$

u = dominant train speed, m/s

D = diameter of vehicle wheels, m

$$\theta = 33 \times 36/(100 \times 0.914) \times 0.057 = 0.7408$$

Determine dynamic load P^d from Equation 7.4:

$$P^d = (1 + \theta)P$$

where

P^d = dynamic wheel load
P = static wheel load

$$P^d = 113400(1 + 0.7408) = 197406 \text{ N}$$

Step 2 Select Materials for Each Structural Component

The appropriate material for each structural component depends on required engineering properties.

Subgrade Engineering Properties

Natural or stabilized soils are normally used as subgrade materials. The engineering properties of any soil deposit are closely related to the specific category of the soil within a soil classification system. Soil classification is a process through which soils are systematically categorized according to their probable engineering characteristics. It therefore serves as a means of identifying suitable subgrade and subbase materials. The two most common classification systems used in the design of travelways are the AASHTO classification system and the Unified Soil Classification System (USCS).

In the AASHTO system, soils are classified into seven groups, A-1 through A-7, with several subgroups as shown in Table 7.5. The classification of a given soil is based on its particle size distribution, the liquid limit, plastic limit, and plasticity index (Atterberg limits).

The particle size distribution is determined by conducting a sieve analysis (sometimes known as a mechanical analysis) on a soil sample if the particles are sufficiently large. This is done by shaking a sample of air-dried soil through a set of sieves with progressively smaller openings. The smallest practical opening of these sieves is 0.075 mm, which is designated as the #200 sieve. Other sieves include #140 (0.106 mm), #60 (0.25mm), #40 (0.425 mm), #20 (0.85 mm), #10 (2.0 mm), #4 (4.75 mm), and several others with openings up to 125 mm (or 5″).

TABLE 7.5
AASHTO Classification System

General Classification	*Granular Materials (35% or Less Passing No. 200)*							*Silt-Clay Materials (More than 35% Passing No. 200)*			
Group Classification	*A-1*			*A-2*							*A-7*
	A-1-a	*A-1-b*	*A-3*	*A-2-4*	*A-2-5*	*A-2-6*	*A-2-7*	*A-4*	*A-5*	*A-6*	*A-7-5, A-7-6*
Sieve analysis, Percent passing											
No. 10	50 max.	—	—	—	—	—	—	—	—	—	—
No. 40	30 max.	50 max.	51 min.	—	—	—	—	—	—	—	—
No. 200	15 max.	25 max.	10 max.	35 max.	35 max.	35 max.	35 max.	36 min.	36 min.	36 min.	36 min.
Characteristics of Fraction passing No. 40:											
Liquid limit			—	40 max.	41 min.	40 max.	41 min.	40 max.	41 min.	40 max.	41 min.
Plasticity index	6 max.		N.P.	10 max.	10 max.	11 min.	11 min.	10 max.	10 max.	11 min.	11 min.*
Usual types of significant constituent materials	Stone fragments, gravel and sand		Fine sand	Silty or clayey gravel and sand				Silty soils		Clayey soils	
General rating as subgrade	Excellent to good							Fair to poor			

*Plasticity index of A-7-5 subgroup is equal to or less than LL minus 30. Plasticity index of A-7-6 subgroup is greater than LL minus 30.

Source: Adapted from *Standard Specifications for Transportation Materials and Methods of Sampling and Testing*, 20th ed., Washington, D.C.: The American Association of State Highway and Transportation Officials, copyright 2000. Used by permission.

The hydrometer test is used to determine the particle sizes that are smaller than the lower sieve limits. This involves suspending a portion of the material that passes through the 2 mm (#10) sieve in water, usually in the presence of a deflocculating agent. This suspension is then left standing until the particles generally settle to the bottom. The specific gravity of the suspension is then determined at different times (t, sec), using a hydrometer. The maximum diameter of the particles in the suspension at depth y is then computed from Equation 7.5:

$$D = \sqrt{\frac{18\eta}{\gamma_s - \gamma_w}\left(\frac{y}{t}\right)} \tag{7.5}$$

where

D = maximum diameter of particles in suspension at depth y—that is, all particles in suspension at depth y have diameters less than D

η = coefficient of viscosity of the suspending medium (in this case water) in poises

γ_s = unit weight of soil particles

γ_w = unit weight of water

The combination of the results of the sieve analysis and the hydrometer test is then used to obtain the particle size distribution of the soil. Figure 7.3 shows examples of the particle size distributions of two soil samples.

The liquid limit (LL) is the moisture content (ratio of the weight of water in the soil mass to the oven-dried weight of soil) at which the soil will flow and close

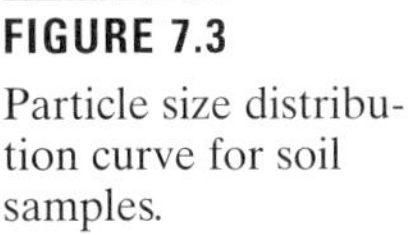

FIGURE 7.3

Particle size distribution curve for soil samples.

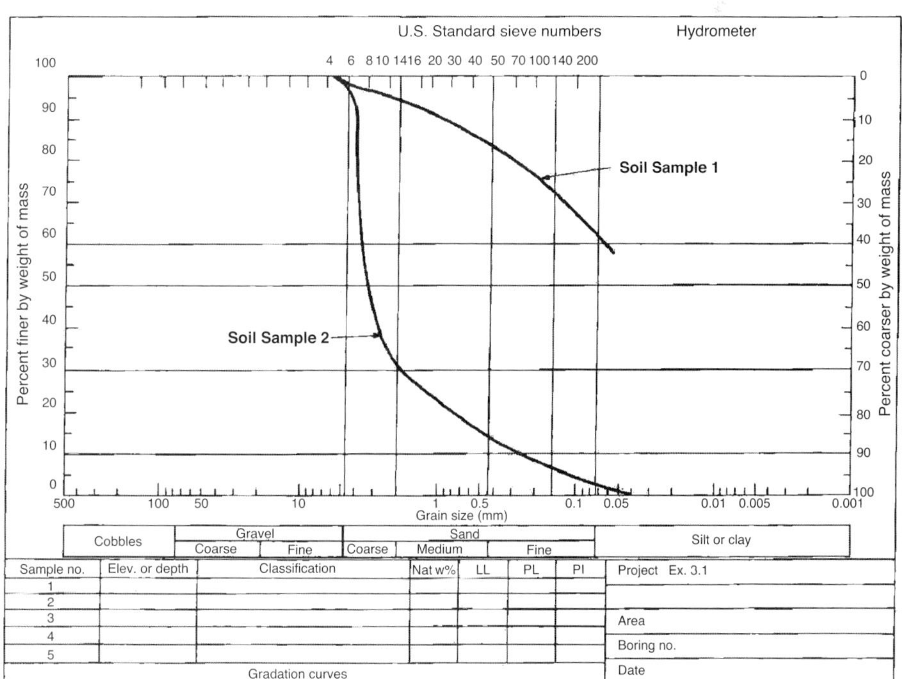

a groove of 12.7 mm within it, after the standard LL equipment has been dropped 25 times. For details of this test, readers may refer to any soil mechanics laboratory manual. The plastic limit (PL) is the moisture content at which the soil crumbles when it is rolled down to a diameter of 3.2 mm. The plasticity index (PI) is the difference between the LL and PL and indicates the range of the moisture content over which the soil is in the plastic state.

Another factor used to evaluate soils within each group in the AASHTO classification system is the group index (GI) of the soil, which is given as

$$GI = (F - 35)[0.2 + 0.005\,(LL - 40)] + 0.01(F - 15)(PI - 10) \tag{7.6}$$

where

GI = group index
F = percent of soil particles passing 0.075 mm (#200) sieve in whole number based on material passing 75 mm (3″) sieve
LL = liquid limit expressed in whole number
PI = plasticity index expressed in whole number

The GI is determined to the nearest whole number. A value of 0 should be recorded when a negative value is obtained for the GI. In determining the GI for A-2-6 and A-2-7 subgroups, the LL part of the equation for GI is not used; that is, only the second term is used. When soils are properly drained and compacted, their value as subgrade material decreases as GI increases. For example, a soil with a GI of 0 (an indication of a good subgrade material) will be better as a subgrade material than one with a GI of 20 (an indication of a poor subgrade material).

EXAMPLE 7.5

Classifying a Soil Sample Using the AASHTO Method

Using the AASHTO soil classification method, determine the classification of the soil having the gradation shown in the following table if the liquid limit (LL) is 45% and the plastic limit (PL) is 25%:

Sieve No.	*Opening (mm)*	*Percent Finer*
4	4.750	95
10	2.000	93
40	0.425	85
100	0.150	75
200	0.075	70

Solution

Since more than 35% passes the 0.075 mm (#200) sieve, the soil is either A-4, A-5, A-6, or A-7. The LL is greater than 40%, therefore the soil cannot be A-4 or A-6; thus it is either A-5 or A-7.

The PI is (45 − 25%) = 20%, which is greater than 10%, thus eliminating A-5. The soil is therefore A-7-5 or A-7-6: (LL − 30) = (45 − 30) = 15, therefore, the soil is A-7-6 since the plasticity index of an A-7-5 soil should not be greater than (LL − 30), but that of an A-7-6 is greater than (LL − 30).
Determine the group index of soil using Equation 7.5:

$$GI = (F - 35)[0.2 + 0.005\,(LL - 40)] + 0.01(F - 15)(PI - 10)$$

$$GI = (70 - 35)[0.2 + 0.005\,(45 - 40)] + 0.01(70 - 15)(20 - 10)$$

$$35(0.225) + 5.5 = 7.88 + 5.50 = 13.38$$

The soil is A-7-6(13).

EXAMPLE 7.6

Classifying a Soil Sample Using the AASHTO Method

The following results were obtained by a mechanical sieve analysis. Classify the soil according to the AASHTO classification system and determine the group index:

Sieve Size	*Opening (mm)*	*Percent Finer*
4	4.750	55
10	2.000	45
40	0.425	40
100	0.150	35
200	0.075	18
Liquid limit	–	16
Plastic limit	–	9

Solution

Since only 18% of the material passes the 0.075 mm (#200) sieve and 45% passes the 2 mm (#10) sieve, the soil is either A-1-b, A-2-4, A-2-5, A-2-6, or A-2-7.

Since the liquid limit (LL) is only 16, the soil is therefore A-1-b, A-2-4, or A-2-6. $PI = (16 - 9) = 7\%$; $6 < PI < 10$; soil is therefore A-2-4.

Determine the GI:

$$GI = (F - 35)[0.2 + 0.005\,(LL - 40)] + 0.01(F - 15)(PI - 10)$$

$$GI = (18 - 35)[0.2 + 0.005\,(16 - 40)] + 0.01(18 - 15)(7 - 10)$$

$$= -1.36 - 0.09$$

$$= -1.45; \text{ therefore } GI \text{ is recorded as } 0$$

The soil is therefore A-2-4(0).

The fundamental premise used in the USCS is that the engineering properties of any coarse-grained soil (soils with more than half of the material larger than the 0.075 mm sieve) depend on its particle size distribution whereas those for a fine-grained soil (soils with less than half of the material smaller than the 0.075 mm sieve) depend on its plasticity. Thus, the system classifies coarse-grained soils on the basis of grain size characteristics and fine-grained soils according to their plasticity characteristics. Other factors used for coarse-grained soils are the coefficient of uniformity (C_u) obtained from Equation 7.7 and the coefficient of curvature (C_c) obtained from Equation 7.8:

$$C_u = \frac{D_{60}}{D_{10}} \tag{7.7}$$

$$C_c = \frac{(D_{30})^2}{D_{10} \times D_{60}} \tag{7.8}$$

where

D_{60} = grain diameter at 60% passing
D_{30} = grain diameter at 30% passing
D_{10} = grain diameter at 10% passing

Fine-grained soils are classified as either silt with low plasticity (ML), silt with high plasticity (MH), clays with high plasticity (CH), clays with low plasticity (CL), or organic silt with high plasticity (OH). Table 7.6 lists the USCS definitions for the four major groups of materials consisting of coarse-grained soils, fine-grained soils, organic soils, and peat. Table 7.7 gives the complete layout of the USCS.

EXAMPLE 7.7

Classifying a Soil Sample Using the Unified Soil Classification System (USCS)

The results obtained from a mechanical analysis and a plasticity test on a soil sample are shown in the following table. Classify the soil using the USCS.

TABLE 7.6
USCS Definition of Particle Sizes

Source: Adapted from *The Unified Soil Classification System,* Annual Book of ASTM Standards, vol.04.08, American Society for Testing and Materials, West Conshohocken, PA, 1996.

Soil Fraction or Component	*Symbol*	*Size Range*
1. Coarse-grained soils		
Gravel	G	75 mm to No. 4 sieve (4.75 mm)
Coarse		75 mm to 19 mm
Fine		19 mm to No. 4 sieve (4.75 mm)
Sand	S	No. 4 (4.75 mm) to No. 200 (0.075 mm)
Coarse		No. 4 (4.75 mm) to No. 10 (2.0 mm)
Medium		No. 10 (2.0 mm) to No. 40 (0.425 mm)
Fine		No. 40 (0.425 mm) to No. 200 (0.075 mm)
2. Fine-grained soils		
Fine		Less than No. 200 sieve (0.075 mm)
Silt	M	(No specific grain size—use Atterberg limits)
Clay	C	(No specific grain size—use Atterberg limits)
3. Organic soils	O	(No specific grain size)
4. Peat	Pt	(No specific grain size)

Gradation Symbols	*Liquid Limit Symbols*
Well graded, W	High LL, H
Poorly graded, P	Low LL, L

Plasticity tests:

LL = nonplastic
PL = nonplastic

Sieve No.	*Opening (mm)*	*Percent Passing (by weight)*
4	4.75	97
10	2.00	33
40	0.425	12
100	0.150	7
200	0.075	4

Solution

Since only 4% of the soil is passing the 0.075 mm sieve, the soil is coarse grained. Since more than 50% of the soil passes the 4.75 mm sieve, the soil is classified as sand.

Determine the coefficient of uniformity (C_u) and coefficient of curvature (C_v). Plot the grain size distribution curve as shown for soil sample 2 in Figure 7.3.

From the grain size distribution curve, determine

$$D_{60} = 3.9 \text{ mm}$$
$$D_{30} = 1.7 \text{ mm}$$
$$D_{10} = 0.28 \text{ mm}$$
$$C_u = D_{60}/D_{10} = 3.9/0.28 = 13.93 > 6$$
$$C_v = (D_{30})^2/(D_{10} \times D_{60}) = (1.7)^2/(0.28 \times 3.9) = 2.65. \; 1 < C_v < 3$$

This sand is well graded with little or no fines and is classified as SW.

TABLE 7.7

Unified Classification System

Source: Joseph E. Bowles, *Foundation Analysis and Design*, McGraw-Hill, New York, 1988.

Major Divisions			Group Symbols	Typical Names	Laboratory Classification Criteria		
Coarse-grained soils (More than half of material is larger than No. 200 sieve size)	Gravels (More than half of coarse fraction is larger than No. 4 sieve size)	Clean gravels (Little or no fines)	GW	Well-graded gravels, gravel-sand mixtures, little or no fines	Determine percentages of sand and gravel from grain-size curve. Depending on percentage of fines (fraction smaller than No. 200 sieve size), coarse-grained soils are classified as follows: Less than 5 per cent GW, GP, SW, SP; More than 12 per cent GM, GC, SM, SC; 5 to 12 per cent *Borderline* cases requiring dual symbols[b]	$C_u = \frac{D_{60}}{D_{10}}$ greater than 4; $C_c = \frac{(D_{30})^2}{D_{10} \times D_{60}}$ between 1	
			GP	Poorly graded gravels, gravel-sand mixtures, little or no fines		Not meeting all gradation requirements for GW	
		Gravels with fines (Appreciable amount of fines)	GM[a] d / u	Silty gravels, gravel-sand-silt mixtures		Atterberg limits below "A" line or P.I. less than 4	Above "A" line with P.I. between 7 and 7 are border-line cases requiring use of dual symbols
			GC	Clayey gravels, gravel-sand-clay mixtures		Atterberg limits below "A" line with P.I. greater than 7	
	Sands (More than half of coarse fraction is smaller than No. 4 sieve size)	Clean sands (Little or no fines)	SW	Well-graded sands, gravelly sands, little or no fines		$C_u = \frac{D_{60}}{D_{10}}$ greater than 6; $C_c = \frac{(D_{30})^2}{D_{10} \times D_{60}}$ between 1 and	
			SP	Poorly graded sands, gravelly sands, little or no fines		Not meeting all gradation requirements for SW	
		Sands with fines (Appreciable amount of fines)	SM[a] d / u	Silty sands, sand-silt mixtures		Atterberg limits above "A" line or P.I. less than 4	Limits plotting in hatched zone with P.I. between 4 and 7 are borderline cases requiring use of dual symbols
			SC	Clayey sands, sand-clay mixtures		Atterberg limits above "A" line with P.I. greater than 7	
Fine-grained soils (More than half material is smaller than No. 200 sieve)	Silts and clays (Liquid limit less than 50)		ML	Inorganic silts and very fine sands, rock flour, silty or clayey fine sands, or clayey silts with slight plasticity	Plasticity Chart (Plasticity index 0–60 vs. Liquid limit 0–100; zones: CL, CL-ML, ML and OL, CH, OH and MH; "A" line)		
			CL	Inorganic clays of low to medium plasticity, gravelly clays, sandy clays, silty clays, lean clays			
			OL	Organic silts and organic silty clays of low plasticity			
	Silts and clays (Liquid limit greater than 50)		MH	Inorganic silts, micaceous or diatomaceous fine sandy or silty soils, elastic silts			
			CH	Inorganic clays of high plasticity, fat clays			
			OH	Organic clays of medium to high plasticity, organic silts			
Highly organic soils			Pt	Peat and other highly organic soils			

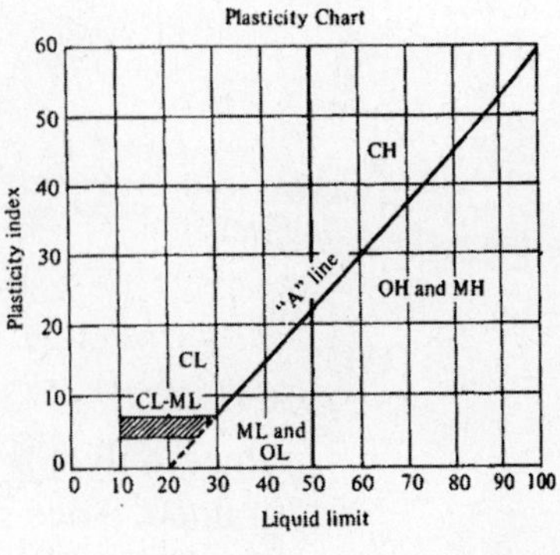

[a]Division of GM and SM groups into subdivisions of d and u are for roads and airfields only. Subdivision is based on Atterberg limits; suffix d used when L.L. is 28 or less and the P.I. is 6 or less; the suffix u used when L.L. is greater than 28.

[b]Borderline classifications, used for soils possessing characteristics of two groups, are designated by combinations of group symbols. For examples GW-GC, well-graded gravel-sand mixture with clay binder.

Subgrade Materials for Highway Flexible Pavements

Soils classified as A-1-a, A-1-b, A-2-4, A-2-5, and A-3 under the AASHTO classification system (see Table 7.5) can be used satisfactorily as subgrade material if properly drained (i.e., low GI values). Materials classified as A-2-6, A-2-7, A-4, A-6, A-7-5, and A-7-6 will require a layer of subbase material if used as subgrade. In addition to the classification of the soil, the strength of the subgrade in terms of its ability to withstand pressure imposed upon it should be known. A measure of the strength of the subgrade is its resilient modulus (M_r), which gives the resilient characteristics of the soil when it is repeatedly loaded with an axle load. It is determined in the laboratory by loading specially prepared samples of soil with a deviator stress of fixed magnitude, frequency, and load duration while the specimen is triaxially loaded in a triaxial chamber. Information on the procedure used to obtain the resilient modulus is given in AASHTO's *Standard Specifications for Transportation Materials and Methods of Testing*. An alternative engineering property is the California Bearing Ratio (CBR), which gives the relative strength of the subgrade with respect to crushed rock. It is determined in the laboratory using standard CBR testing equipment. Information on this test is also given in AASHTO's *Standard Specifications for Transportation Materials and Methods of Testing*.

It should be noted that both the M_r and CBR depend on the moisture content of the soil. However, the moisture content of the subgrade varies from one season to the other. For example, the moisture content tends to be highest during the spring thaw period, resulting in the subgrade having the least strength during the spring. An equivalent value of the M_r or the CBR is therefore first determined to account for this variation and then used in the design procedure. The method used to determine this equivalent M_r is discussed later in this chapter in the section on AASHTO method for the design of flexible highway pavements. To facilitate the use of either of these properties when the other is known, AASHTO recommends the use of the conversion factor shown in Equation 7.9:

$$M_r(\text{N/mm}^2) = 10.5 \text{ CBR (for soils having a CBR of up to 10)} \tag{7.9}$$

Subgrade Materials for Highway Rigid Pavements

Soils having the same classification as those suitable for flexible highway pavements' subgrades are also suitable for rigid highway pavements. However, the main strength property of the subgrade for rigid highway pavements is the modulus of subgrade reaction (k). This is the stress (N/mm^2) that will cause a 25.4 mm deflection of the underlying soil. This is obtained by conducting a plate bearing test, which measures the bearing capacity of the pavement foundation. However, the effective k value used for design is influenced by several factors, such as density, moisture content (seasonal effect), the type and thickness of the subbase material used in the pavement, the effect of potential erosion of the subbase, and whether rock lies within 3 m of the subgrade surface. The effective modulus of subgrade reaction (k) used in design is determined through a process that adjusts the measured k for each of these factors. A brief description for determining the effective modulus of subgrade reaction (k) is given later in this chapter in the section on the AASHTO methods for the design of highway pavements.

Subgrade Materials for Airport Flexible Pavements

The USCS is recommended for use in determining the suitability of a soil deposit as the subgrade for a flexible airport pavement. Table 7.8 shows the characteristics of soils that can be used as subgrade materials based on the USCS. In addition, the FAA recommends that consideration be given to the protection of pavements in areas where adverse effects due to seasonal frost and permafrost may occur. One of two approaches is used to compensate for the effect of seasonal frost or permafrost. The first approach controls the deformation caused by the frost action, and the second provides adequate pavement load-carrying capacity during the critical frost-melting period. In the first approach, a combined thickness of pavement and non-frost-susceptible material is provided to eliminate, or limit to an acceptable amount, the frost penetration into the subgrade. In the second approach, reduced subgrade strength is used. The susceptibility of soils to frost penetration depends on the size and distribution of voids in the soil mass, which is reflected in the classification of the soil. Table 7.9 shows the relative susceptibility of different soils based on their USCS category. The soils are divided into four different groups (FG-1 to FG-4) with those with higher frost group numbers being more susceptible to frost. For example, frost group 4 (FG-4) is more susceptible than frost group 1 (FG-1). Three design methods have been developed to incorporate these considerations: complete frost protection, limited subgrade frost penetration, and reduced subgrade strength.

The *Complete Frost Penetration* method provides a sufficient thickness of pavement and non-frost-susceptible material to totally contain frost penetration. The depth of frost penetration depends on the Air Freezing Index and the dry unit weight of the subgrade soil. The Air Freezing Index is defined by the FAA as a measure of the combined duration and magnitude below freezing temperatures occurring during any given freezing season. It is determined as the product of the average daily temperature below freezing by the number of days during which the average daily temperature is below freezing. The FAA recommends that the Air Freezing Index used for design should be based on the average of the three coldest winters in a 30-year period, if available, or the coldest winter observed in a 10-year period. Figures 7.4 and 7.5 give distributions of the air freezing indices in the continental United States and Alaska respectively. Figure 7.6 gives the depth of frost penetration for different air freezing indices and unit weights. The depth of frost penetration is then compared

TABLE 7.8

FAA Soil Characteristics Pertinent to Pavement Foundation

Major Divisions (1)	(2)	Letter (3)	Name (4)	Value as Foundation When Not Subject to Frost Action (5)	Value as Base Directly under Wearing Surface (6)	Potential Frost Action (7)	Compressibility and Expanston (8)	Drainage Characteristics (9)	Compaction Equipment (10)	Unit Dry Weight (pcf) (11)	Field CBR (12)	Subgrade Modulus *k* (pci) (13)
Coarse-grained soils	Gravel and gravelly soils	GW	Gravel or sandy gravel. well graded	Excellent	Good	None to very slight	Ahnost none	Excellent	Crawler-type tractor. rubber-tired equipment. steel-wheeled roller	125-140	60-80	300 or more
		GP	Gravel or sandy gravel, poorly graded	Good to excellent	Poor to fair	None to very slight	Almost none	Excellent	Crawler-type tractor, rubber-tired equipment. steel-wheeled roller	120-130	35-60	300 or more
		GU	Gravel or sandy gravel, uniformly graded	Good	Poor	None to very slight	Almost none	Excellent	Crawler-type tractor, rubber-tired equipment	115-125	25-50	300 or more
		GM	Silty gravel or silty sandy gravel	Good to excellent	Fair to good	Slight to medium	Very slight	Fair to poor	Rubber-tired equipment, sheepsfoot roller, close control of moisture	130-145	40-80	3tJtJ or more
		GC	Clayey gravel or clayey sandy gravel	Good	Poor	Slight to medium	Slight	Poor to practically impervious	Rubber-tired equipment, sheepsfoot roller	120-140	20-40	200-300
	Sand and sandy soils	SW	Sand or gravelly sand, well graded	Good	Poor	None to very slight	Almost noue	Excellent	Crawler-type tractor. rubber-tired equipmeut	I IO-130	20-40	200-300
		SP	Sand or gravelly sand. poorly graded	Fair to good	Poor to not suitable	None to very slight	Almost none	Excellent	Crawler-type tractor. rubber-tired equipment	105-120	15-25	200-300
		SU	Sand or gravelly sand, uniformly graded	Fair to good	Not suitable	None to very slight	Almost none	Excellent	Crawler-type tractor, rubber-tired equipment	100-115	IO-20	200-300
		SM	Silty sand or silty gravelly sand	Good	Poor	Slight to high	Very slight	Fair to poor	Rubber-tired equipment, sheepsfoot roller. close control of moisture	120-135	20-40	200-300
		SC	Clayey sand or clayey gravelly sand	Fair to good	Not suitable	Slight to high	Slight to medium	Poor to practically impervious	Rubber-tired quipment. sheepsfoot roller	105-130	JO-20	200-300
Fine-grained Soils	Low compressibility LL < 50	ML	Silts, sandy silts, gravelly silts, or diatomaceous soils	Fair to good	Not suitable	Medium to very high	Slight to medium	Fair to poor	Rubber-tired equipment, sheepsfoot roller, close control of moisture	100-125	5-15	IOO-200
		CL	Lean clays, sandy clays, or gravelly clays	Fair to good	Not suitable	Medium to high	Medium	Practically impervious	Rubber-tired equipment, sheepsfoot roller	It-IO-125	5-15	100-200
		OL	Organic silts or lean organic clays	Poor	Not suitable	Medium to high	Medium to high	Poor	Rubber-tired quipmeut. sheepsfoot roller	90-105	4-8	IOO-200
	High compressibility LL > 50	MH	Micaceous clays or diatomaceous soils	Poor	Not suitable	Medium to very high	ffigh	Fair to poor	Rubber-tired quipment, sheepsfoot roller	80-100	4-g	IOO-200
		CH	Fat clays	Poor to very poor	Not suitable	Medium	ffigh	Practically impervious	Rubber-tired equipment. sheepsfoot roller	90-I IO	3-5	50-100
		OH	Fat organic clays	Poor to very poor	Not suitable	Medium	High	Practically impervious	Rubber-tired quipmeut, sheepsfoot roller	SO-105	3-5	50-100
Peat and other fibrous organic soils		Pt	Peat, humus and other	Not suitable	Not suitable	Slight	Very high	Fair to poor	Compaction not practical			

Source: Airport Pavement Design and Evaluation, Advisory Circular AC No 150/5320-6D, Federal Aviation Administration, U.S. Department of Transportation, (Incorporating Changes 1 through 5), Washington D.C., April 2004.

Note: 1 pcf = 16 kg/m^3, 1 pci = 2.72×10^{-4} N/mm^3

TABLE 7.9

Soil Frost Groups

Source: Airport Pavement Design and Evaluation, Advisory Circular AC No 150/5320 – 6D, Federal Aviation Administration, U.S. Department of Transportation, (Incorporating Changes 1 through 5), Washington D.C., April 2004.

Frost Group	Kind of Soil	Percentage Finer Than 0.02 mm By Weight	Soil Classification
FG-1	Gravelly soils	3to 10	GW, GP, GW-GM, GP-GM
FG-2	Gravelly soils	10 to 20	GM, GW-GM, GP-GM,
	Sands	3to 15	SW, SP, SM, SW-SM SP-SM
FG-3	Gravelly soils	Over 20	GM, GC
	Sands, except very fine silty sands	Over 15	SM, SC
	Clays, PI above 12		CL, CH
FG-4	Very fine silty sands	Over 15	SM
	All silts		ML, MH
	Clays, PI = 12 or less		CL, CL-ML
	Varied clays and other fine grained banded sediments.	-	CL, CH, ML, SM

with the thickness for structural design and the difference between the two depths is made up with non-frost-susceptible material. This procedure is illustrated later under the Federal Aviation Administration Design Method for Flexible Airport Pavements section.

The difference between *Limited Subgrade Frost Penetration* method and the complete frost penetration method is that here a limited amount of frost penetration into the underlying frost susceptible subgrade is allowed. When the

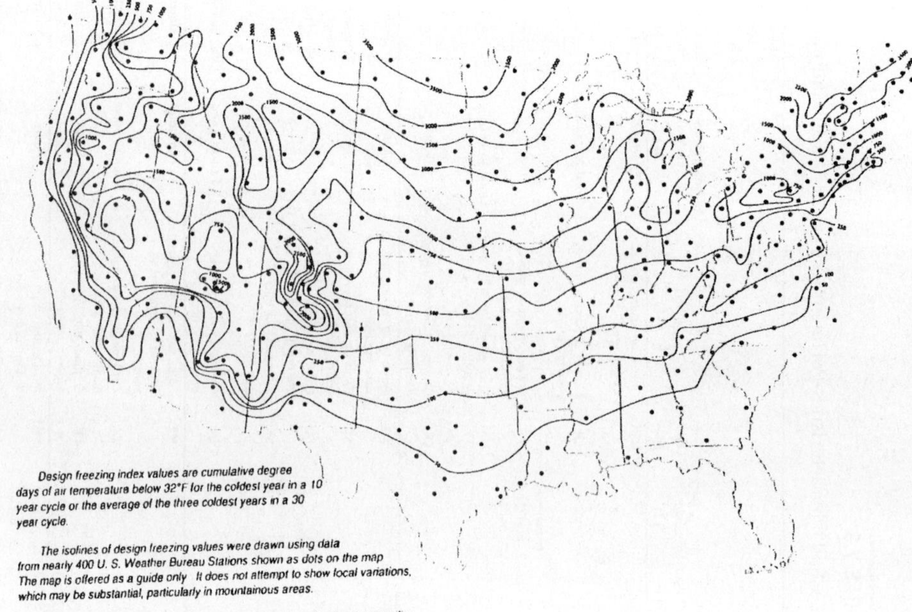

FIGURE 7.4

Distribution of design air freezing indexes in the continental U.S.

Source: Airport Pavement Design and Evaluation, Advisory Circular AC 150/5320-6D (Incorporating Changes 1 through 5), Federal Aviation Administration, Department of Transportation, Washington, D.C., April, 2004.

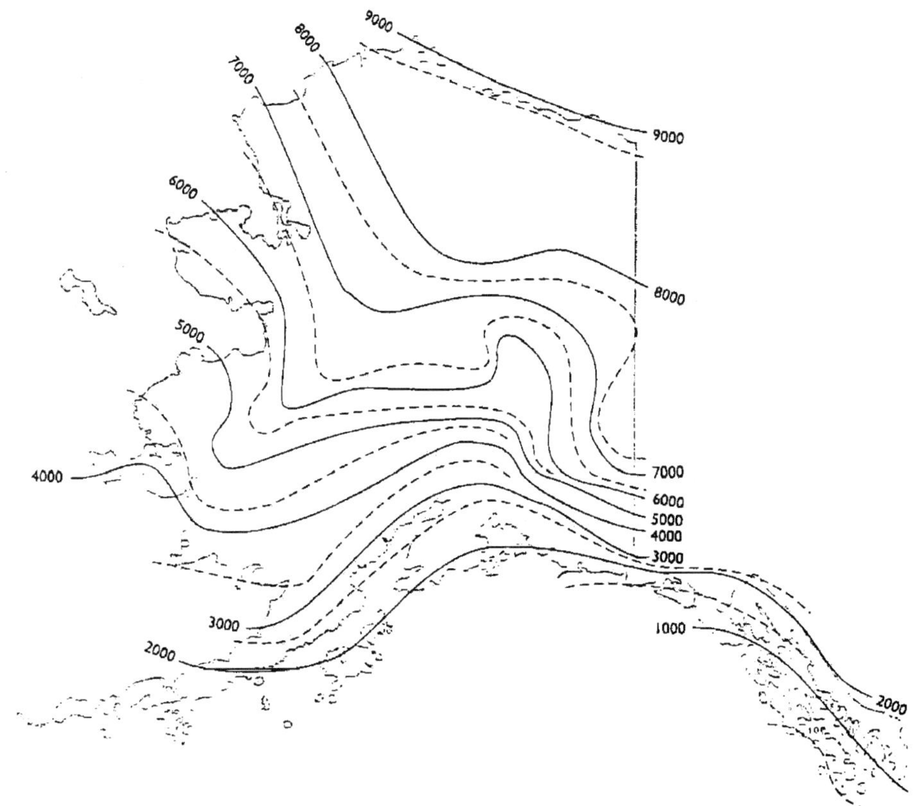

FIGURE 7.5

Distribution of design air freezing index values in Alaska.

Source: Airport Pavement Design and Evaluation, Advisory Circular AC 150/5320-6D (Incorporating Changes 1 through 5), Federal Aviation Administration, Department of Transportation, Washington, D.C., April, 2004.

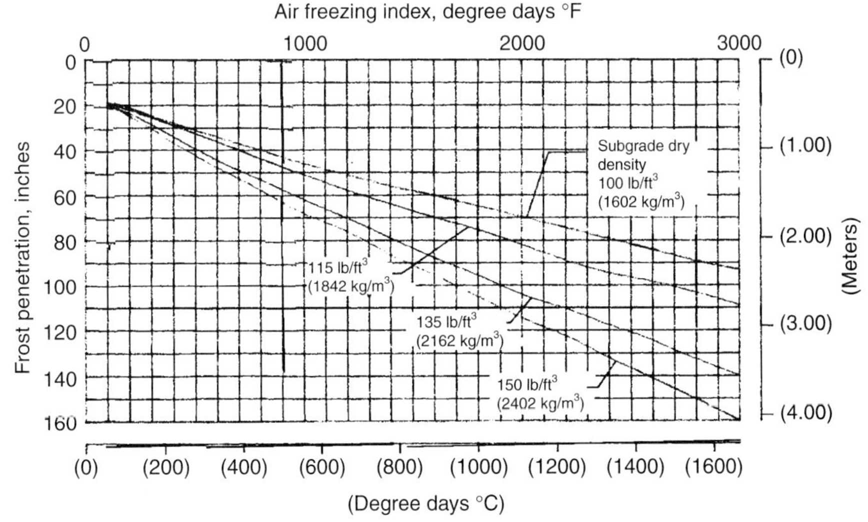

FIGURE 7.6

Depth of frost penetration.

Source: Airport Pavement Design and Evaluation, Advisory Circular AC 150/5320-6D (Incorporating Changes 1 through 5), Federal Aviation Administration, Department of Transportation, Washington, D.C., April, 2004.

TABLE 7.10

Reduced Subgrade Strength Ratings

Source: Airport Pavement Design and Evaluation, Advisory Circular AC No 150/5320-6D, Federal Aviation Administration, U.S. Department of Transportation, (Incorporating Changes 1 through 5), Washington D.C., April 2004.

Frost Group	Flexible Pavement CBR Value	Rigid Pavement k-Value (pci)
FG-1	9	50
FG-2	7	40
FG-3	4	25
FG-4	Reduced Subgrade Strength Method Does Not Apply	

Note: 1 pci = 2.72×10^{-4} N/mm^3

thickness of the structural section is less than 65% of the frost penetration, additional frost protection is provided.

In the *Reduced Subgrade Strength* method, a subgrade strength rating is assigned to the pavement for the frost-melting period. This is based on the frost group of the subgrade soil. Table 7.10 shows reduced subgrade strength ratings for different frost groups.

In addition to the soil classification characteristics, the FAA method uses the CBR as the main engineering property for the subgrade of a flexible airport runway. The CBR gives the relative strength of the soil with respect to crushed rock, which is considered an excellent coarse base material. The CBR Information on the Procedures for Determining the CBR can also be found in standard specifications for Transportation Materials and Methods of Testing.

The FAA recommends that the CBR used for design should not be greater than 85% of all the subgrade CBR values obtained from test results. This value should be adjusted for the effect of subbase, as discussed later in the section dealing with subbase materials for airport flexible pavements.

Subgrade Materials for Rigid Airport Runway Pavements

Soil classification requirements for suitable subgrade material for airport rigid pavements are similar to those for flexible airport runway pavements. Also, the main strength property for the subgrade of rigid airport runways is the same as that for rigid highway pavements; that is, the modulus of subgrade reaction (k).

Subgrade Materials for Rail Tracks

The classification characteristics of acceptable subgrade materials for a rail track subgrade are similar to those for airport runway pavements as the USCS is also used. Table 7.11 gives the relative suitability of different types of materials for use as subgrade for rail tracks. The main strength property of the subgrade used by AREMA is the bearing capacity of the subgrade. This is usually obtained from an Unconfined Compression Test. AREMA suggests that an allowable bearing pressure of 0.175 N/mm^2 may be used, but caution is advised in applying this value. AREMA also suggests that the level of stress applied on the subgrade should not be greater than an allowable pressure that includes a safety factor. The safety factor should be at least 2 but can be as high as 5 in order to avoid bearing capacity failure or undue long-term deformation of the subgrade.

TABLE 7.11

Soil Groups, Their Characteristics and Uses

(1)		(2)	(3)	(4)	(5)	(6)	(7)	(8)	(9)	(10)	(11)	(12)
Symbol		Soil Group	Field Identification	Frost Heaving	Drainage	Value as Filter Layer	Erosion on Exposed Slope	Value as Subgrade	Pumping Action	Stability in Compacted Fills	Compaction Characteristics	Typical Duty Type Geotextile Fabric Use
Gravels	GW	Well-graded GRAVELS and well-graded GRAVELS with SAND mixtures, trace to no silt or clay	Wide range in grain sizes, substantial amounts of all intermediate sizes, no dry strength	None to very slight	Excellent	Fair	None*	Excellent	None	Very Good	Excellent; crawler-type tractor, rubber-tired roller, steel-wheeled roller	None required
	GP	Poorly-graded GRAVELS and poorly-graded GRAVEL with SAND mixtures, trace to no silt or clay	Predominantly one size, or a range of sizes with some missing, no dry strength	None to very slight	Excellent	Fair to poor	None*	Excellent	None	Reasonably good	Good; crawler-type tractor, rubber-tired roller, steel-wheeled roller	None required
	GM	SILTY GRAVEL and SILTY GRAVEL with sand mixture	Fines with low or no plasticity, slight to no dry strength	Slight to medium	Fair to very poor	Very poor	None to slight	Good	None	Reasonably good	Good with close moisture control; rubber-tired roller, sheepfoot roller	None required
	GC	CLAYEY GRAVEL and CLAYEY GRAVEL with sand mixture	Plastic fines, medium to high dry strength	Slight to medium	Poor to very poor	Not to be used	None to slight	Good	Slight	Fair	Excellent; rubber-tired roller, sheepfoot roller	None required
Sands	SW	Well-graded SAND and well-graded SAND with GRAVEL mixtures, trace to no silt or clay	Wide range in grain sizes, substantial amounts of all intermediate sizes, no dry strength	None to very slight	Excellent	Excellent	Slight to high with decreasing gravel content	Excellent	None	Very good	Excellent; crawler-type tractor, rubber-tired roller	None required
	SP	Poorly graded SAND and poorly graded SAND with GRAVEL mixtures, trace to no silt or clay	Predominantly one size, or a range of sizes with some missing, no dry strength	None to very slight	Excellent	Fair to poor	High	Good	None	Reasonably good with flat slopes	Good; crawler-type tractor, rubber-tired roller	None required
	SM	SILTY SAND and SILTY SAND with GRAVEL mixtures	Fines of low to no plasticity, slight to no dry strength	Slight to high	Fair to very poor	Very poor	High	Poor	None to slight	Fair	Good with close moisture control; rubber-tired roller, sheepfoot roller	Slight regular
	SC	CLAYEY SAND and CLAYEY SAND with GRAVEL mixture	Plastic fines, medium to high dry strength	Slight to high	Very poor	Not to be used	Slight	Poor	Slight	Fair	Excellent; rubber-tired roller, sheepfoot roller	Slight regular

(continued)

TABLE 7.11

Soil Groups, Their Characteristics and Uses (*continued*)

(1)			(2)	(3)	(4)	(5)	(6)	(7)	(8)	(9)	(10)	(11)	(12)
Symbol			Soil Group	Field Identification	Frost Heaving	Drainage	Value as Filter Layer	Erosion on Exposed Slope	Value as Subgrade	Pumping Action	Stability in Compacted Fills	Compaction Characteristics	Typical Duty Type Geotextile Fabric Use
Silts and Clays	Of Low Plasticity	ML	SILT or SILT with SAND or GRAVEL; SANDY SILT or SANDY SILT with GRAVEL; GAVELLY SILT or GRAVELLY SILT with SAND mixture	Fine grained, slight to no dry strength	Medium to very high	Fair to very poor	Not to be used	Very high	Poor	Slight to bad	Poor	Poor to good with close control of moisture; rubber-tired roller; sheepfoot roller	Yes regular
Silts and Clays	Of Low Plasticity	CL	Lean CLAY or Lean CLAY with SAND or GRAVEL; SANDY LEAN CLAY or SANDY Lean CLAY with GRAVEL; GRAVELLY Leam CLAY or GRAVELLY Lean CLAY with SAND mixtures	Medium to high dry strength	Medium to high	Very poor	Not to be used	None to slight	Bad	Bad	Reasonable	Fair to good; rubber-tired roller, sheepfoot roller	Yes heavy
Silts and Clays	Of High Plasticity	MH	Elastic SILT or elastic SILT with SAND or GRAVEL; SANDY elastic SILT or SANDY elastic SILT with GRAVEL; GRAVELLY elastic SILT or GRAVELLY elastic SILT with SAND mixtures	Slight to medium dry strength	Medium to very high	Poor to very poor	Not to be used	None to slight	Bad	Very bad	Poor	Poor to very poor; sheepfoot roller	Yes heavy
Silts and Clays	Of High Plasticity	CH	Fat CLAY or fat CLAY with SAND or GRAVEL; SANDY fat CLAY or SANDY fat CLAY with GRAVEL; GRAVELLY fat CLAY or GRAVELLY fat CLAY with SAND mixtures	Sticky when wet, high dry strength	Medium	Very poor	Not to be used	None	Bad	Very bad	Fair with flat slopes	Fair to poor; sheepfoot roller	Yes extra heavy
ORGANIC		OH	Organic SILT or CLAY and with SAND or GRAVEL; SANDY or GRAVELLY organic SILT or CLAY with GRAVEL or SAND respectively	High smell, dark colour, mottled appearance, slight to high dry strength	Medium to high	Poor to very poor	Not to be used	Variable	Bad	Very bad	Not to be used	Poor to very poor	Yes extra heavy
ORGANIC		PT	PEAT	Dark colour, spongy feel and fibrous texture	Slight to high	Poor	Not to be used	Not applicable	Remove completely	Very bad	Not to be used	Compaction not possible	Yes extra heavy

Adapted from ASTM Method D 2487T

NOTES:

Column 2: Soil types in capitals and underlined make up more than 50% of sample. Other soil types in capitals make up more than 5%.

Column 4: Tendency of soil to frost heave.

Column 5: Ability of soil to drain water by gravity. Drainage ability decreases with decreasing average grain size.

Column 6: Value of soil as filter backfill around subdrain pipes to prevent clogging with fines, and as filter layer to prevent migration of fines from below.

Column 7: Ability of natural soil to resist erosion on an exposed slope. Soils marked * may be used to protect eroding slopes of other materials.

Column 8: Value as stable subgrade for roadbed, when protected by suitable ballast and sub-ballast material. Good soils may be used to protect poorer soils in subgrade.

Source: Manual for Highway Engineering, American Railway Engineering and Maintenance-of-Way Association, 2005.

Subbase and Base Materials Engineering Properties

Materials used for subbase and base in highway and airport runway pavements should satisfy certain gradation and plasticity requirements. They may be untreated granular material or stabilized material.

Subbase and Base Materials for Highway Pavements

The AASHTO guide for highway pavement design gives recommended particle size distributions for acceptable types of subbase materials, and these are given in Table 7.12. AASHTO suggests that the first five types, A through E, can be used within the upper four in layers, whereas type F can be used below the uppermost four in layers. AASHTO also suggests that in cases where the pavement is subjected to frost action, the percent of fines in the A, B, and F materials should be reduced to a minimum. The subbase thickness is usually not less than 150 mm and should be extended 300 to 900 mm outside the edge of the pavement

TABLE 7.12
Recommended Particle Size Distribution for Different Types of Subbase Materials for Rigid Highway Pavements

Source: Adapted with permission from *Standard Specifications for Transportation Materials and Methods of Sampling and Testing,* 20th ed., American Association of State Highway and Transportation Officials, Washington, D.C., 2000.

	Types of Subbase					
Sieve Designation	*Type A*	*Type B*	*Type C (Cement Treated)*	*Type D (Lime Treated)*	*Type E (Bituminous Treated)*	*Type F (Granular)*
Sieve analysis percent passing						
2 in.	100	100	—	—	—	—
1 in.	—	75–95	100	100	100	100
3/8 in.	30–65	40–75	50–85	60–100	—	—
No. 4	25–55	30–60	35–65	50–85	55–100	70–100
No. 10	15–40	20–45	25–50	40–70	40–100	55–100
No. 40	8–20	15–30	15–30	25–45	20–50	30–70
No. 200	2–8	5–20	5–15	5–20	6–20	8–25
(The minus No. 200 material should be held to a practical minimum.)						
Compressive strength lb/in² at 28 days			400–750	100		
Stability						
Hveem Stabilometer					20 min.	
Hubbard field					1000 min.	
Marshall stability					500 min.	
Marshall flow					20 max.	
Soil constants						
Liquid limit	25 max.	25 max.				25 max.
Plasticity index[a]	N.P.	6 max.	10 max.[b]		6 max.[b]	6 max.

[a]As performed on samples prepared in accordance with AASHTO Designation T87.
[b]These values apply to the mineral aggregate prior to mixing with the stabilizing agent.

Note: 1″ = 25.4 mm

TABLE 7.13

Specifications for Subbase Materials for Flexible Airport Runway Pavement

Source: Airport Pavement Design and Evaluation, Advisory Circular AC No 150/5370 – 10B, Federal Aviation Administration, U.S. Department of Transportation, Washington D.C., April 2004.

Sieve Designation (Square Openings) as per ASTM C 136	*Percentage by Weight Passing Sieves*
3 inch (75.0 mm)	100
No. 10 (2.0 mm)	20–100
No. 40 (0.450 mm)	5–60
No. 200 (0.075 mm)	0–15

structure. The subbase material is defined in terms of its elastic modulus E_{SB}. The type of material used is an important input in the determination of the effective modulus subgrade reaction of the subgrade. It should be noted that rigid highway pavements may or may not have a base course between the subgrade and the concrete surface. When a base course is used, it is usually referred to as the subbase.

Subbase and Base Materials for Airport Runway Pavements

The Federal Aviation Administration recommends that a subbase layer should be included as part of the structure of all airport flexible pavements, unless the subgrade CBR value is 20 or higher. The gradation requirements for suitable subbase materials for flexible highway pavements are given in Table 7.13. The Federal Aviation Administration also stipulates that the Liquid Limit and Plasticity Index of the portion of the material passing 0.450 mm sieve should not be greater than 25 and 6%, respectively. Also, at locations where frost penetration may be a problem, the maximum amount of material finer than 0.02 mm in diameter should be less than 3%.

Base course materials for flexible airport runway pavements are usually composed of selected durable aggregates, crushed aggregates, lime rock, cement-treated soil, or a plant mix bituminous mixture. A minimum CBR value of 80 is assumed for these materials. Table 7.14 gives the specifications for crushed aggregates that could be used as base course for a flexible airport runway. The material used should be well graded from coarse to fine and should not vary from the high limit on one sieve to the low limit on the next sieve or vice versa.

Materials usually accepted as suitable for use in subbase and base courses of rigid airport pavements are similar to those used as subbase and base materials in flexible airport pavements. The gradation requirements shown in Tables 7.13 and 7.14 are therefore applicable. Other materials that may be used include crushed aggregate, lime rock, cement-treated soil, and plant mix bituminous asphalt concrete. A minimum depth of 100 mm of subbase is recommended for airport rigid pavements. A subbase layer may, however, not be needed if the subgrade is classified as GW, GP, GM, GC, and SW, with good drainage, and is not susceptible to frost action.

TABLE 7.14

Specification for Base Materials for Flexible Airport Runway Pavements (a) Requirements for Gradation of Aggregate

Sieve Designation (mm)	Percentage by Weight Passing Sieves: 51 mm Maximum	38 mm Maximum	25 mm Maximum
50	100	—	—
37	70–100	100	—
25	55–85	70–100	100
19	50–80	55–85	70–100
4.75	30–60	30–60	35–65
0.45	10–30	10–30	10–25
0.075	5–15	5–15	5–15

(b) Requirements for Gradation of Aggregate[a]

Source: Airport Pavement Design and Evaluation, Advisory Circular AC No 150/5370 – 10B, Federal Aviation Administration, U.S. Department of Transportation, Washington D.C., April 2004.

Sieve Size (mm)	Design Range Percentage by Weight	Job Mix Tolerances Percent
50	100	0
37	95–100	+/− 5
25	70–95	+/− 8
19	55–85	+/− 8
4.75	30–60	+/− 8
0.60	12–30	+/− 5
0.075	0–5	+/− 3

Note:
a. Where environmental conditions (temperature and availability of free moisture) indicate potential damage due to frost action, the maximum percent of material, by weight, of particles smaller than 0.02 mm shall be 3 percent. It also may be necessary to have a lower percentage of material passing the No. 200 sieve to help control the percentage of particles smaller than 0.02 mm.

Ballast Materials for Railway Tracks

It is recommended by AREMA that any material used as ballast for a railway track should have not more than 1% passing the 0.075 mm sieve. Recommended gradation for these materials is shown in Table 7.15. AREMA also recommends that, in order to provide suitable support for the ties of a main line track, the depth of the ballast should be at least 305 mm, and that of the subballast at least 150 mm. A subballast of 305 mm compacted depth is commonly used for standard gauge construction in main track service. It should be emphasized that greater depths than the minimum specified may be needed, depending on the strength of the subgrade.

Surface Materials' Engineering Properties

Materials used in the construction of highway and airport surface courses, and the rail road track superstructure should also satisfy certain engineering properties. For example, the surface course of highway pavements should be capable of withstanding high tire pressure, resisting the abrasive forces due to traffic, and providing resistance to skidding and preventing surface water from

TABLE 7.15
Recommended Ballast Gradations

Source: Manual for Highway Engineering, American Railway Engineering and Maintenance-of-Way Association, 2005.

Size No. (See Note 1)	Nominal Size Square Opening	Percent Passing									
		3″	2½″	2″	1½″	1″	¾″	½″	⅜″	No.4	No. 8
24	2½″- ¾″	100	90-100		25-60		0-10	0-5	–	–	–
25	2½″ - ⅜″	100	80-100	60-85	50-70	25-50	–	5-20	0-10	0-3	–
3	2″ - 1″	–	100	95-100	35-70	0-15	–	0-5	–	–	–
4A	2″ - ¾″	–	100	90-100	60-90	10-35	0-10	–	0-3	–	–
4	1½″ - ¾″	–	–	100	90-100	20-55	0-15	–	0-5	–	–
5	1″ - ⅜″	–	–	–	100	90-100	40-75	15-35	0-15	0-5	–
57	1″ - No. 4	–	–	–	100	95-100	–	25-60	–	0-10	0-5

Note 1: Gradation Numbers 24, 25, 3, 4A and 4 are main line ballast materials. Gradation Numbers 5 and 57 are yard ballast materials.

Note 2: 1″ = 25.4 mm

penetrating into the underlying layers. A description of commonly used surface materials for flexible and rigid pavements and their required engineering properties are presented.

Surface Materials for Highway and Airport Runway

Flexible Pavements

The material used as the surface course in flexible pavements is asphalt concrete. This is a uniformly mixed combination of asphalt cement, coarse aggregate, fine aggregate, and other materials depending on the type of asphalt concrete.

Asphalt cements are obtained from the fractional distillation of natural deposits of asphalt materials. This is a process through which the different volatile materials in the crude oil are removed at successively higher temperatures until the petroleum asphalt is obtained as residue. They are semisolid hydrocarbons with certain physiochemical characteristics that make them good cementing agents. They are also very viscous and, when used as a binder for aggregates in pavement construction, it is necessary to heat both the aggregates and the asphalt cement prior to mixing the two materials. Several types of asphalt cement can be produced depending on the treatment adopted. The residual asphalt obtained directly from the distillation process is asphalt cement. When the residue is mixed (cut back) with a heavy distillate such as diesel oil, it is known as slow-curing asphalt; when it is cut back with light fuel oil or kerosene, it is known as medium-curing asphalt, and when it is cut back with a petroleum distillate that will easily evaporate, thereby facilitating a quick change from the liquid form to the original asphalt cement, it is known as rapid-curing cutback asphalt. The particular grade of asphalt cement is designated by its penetration and viscosity, both of which give an indication of the consistency of the material at a given temperature. The penetration is the distance in 0.1 mm that a standard needle will penetrate a given sample under specific conditions of loading, time, and temperature. The viscosity can be determined by conducting either the Saybolt Furol viscosity test or the kinematic viscosity test. The Saybolt Furol viscosity is given as the time exactly

60 mL of the asphalt material takes in seconds to flow through the orifice of the Saybolt Furol viscometer at a specified temperature. Temperatures at which asphalt materials for highway construction are tested include 25°C, 50°C, and 60°C. The kinematic viscosity is defined as the absolute viscosity divided by the density and is given in units of centistokes. It is obtained as the product of the time in seconds it takes the material to flow between two timing marks in a kinematic viscometer tube, and a calibration factor for the viscometer used. The manufacturer of the viscometer gives the calibration factor. Standard calibrating oils with known viscosity characteristics are used to calibrate each viscometer. Each asphalt material is designated in terms of the treatment used in its production and its viscosity. For example, an RC-70 is a rapid-curing cutback asphalt with a minimum kinematic viscosity of 70 centistokes at 60°C. It is important that the temperature at which the consistency is determined be specified, since temperature significantly affects the consistency of the asphalt material. It should also be noted that specifications given for asphalt materials usually indicate minimum and maximum values for viscosity. For example, while the minimum value for RC-70 is 70 centistokes at 60°C, the maximum value acceptable is 140 centistokes. Although viscosity is an important parameter, several other parameters are also included in specifying suitable asphalt materials for highway construction. For example, Table 7.16 shows the different parameters used for specifying rapid-curing cutback asphalts.

Another type of asphalt material used in highway construction is asphalt emulsion. Asphalt emulsions are produced by breaking asphalt cement, usually 100–250 penetration range, into minute particles and dispersing them in water with an emulsifier. They remain in suspension in the liquid phase as long as the water does not evaporate or the emulsifier does not break. These minute particles have like electrical charges and therefore do not coalesce. Asphalt emulsions are classified as anionic, cationic, or nonionic. Particles of the anionic and cationic types are surrounded by electrical charges, while the nonionic type is neutral. When the surrounding electrical charge is negative, the emulsion is anionic, and when it is positive, the emulsion is cationic. The anionic and cationic emulsions are commonly used in asphalt construction, particularly as base and subbase courses. The emulsions are further classified in a way similar to that of the asphalt cements. They are classified as rapid setting (RS), medium setting (MS), or slow setting (SS), depending on how rapidly the material will return to the state of the original asphalt cement. These classifications are used to designate the specific type of emulsion. For example, CRS-2 denotes a cationic rapid-setting emulsion. Specifications for the use of emulsified asphalts are given in AASHTO M140 as described in the Standard Specifications for Transportion Materials and Methods of Sampling and Testing.

The aggregates used in asphalt concrete are usually crushed rock, sand, and filler. Coarse aggregates retained in the 2.36 mm sieve are the predominant rock materials, while sand is predominantly material that passes the 2.36 mm sieve. The

TABLE. 7.16
Specification for Rapid Curing of Cutback Asphalts

Source: Used with permission from *Standard Specifications for Transportation Materials and Methods of Sampling and Testing,* 20th ed., Washington, D.C., American Association of State Highway and Transportation Officials, copyright 2000.

	RC-70		*RC-250*		*RC-800*		*RC-3000*	
	Min.	*Max.*	*Min.*	*Max.*	*Min.*	*Max.*	*Min.*	*Max.*
Kinematic viscosity at 60 C (140 F) (See Note 1) centistokes	70.	140.	250.	500.	800.	1600.	3000.	6000.
Flash point (Tag, open-cup), degrees C (F)	...	...	27. (80).	...	27. (80).	...	27. (80).	...
Water, percent	...	0.2	...	0.2	...	0.2	...	0.2
Distillation test:								
Distillate, percentage by volume of total distillate to 360 C (680 F)								
to 190 C (374 F)	10.	...	...	...	...	...	...	...
to 225 C (437 F)	50.	...	35.	...	15.	...	...	...
to 260 C (500 F)	70.	...	60.	...	45.	...	25.	...
to 315 C (600 F)	85.	...	80.	...	75.	...	70.	...
Residue from distillation to 360 C (680 F) volume percentage of sample by difference	55.	...	65.	...	75.	...	80.	...
Tests on residue from distillation:								
Absolute viscosity at 60 C (140 F) (See Note 3) poises	600.	2400.	600.	2400.	600.	2400.	600.	2400.
Ductility, 5 cm/min. at 25 C (77 F) cm	100.	...	100.	...	100.	...	100.	...
Solubility in trichloroethylene, percent	99.0	...	99.0	...	99.0	...	99.0	...
Spot test (See Note 2) with:								
Standard naphtha	Negative for all grades							
Naphtha - xylene solvent, -percent xylene	Negative for all grades							
Heptane - xylene solvent, -percent xylene	Negative for all grades							

NOTE 1—As an alternate, Saybolt Furol viscosities may be specified as follows:
Grade RC-70—Furol viscosity at 50 C (122 F)—60 to 120 sec.
Grade RC-250—Furol viscosity at 60 C (140 F)—125 to 250 sec.
Grade RC-800—Furol viscosity at 82.2 C (180 F)—100 to 200 sec.
Grade RC-3000—Furol viscosity at 82.2 C (180 F)—300 to 600 sec.

NOTE 2—The use of the spot test is optional. When specified, the Engineer shall indicate whether the standard naphtha solvent, the naphtha xylene solvent or the heptane xylene solvent will be used in determining compliance with the requirement, and also, in the case of the xylene solvents, the percentage of xylene to be used.

NOTE 3—In lieu of viscosity of the residue, the specifying agency, at its option, can specify penetration at 100 g; 5s at 25 C (77 F) of 80-120 for Grades RC-70, RC-250, RC-800, and RC-3000. However, in no case will both be required.

filler is predominantly mineral dust that passes the 0.075 mm sieve. Specifications have been developed for the combined aggregates. Table 7.17 gives suggested grading requirements of aggregates based on ASTM designation 3515.

The mixture of asphalt cement, coarse aggregates, and filler materials to form asphalt concrete should be capable of resisting imposed traffic loads, be skid-resistant even when wet, and not be easily affected by weathering forces. The design of the mix used in producing the asphalt concrete determines the degree to which it achieves these characteristics. There are mainly three different

TABLE 7.17
Suggested Aggregate Grading Requirements for Asphalt Concrete

				Dense Mixtures					
				Mix Designation and Nominal Size of Aggregate					
Sieve Size	***2 in. (50 mm)***	***1½ in. (37.5 mm)***	***1 in. (25.0 mm)***	***¾ in. (19.0 mm)***	***½ in. (12.5 mm)***	***⅜ in. (9.5 mm)***	***No. 4 (4.75 mm) (Sand Asphalt)***	***No. 8 (2.36 mm)***	***No. 16 (1.18 mm) (Sheet Asphalt)***
				Grading of Total Aggregate (Coarse Plus Fine, Plus Filler if Required)					
				Amounts Finer Than Each Laboratory Sieve (Square Opening), Weight %					
2 ½ in. (63-mm)	100	...	...	...	...	...	...	...	...
2 in. (50-mm)	90 to 100	100	...	...	...	...	...	...	...
1 ½ in. (37.5-mm)	...	90 to 100	100	...	...	...	...	...	...
1 in. (25.0-mm)	60 to 80	...	90 to 100		...	...	...	...	...
¾ in. (19.0-mm)	...	56 to 80	...	90 to 100	100	...	...	...	...
½ in. (12.5-mm)	35 to 65	...	56 to 80	...	90 to 100	100	...	...	...
⅜ in. (9.5-mm)	...	...	...	56 to 80	...	90 to 100	100	...	...
No. 4 (4.75-mm)	17 to 47	23 to 53	29 to 59	35 to 65	44 to 74	55 to 85	80 to 100	...	100
No. 8 (2.36-mm)[A]	10 to 36	15 to 41	19 to 45	23 to 49	28 to 58	32 to 67	65 to 100	...	95 to 100
No. 16 (1.18-mm)	...	...	...	...	...	...	40 to 80	...	85 to 100
No. 30 (600-μm)	...	...	...	...	...	...	25 to 65	...	70 to 95
No. 50 (300-μm)	3 to 15	4 to 16	5 to 17	5 to 19	5 to 21	7 to 23	7 to 40	...	45 to 75
No. 100 (150-μm)	...	...	...	...	...	...	3 to 20	...	20 to 40
No. 200 (75-μm)[B]	0 to 5	0 to 6	1 to 7	2 to 8	2 to 10	2 to 10	2 to 10	...	9 to 20
				Open Mixtures					
				Mix Designation and Nominal Maximum Size of Aggregate					
Sieve Size	***2 in. (50 mm)***	***1½ in. (37.5 mm)***	***1 in. (25.0 mm)***	***¾ in. (19.0 mm)***	***½ in. (12.5 mm)***	***⅜ in. (9.5 mm)***	***No. 4 (4.75 mm) (Sand Asphalt)***	***No. 8 (2.36 mm)***	***No. 16 (1.18 mm) (Sheet Asphalt)***
	Base and Binder Courses						***Surface and Leveling Courses***		
2 ½ in. (63-mm)	100	...	...	...	...	...	...	...	...
2 in. (50-mm)	90 to 100	100	...	...	...	...	...	...	...
1 ½ in. (37.5-mm)	...	90 to 100	100	...	...	...	...	...	...

(continued)

TABLE 7.17

Suggested Aggregate Grading Requirements for Asphalt Concrete (*continued*)

	Open Mixtures								
	Mix Designation and Nominal Maximum Size of Aggregate								
Sieve Size	*2 in. (50 mm)*	*1½ in. (37.5 mm)*	*1 in. (25.0 mm)*	*¾ in. (19.0 mm)*	*½ in. (12.5 mm)*	*⅜ in. (9.5 mm)*	*No. 4 (4.75 mm) (Sand Asphalt)*	*No. 8 (2.36 mm)*	*No. 16 (1.18 mm) (Sheet Asphalt)*
	Base and Binder Courses					*Surface and Leveling Courses*			
1 in. (25.0-mm)	40 to 70	...	90 to 100	100	...	...	...	...	...
¾ in. (19.0-mm)	...	40 to 70	...	90 to 100	100	...	...	...	...
½ in. (12.5-mm)	18 to 48	...	40 to 70	...	85 to 100	100	...	...	...
⅜ in. (9.5-mm)	...	18 to 48	...	40 to 70	60 to 90	85 to 100	...	...	...
No. 4 (4.75-mm)	5 to 25	6 to 29	10 to 34	15 to 39	20 to 50	40 to 70	...	100	...
No. 8 (2.36-mm)[A]	0 to 12	0 to 14	1 to 17	2 to 18	5 to 25	10 to 35	...	75 to 100	...
No. 16 (1.18-mm)	...	...	...	...	3 to 19	5 to 25	...	50 to 75	...
No. 30 (600-m)	0 to 8	0 to 8	0 to 10	0 to 10	...	...	...	28 to 53	...
No. 50 (300-m)	...	...	...	...	0 to 10	0 to 12	...	8 to 30	...
No. 100 (150-m)	...	...	...	...	...	...	...	0 to 12	...
No. 200 (75-m)[B]	...	...	...	...	...	...	...	0 to 5	...
	Bitumen, Weight % of Total Mixture[C]								
	2 to 7	3 to 8	3 to 9	4 to 10	4 to 11	5 to 12	6 to 12	7 to 12	8 to 12
	Suggested Coarse Aggregate Sizes								
	3 and 57	4 and 67 or 4 and 68	5 and 7 or 57	67 or 68 or 6 and 8	7 or 78	8			

[A]In considering the total grading characteristics of a bituminous paving mixture, the amount passing the No. 8 (2.36-mm) sieve is a significant and convenient field control point between fine and coarse aggregate. Gradings approaching the maximum amount permitted to pass the No. 8 sieve will result in pavement surfaces having comparatively fine texture, while coarse gradings approaching the minimum amount passing the No. 8 sieve will result in surfaces with comparatively coarse texture.

[B]The material passing the No. 200 (75-μm) sieve may consist of fine particles of the aggregates or mineral filler, or both, but shall be free of organic matter and clay particles. The blend of aggregates and filler, when tested in accordance with Test Method D 4318, shall have a plasticity index of not greater than 4, except that this plasticity requirement shall not apply when the filler material is hydrated lime or hydraulic cement.

[C]The quantity of bitumen is given in terms of weight % of the total mixture. The wide difference in the specific gravity of various aggregates, as well as a considerable difference in absorption, results in a comparatively wide range in the limiting amount of bitumen specified. The amount of bitumen required for a given mixture should be determined by appropriate laboratory testing or on the basis of past experience with similar mixtures, or by a combination of both.

Source: Annual Book of ASTM Standards, Section 4, Construction, Vol. 04.03, Road and Paving Materials; Pavement Management Technologies, American Society for Testing and Materials, Philadelphia, PA, 1996.

kinds of asphalt concrete used in highway pavement construction: hot-mix-hot-laid, hot-mix-cold-laid, and cold-mix-cold-laid.

Hot-mix-hot-laid asphalt concrete is a properly produced mixture of asphalt cement, coarse aggregate, fine aggregate, and filler (dust) at temperatures ranging from 80°C to 163°C depending on the type of asphalt cement used. Suitable types of asphalt materials include AC-20, AC-10, and AR-8000. Hot-mix-hot-laid asphalt concrete can also be classified as open graded, coarse graded, dense graded, or fine graded, depending on the maximum size of aggregates used in the mixture and the use of the mixture. For example, when the mixture is for high-type surfacing, the maximum aggregate size is between 12.7 mm and 19 mm, for open graded; between 12.7 mm and 19 mm, for coarse graded; between 12.7 mm and 25.4 mm for dense graded. It is important that the asphalt concrete be an optimum blend of the different components that will satisfy the specified requirements for stability and durability. Two mix procedures for obtaining this objective are the Marshall Method described in detail in ASTM D1559 and the *Su*perior *per*forming asphalt *pave*ment (*Superpave*), which was developed as part of the Strategic Highway Research Program (SHRP) and is described in *Superpave Mix Design* (SP-2).

Hot-mix-cold-laid asphalt concrete is manufactured hot and then shipped and either immediately laid or stockpiled for use at a future date. Asphalt cements with high penetration with lower limits of the 200–300 penetration grade have been found suitable for this type of asphalt concrete.

Cold-mix-cold-laid asphalt concrete is usually manufactured with emulsified asphalt or a low-viscosity cutback asphalt as the binder. It can also be used immediately after production or stockpiled for later use. The type and grade of asphalt material used depends on whether the material is to be stored for a long time, the use of the material, and the gradation of the aggregates.

The most commonly used material in flexible pavements is a hot plant mix of asphalt cement and dense-graded aggregates with a maximum size of 1″. Details of different methods for the mix design of this material are given in *Traffic and Highway Engineering* by Garber and Hoel.

The Federal Aviation Administration recommends the use of a dense-graded hot-mix asphalt concrete for use as surface material for flexible airport runway pavements. Detailed specifications for the composition of the asphalt concrete are given in Part V of the FAA *Advisory Circular 150/5370-10A*.

Surface Materials for Highway and Airport Runway Rigid Pavements

Portland cement concrete is commonly used as the surface material for rigid highway pavements. It is a mixture of Portland cement, coarse aggregate, fine aggregate, and water. Steel reinforcement is sometimes used depending on the type of pavement being constructed.

Portland cement is manufactured from a carefully prepared mix of limestone, marl, and clay or shale. The mixture is crushed and pulverized and then burnt at a high temperature (about 1540°C) to form a clinker. This is then allowed to cool and a small amount of gypsum added, and the mixture is ground until more than 90% of the material passes the 0.075 mm sieve. AASHTO has specified five main types of Portland cement:

Type I is suitable for general concrete construction, where no special properties are required.

Type II is suitable for use in general construction, where the concrete will be exposed to moderate action of sulphate or where moderate heat of hydration is required.

Type III is suitable for concrete construction that requires a high concrete strength in a relatively short time. It is sometimes referred to as high early strength concrete.

Types IA, IIA, and IIIA are similar to types I, II, and III, respectively, but contain a small amount (4–8% of total mix) of entrapped air. In addition to the properties listed for types I, II, and III, types IA, IIA, and IIIA are also more resilient to calcium chloride and de-icing salts and are therefore more durable.

Type IV is suitable for projects where low heat of hydration is necessary.

Type V is suitable for concrete construction projects where the concrete will be exposed to high sulphate action.

The recommended proportions of the different chemical constituents for the different types are shown in Table 7.18. Inert materials that do not react with the cement are used as the coarse aggregates in Portland cement. They usually consist of one or a combination of two or three of the following: crushed gravel, stone, or blast furnace slag.

The fine aggregate in Portland cement is mainly sand. Tables 7.19 and 7.20 show AASHTO-recommended gradation for fine and coarse aggregates used in Portland cement. Apart from the gradation requirements, AASHTO also recommends minimum standards for soundness and cleanliness. The soundness requirement is usually given in terms of the maximum permitted loss in the material after five alternate cycles of wetting and drying in the soundness test. A maximum of 10% weight loss is usually specified. The maximum amount of different types of deleterious materials contained in the fine aggregate is often used to specify the cleanliness requirement. For example, the maximum amount of silt (material passing 0.075 mm sieve) should not be greater than 5% of the total fine aggregates.

The main requirement usually specified for the water is that it should be suitable for drinking; that is, the quantity of organic matter, oil, acids, and alkalines should not be greater than the allowable amount in drinking water.

In order to control cracking of concrete pavement, reinforcing steel may be used in the form of a bar mat or wire mesh, placed about 75 mm below the surface

TABLE 7.23
Tensile Properties Requirements for Railway Steel Tracks

Source: Manual for Highway Engineering, American Railway Engineering and Maintenance-of-Way Association, 2005.

Description	*Standard*	*High-Strength*
Yield Strength, ksi, minimum	70	110
Tensile Strength, ksi, minimum	140	170
Elongation in 2″, percent, minimum	9	10

Note: 1 ksi = 6.9 N/mm^2
1″ = 25.4 mm

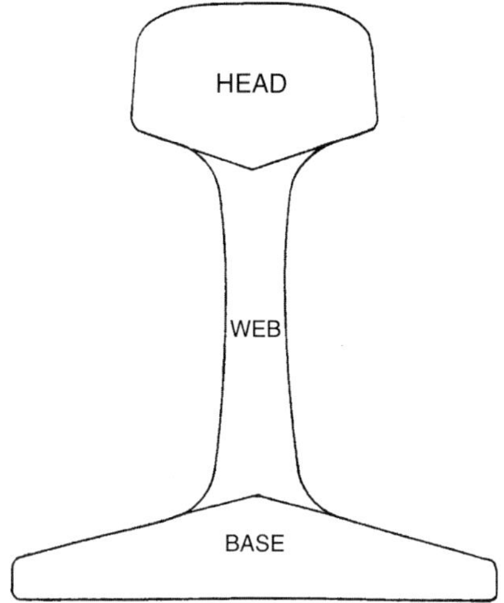

FIGURE 7.8
Definition of rail cross-sectional area.

Source: Manual for Highway Engineering, American Railway Engineering and Maintenance-of-Way Association, Landover, MD, 2005.

Step 3 Determine Minimum Size and/or Minimum Thickness for Each Structural Component

This step involves the determination of the minimum size and/or minimum thickness of each structural component such that the stress and/or strain within any component does not exceed the maximum allowable. This depends on the load transmitted from the wheels of the vehicles using the travelway and the strength of the subgrade. For example, the thickness of the surface, base, or subbase course of a flexible highway pavement depends on the accumulated equivalent axle load for the design period and the subgrade resilient modulus. Similarly, the total thickness of the ballasts and subballast of a railway track depends on the wheel load of the train for which the track is being designed and the permissible bearing pressure on the subgrade. The design procedures presented are the AASHTO methods for flexible and rigid highway pavements, the Federal Aviation Administration methods for flexible and rigid airport pavements, and the AREMA method for railway tracks.

AASHTO Method for the Design of Flexible Highway Pavements

Highway flexible pavements are divided into three subgroups: high type, intermediate type, and low type. High-type pavements should not be susceptible to weather conditions and should be capable of adequately supporting the expected traffic load without visible distress due to fatigue. The quality of an intermediate-type pavement could range from as high as that of high-type pavements to that for surface-treated roads. Low-type pavements usually have wearing surfaces ranging from untreated loose natural materials to surface-treated earth. These are used mainly for low-cost roads. The method presented here is for the design of high-type pavements, although it can also be used for some intermediate-type pavements.

This AASHTO method is based on the results obtained from the American Association of Highway Officials (AASHO) (now AASHTO) road test that was conducted in Ottawa, Illinois, a cooperative effort carried out under the auspices of 49 states, the District of Columbia, Puerto Rico, the Bureau of Public Roads, and several industry groups. Data were collected from the application of several thousand single- and tandem-axle loads on flexible and rigid pavements with different combinations of subbase, base, and surface thicknesses on A-6 subgrade material. The loads ranged from 900 to 13600 kg, and 10890 to 21780 kg for single- and tandem axles, respectively. Data collected included the extent of cracking and the amount of patching required to maintain the section in service, the effect of the load applications on the longitudinal and transverse profiles, the extent of rutting and surface deflection, pavement curvature at different vehicle speeds, stresses imposed on the subgrade surface, and the temperature distribution in the pavement layers. A thorough analysis of these data formed the basis for the AASHTO method of flexible pavement.

The first interim guide for the design of pavement structures was published by AASHTO in 1961. Revisions were then made in 1972, 1986, and 1993. The 1993 edition includes a procedure for overlay design. It should, however, be noted that AASHTO clearly indicated in each of these editions that the design procedures presented do not necessarily cover all conditions that may exist in any one specific site. AASHTO therefore recommends that, in using the guide, local experience should be used to augment the procedures given in the guide.

The method is presented by first discussing the specific factors used in the procedure and then the equations for determining the thicknesses of the pavement layers are presented.

Factors Used in AASHTO'S Method for Flexible Highway Pavement Design

The factors used include the accumulated equivalent axle load (ESAL) for the design period, the resilient modulus (M_r) of the subgrade, the quality of materials used to construct the base and surface layers, the impact of variable environmental conditions during a year, drainage characteristics, and the reliability of predicted traffic.

The design ESAL is computed as discussed earlier in this chapter. AASHTO has provided tables for equivalent axle loads for different terminal serviceability indices (P_t) (see below for definition). The values given in Tables 7.1a and 7.1b are for single and tandem axles, respectively, and for a terminal serviceability index (P_t) of 2.5 as this is commonly used in the AASHTO flexible pavement design method.

Although the guide uses the resilient modulus to indicate the quality of the subgrade in the design procedure, it allows for the conversion of the CBR value of the soil to an equivalent M_r value using the following conversion factor:

$$M_r \text{ (N/mm}^2\text{)} = 10.5 \text{ CBR (for fine-grain soils with soaked CBR of 10 or less)}$$

Because the strength of the subgrade varies from one season to another during the year, an effective M_r for the whole year is determined using the procedure discussed later in the section on the effect of the environment.

Pavement performance is based on the structural and functional performance of the pavement. The structural performance reflects the physical condition of the pavement with respect to cracking, faulting, raveling, and so on. These factors negatively impact the ability of the pavement to carry the accumulated ESAL used for design. Functional performance reflects the ability of the roadway to provide a comfortable ride. A concept known as the *serviceability index* is used to quantify the pavement performance. The roughness and distress, which are quantified in terms of the extent of cracking and patching of the pavement, are used to determine its present serviceability index (PSI). It is given as a function of the extent and type of cracking and the slope variance of the two wheel paths, which is a measure of the variations in the longitudinal profile. Individual ratings were assigned by experienced engineers to different pavements with varying conditions, and the mean of these ratings used to relate the PSI to the factors considered. The lowest PSI is 0 and the highest is 5.

The serviceability index immediately after the construction of a new pavement is the initial serviceability index (P_i), and the minimum acceptable value is the terminal serviceability index (P_t). Recommended values for the terminal serviceability index for flexible pavements are 2.5 or 3.0 for major highways and 2.0 for highways with a lower classification. A P_t value of 1.5 has been used in cases where economic constraints restrict capital expenditures for construction. This low value should, however, be used only in special cases on selected classes of highways.

The materials used for construction can be classified under three general groups: those used for subbase construction, those used for base construction, and those used for surface construction.

The quality of *subbase construction materials* is given in terms of the layer coefficient, a_3, which is used to convert the actual thickness of the subbase to an equivalent SN. For example, a value of 0.11 is assumed for the sandy gravel subbase course material used in the AASHTO road test. Figure 7.9 gives values for different granular subbase materials. Because different environmental, traffic,

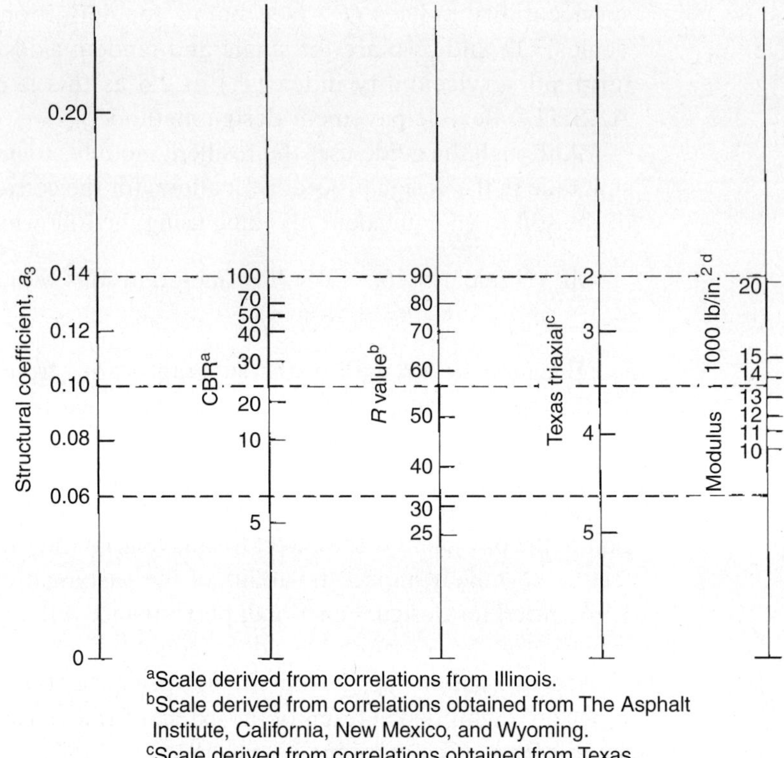

FIGURE 7.9
Variation in granular subbase layer coefficient, a_3, with various subbase strength parameters.

Source: AASHTO Guide for Design of Pavement Structures, American Association of State Highway and Transportation Officials, Washington, D.C., 1993. Used by permission.

[a]Scale derived from correlations from Illinois.
[b]Scale derived from correlations obtained from The Asphalt Institute, California, New Mexico, and Wyoming.
[c]Scale derived from correlations obtained from Texas.
[d]Scale derived on NCHRP project 128, 1972.

Note: 1000 lb/in^2 = 6.9 N/mm^2

and construction conditions may exist, AASHTO suggests that each design agency develop appropriate layer coefficients that reflect the conditions that exist in its location.

Base course construction materials should satisfy the general requirements for base course materials given earlier in this chapter. Figure 7.10 gives structural layer coefficient, a_2, for different materials that can be used as base construction materials.

The material commonly used for *surface course construction* is a hot plant mix of asphalt cement and dense-graded aggregates with a maximum size of 25.4 m. The structural layer coefficient (a_1) for this material depends on its resilient modulus and can be obtained from Figure 7.11.

The AASHTO design procedure for flexible pavements considers temperature and rainfall as the two main *environmental factors*. The factors that are related to the temperature effect include stresses induced by thermal action, changes in the creep properties, and the effect of freezing and thawing of the subgrade soil. The rainfall effect takes into consideration the possibility of surface water penetrating into the underlying material. When this occurs, the properties of the underlying materials may be significantly altered. Although there are several ways of preventing this (see *Traffic and Highway Engineering* by

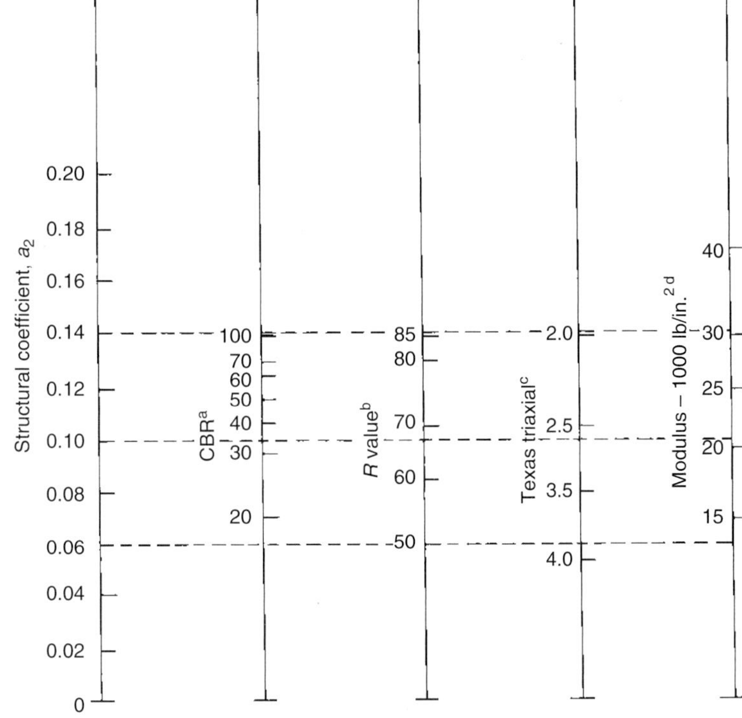

FIGURE 7.10

Variation in granular base layer coefficient, a_2, with various subbase strength parameters.

Source: AASHTO Guide for Design of Pavement Structures, American Association of State Highway and Transportation Officials, Washington, D.C., 1993. Used by permission.

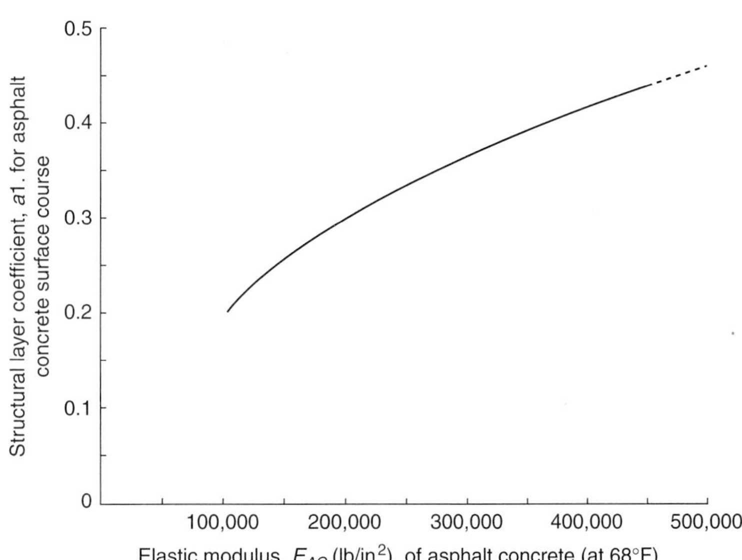

FIGURE 7.11

Chart for estimating structural layer coefficient of dense-graded/asphalt concrete based on the elastic (resilient) modulus.

Source: AASHTO Guide for Design of Pavement Structures, American Association of State Highway and Transportation Officials, Washington, D.C., 1993. Used by permission.

Garber and Hoel), the AASHTO design procedure corrects for this effect, as mentioned later in our discussion of drainage.

The *effect of temperature* considers the weakening of the underlying material during the thaw period. Test results have shown that when materials are susceptible to frost action, the modulus during the thaw period can be as low as 50–80% of the modulus during summer and fall seasons. Also, in areas with potential for heavy rains during specific periods of the year, the strength of the material may reduce during these periods, resulting in the variation of its strength during the year, even when there is no specific thaw period. In order to compensate for this variation of material strength during the year, an *effective resilient* modulus that is equivalent to the combined effect of the different seasonal moduli during the year is determined. This effective resilient modulus will result in a PSI of the pavement during a full 12-month period that is equivalent to that obtained by using the appropriate resilient modulus for each season.

The first of the two methods suggested by AASHTO to determine the effective resilient modulus is presented here. This method uses results of laboratory tests to develop a mathematical relationship between the resilient modulus of the soil material and its moisture content. The resilient modulus is then determined for the estimated moisture content during each season. It is necessary to divide the whole year into the different time intervals that correspond to the different seasonal resilient moduli. The minimum time interval suggested by AASHTO is one-half month. Equation 7.10, suggested by AASHTO, can be used to develop the relative damage u_f for each time period. The mean relative damage u_f is then computed, and the effective subgrade resilient modulus is then determined using Equation 7.10 or the chart shown in Figure 7.12.

$$u_f = 1.18 \times 10^8 \times M_r^{-2.32} \tag{7.10}$$

EXAMPLE 7.8

Computing Effective Resilient Modulus

Figure 7.12 shows roadbed soil resilient modulus M_r for each month estimated from laboratory results correlating M_r with moisture content. Determine the effective resilient modulus of the subgrade.

Solution

Note that in this case the moisture content does not vary within any one month. The solution of the problem is given in Figure 7.12.

FIGURE 7.12

Chart for estimating effective roadbed soil layer resilient modulus for flexible pavement designed using the serviceability criteria.

Source: AASHTO Guide for Design of Pavement Structures, American Association of State Highway and Transportation Officials, Washington, D.C., 1993. Used by permission.

Month	Roadbed soil modulus M_r (lb/in.^{2})	Relative damage u_f
Jan.	22000	0.01
Feb.	22000	0.01
Mar.	5500	0.25
Apr.	5000	0.30
May	5000	0.30
June	8000	0.11
July	8000	0.11
Aug.	8000	0.11
Sept.	8500	0.09
Oct.	8500	0.09
Nov.	6000	0.20
Dec.	22000	0.01
Summation: $\Sigma u_f =$		1.59

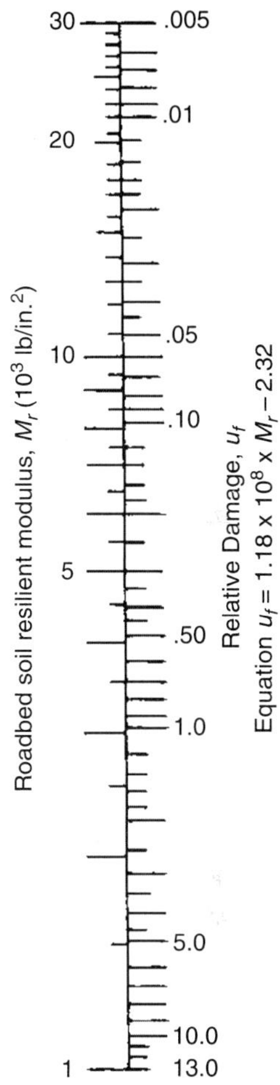

Average $\bar{u}_f = \frac{\Sigma u_f}{n} = \frac{1.59}{12} = 0.133$

Effective roadbed soil resilient modulus, M_r (lb/in.^{2}) = 7250 (corresponds to $\bar{u}_f$)

Note: 1 lb/in.^{2} = 7 kPa

Determine the value of u_f for each M_r—use Equation 7.10:

$$u_f = 1.18 \times 10^8 \times M_r^{-2..32}$$

For example, for the month of May, $u_f = 1.18 \times 10^8 \times 5000^{-2.32} \cong 0.30$
Determine the mean relative damage:

$$u_f = 0.133$$

Determine effective resilient modulus—use Equation 7.10 or the chart shown in Figure 7.12, which in turn gives an effective resilient modulus of 50.6 N/mm^2.

Drainage effect on the performance of flexible highway pavements is considered in the AASHTO design procedure by first providing a suitable

FIGURE 7.13

Example of drainage layer in pavement structure.

Source: AASHTO Guide for Design of Pavement Structures, American Association of State Highway and Transportation Officials, Washington, D.C., 1993. Used by permission.

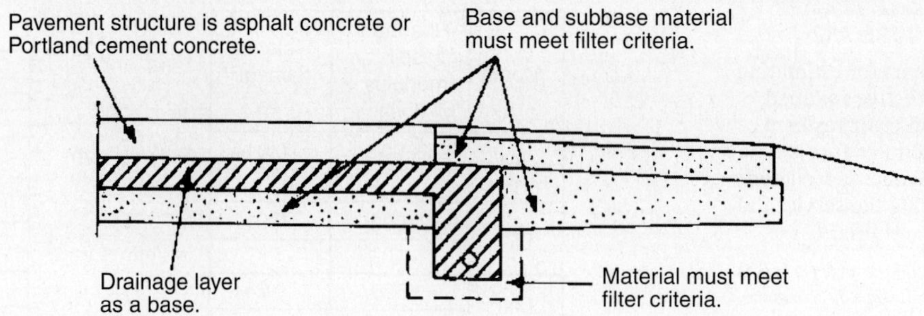

(a) Base is used as the drainage layer.

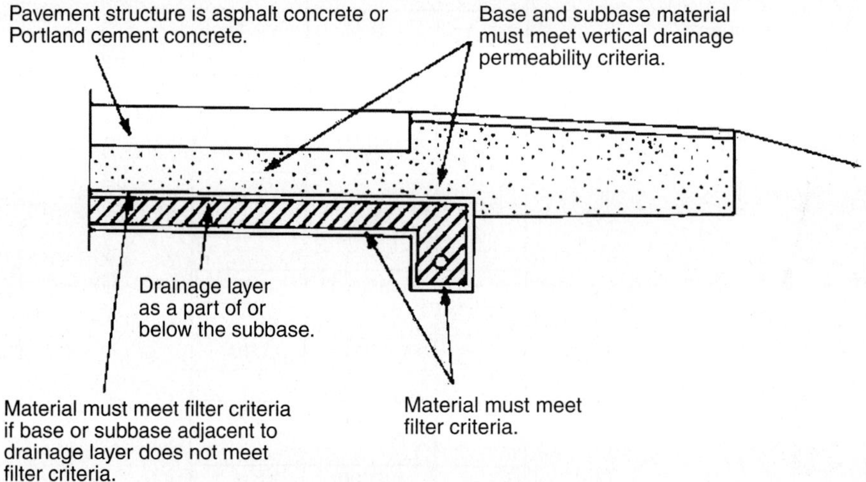

(b) Drainage layer is part of or below the subbase.

Note: Filter fabrics may be used in lieu of filter material, soil, or aggregate, depending on economic considerations.

drainage layer, as shown in Figure 7.13, and modifying the structural layer coefficient by incorporating a factor m_i for the base and subbase layer coefficients a_2 and a_3. The value of m_i is based on the percentage of time during which the pavement structure will be nearly saturated and on the quality of drainage, which is dependent on the time it takes to drain the base layer to 50% of saturation. AASHTO-suggested general definitions of the different levels of drainage quality are given in Table 7.24 and the recommended m_i values are given in Table 7.25.

AASHTO has proposed the use of a reliability factor in the design of highway pavements to take into consideration the uncertainty associated with the determination of the design ESAL, particularly with respect to the use of growth factors that may not be accurate. A detailed discussion of the development

TABLE 7.24
Definition of Drainage Quality

Source: Adapted with permission from *AASHTO Guide for Design of Pavement Structures,* American Association of State Highway and Transportation Officials, Washington, D.C., 1993. Used by permission.

Quality of Drainage	*Water Removed Within*[*]
Excellent	2 hours
Good	1 day
Fair	1 week
Poor	1 month
Very poor	(water will not drain)

[*]Time required to drain the base layer to 50 percent saturation.

of the approach used is beyond the scope of this book. However, a general description is presented to facilitate the understanding of the design equation and the associated charts. First, a reliability design level (R%), which represents the assurance that the designed pavement section will survive for its design life, is selected. For example, a 60% reliability design level implies that the probability of design performance success is 60%. Table 7.26 gives reliability levels suggested by AASHTO and based on a survey of the AASHTO design task force. A reliability factor is then determined based on the reliability level and the overall variation S_0^2 using Equation 7.11. The overall variation accounts for the chance variation of the traffic forecast and the chance variation in actual pavement performance for a given design period traffic, W_{18}.

$$\log_{10} FR = - Z_R S_0 \tag{7.11}$$

where

F_r = reliability factor for a reliability design level of R%
Z_R = standard normal variant for a given reliability (R%)
S_0 = estimated overall standard deviation

Table 7.27 gives values of Z_R for different reliability (R%).

AASHTO also recommends an overall standard deviation range of 0.30–0.40 for rigid pavements and 0.4–0.5 for flexible pavements. Although

TABLE 7.25
Recommended m_i Values

Source: Adapted from *AASHTO Guide for Design of Pavement Structures,* American Association of State Highway and Transportation Officials, Washington, D.C., 1993. Used by permission.

	Percent of Time Pavement Structure Is Exposed to Moisture Levels Approaching Saturation			
Quality of Drainage	*Less Than 1 Percent*	*1–5 Percent*	*5–25 Percent*	*Greater Than 25 Percent*
Excellent	1.40–1.35	1.35–1.30	1.30–1.20	1.20
Good	1.35–1.25	1.25–1.15	1.15–1.00	1.00
Fair	1.25–1.15	1.15–1.05	1.00–0.80	0.80
Poor	1.15–1.05	1.05–0.80	0.80–0.60	0.60
Very poor	1.05–0.95	0.95–0.75	0.75–0.40	0.40

TABLE 7.26
Suggested Levels of Reliability for Various Functional Classifications

Source: Adapted with permission from *AASHTO Guide for Design of Pavement Structures,* American Association of State Highway and Transportation Officials, Washington, D.C., 1993. Used by permission.

Recommended Level of Reliability		
Functional Classification	***Urban***	***Rural***
Interstate and other freeways	85–99.9	80–99.9
Other principal arterials	80–99	75–95
Collectors	80–95	75–95
Local	50–80	50–80

Note: Results based on a survey of the AASHTO Pavement Design Task Force.

these values are based on a detailed analysis of existing data, very little data presently exist for certain design components such as drainage. A methodology for improving these estimates is presented in the AASHTO guide which may be used when additional data are available.

AASHTO Design Equations for Flexible Highway Pavements

There are two equations used in this procedure. Equation 7.12 is the first equation, and it gives the relationship between the overall required SN as the dependent variable and several input variables that include the design ESAL, the difference between the initial and the terminal serviceability indices, and the resilient modulus of the subgrade. The determined SN is capable of carrying the projected design ESAL. The second is Equation 7.13, which gives the required

TABLE 7.27
Standard Normal Deviation (Z_R) Values Corresponding to Selected Levels of Reliability

Source: Adapted with permission from *AASHTO Guide for Design of Pavement Structures,* American Association of State Highway and Transportation Officials, Washington, D.C., 1993. Used by permission.

Reliability (R%)	*Standard Normal Deviation,* Z_R
50	−0.000
60	−0.253
70	−0.524
75	−0.674
80	−0.841
85	−1.037
90	−1.282
91	−1.340
92	−1.405
93	−1.476
94	−1.555
95	−1.645
96	−1.751
97	−1.881
98	−2.054
99	−2.327
99.9	−3.090
99.99	−3.750

structural number based on the drainage coefficient, m_i, for each layer coefficient for the surface, base, and subbase layers and the actual depth of each layer. This design procedure is not used for ESAL less than 50,000 for the performance period as these roads are usually considered as low-volume loads. The design equation for the equivalent structural number is given as:

$$\log_{10} W_{18} = Z_R S_o + 9.36 \log_{10}(\mathrm{SN}+1) - 0.20 + \frac{\log_{10}[\Delta \mathrm{PSI}/(4.2-1.5)]}{0.40 + [1094/(\mathrm{SN}+1)^{5.19}]} + 2.32 \log_{10} M_r - 8.07 \quad (7.12)$$

where

W_{18} = predicted number of 18,000-lb (80 kN) single-axle load applications
Z_R = standard normal deviation for a given reliability
S_o = overall standard deviation
SN = structural number indicative of the total pavement thickness
$\Delta\mathrm{PSI} = P_i - P_t$
P_i = initial serviceability index
P_t = terminal serviceability index
M_r = resilient modulus in lb/in^2

$$\mathrm{SN} = a_1 D_1 + a_2 D_2 m_2 + a_3 D_3 m_3 \quad (7.13)$$

where

m_i = drainage coefficient for layer i
a_1, a_2, a_3 = layer coefficients representative of surface, base, and subbase course, respectively
D_1, D_2, D_3 = actual thickness in inches of surface, base, and subbase courses, respectively

Equation 7.12 can be solved for SN using a computer program or the chart in Figure 7.14. The use of the chart is demonstrated by the example solved on the chart and in the solution of Example 7.9. The designer selects the type of surface to be used, which can be either asphalt concrete, a single surface treatment, or a double surface treatment. Table 7.28 gives AASHTO minimum thicknesses for surface and base materials.

TABLE 7.28

AASHTO-Recommended Minimum Thicknesses of Highway Layers

Source: Adapted with permission from *AASHTO Guide for Design of Pavement Structures,* American Association of State Highway and Transportation Officials, Washington, D.C., 1993. Used by permission.

	Minimum Thickness (in.)	
Traffic, ESALs	*Asphalt Concrete*	*Aggregate Base*
Less than 50,000	1.0 (or surface treatment)	4
50,001–150,000	2.0	4
150,001–500,000	2.5	4
500,001–2,000,000	3.0	6
2,000,001–7,000,000	3.5	6
Greater than 7,000,000	4.0	6

Note: 1 inch = 25.4 mm

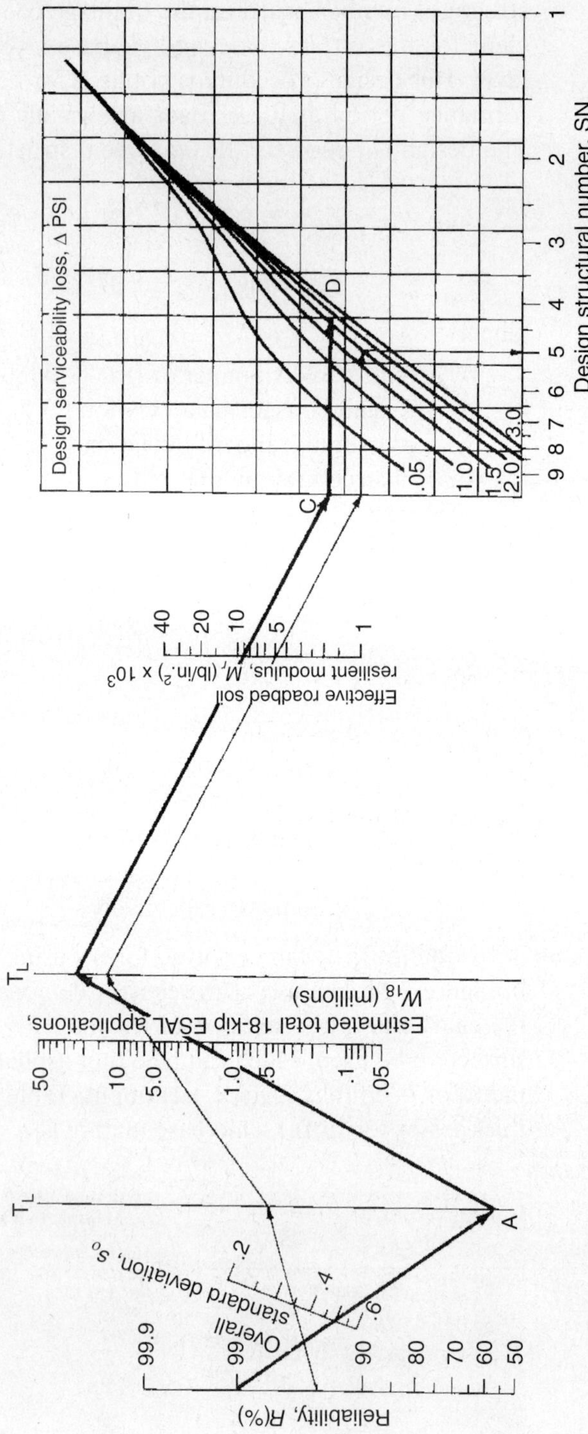

Note: 1000 lb/in² = 6.9 N/mm²

FIGURE 7.14

Design chart for flexible pavements based on using mean values for each input.

Source: AASHTO Guide for Design of Pavement Structures, American Association of State Highway and Transportation Officials, Washington, D.C., 1993. Used by permission.

Example 7.9

Designing a Flexible Pavement Using the AASHTO Method

A flexible pavement for an urban interstate highway is to be designed using the 1993 AASHTO guide procedure to carry a design ESAL of 3.5×10^6. It is estimated that it takes about one week for water to be drained from within the pavement and the pavement structure will be exposed to moisture levels approaching saturation for 26% of the time. The following additional information is available:

Resilient modulus of asphalt concrete at 68°F = 450,000 lb/in^2 (3105 N/mm^2)

CBR value of base course material = 100, M_r = 35,000 lb/in^2 (242 N/mm^2)

CBR value of subbase course material = 25, M_r = 14,500 lb/in^2 (100 N/mm^2)

CBR value of subgrade material = 6

Solution

Determine a suitable pavement structure, M_r, of subgrade = 6×1500 lb/in^2 = 9,000 lb/in^2

Since the pavement is to be designed for an interstate highway, the following assumptions are made:

Reliability level (R) = 99% (range is 80–99.9 from Table 7.26)

Standard deviation (S_o) = 0.49 (range is 0.4–0.5)

Initial serviceability index P_i = 4.5

Terminal serviceability index P_t = 2.5

The nomograph in Figure 7.14 is used to determine the design SN through the following steps:

Step i. Draw a line joining the reliability level of 99% and the overall standard deviation S_o of 0.49, and extend this line to intersect the first T_L line at point A.

Step ii. Draw a line joining point A to the ESAL of 3.5×10^6, and extend this line to intersect the second T_L line at point B.

Step iii. Draw a line joining point B and resilient modulus (M_r) of the roadbed soil, and extend this line to intersect the design serviceability loss chart at point C.

Step iv. Draw a horizontal line from point C to intersect the design serviceability loss (ΔPSI) curve at point D. In this problem $\Delta PSI = 4.5 - 2.5 = 2$.

Step v. Draw a vertical line to intersect the design SN, and read this value. SN = 4.4

Step vi. Determine the appropriate structure layer coefficient for each construction material.

(a) Resilient value of asphalt cement = 450,000 lb/in² (3105 N/mm²). From Figure 7.11, $a_1 = 0.44$.

(b) CBR of base course material = 100. From Figure 7.10, $a_2 = 0.14$.

(c) CBR of subbase course material = 22. From Figure 7.9, $a_3 = 0.10$.

Step vii. Determine appropriate drainage coefficient m_i. Since only one set of conditions is given for both the base and subbase layers, the same value will be used for m_1 and m_2. The time required for water to drain from within pavement is one day, and based on Table 7.23, drainage quality is good. The percentage of time pavement structure will be exposed to moisture levels approaching saturation = 26, and from Table 7.24, $m_i = 0.80$.

Step viii. Determine appropriate layer thicknesses from Equation 7.13:

$$SN = a_1D_1 + a_2D_2m_2 + a_3D_3m_3$$

It can be seen that several values of D_1, D_2, and D_3 values can be obtained to satisfy the SN value of 4.4. Layer thicknesses, however, are usually rounded up to the nearest 0.5″ (12.7 mm).

The selection of the different layer thicknesses should also be based on constraints associated with maintenance and construction practices so that a practical design is obtained. For example, it is normally impractical and uneconomical to construct any layer with a thickness less than some minimum value, as shown in Table 7.28.

Taking into consideration that a flexible pavement structure is a layered system, the determination of the different thicknesses should be carried out as indicated in Figure 7.15. The required SN above the subgrade is first determined, and then the required SNs above the base and subbase layers are determined using the appropriate strength of each layer. The minimum allowable thickness of each layer can then be determined using the differences of the computed SNs.

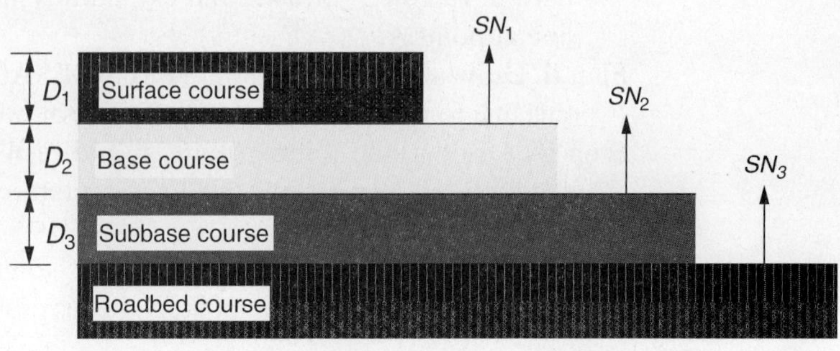

FIGURE 7.15 Procedure for determining thicknesses of layers using a layered analysis approach.

Using the appropriate values for M_r in Figure 7.14, we obtain $SN_3 = 4.4$ and $SN_2 = 3.8$. Note that when SN is assumed to compute ESAL, the assumed and computed SN values must be approximately equal. If these are significantly different, the computation must be repeated with a new assumed SN.
We know

M_r for base course $-$ 31,000 (214 N/mm²)

Using this value in Figure 7.14, we obtain:

$SN_1 = 2.6$

giving

$$D_1 = \frac{2.6}{0.44} = 5.9'' \text{ (149.8 mm)}$$

Using 6" for the thickness of the surface course D^*.

$D_1^* = 6''$ (152.4 mm).

$SN_1^* = a_1 D_1^* = 0.44 \times 6 = 2.64$

$$D_2^* \geq \frac{SN_2 - SN_1^*}{a_2 m_2} \geq \frac{3.8 - 2.64}{0.14 \times 0.8} \geq 10.36'' \quad \text{(use 12'' (304.8 mm))}$$

$SN_2^* = 0.14 \times 0.8 \times 12 + 2.64 = 1.34 + 2.64$

$$D_3^* = \frac{SN_3 - SN_2^*}{a_3 m_2} = \frac{4.4 - (2.64 + 1.34)}{0.1 \times 0.8} = 5.25'' \quad \text{(use 6'' (152.4 mm))}$$

$SN_3^* = 2.64 + 1.34 + 6 \times 0.8 \times 0.1 = 4.46$

where * denotes actual values used.

The Federal Aviation Administration Design Method for Flexible Airport Runway Pavements

The design inputs in this procedure are the CBR value for the subgrade material, the CBR value for the subbase material, the gross weight of the design aircraft, and the number of departures of the design aircraft. Although this design method is basically empirical, it is based on extensive research, and reliable correlations have been developed. The FAA has developed generalized design curves that apply to families of aircraft for determining the required total pavement thickness and the thickness of the hot-mix asphalt surfacing, for single, dual, and tandem main-gear landing assemblies as shown in Figures 7.16, 7.17, and 7.18. It has also developed design curves for specific aircraft examples of charts given in Figures 7.19, 7.20, and 7.21. A minimum base course thickness is also specified for each family of aircraft and for each specific aircraft, as shown in Table 7.29. The thicknesses provided on these charts are adequate for annual departures of 25,000 or less and should be adjusted by the percentages shown in Table 7.30 for

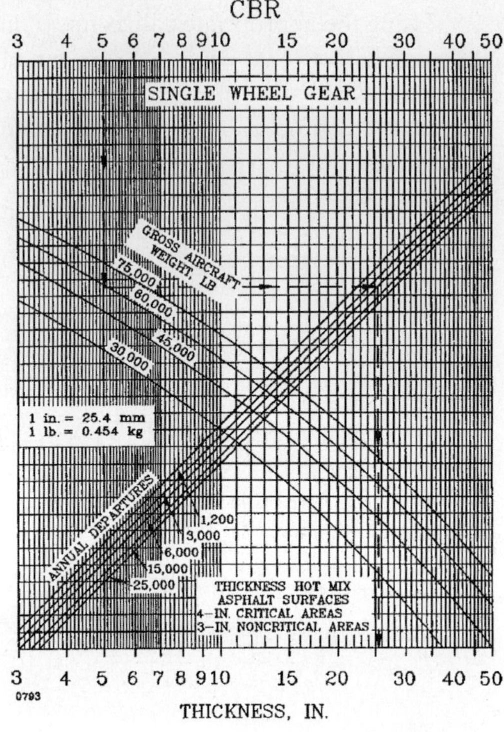

FIGURE 7.16

Design chart for airport flexible pavements serving single-wheel gear airplanes.

Source: Airport Pavement Design and Evaluation, Advisory Circular AC 150/5320-6D (Incorporating Changes 1 through 5), Federal Aviation Administration, Department of Transportation, Washington, D.C., April, 2004.

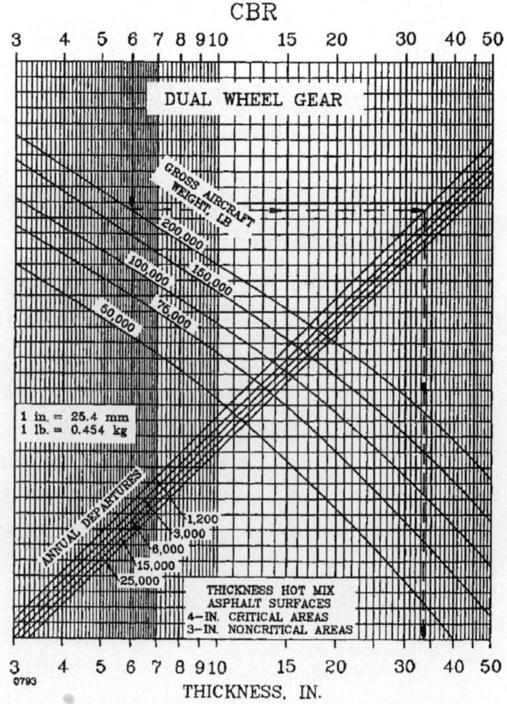

FIGURE 7.17

Design charts for airport flexible pavements serving dual-wheel gear airplanes.

Source: Airport Pavement Design and Evaluation, Advisory Circular AC 150/5320-6D (Incorporating Changes 1 through 5), Federal Aviation Administration, Department of Transportation, Washington, D.C., April, 2004.

FIGURE 7.18

Design charts for airport flexible pavements serving dual-tandem gear airplanes.

Source: Airport Pavement Design and Evaluation, Advisory Circular AC 150/5320-6D (Incorporating Changes 1 through 5), Federal Aviation Administration, Department of Transportation, Washington, D.C., April, 2004.

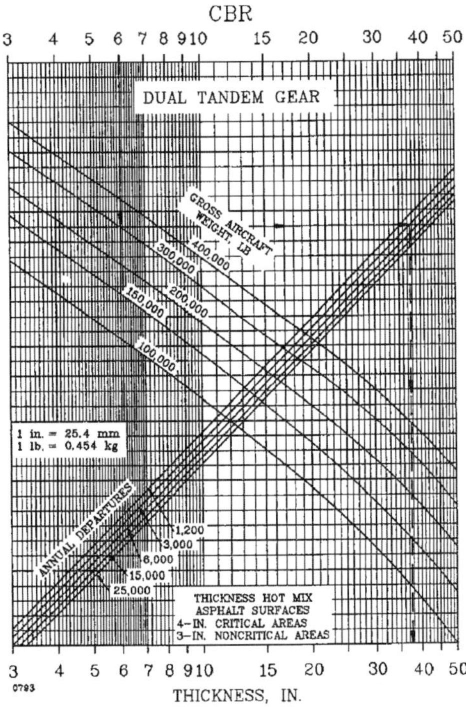

FIGURE 7.19

Design charts for airport flexible pavements serving A-300 Model B2 airplanes.

Source: Airport Pavement Design and Evaluation, Advisory Circular AC 150/5320-6D (Incorporating Changes 1 through 5), Federal Aviation Administration, Department of Transportation, Washington, D.C., April, 2004.

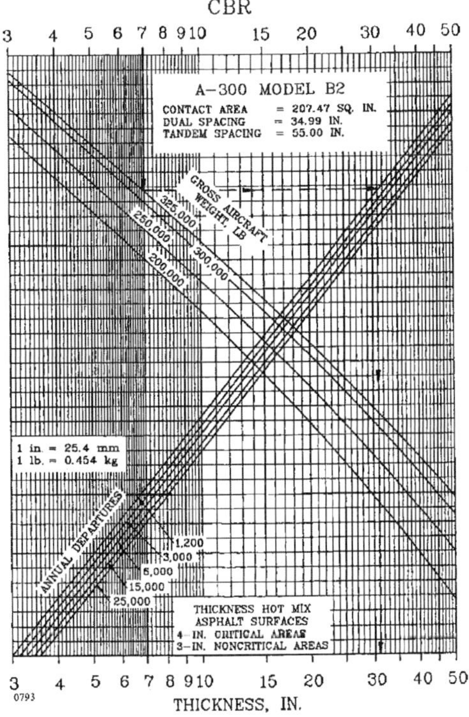

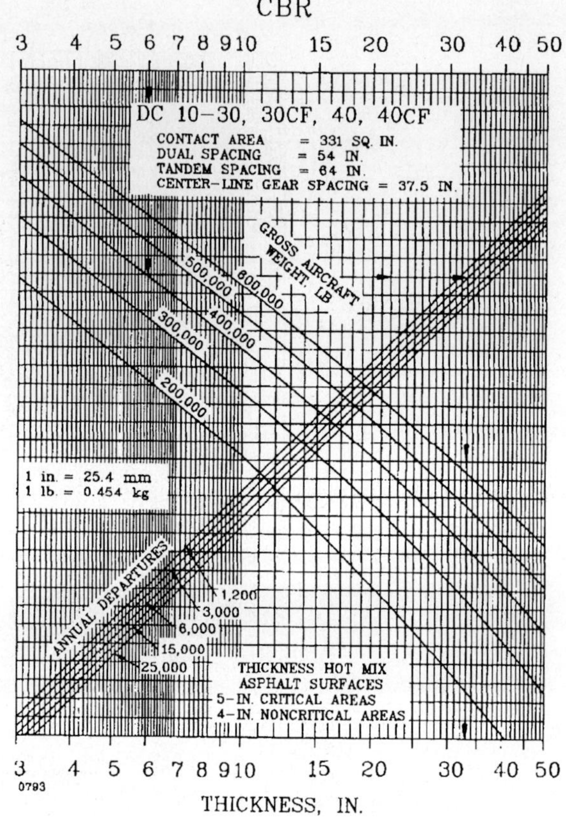

FIGURE 7.20
Design charts for airport flexible pavements serving DC 10-30, 30 CF, 40, and 40 CF airplanes.

Source: Airport Pavement Design and Evaluation, Advisory Circular AC 150/5320-6D (Incorporating Changes 1 through 5), Federal Aviation Administration, Department of Transportation, Washington, D.C., April, 2004.

departures greater than 25,000. One inch (25.4 mm) of the thickness increase should be hot-mix asphalt surfacing and the remainder should be proportioned between the base and subbase courses. Although the thicknesses given by these charts are based on the use of subbase materials having the quality given in Table 7.8, the design procedure also provides for use of higher-quality materials. When higher-quality materials are used, the equivalent thicknesses are obtained by dividing the thicknesses obtained from the charts by the equivalency factors shown in Table 7.31. Also, note that although airport pavements are usually constructed in uniform full-depth sections, they may sometimes be constructed with a transversely variable section, which permits a reduction of the total thickness (T) at noncritical areas. The critical section is the area used by departing traffic; the noncritical area is the area where traffic is arrival, such as high-speed turn-offs, and the edge, where traffic is unlikely; for example, along the outer edges of the runway. As a general rule of thumb, the full thickness (T) obtained from the chart is specified for the critical area, a thickness of $0.9T$ is specified for the noncritical area (arrival area) and a thickness of $0.7T$ is used for the outer edge of the pavement. Note, however, that the thickness of

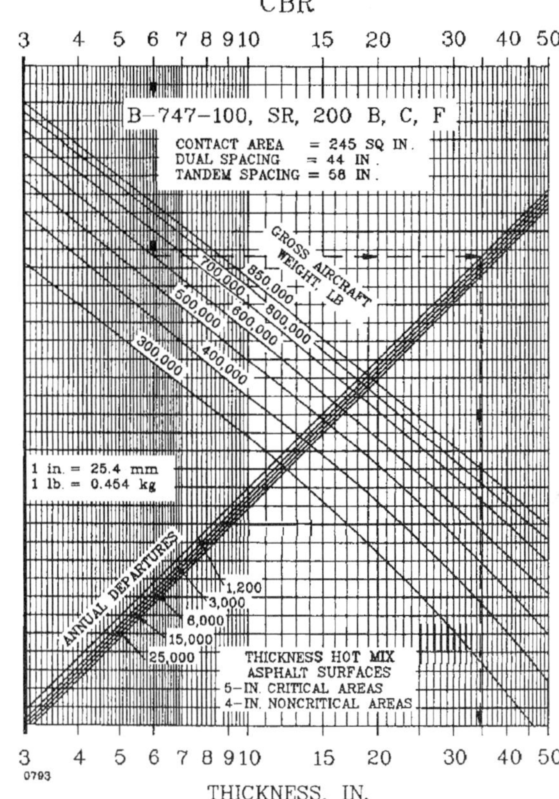

FIGURE 7.21

Design charts for airport flexible pavements serving B-747-100, SR, and 200 B, C, and F airplanes.

Source: Airport Pavement Design and Evaluation, Advisory Circular AC 150/5320-6D (Incorporating Changes 1 through 5), Federal Aviation Administration, Department of Transportation, Washington, D.C., April, 2004.

TABLE 7.29

Minimum Base Course Thickness for Flexible Airport Runway Pavements

Source: Airport Pavement Design and Evaluation, Advisory Circular AC No 150/5320 – 6D, Federal Aviation Administration, U.S. Department of Transportation, (Incorporating Changes 1 through 5), Washington D.C., April 2004.

Design Aircraft	Design Load Range		Minimum Base Course Thickness	
	lbs.	(kg)	in.	(mm)
Single Wheel	30,000 - 50,000	(13600 - 22 700)	4	(100)
	50,000 - 75,000	(22700 - 34 000)	6	(150)
Dual Wheel	50,000 - 100,000	(22700 - 45 000)	6	(150)
	100,000 - 200,000	(45 000 - 90 700)	8	(200)
Dual Tandem	100,000 - 250,000	(45 000 - 113 400)	6	(150)
	250,000 - 400,000	(113400 - 181 000)	8	(200)
757 767	200,000 - 400,000	(90700 - 181000)	6	(150)
DC-10 L1011	400,000 - 600,000	(181 000 - 272000)	8	(200)
B-747	400,000 - 600,000	(181 000 - 272000)	6	(150)
	600,000 - 850,000	(272 000 - 385 700)	8	(200)
c-130	75,000 - 125,000	(34 000 - 56 700)	4	(100)
	125,000 - 175,000	(56700 - 79 400)	6	(150)

Note: The calculated base course thickness should be compared with the minimum base course thickness listed above. The greater thickness, calculated or minimum, should be specified in the design section.

TABLE 7.30

Percentages for Pavement Thickness Adjustment for High Departure Levels

Source: Airport Pavement Design and Evaluation, Advisory Circular AC No 150/5320 – 6D, Federal Aviation Administration, U.S. Department of Transportation, (Incorporating Changes 1 through 5), Washington D.C., April 2004.

Annual Departure Level	*Percent of 25,000 Departure Thickness*
50,000	104
100,000	108
150,000	110
200,000	112

the surface course obtained from the design should be used throughout the pavement width. The factor of $0.9T$ for the noncritical area should be applied only to the base and subbase courses and that of $0.7T$ for the edge to only the base course.

TABLE 7.31

Recommended Equivalency Factor Ranges High Quality Base and Subbase

Source: Airport Pavement Design and Evaluation, Advisory Circular AC No 150/5320 – 6D, Federal Aviation Administration, U.S. Department of Transportation, (Incorporating Changes 1 through 5), Washington D.C., April 2004.

Material	Equivalency Factor Range
P-208, Aggregate Base Course	1.0 - 1.5
P-209, Crushed Aggregate Base Course	1.2 - 1.8
P-2 I 1, Lime Rock Base Course	1.0 - 1.5

(a) Granular Subbase

Material	Equivalency Factor Range
P-301, Soil Cement Base Course	1.0 - 1.5
P-304, Cement Treated Base Course	1.6 - 2.3
P-306, Econocrete Subbase Course	1.6 - 2.3
P-401, Plant Mix Bituminous Pavements	1.7 - 2.3

(b) Stabilized Subbase

Material	Equivalency Factor Range
P-208, Aggregate Base Course	1.0'
P-21 1, Lime Rock Base Course	1.0

'Substitution of P-208 for P-209 is permissible only if the gross weight of the design aircraft is 60,000 lbs (27 000 kg) or less. In addition, if P-208 is substituted for P-209, the required thickness of hot-mix asphalt surfacing shown on the design curves should be increased 1 inch (25 mm).

(c) Granular Base

Material	Equivalency Factor Range
P-304, Cement Treated Base Course	1.2 - 1.6
P-306, Econocrete Subbase Course	1.2 - 1.6
P-401, Plant Mix Bituminous Pavements	1.2 - 1.6

Note: Reflection cracking may be encountered when P-304 or P-306 is used as base for a flexible pavement. The thickness of the hot-mix asphalt surfacing course should be at least 4 inches (100 mm) to minimize reflection cracking in these instances.

(d) Stabilized Base

The procedure for using the design curves consists of the following steps:

(i) Determine design aircraft.

(ii) Determine equivalent annual departures based on design aircraft.

(iii) Determine total pavement thickness required based on the equivalent annual departures by the design aircraft and the CBR of the subgrade.

(iv) Determine total thickness required based on the equivalent annual departures by the design aircraft and the CBR of the subbase. This gives the combined thickness of hot-mix asphalt and the base course needed above the subbase. Subtract this thickness value from that obtained from (i) to obtain the thickness of the subbase.

(v) Select minimum thickness of hot-mix asphalt thickness. Note that the minimum thickness of hot-mix asphalt for critical areas is 4″ (100 mm) and for noncritical 3″ (76 mm). Determine the minimum thickness of the base course by subtracting the thickness of the hot-mix asphalt layer from that obtained for the combined thickness of the hot-mix asphalt surface and the subbase. Compare the thickness obtained for the base course layer with the minimum required as given in Table 7.28. Use the greater value as the required thickness of the base course.

(vi) Adjust each thickness obtained for high departure levels (i.e., for annual departure levels higher than 25,000) using percentages given in Table 7.29.

Note that the thickness (T) obtained from the charts should be rounded up to the higher whole number for fractions of 0.5″ or more, and be rounded down to the next lower whole number for fractions less than 0.5″.

EXAMPLE 7.10

Designing a Flexible Airport Runway Pavement

If the average annual departures and maximum take-off weight of each aircraft type expected to use an airport runway are those given in Example 7.3 and as shown in the following table, determine the minimum thickness for each of the hot mix asphalt surface course, the base course, and the subbase course. The CBR values of the subbase and subgrade are 20 and 6, respectively.

Aircraft	*Gear Type*	*Average Annual Departures*	*Max. Take-Off Weight*
727-100	Dual	3500	68038 kg (150,000 lb)
727-200	Dual	9100	86409 kg (190,500 lb)
707-320B	Dual tandem	3000	148324 kg (327,000 lb)
DC-10-30	Dual	5800	48988 kg (108,000 lb)
737-200	Dual	2650	52390 kg (115,500 lb)
747-100	Double dual tandem	80	317513 kg (700,000 lb)

Solution

Determine the total pavement thickness required for each aircraft type for the associated average annual departures, using the appropriate figure.

727-100 dual gear = 34″ (863.6 mm) from Figure 7.17

727-200 dual gear = 40″ (1016 mm) from Figure 7.17

707-320B dual tandem = 38″ (965.2 mm) from Figure 7.18

DC-10-30 dual = 19″ (482.6 mm) from Figure 7.20

747-100 double dual tandem = 30″ (762 mm) from Figure 7.21

Determine the design aircraft. The greatest thickness is for the 727-200 and is therefore the design aircraft for this airport runway pavement. The assumption made in Problem 7.3 is therefore correct.

Determine the equivalent annual departures based on design aircraft. This has been done in Problem 7.3 and the results are repeated here:

Aircraft	*Equivalent Dual Gear Departures (R_2)*	*Wheel Load (W_2)*	*Wheel Load of Design Aircraft (lb) (W_1)*	*Equivalent Annual Departures for Design Aircraft (R_1)*
727-100	3500	(35,625 lb) 16159 kg	45,244	1396
727-200	9100	(45,244 lb) 20522 kg	45,244	9100
707-320B	5100	(38,831 lb) 17613 kg	45,244	2721
DC-10-30	5800	(25,650 lb) 11635 kg	45,244	682
737-200	2650	(27,431 lb) 12442 kg	45,244	462
747-100	136	(35,625 lb) 16159 kg	45,244	78

Total annual departures based on design aircraft = (1396 + 9100 + 2721 + 682 + 462 + 78)

= 14,439

Determine the total pavement thickness required for a 727-200 (dual-wheel gear) with an annual departure of 14,500 and a gross weight of 190,500 lb, using Figure 7.17:

Total pavement thickness = 40″ (1016 mm)

Determine the total thickness based on the CBR value of 20 for the subbase, using Figure 7.17:

Total thickness required over the subbase = 18″ (457.2 mm)

Determine the thickness of subbase:

Thickness of subbase = (40 − 18) = 22″ (558.8 mm)

Determine the thickness of the base course:

Thickness of base course = (thickness above base course − minimum required hot-mix asphalt surface)

Minimum required hot-mix asphalt = 4″ (101.6 mm) (from Figure 7.17)

Thickness of base course = (18 − 4)″ = 14″ (355.6 mm)

Compare the thickness of base course with minimum required:

Minimum required base course thickness = 8″ (203.2 mm) (from Table 7.29)

Computed value is higher than minimum required.

Adjust for high departure levels. No adjustment is required as equivalent annual departure level is less than 25,000.

The thickness requirements for this design are

Hot-mix asphalt surface thickness = 4″ (101.6 mm)

Base course thickness = 14″ (355.6 mm)

Subbase thickness = 22″ (558.8 mm)

Design of Rigid Highway and Airport Runway Pavements

Rigid pavements can be divided into four general types: jointed plain concrete pavements, jointed reinforced concrete pavements, continuously reinforced concrete pavements, and prestressed concrete pavements. Prestressed concrete pavements are not discussed as they are outside the scope of this book, but interested readers may refer to the AASHTO *Guide for the Design of Pavement Structures*. The amount of reinforcement used in the pavement determines its type. The reinforcement does not prevent cracking, but it keeps the cracks that are formed tightly closed, such that the structural integrity of the slab is maintained by the interlocking of the irregular faces of the coarse aggregates. The determination of the thickness of the slab is the same for all pavement types when using the design procedures described below.

Jointed plain concrete pavements do not have any steel or dowels for load transfer. These pavements are mainly used on low-volume highways or when cement-stabilized soils are used as the subbase. In order to reduce the amount of cracking in these pavements, transverse joints are provided at relatively shorter distances than those for other types, usually at distances between 10 and 20 ft (254 mm and 508 mm). The transverse joints of plain concrete are sometimes skewed so that one wheel of a vehicle passes through the joint at a time, which enhances a smooth drive.

Jointed reinforced concrete pavements have dowels for the transfer of traffic loads across the joints, with these joints spaced at larger distances, ranging from 30 to 100 ft (9.1 m to 30.5 m). Tie bars are often commonly used across the longitudinal joints and temperature steel is placed throughout the slab. The amount of temperature steel used depends on the length of the slab.

Continuously reinforced concrete pavements have no joints, except construction joints or expansion joints when they are necessary at specific locations such as at bridges. A relatively high percentage of steel is used in these pavements, with at least a longitudinal steel cross-sectional area of 0.6% of highway slab cross-sections and between 0.5 and 1% of airport runway slab cross-sections.

AASHTO Method for the Design of Highway Rigid Pavements

The AASHTO method for the design of rigid highway pavements is also based on the results obtained from the AASHTO road test that was conducted in Ottawa, Illinois. The method is presented by first discussing the specific factors used in the procedure and then the equation for determining the thickness of the pavement is presented.

Factors Used in AASHTO'S Method for the Design of Rigid Pavements The *AASHTO Design Factors for Rigid Highway Pavements* include pavement performance, the design equivalent axle load, resilient modulus (k) of the subgrade, quality of the base and surface materials, the environment, drainage, and reliability.

Pavement performance is also based on the structural and functional performance of the pavement. In this case, however, AASHTO recommends the use of 4.5 for the initial serviceability index (P_i) for a new rigid pavement and 2.5 for the terminal serviceability index (P_t), although the designer is free to select a different value.

The design ESAL is computed in a way similar to that for flexible pavements in that the traffic load application is given in terms of the number of 8.16×10^3 kg (18,000-lb) equivalent axle loads (*ESALs*). However, in this method, the ESAL factors depend on the slab thickness and the terminal serviceability index of the pavement. Tables 7.32 and 7.33 give ESAL factors for rigid pavements with a terminal serviceability index of 2.5. Since the ESAL factors depend on the thickness of the slab, it is necessary to assume the thickness of the slab at the start of the computation. This assumed value is used to compute the number of accumulated ESALs, which in turn is used to compute the required thickness. If this is significantly different from the assumed thickness, the accumulated ESALs should be recomputed. This procedure is repeated until the assumed and computed thicknesses are approximately the same.

The strength characteristic of the subgrade used in the design of rigid pavements is the Westergaard modulus of subgrade reaction (k). It is, however, necessary to determine the effective value of k as this value depends on several

TABLE 7.32

ESAL Factors for Rigid Pavements, Single Axle, P_t of 2.5

Source: Adapted with permission from *AASHTO Guide for Design of Pavement Structures,* American Association of State Highway and Transportation Officials, Washington, D.C., 1993. Used by permission.

	Slab Thickness, D (in.)								
Axle Load (kip)	*6*	*7*	*8*	*9*	*10*	*11*	*12*	*13*	*14*
2	.0002	.0002	.0002	.0002	.0002	.0002	.0002	.0002	.0002
4	.003	.002	.002	.002	.002	.002	.002	.002	.002
6	.012	.011	.010	.010	.010	.010	.010	.010	.010
8	.039	.035	.033	.032	.032	.032	.032	.032	.032
10	.097	.089	.084	.082	.081	.080	.080	.080	.080
12	.203	.189	.181	.176	.175	.174	.174	.173	.173
14	.376	.360	.347	.341	.338	.337	.336	.336	.336
16	.634	.623	.610	.604	.601	.599	.599	.599	.598
18	1.00	1.00	1.00	1.00	1.00	1.00	1.00	1.00	1.00
20	1.51	1.52	1.55	1.57	1.58	1.58	1.59	1.59	1.59
22	2.21	2.20	2.28	2.34	2.38	2.40	2.41	2.41	2.41
24	3.16	3.10	3.22	3.36	3.45	3.50	3.53	3.54	3.55
26	4.41	4.26	4.42	4.67	4.85	4.95	5.01	5.04	5.05
28	6.05	5.76	5.92	6.29	6.61	6.81	6.92	6.98	7.01
30	8.16	7.67	7.79	8.28	8.79	9.14	9.35	9.46	9.52
32	10.8	10.1	10.1	10.7	11.4	12.0	12.3	12.6	12.7
34	14.1	13.0	12.9	13.6	14.6	15.4	16.0	16.4	16.5
36	18.2	16.7	16.4	17.1	18.3	19.5	20.4	21.0	21.3
38	23.1	21.1	20.6	21.3	22.7	24.3	25.6	26.4	27.0
40	29.1	26.5	25.7	26.3	27.9	29.9	31.6	32.9	33.7
42	36.2	32.9	31.7	32.2	34.0	36.3	38.7	40.4	41.6
44	44.6	40.4	38.8	39.2	41.0	43.8	46.7	49.1	50.8
46	54.5	49.3	47.1	47.3	49.2	52.3	55.9	59.0	61.4
48	66.1	59.7	56.9	56.8	58.7	62.1	66.3	70.3	73.4
50	79.4	71.7	68.2	67.8	69.6	73.3	78.1	83.0	87.1

Note: 1 inch = 25.4 mm
1 kip = 4.5 kN

different factors, such as (1) the seasonal effect of the resilient modulus of the subgrade; (2) the elastic modulus and thickness of the subbase; (3) whether bedrock lies within 3 m of the subgrade surface; and (4) the effect of potential erosion of the subbase. Detailed discussion on the methodology to determine the effective value of k is beyond the scope of this book as geotechnical engineers normally carry out this task. Interested readers may refer to the AASHTO *Guide for the Design of Pavement Structures.* A brief description of the methodology is, however, presented to facilitate the understanding of the overall AASHTO design method.

The procedure for adjusting for seasonal effect is similar to that for flexible pavements. The year is therefore divided into time intervals and an appropriate value of M_r used for each time interval. AASHTO suggests that a division of less than one half-month for any given season is not necessary, as shown in Table 7.34. Similarly, it is necessary to obtain seasonal elastic moduli (E_{SB}) for the subbase corresponding to the selected time intervals.

TABLE 7.33

ESAL Factors for Rigid Pavements, Tandem Axle, P_t of 2.5

Source: Adapted with permission from *AASHTO Guide for Design of Pavement Structures,* American Association of State Highway and Transportation Officials, Washington, D.C., 1993. Used by permission.

	Slab Thickness, D (in.)								
Axle Load (kip)	*6*	*7*	*8*	*9*	*10*	*11*	*12*	*13*	*14*
2	.0001	.0001	.0001	.0001	.0001	.0001	.0001	.0001	.0001
4	.0006	.0006	.0005	.0005	.0005	.0005	.0005	.0005	.0005
6	.002	.002	.002	.002	.002	.002	.002	.002	.002
8	.007	.006	.006	.005	.005	.005	.005	.005	.005
10	.015	.014	.013	.013	.012	.012	.012	.012	.012
12	.031	.028	.026	.026	.025	.025	.025	.025	.025
14	.057	.052	.049	.048	.047	.047	.047	.047	.047
16	.097	.089	.084	.082	.081	.081	.080	.080	.080
18	.155	.143	.136	.133	.132	.131	.131	.131	.131
20	.234	.220	.211	.206	.204	.203	.203	.203	.203
22	.340	.325	.313	.308	.305	.304	.303	.303	.303
24	.475	.462	.450	.444	.441	.440	.439	.439	.439
26	.644	.637	.627	.622	.620	.619	.618	.618	.618
28	.855	.854	.852	.850	.850	.850	.849	.849	.849
30	1.11	1.12	1.13	1.14	1.14	1.14	1.14	1.14	1.14
32	1.43	1.44	1.47	1.49	1.50	1.51	1.51	1.51	1.51
34	1.82	1.82	1.87	1.92	1.95	1.96	1.97	1.97	1.97
36	2.29	2.27	2.35	2.43	2.48	2.51	2.52	2.52	2.53
38	2.85	2.80	2.91	3.03	3.12	3.16	3.18	3.20	3.20
40	3.52	3.42	3.55	3.74	3.87	3.94	3.98	4.00	4.01
42	4.32	4.16	4.30	4.55	4.74	4.86	4.91	4.95	4.96
44	5.26	5.01	5.16	5.48	5.75	5.92	6.01	6.06	6.09
46	6.36	6.01	6.14	6.53	6.90	7.14	7.28	7.36	7.40
48	7.64	7.16	7.27	7.73	8.21	8.55	8.75	8.86	8.92
50	9.11	8.50	8.55	9.07	9.68	10.14	10.42	10.58	10.66
52	10.8	10.0	10.0	10.6	11.3	11.9	12.3	12.5	12.7
54	12.8	11.8	11.7	12.3	13.2	13.9	14.5	14.8	14.9
56	15.0	13.8	13.6	14.2	15.2	16.2	16.8	17.3	17.5
58	17.5	16.0	15.7	16.3	17.5	18.6	19.5	20.1	20.4
60	20.3	18.5	18.1	18.7	20.0	21.4	22.5	23.2	23.6
62	23.5	21.4	20.8	21.4	22.8	24.4	25.7	26.7	27.3
64	27.0	24.6	23.8	24.4	25.8	27.7	29.3	30.5	31.3
66	31.0	28.1	27.1	27.6	29.2	31.3	33.2	34.7	35.7
68	35.4	32.1	30.9	31.3	32.9	35.2	37.5	39.3	40.5
70	40.3	36.5	35.0	35.3	37.0	39.5	42.1	44.3	45.9
72	45.7	41.4	39.6	39.8	41.5	44.2	47.2	49.8	51.7
74	51.7	46.7	44.6	44.7	46.4	49.3	52.7	55.7	58.0
76	58.3	52.6	50.2	50.1	51.8	54.9	58.6	62.1	64.8
78	65.5	59.1	56.3	56.1	57.7	60.9	65.0	69.0	72.3
80	73.4	66.2	62.9	62.5	64.2	67.5	71.9	76.4	80.2
82	82.0	73.9	70.2	69.6	71.2	74.7	79.4	84.4	88.8
84	91.4	82.4	78.1	77.3	78.9	82.4	87.4	93.0	98.1
86	102.0	92.0	87.0	86.0	87.0	91.0	96.0	102.0	108.0
88	113.0	102.0	96.0	95.0	96.0	100.0	105.0	112.0	119.0
90	125.0	112.0	106.0	105.0	106.0	110.0	115.0	123.0	130.0

Note: 1 inch = 25.4 mm
1 kip = 4.5 kN

TABLE 7.34

Table for Estimating Effective Modulus of Subgrade Reaction

Source: AASHTO Guide for Design of Pavement Structures, American Association of State Highway and Transportation Officials, 1993. Used by permission.

Trial Subbase: Type ____________ Depth to Rigid Foundation (feet) ________

Thickness (inches) ________ Projected Slab Thickness (inches) ________

Loss of Support, LS ________

(1) Month	*(2)* Roadbed Modulus M_r (psi)	*(3)* Subbase Modulus E_{SB} (psi)	*(4)* Composite k-Value (pci) (Fig. 7.26)	*(5)* k-Value (pci) on Rigid Foundation (Fig. 7.27)	*(6)* Relative Damage, u_r (Fig. 7.28)
Jan.					
Feb.					
Mar.					
Apr.					
May					
June					
July					
Aug.					
Sept.					
Oct.					
Nov.					
Dec.					
				Summation: $\sum u_r =$	

Average: $\bar{u}_r = \dfrac{\sum u_r}{n} =$ ________

Effective Modulus of Subgrade Reaction, k(pci) = ________

Corrected for Loss of Support: k(pci) = ________

Note: 1000 psi = 6.9 N/mm^2, 1000 pci = 0.272 N/mm^3

Assuming a semi-infinite depth (greater than 3 m) of the subgrade, a composite modulus of subgrade reaction is then determined for each season, based on the elastic modulus of the subbase, the depth of the subbase, and the resilient modulus of the subgrade, using the chart shown in Figure 7.22.

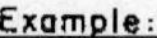

Example:

D_{SB} = 6 inches

E_{SB} = 20,000 psi

M_R = 7,000 psi

Solution: k_∞ = 400 pci

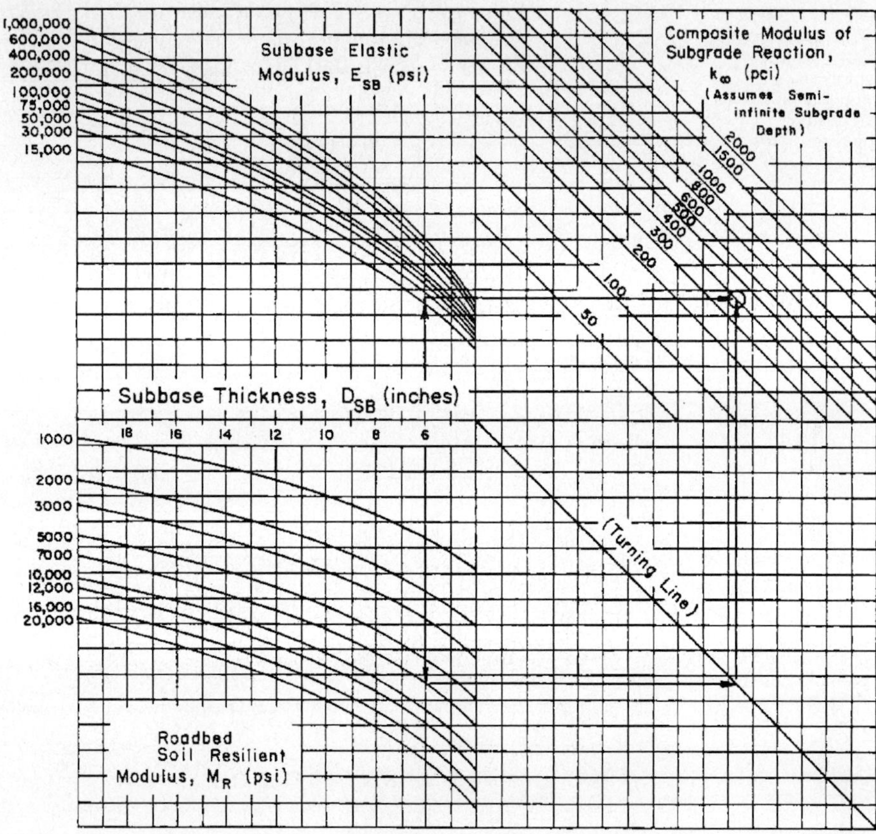

Note: 1 inch = 25.4 mm, 1000 psi = 6.9 N/mm², 1000 pci = 0.272 N/mm³

FIGURE 7.22

Chart for estimating composite modulus of subgrade reaction k_∞, assuming a semi-infinite subgrade depth.

Source: AASHTO Guide for Design of Pavement Structures, American Association of State Highway and Transportation Officials, Washington, D.C., 1993. Used by permission.

It should be noted that a subbase thickness is required for this chart to be used. In cases where there is no subbase (i.e., the concrete slab is placed directly over the subbase), the composite modulus of subgrade (k_c) reaction is obtained from the elastic modulus of the subgrade (M_r), using the theoretical expression:

$$k_c^{(in\ pci)} = M_r^{(in\ psi)}/19.4$$

Also, the presence of bedrock within 3 m of the subgrade surface and extending over a significant length along the highway alignment may result in an increase of the overall modulus of subgrade reaction. This effect is taken into consideration by adjusting the effective modulus of subgrade reaction using the chart shown in Figure 7.23. Using the assumed thickness of the slab,

FIGURE 7.23

Chart to modify modulus of subgrade reaction to consider effects of rigid foundation near surface (within 10ft).

Source: AASHTO Guide for Design of Pavement Structures, American Association of State Highway and Transportation Officials, Washington, D.C., 1993. Used by permission.

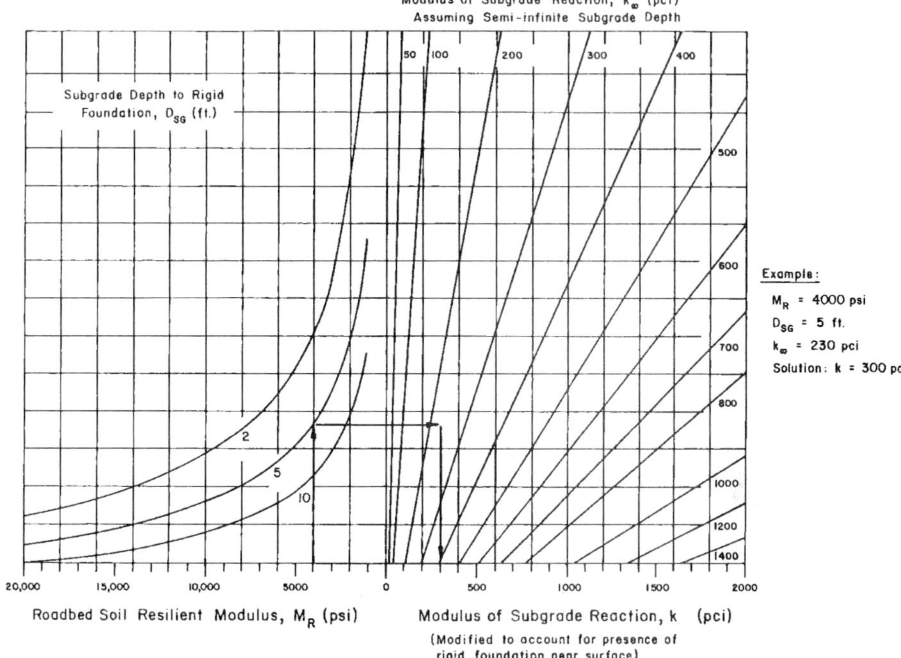

Note: 1 ft = 304.8 mm, 1000 psi = 6.9 N/mm^2, 1000 pci = 0.272 N/mm^3

the relative damage for each season is determined using Figure 7.24. The mean of the relative damages for all seasons is then determined and used to obtain the effective modulus of subgrade reaction from Figure 7.24.

The effect of potential erosion of the subbase is then considered by incorporating a loss of support (LS) factor to take into consideration the potential of loss of support due to erosion of the subbase and/or differential vertical movements of the soil. This factor depends on the type of material used as the subbase and its elastic or resilient modulus, as shown in Table 7.35. The LS value increases with an increase in the potential of the subbase to erode, resulting in a higher reduction of the effective modulus of subgrade reaction, as shown in Figure 7.25.

The effect of drainage on the performance of rigid pavements is considered by incorporating a drainage coefficient (C_d) in the equation used for design. This factor is based on the drainage quality of the subbase material, which depends on the time it takes to drain the subbase layer to 50% of saturation, and the length of time during which the pavement structure will be nearly saturated. Table 7.24 gives the general definition of different levels of drainage quality, and Table 7.36 gives AASHTO-recommended values for C_d.

The same procedure used for reliability in the flexible pavement design procedure is used for the rigid pavement design. Reliability factors for rigid pavements are the same as for flexible pavements. AASHTO, however, recommends an overall standard deviation range of 0.30–0.40 for rigid pavements.

FIGURE 7.24

Chart for estimating relative damage to rigid pavements based on slab thickness and underlying support.

Source: AASHTO Guide for Pavement Structures, American Association of State Highway and Transportation Officials, Washington, D.C., 1993. Used by permission.

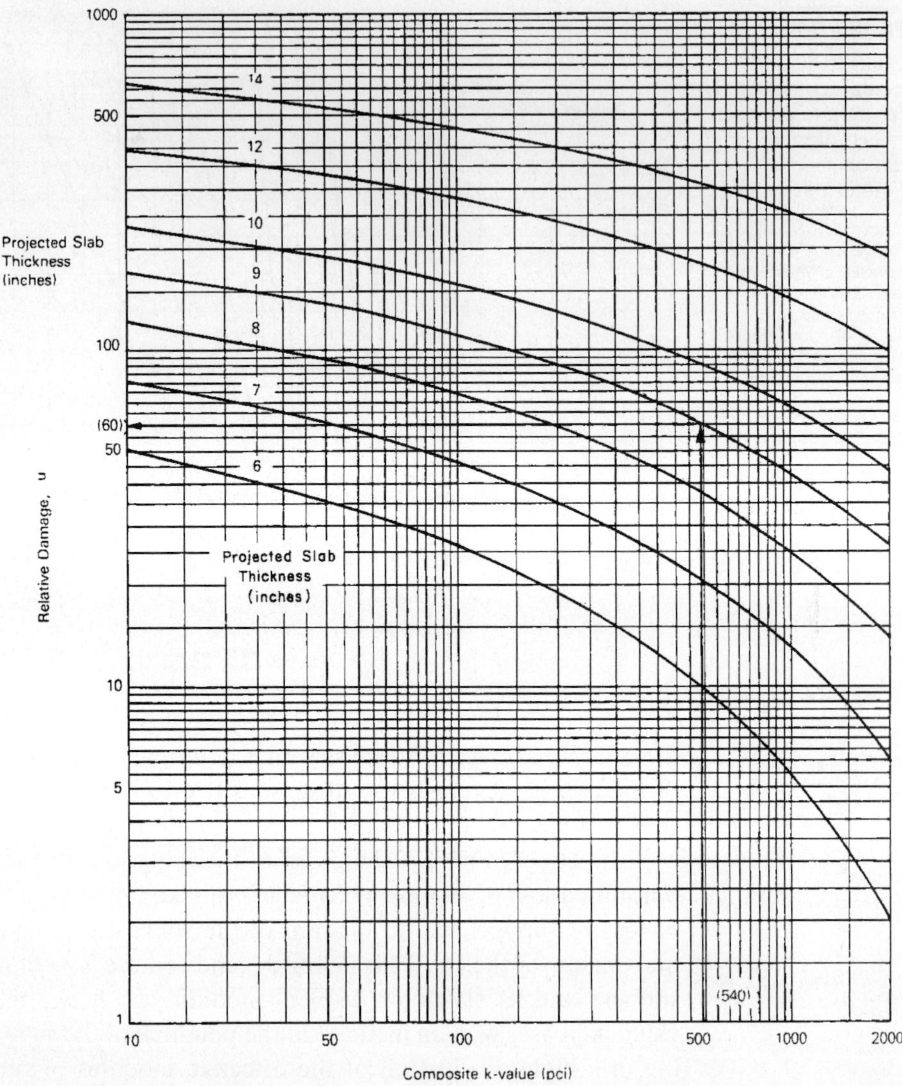

Note: 1 inch = 25.4 mm, 1000 pci = 0.272 N/mm³

EXAMPLE 7.11

Computing Effective Modulus of Subgrade Reaction for a Rigid Pavement Using the AASHTO Method

An 8-″(203.2 mm) layer of cement-treated granular material is to be used as subbase for a rigid pavement. The seasonal values for the roadbed resilient modulus and the subbase elastic modulus are given in columns 2 and 3 of Table 7.37. If the rock depth is located 5 ft (1524 mm) from the subgrade surface and the projected slab thickness is 8″(203.2 mm), estimate the effective modulus of

TABLE 7.35

Typical Range of Loss of Support Factors for Various Types of Materials

Source: Adapted from B.F. McCullough and Gary E. Elkins, *CRC Pavement Design Manual*, Austin Research Engineers, Inc., Austin, TX, October 1979.

Type of Material	*Loss of Support (LS)*
Cement-treated granular base (E = 1,000,000 to 2,000,000 lb/in.^{2})	0.0 to 1.0
Cement aggregate mixtures (E = 500,000 to 1,000,000 lb/in.^{2})	0.0 to 1.0
Asphalt-treated base (E = 350,000 to 1,000,000 lb/in.^{2})	0.0 to 1.0
Bituminous stabilized mixtures (E = 40,000 to 300,000 lb/in.^{2})	0.0 to 1.0
Lime-stabilized mixtures (E = 20,000 to 70,000 lb/in.^{2})	1.0 to 3.0
Unbound granular materials (E = 15,000 to 45,000 lb/in.^{2})	1.0 to 3.0
Fine-grained or natural subgrade materials (E = 3,000 to 40,000 lb/in.^{2})	2.0 to 3.0

Note:
E in this table refers to the general symbol for elastic or resilient modulus of the material.
1000 lb/in^2 = 6.9 N/mm^2
1000 lb/in^3 = 0.272 N/mm^3

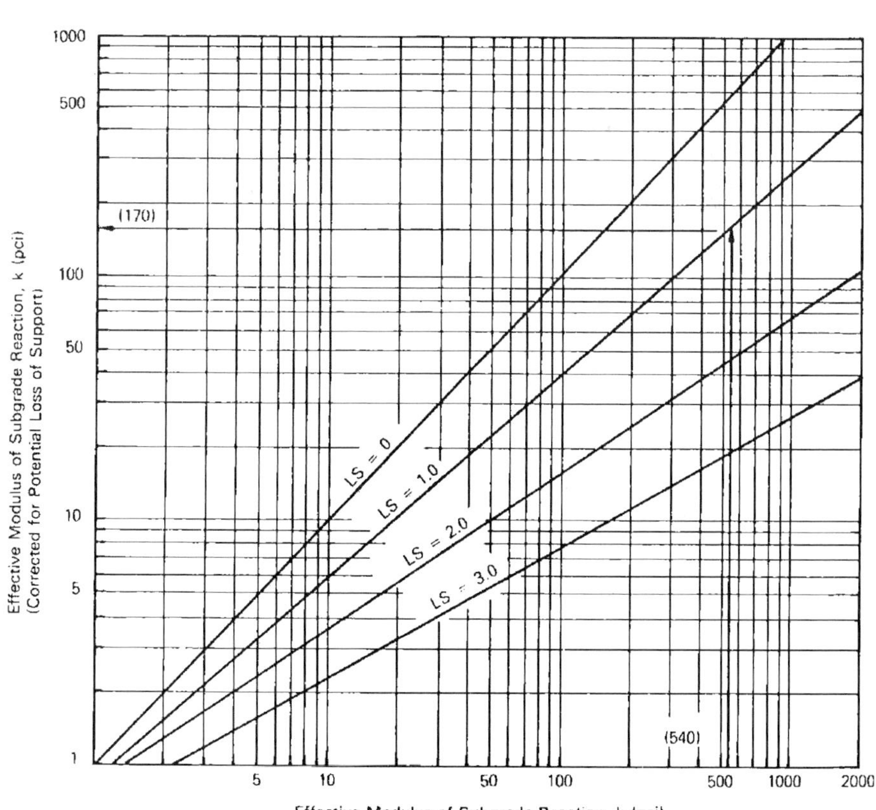

Note: 1000 pci = 0.272 N/mm^3

FIGURE 7.25

Correction of effects modulus of subgrade reaction for potential loss of support

Source: AASHTO Guide for Pavement Structures, American Association of State Highway and Transportation Officials, Washington, D.C., 1993. Used by permission.

TABLE 7.36

Recommended Values for Drainage Coefficient C_d for Rigid Pavements

Source: Adapted from *AASHTO Guide for Design of Pavement Structures,* American of State Highway and Transportation Offices, Washington, D.C., 1973. Used by permission.

	Percent of Time Pavement Structure is Exposed to Moisture Levels Approaching Saturation			
Quality of Drainage	*Less Than 1 Percent*	*1–5 Percent*	*5–25 Percent*	*Greater Than 25 Percent*
Excellent	1.25–1.20	1.20–1.15	1.15–1.10	1.10
Good	1.20–1.15	1.15–1.10	1.10–1.00	1.00
Fair	1.15–1.10	1.10–1.00	1.00–0.90	0.90
Poor	1.10–1.00	1.00–0.90	0.90–0.80	0.80
Very poor	1.00–0.90	0.90–0.80	0.80–0.70	0.70

TABLE 7.37

Illustrative Example for Determining Effective Modulus of Subgrade Reaction

(1) Month	*(2) Roadbed Modulus, M_r (psi)*	*(3) Subbase Modulus, E_{SB} (psi)*	*(4) Composite k_∞ Value (pci) (Figure 7.22)*	*(5) k Value (pci) on Rigid Foundation (Figure 7.23)*	*(6) Relative Damage, u_r (Figure 7.24)*
January	20,000	50,000	1100	1350	0.20
	20,000	50,000	1100	1350	0.20
February	20,000	50,000	1100	1350	0.20
	20,000	50,000	1100	1350	0.20
March	3000	20,000	190	290	0.50
	3000	20,000	190	290	0.50
April	4000	20,000	260	370	0.45
	4000	20,000	260	370	0.45
May	4000	20,000	260	370	0.45
	4000	20,000	260	370	0.45
June	8000	25,000	500	810	0.28
	8000	25,000	500	810	0.28
July	8000	25,000	500	810	0.28
	8000	25,000	500	810	0.28
August	8000	25,000	500	810	0.28
	8000	25,000	500	810	0.28
September	8000	25,000	500	810	0.28
	8000	25,000	500	810	0.28
October	8000	25,000	500	810	0.28
	8000	25,000	500	810	0.28
November	8000	25,000	500	810	0.28
	8000	25,000	500	810	0.28
December	20,000	50,000	1100	1350	0.20
	20,000	50,000	1100	1350	0.20

Note: 1000 psi = 6.9 N/mm^2, 1000 pci = 0.272 N/mm^3

subgrade reaction using the AASHTO method. The LS factor is 1. Note that in practice the values for the seasonal moduli of the roadbed and subbase materials are determined by using the appropriate test.

Solution

Determine the composite modulus of subgrade reaction $k_{\propto}$ value for each seasonal period for the corresponding M_r and E_{SB} values using Figure 7.22 and assuming a semi-infinite subgrade depth. These values are shown in column 4.

Modify $k_{\propto}$ to take into consideration the presence of a rigid foundation within 10 ft (3.05 m) of the subgrade surface using Figure 7.23. These values are shown in column 5.

Determine the relative damage for each seasonal period, using the modified k value in Figure 7.24. These values are shown in column 6.

Summation: $\Sigma u_r = 7.36$.

Determine the average $u_r = 7.36/24 = 0.31$.

Determine the overall effective modulus of subgrade reaction $(k) = 750$ pci ($0.204\ \text{N/mm}^3$) (obtained from Figure 7.24).

Determine the corrected k for loss of support, $k = 210\ \text{lb/in}^3$ ($0.057\ \text{N/mm}^3$) (obtained from Figure 7.25).

AASHTO Design Equation for Determining the Thickness of a Rigid Highway Pavement The minimum thickness of the concrete pavement that is adequate to carry the design ESAL is obtained from Equation 7.14:

$$\log_{10} W_{18} = Z_R S_o + 7.35 \log_{10}(D + 1) - 0.06 + \frac{\log_{10}\left[\Delta PSI/(4.5 - 1.5)\right]}{1 + \left[(1.624 \times 10^7)/(D + 1)^{8.46}\right]} + (4.22 - 0.32P_t)\log_{10} \times \left[\frac{S_c C_d}{215.63J}\left(\frac{D^{0.75} - 1.132}{D^{0.75} - [18.42/(E_c/k)^{0.25}]}\right)\right] \tag{7.14}$$

where

Z_R = standard normal variant corresponding to the selected level of reliability
S_o = overall standard deviation
W_{18} = predicted number of 18-kip (81.6 kN) ESAL application that can be carried by the pavement structure after construction
D = thickness of concrete pavement to the nearest half-inch
ΔPSI = design serviceability loss = $P_i - P_t$
P_i = initial serviceability index
P_t = terminal serviceability index
E_c = elastic modulus of the concrete to be used in construction (lb/in^2)
S_c' = modulus of rupture of the concrete to be used in construction (lb/in^2)

J = load transfer coefficient = 3.2 (assumed)
C_d = drainage coefficient
k = corrected effective modulus of subgrade reaction
E_c = elastic modulus of the concrete

The thickness of the concrete pavement (D) can be determined by using a computer program or a set of two charts, as shown in Figures 7.26 and 7.27.

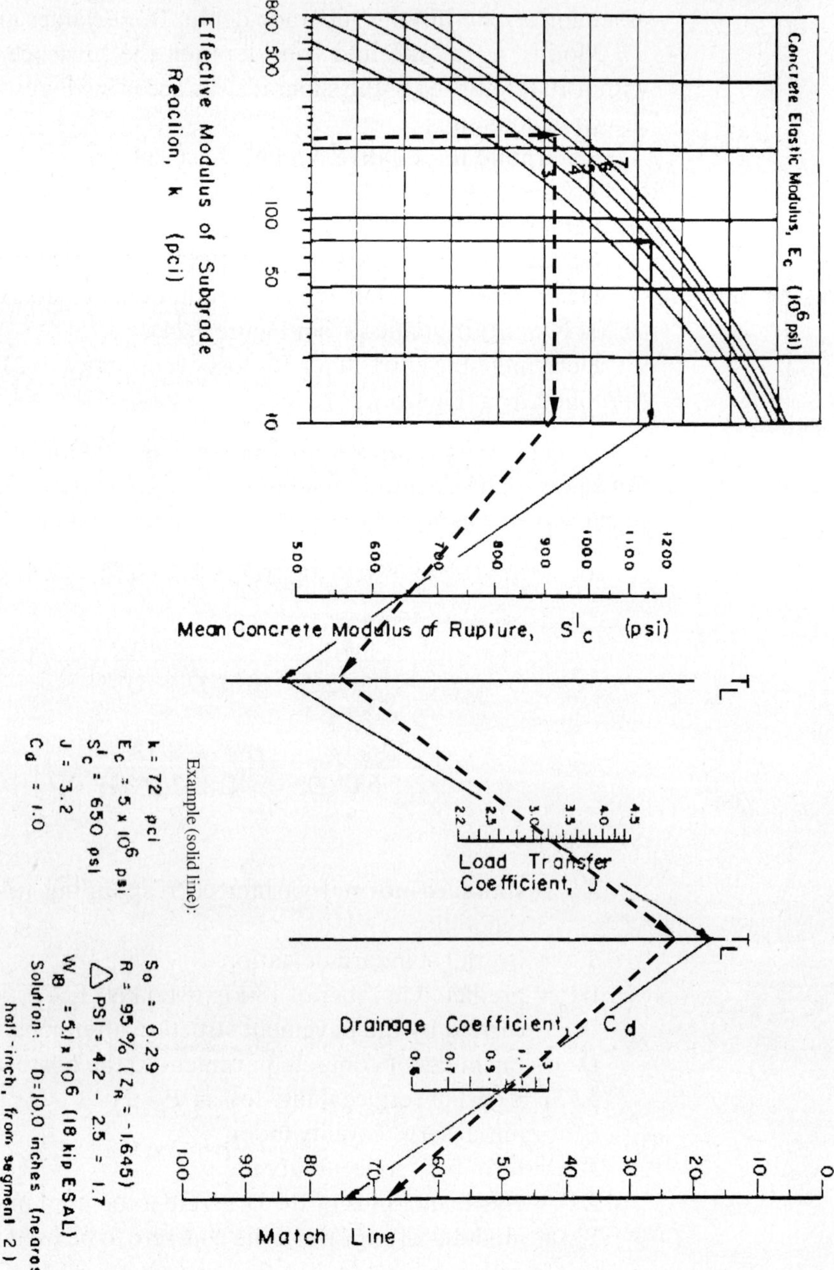

FIGURE 7.26
AASHTO design chart for rigid pavements based on using mean values for each input variables (Segment 1).

Source: AASHTO Guide for Pavement Structures, American Association of State Highway and Transportation Officials, Washington, D.C., 1993. Used by permission.

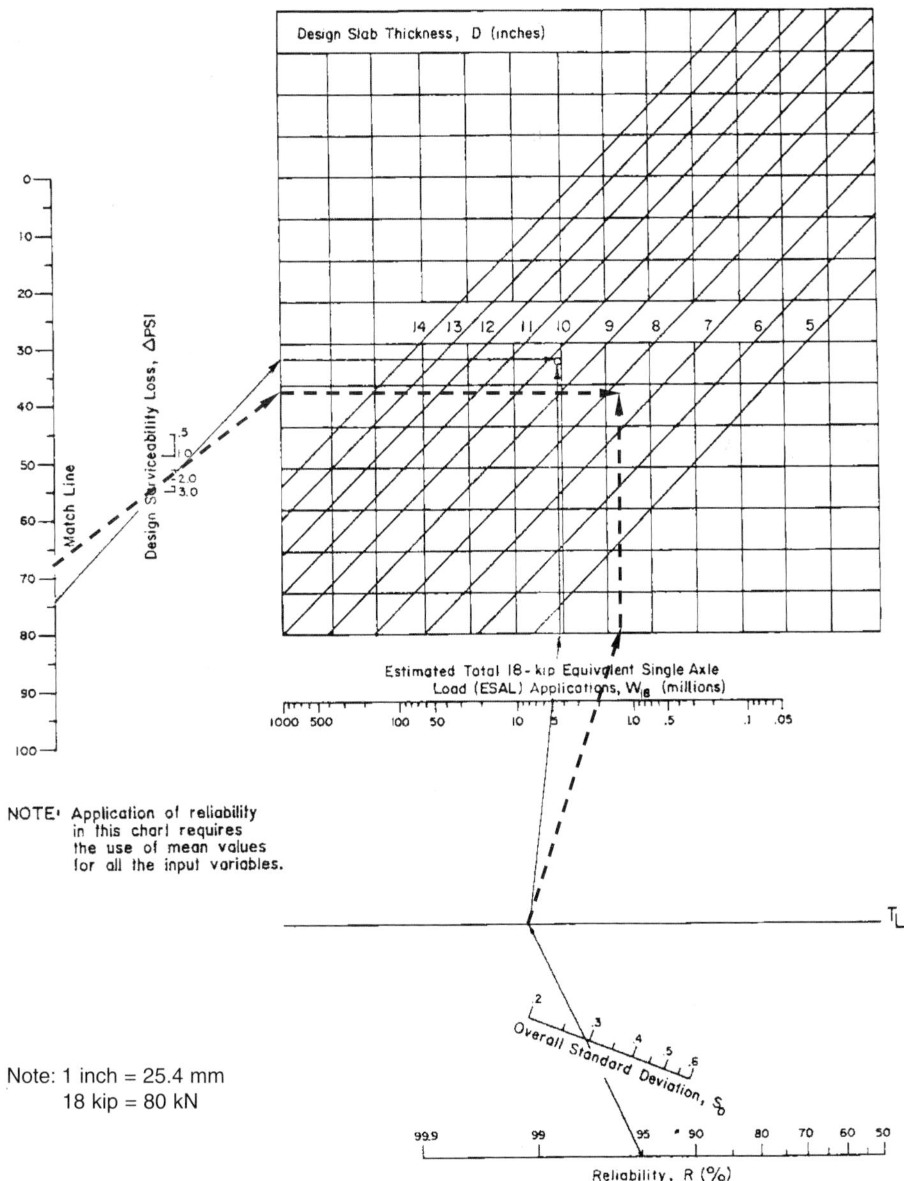

FIGURE 7.27

AASHTO design chart for rigid pavements based on using mean values for each input variables (Segment 2).

Source: AASHTO Guide for Pavement Structures, American Association of State Highway and Transportation Officials, Washington, D.C., 1993. Used by permission.

EXAMPLE 7.12

Determining the Slab Thickness of a Rigid Concrete Highway Pavement Using the AASHTO Method

A rigid highway pavement is to be constructed to carry an accumulative single-axle load of 1.05×10^6. The subbase is a 152.4 mm(6″) layer of cement-treated granular material, the seasonal values for the roadbed resilient modulus and the subbase elastic modulus are as given in columns 2 and 3 of Table 7.37, and rock is located 1524 m(5 ft) from the subgrade surface. Using the AASHTO design procedure, determine the required thickness of the slab for the following values for the input variables:

Loss of support factor (LS) = 1

Elastic modulus of concrete (E_c) = 5 × 10^6 lb/in^2 (3.45 × 10^4 N/mm^2)

Modulus of rupture of the concrete to be used in construction (S'_c) = 650 lb/in^2 (4.485 N/mm^2)

Load transfer coefficient (J) = 3.2

Drainage coefficient (C_d) = 1.0

Overall standard deviation (S_o) = 2.9

Reliability level = 95% (Z_R = 1.645)

Initial serviceability index (P_i) = 4.5

Terminal serviceability index(P_t) = 2.5

Equivalent single-axle load application ($ESAL$) = 2.0 × 10^6

Solution

Determine the overall effective modulus of subgrade reaction (k). Since the subgrade characteristics are the same as those given in Table 7.36, let us assume a slab thickness of 8″ (203.2 mm), which gives an overall k value of 210 lb/in^3 (0.057 N/mm^3), as shown in Example 7.11.

Determine the required depth using Figures 7.26 and 7.27 (dashed lines):

$$\Delta PSI = \text{design serviceability loss} = P_i - P_t = 4.5 - 2.5 = 2.0$$

The required thickness of the concrete slab is 8″ ≈ 200 mm, as shown in Figure 7.27 (dashed lines).

Note that if the thickness obtained is significantly different from the assumed thickness of 8, the whole procedure should be repeated including the computation of k using another assumed value for the thickness.

The FAA Method for the Design of Rigid Airport Runway Pavements

The design input parameters used in this method are (1) the concrete flexural strength; (2) the subgrade modulus (k); (3) the gross weight of the design aircraft; and (4) the annual departure of the design aircraft. The flexural strength that should be used in design should be based on the strength requirement at the time the pavement is opened to traffic. The modulus k on top of the subbase is determined from the subgrade modulus, the subbase material, and the depth of the subbase layer using Figures 7.28, 7.29, or 7.30, depending on the material used for the subbase layer. Figure 7.30 can be used for cement-stabilized and bituminous-stabilized materials.

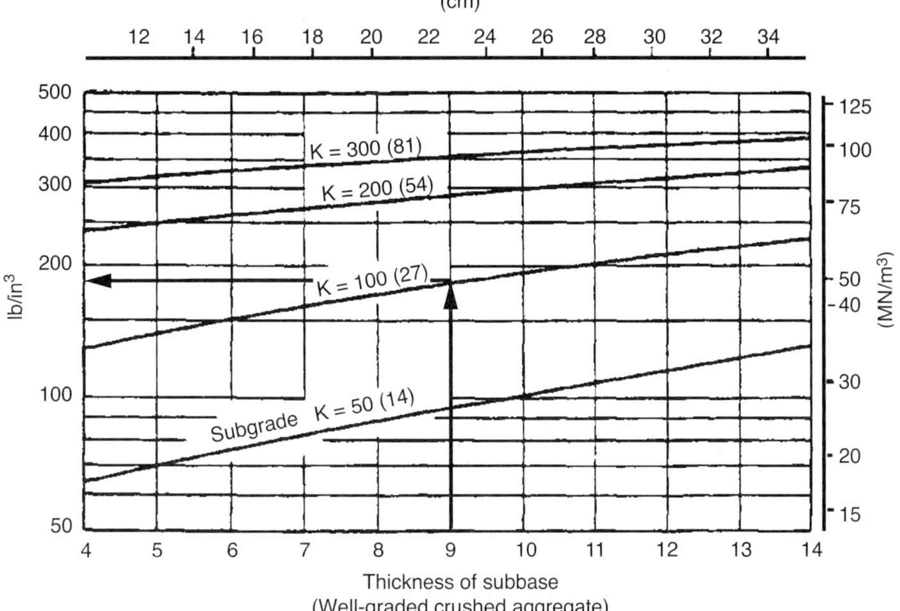

FIGURE 7.28

Effect of subbase on modulus of subgrade reaction for well-graded crushed aggregate.

Source: Airport Pavement Design and Evaluation, Advisory Circular AC 150/5320-6D (Incorporating Changes 1 through 5), Federal Aviation Administration, Department of Transportation, Washington, D.C., April, 2004.

The FAA has also developed design charts for a variety of landing gear types and specific aircraft. In developing these charts, it was assumed that the wheel load is located at a joint in a direction that is either perpendicular or tangential to the joint. Figures 7.31 through 7.33 show design curves for the different landing gear types and Figures 7.34 through 7.39 give examples of design curves for specific aircraft. The use of these charts is illustrated in the following examples.

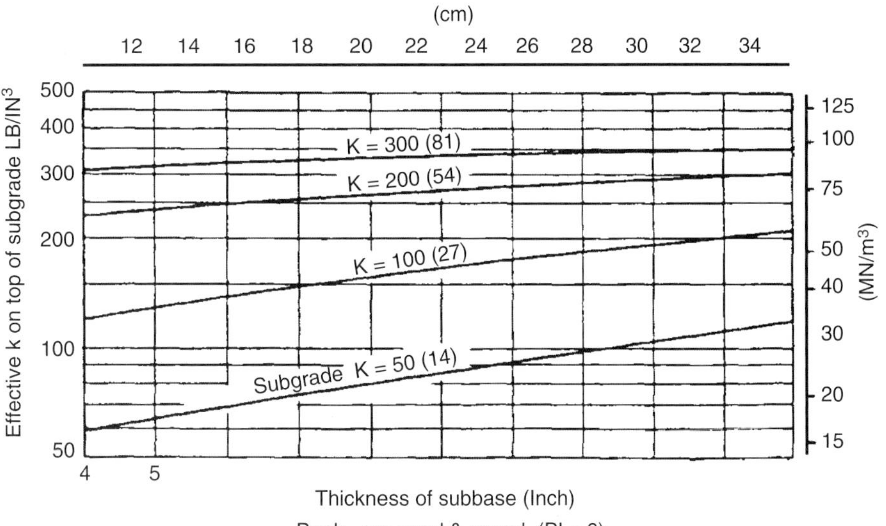

FIGURE 7.29

Effect of subbase on modulus of subgrade reaction for bank-run sand and gravel.

Source: Airport Pavement Design and Evaluation, Advisory Circular AC 150/5320-6D (Incorporating Changes 1 through 5), Federal Aviation Administration, Department of Transportation, Washington, D.C., April, 2004.

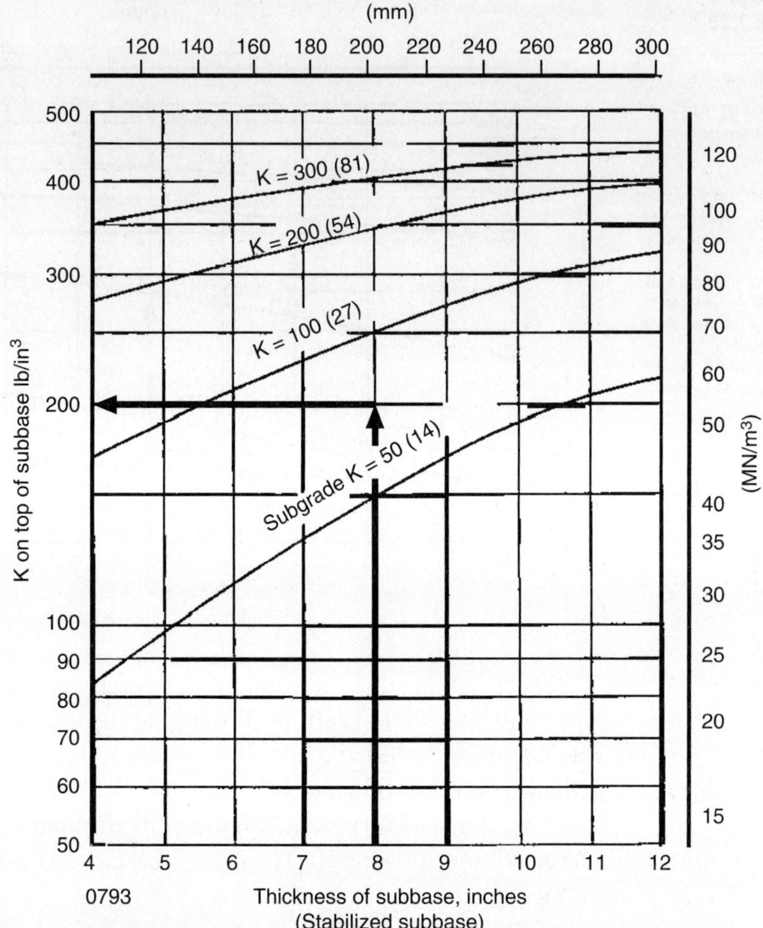

FIGURE 7.30
Effects of stabilized subbase on subgrade modulus.

Source: Airport Pavement Design and Evaluation, Advisory Circular AC 150/5320-6D (Incorporating Changes 1 through 5), Federal Aviation Administration, Department of Transportation, Washington, D.C., April, 2004.

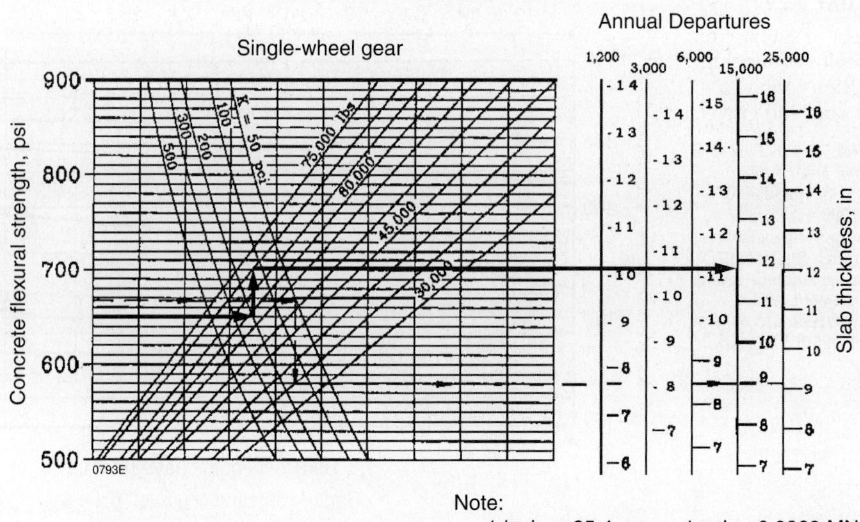

FIGURE 7.31
FAA rigid pavement design curve (single-wheel gear).

Source: Airport Pavement Design and Evaluation, Advisory Circular AC 150/5320-6D (Incorporating Changes 1 through 5), Federal Aviation Administration, Department of Transportation, Washington, D.C., April, 2004.

Note:
1 inch = 25.4 mm   1 psi = 0.0069 MN/m^2
1 lb = 0.454 kg   1 pci = 0.272 MN/m^3

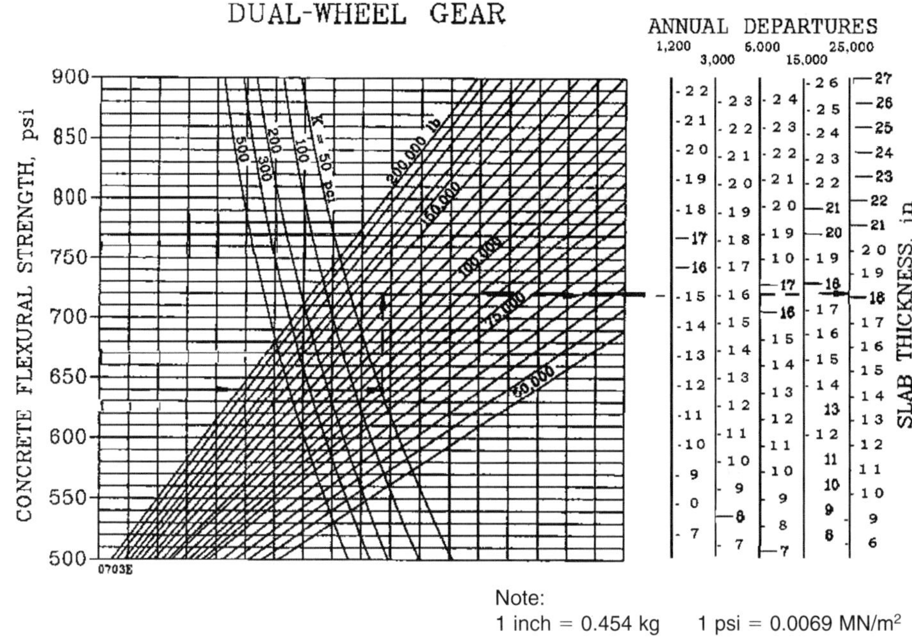

FIGURE 7.32

FAA rigid pavement design curve (dual-wheel gear).

Source: Airport Pavement Design and Evaluation, Advisory Circular AC 150/5320-6D (Incorporating Changes 1 through 5), Federal Aviation Administration, Department of Transportation, Washington, D.C., April, 2004.

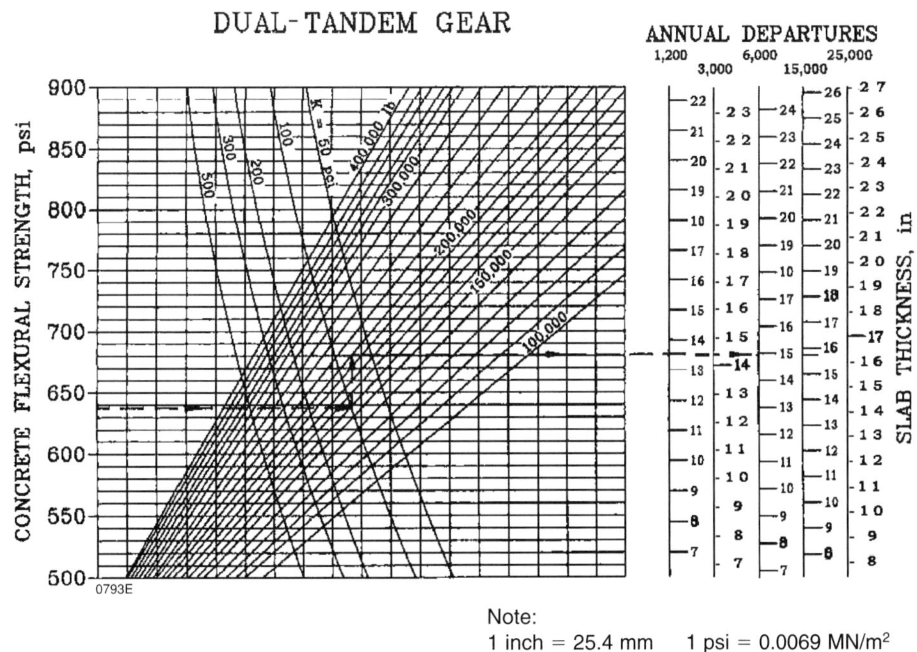

FIGURE 7.33

FAA rigid pavement design curve (dual-tandem gear).

Source: Airport Pavement Design and Evaluation, Advisory Circular AC 150/5320-6D (Incorporating Changes 1 through 5), Federal Aviation Administration, Department of Transportation, Washington, D.C., April, 2004.

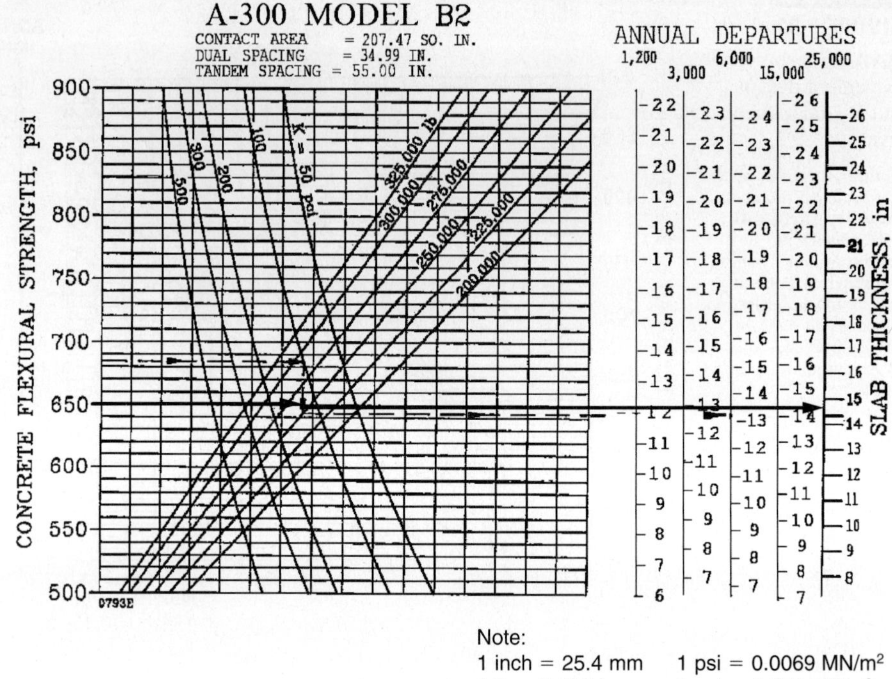

FIGURE 7.34

FAA rigid pavement design curve for specific aircraft (A-300 model B2).

Source: Airport Pavement Design and Evaluation, Advisory Circular AC 150/5320-6D (Incorporating Changes 1 through 5), Federal Aviation Administration, Department of Transportation, Washington, D.C., April, 2004.

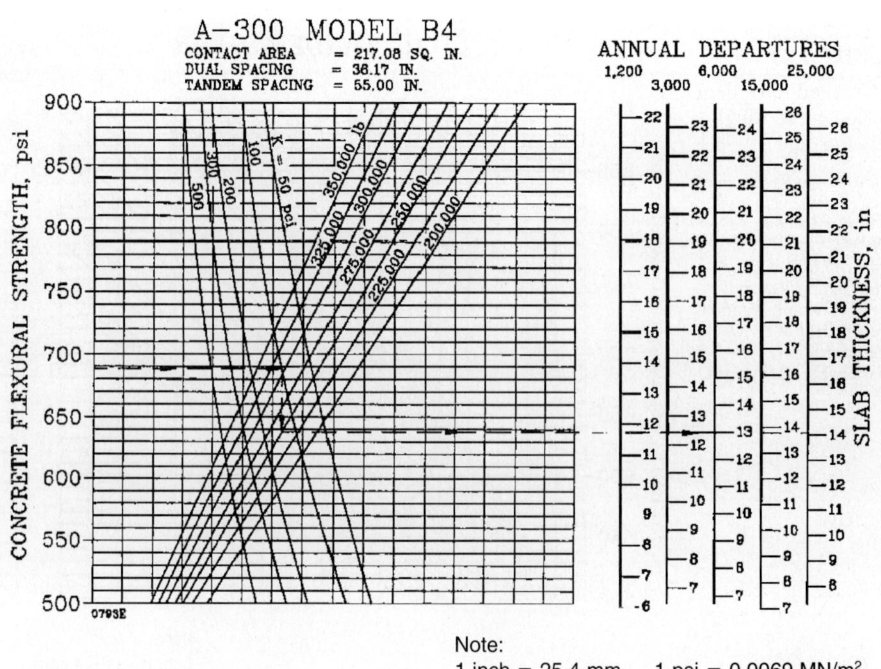

FIGURE 7.35

FAA rigid pavement design curve for specific aircraft (A-300 model B4).

Source: Airport Pavement Design and Evaluation, Advisory Circular AC 150/5320-6D (Incorporating Changes 1 through 5), Federal Aviation Administration, Department of Transportation, Washington, D.C., April, 2004.

FIGURE 7.36

FAA rigid pavement design curve for specific aircraft (B-747-100, SR, and 200 B, C, and F).

Source: Airport Pavement Design and Evaluation, Advisory Circular AC 150/5320-6D (Incorporating Changes 1 through 5), Federal Aviation Administration, Department of Transportation, Washington, D.C., April, 2004.

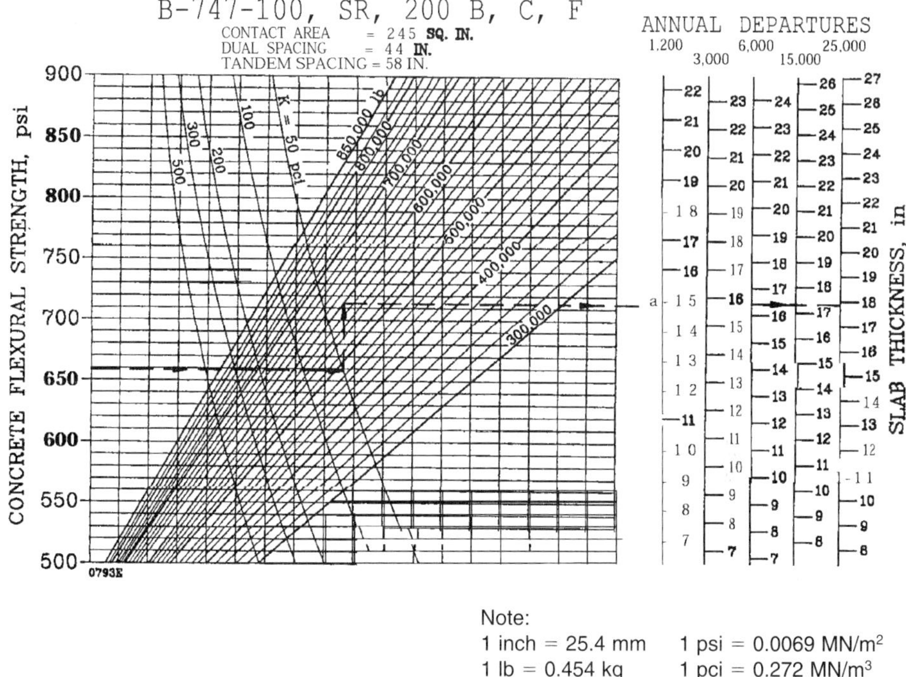

FIGURE 7.37

FAA rigid pavement design curve for specific aircraft (B-747 SP).

Source: Airport Pavement Design and Evaluation, Advisory Circular AC 150/5320-6D (Incorporating Changes 1 through 5), Federal Aviation Administration, Department of Transportation, Washington, D.C., April, 2004.

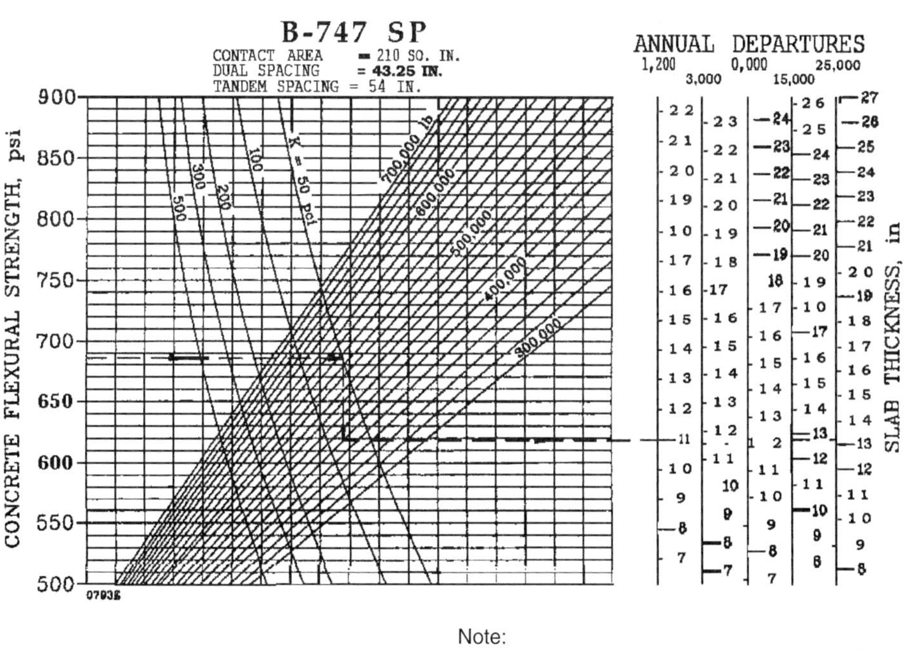

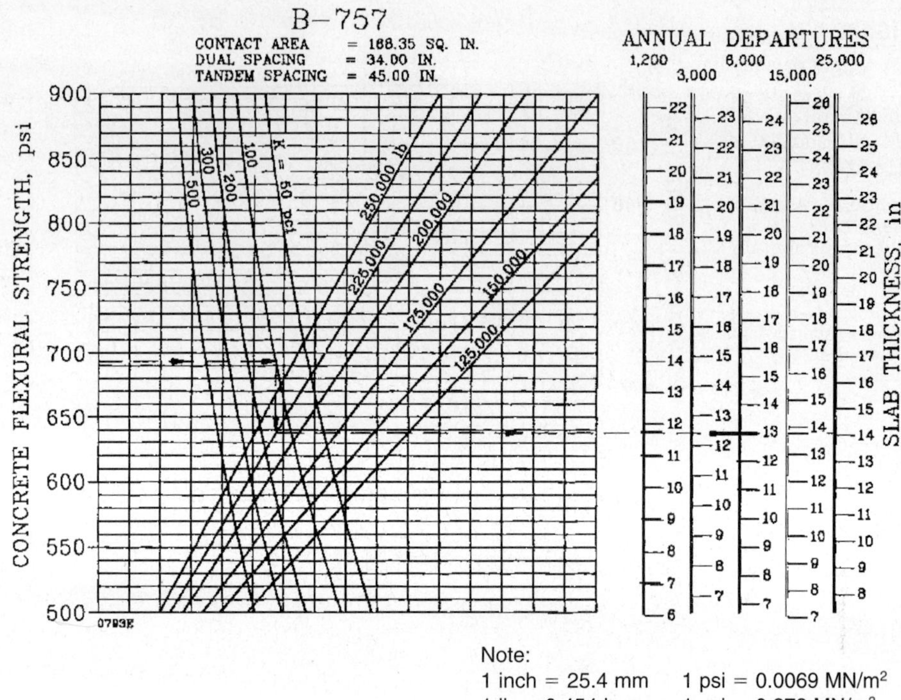

FIGURE 7.38

FAA rigid pavement design curve for specific aircraft (B-757).

Source: Airport Pavement Design and Evaluation, Advisory Circular AC 150/5320-6D (Incorporating Changes 1 through 5), Federal Aviation Administration, Department of Transportation, Washington, D.C., April, 2004.

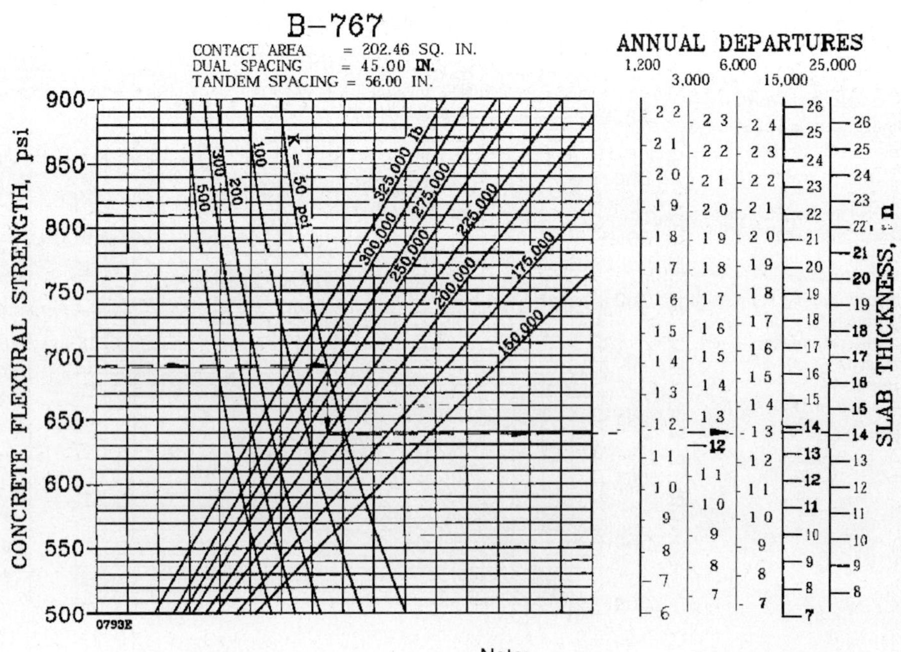

FIGURE 7.39

FAA rigid pavement design curve for specific aircraft (B-767).

Source: Airport Pavement Design and Evaluation, Advisory Circular AC 150/5320-6D (Incorporating Changes 1 through 5), Federal Aviation Administration, Department of Transportation, Washington, D.C., April, 2004.

EXAMPLE 7.13

Determining the Required Slab Thickness for an Airport Rigid Pavement Serving a Given Landing Gear Type

Determine the required thickness for the concrete pavement slab of a rigid airport pavement that will be serving single-wheel gear aircraft, with a gross weight of 317510 N and 15,000 annual departures. The subbase will consist of an 8-″ (= 203.2 mm) cement-stabilized layer, and the subgrade modulus is 75 lb/in^3(0.02 N/mm^3). Assume the flexural strength of the concrete is 650 lb/in^2 (4.485 N/mm^2).

Solution

Determine the equivalent k value. Use Figure 7.30 to determine the effect of the stabilized subbase on the subgrade modulus (solid lines).

For an 8-″(203.2 mm) stabilized subbase and a subgrade modulus of 75 lb/in^3 (0.02 N/mm^3), the equivalent k value is 200 lb/in^3 (= 0.054 N/mm^3).

Determine the slab thickness from Figure 7.31. The thickness of the slab is 11.75″ (298.5 mm), and the required thickness is 12″ (304.8 mm).

Note that since the number of annual departures is less than 25,000, there is no need to correct for high departure levels.

EXAMPLE 7.14

Determining the Required Slab Thickness for an Airport Rigid Pavement Serving Specific Aircraft

Determine the required thickness for a rigid concrete pavement slab serving a fleet of specific aircraft with annual departures and gross weights such that the A-300 Model B2 is the design aircraft with a gross weight of 1224700 N and equivalent annual departures of 25,000. The subbase of the pavement will be a 9-″ (228.6 mm) well-graded crushed aggregate, and the subgrade modulus is 100 lb/in^3 Assume that the flexural strength of the concrete is 650 lb/in^2 (4.485 N/mm^2). *Note:* This implies that the design aircraft and equivalent annual departures have been determined through a procedure similar to that in Example 7.10.

Solution

Determine the equivalent k value. Use Figure 7.28 to determine the effect of well-graded crushed aggregate on modulus of subgrade reaction.

For a 9-″ (228.6 mm) well-graded crushed aggregate and a subgrade value of 100 lb/in^3 (0.0272 N/mm^3), the effective k is about 175 lb/in^3 (0.0476 N/mm^3).

Determine slab thickness. Use Figure 7.34 (see solid line) with a concrete flexural strength of 650 lb/in^2 (4.485 N/mm^2), an effective k of 175 lb/in^3 (0.0476 N/mm^3), a gross weight of 1224700 N, and annual equivalent departures of 25,000. The thickness is 15″ (381 mm).

Note that since the number of annual departures is not higher than 25,000, there is no need to correct for higher departures.

Other Related Design Factors for Rigid Highway and Airport Runway Pavements

In addition to determining the concrete slab thickness for rigid pavements, consideration must also be given to the design of the transverse and longitudinal joints, the computation of the minimum reinforcing steel, and the effect of pumping.

Types of Rigid Pavement Joints Rigid pavement joints are categorized with respect to their function, and can be divided into four basic categories:

- Expansion joints
- Contraction joints
- Hinge joints
- Construction joints

Expansion joints are usually placed transversely at regular intervals to provide adequate space for the slab to expand when subjected to a sufficiently high increased temperature and they should create a distinct brake throughout the depth of the slab. They are therefore placed across the full width of the slab and are 19–25.4 mm (¾ to 1″) wide. An expansion joint usually contains nonextruding compressible material and may be provided with dowel reinforcing bars that are lubricated at one side to form a load transfer mechanism. At locations where a load transfer across the joint is not feasible, such as where the pavement abuts a structure, dowel bars are not provided but the thickness of the slab along the edge may be increased. Expansion joints are placed in both highway and airport rigid pavements. Figures 7.40a and 7.40b show different types of expansion joints.

Contraction joints are used to control the amount of cracking of the pavement due to reduction in the moisture content or the temperature. These joints are placed transversely across the width of the pavement and at regular intervals along the pavement's length. Although it is not usually necessary to install a load-transfer mechanism in the form of a dowel bar at these joints, it may be necessary to do so when there is doubt that the interlocking grains will adequately transfer the load. Figure 7.40 also shows examples of contracting joints.

Hinge joints are mainly used to reduce cracking along the centerline of rigid pavements, although they are sometimes used as construction joints. Figure 7.40 also shows a typical hinged joint.

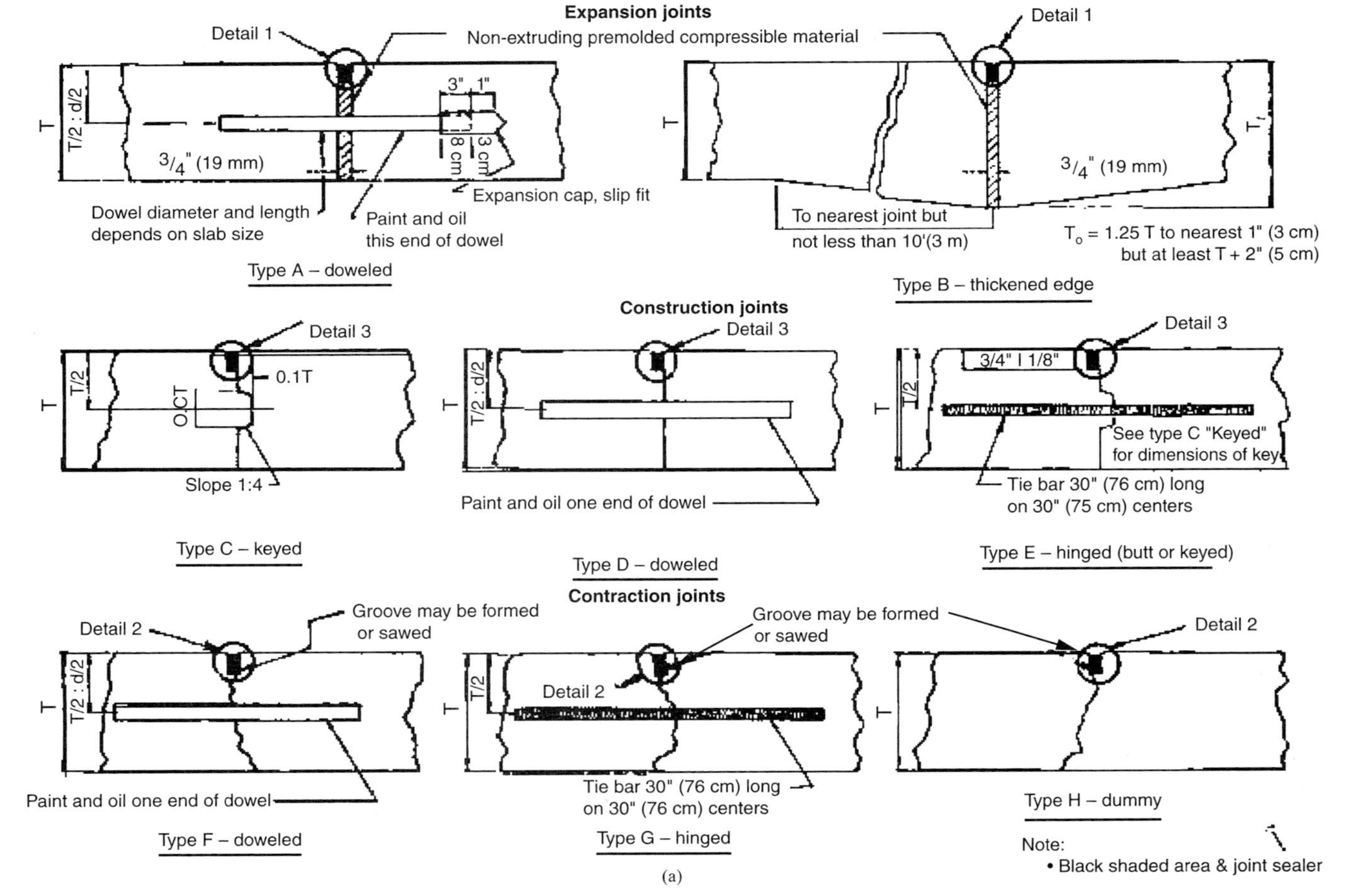

FIGURE 7.40a

Rigid pavement joint types and details.

Source: Airport Pavement Design and Evaluation, Advisory Circular AC 150/5320-6D (Incorporating Changes 1 through 5), Federal Aviation Administration, Department of Transportation, Washington, D.C., April, 2004.

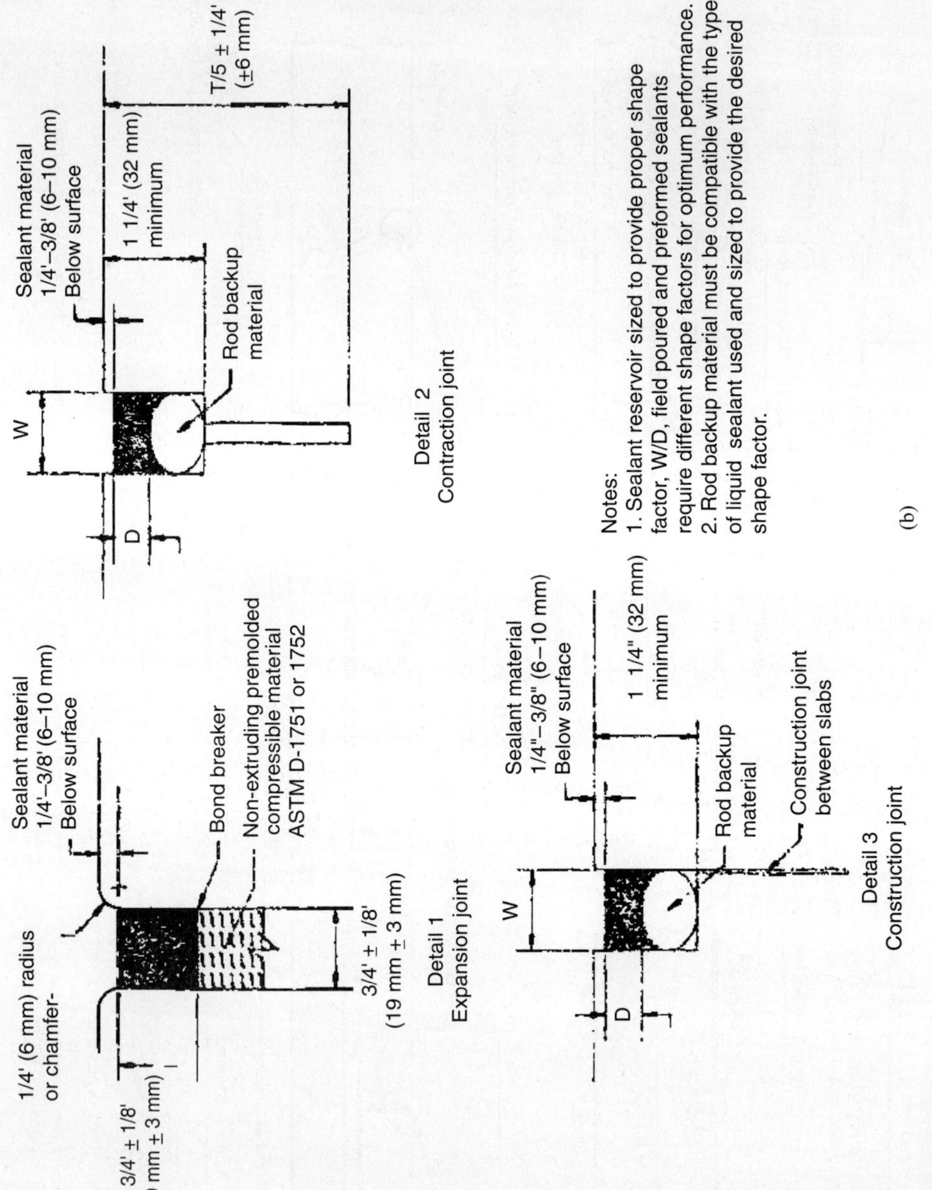

FIGURE 7.40b

Rigid pavement joint types and details.

Source: Airport Pavement Design and Evaluation, Advisory Circular AC 150/5320-6D (Incorporating Changes 1 through 5), Federal Aviation Administration, Department of Transportation, Washington, D.C., April, 2004.

TABLE 7.38
Recommended Maximum Joint Spacing for Rigid Pavements Without Stabilized Subbase

Source: Airport Pavement Design and Evaluation, Advisory Circular AC No 150/5320 – 6D, Federal Aviation Administration, U.S. Department of Transportation, (Incorporating Changes 1 through 5), Washington D.C., April 2004.

Slab Thickness		Transverse		Longitudinal	
Inches	Millimeters	Feet	Meters	Feet	Meters
6	150	12.5	3.8	12.5	3.8
7-9	175-230	15	4.6	15	4.6
9-12	230-305	20	6.1	20	6.1
> 12	>305	25	7.6	25	7.6

Note: Joint spacings shown in this table are maximum values that may be acceptable under ideal conditions. Smaller joint spacings should be used if indicated by past experience. Pavements subject to extreme seasonal temperature differentials or extreme temperature differentials during placement may require smaller joint spacings.

Construction joints are placed between two abutting slabs when they are placed at different times; for example, at the end of a day's work. These joints provide suitable bonding of the abutting slabs. Examples of these joints are also shown in Figure 7.40.

Spacing of Rigid Pavement Joints Rigid pavement joints should be spaced at distances that allow them to adequately perform their functions. AASHTO suggests that local experience could be used in specifying joint spacings for rigid highway pavements. However, consideration should be given to whether the coarse aggregate used is different from that for which experience was gained, as this may have a significant impact on the maximum joint spacings. The reason is that differences may exist in the thermal coefficients of the concrete with different coarse aggregates. AASHTO also suggests that a general rule for determining surface joint spacing for plain concrete pavements is that the spacing (in feet) should not greatly exceed twice the slab thickness (in inches).

FAA-recommended maximum values for joint spacings on rigid airport runway pavements without stabilized subbases are shown in Table 7.38. These values are based on the same rule of thumb as that suggested by AASHTO, which was originally given by the Portland Cement Association, and stipulates that the joint spacing (in feet) should not greatly exceed twice the slab thickness (in inches). It should be noted that these are maximum values, and shorter spacings may be more appropriate in some instances. The FAA recommends a different procedure for determining joint spacings on rigid pavements with stabilized subbases, as these pavements are subjected to higher warping stresses than those with unstabilized subbases. For these pavements, the FAA recommends that the ratio of the joint spacing to the radius of relative stiffness of the concrete slab should be between 4 and 6. The radius of relative stiffness of the slab is given as

$$l = \left[\frac{Eh^3}{12(1 - \mu^2)k}\right]^{\frac{1}{4}} \tag{7.15}$$

where

l = radius of relative stiffness, inches
E = modulus of elasticity of the concrete
h = slab thickness
μ = Poisson's ratio for the concrete, usually 0.15
k = modulus of subgrade reaction

In addition to recommendations given for joint spacing, the Federal Aviation Administration has recommended several other factors that should be considered in the use of joints. First, the FAA suggests that key joints should not be used for slabs less than 230 mm thick as this results in keyways with limited strengths. Second, special consideration should be given to the type of longitudinal joints used for wide-body jet aircraft as experience has shown that the use of inappropriate joints will result in poor performance of the joints. For example, when the modulus of subgrade reaction is 0.055 N/mm^2 or less, keyed joints should not be used and a doweled or thickened joint should be used. When the modulus of subgrade reaction is between 0.055 N/mm^2 and 0.11 N/mm^2, hinged, dowel, or thickened joints may be used, and when the modulus of subgrade reaction is 400 lb/in^3 or greater, a conventional keyed joint may be used.

Type, Area, and Spacing of Steel This should also be considered in the design of rigid pavements. The types of reinforcement are welded wire fabric or bar mats. The wire fabric consists of longitudinal and transverse steel wires welded at regular intervals and is typically used in jointed reinforced concrete pavements, while the bar mats consist of longitudinal and transverse reinforcing rods at regular intervals forming a mat and is typically used in continuously reinforced concrete pavements.

Area and Spacing of Temperature Steel in Jointed Reinforced Highway Pavements The AASHTO procedure provides for the estimation of the percent of steel reinforcement in a jointed reinforcement concrete pavement. In addition to the length of the pavement slab (joint spacing), other factors considered are the friction factor and the reinforcing steel working stress. The friction factor is the coefficient of friction between the subbase or subgrade and the bottom of the slab. Recommended friction factors for different subbase materials and natural subgrade are given in Table 7.39. The working stress is usually 75% of the steel yield strength, with typical values of 210 and 310 N/mm^2 for Grade 40 and Grade 60 steel, respectively, and 335 N/mm^2 for welded wire fabric (WWF) and deformed wire fabric (DWF). In order to reduce the impact of potential corrosion on the cross-sectional area of the pavement, it is suggested that the minimum acceptable wire size be used. Equation 7.16 gives the percentage of steel that is required:

$$p_s = 1.1314 \times 10^{-3} \times \frac{LF}{f_s} \tag{7.16}$$

TABLE 7.39

Recommended Values for Friction Factors for Different Subbase Materials and Natural Subgrade

Source: AASHTO Guide for Design of Pavement Structures, American Association of State Highway and Transportation Officials, 1993. Used by permission.

Type of Material Beneath Slab	*Friction Factor*
Surface treatment	2.2
Lime stabilization	1.8
Asphalt stabilization	1.8
Cement stabilization	1.8
River gravel	1.5
Crushed stone	1.5
Sandstone	1.2
Natural subgrade	0.9

where

p_s = percent of steel required (percent of cross-sectional area of slab)
f_s = working stress of the steel used, N/mm^2
L = length of slab (joint spacing), mm

Equation 7.16 is used in the AASHTO procedure to estimate the required steel in the transverse and longitudinal directions for jointed reinforced concrete pavement. Table 7.40a shows sectional areas of welded fabrics that can be used to select appropriate fabrics.

TABLE 7.40a

Sectional Areas of Welded Fabric (mm^2)

Wire size smooth	Number deformed	Nominal diameter mm	Nominal weight gm/lin mm	Center-to-Center Spacing(mm) 102	152	203	254	305
W31	D31	15.95	1.57	600.00	400.00	300.00	240.00	200.00
W30	D30	15.70	1.52	580.64	387.10	290.32	232.26	193.55
W28	D28	15.16	1.42	541.93	361.29	270.97	216.77	180.64
W26	D26	14.61	1.39	503.22	335.48	251.61	201.29	167.74
W24	D24	14.05	1.21	464.52	309.68	232.26	185.81	154.84
W22	D22	13.44	1.11	425.81	283.87	212.90	170.32	141.94
W20	D20	12.80	1.01	387.10	258.06	193.55	154.84	129.03
W18	D18	12.14	0.91	348.39	232.26	174.19	139.35	116.13
W16	D16	11.46	0.81	309.68	206.45	154.84	123.87	103.23
W14	D14	10.72	0.71	270.97	180.64	135.48	108.39	90.32
W12	D12	9.91	0.61	232.26	154.84	116.13	92.90	77.42
W11	D11	9.50	0.56	212.90	141.94	106.45	85.16	70.97
W10.5		9.30	0.53	203.23	135.48	101.29	81.29	67.74
W10	D10	9.04	0.51	193.55	129.03	96.77	77.42	64.52
W9.5		8.84	0.48	183.87	122.58	91.61	73.55	61.29
W9	D9	8.59	0.46	174.19	116.13	87.10	69.68	58.06
W8.5		8.36	0.43	164.52	109.68	81.94	65.81	54.84
W8	D8	8.10	0.40	154.84	103.23	77.42	61.29	51.61
W7.5		7.85	0.38	164.52	96.77	72.26	58.06	48.39
W7	D7	7.57	0.35	135.48	90.32	67.74	54.19	45.16
W6.5		7.32	0.33	125.81	83.87	62.58	50.32	41.94
W6	D6	7.01	0.30	116.13	77.42	58.06	46.45	38.71
W5.5		6.71	0.28	106.45	70.97	52.90	42.58	35.48
W5	D5	6.40	0.25	96.77	64.52	48.39	38.71	32.26
W4.5		6.10	0.23	87.10	58.06	43.23	34.84	29.03
W4	D4	5.72	0.20	77.42	51.61	38.71	30.97	25.81

TABLE 7.40b

Dimensions and Unit Weights of Deformed Steel Reinforcing Bars

Number	Diameter in. (mm)	Area in.^{2} (cm^2)	Perimeter in. (cm)	Unit Weight lbs./ft. (kg/m)
3	0.375 (9.5)	0.11 (0.71)	1.178 (3.0)	0.376 (0.56)
4	0.500 (12.7)	0.20 (1.29)	1.571 (4.0)	0.668 (1.00)
5	0.625 (15.9)	0.31 (2.00)	1.963 (5.0)	1.043 (1.57)
6	0.750 (19.1)	0.44 (2.84)	2.356 (6.0)	1.502 (2.26)
7	0.875 (22.2)	0.60 (3.86)	2.749 (7.0)	2.044 (3.07)

NOMINAL DIMENSIONS

Source: Airport Pavement Design and Evaluation, Advisory Circular AC No 150/5320 – 6D, Federal Aviation Administration, U.S. Department of Transportation, (Incorporating Changes 1 through 5), Washington D.C., April 2004.

EXAMPLE 7.15

Estimating the Required Temperature Steel for a Jointed Reinforced Concrete Highway Pavement

The highway rigid slab designed in Example 7.12 is to be constructed as a jointed reinforced concrete slab. If the slab will be constructed in lengths of 15 m and widths of 7.5 m, determine

(a) Required steel area in each direction
(b) A suitable welded fabric that can be used

Solution

Determine the steel percentage in the longitudinal direction. Use Equation 7.16:

$$p_s = \frac{LF}{2f_s}100$$

$$L = 15 \text{ m}$$

$F = 1.8$ (from Table 7.39, for cement-stabilized material beneath slab)

$f_s = 335$ N/mm^2 (for welded fabric having a yield strength of 414 N/mm^2)

$$p_s = 1.1314 \times 10^{-3} \times \frac{15 \times 10^3 \times 1.8}{335}$$

$$= 0.091$$

Determine the area of steel/ft width in the longitudinal direction:

Depth of pavement = 200 mm (from Example 7.12)

$$\text{Area of steel/m width} = \frac{0.091}{100} \times 200 \times 1000 = 182 \text{ mm}^2$$

Determine the wire size and center-to-center spacing:
From Table 7.40a, suitable wire sizes and spacings are

(i) W9 at 305 mm spacing or
(ii) W6 at 203 mm spacing

In order to reduce the impact of corrosion on the cross-sectional area of the pavement, use W6 at 203 mm spacing.
Determine the steel percentage in the transverse direction:

$$p_s = 1.1314 \times 10^{-3} \frac{7.5 \times 10^3 \times 1.8}{335} = 0.0455$$

Determine the area of steel/m width in the transverse direction:

Pavement depth = 200 mm

$$\text{Area of steel/ft width} = \frac{0.0455}{100} \times 200 \times 1000 = 91 \text{ mm}^2$$

Determine wire size and center-to-center spacing:
From Table 7.40a, the wire size is W4 at 254 mm spacing.

Area and Spacing of Temperature Steel in Jointed Reinforced Airport Runway Pavements The equation given by the FAA to determine the steel area for a jointed reinforced concrete pavement is obtained from the subgrade drag formula and the coefficient of friction formula and is given as

$$A_s = 2.0141 \times 10^{-3} \times \frac{L\sqrt{Lt}}{f_s} \tag{7.17}$$

where

A_s = area of steel/m of width or length, mm²
L = length or width of slab, mm
t = thickness of slab, mm
f_s = allowable tensile strength in steel, N/mm² ($\frac{2}{3}$ of yield strength)

The FAA recommends that the minimum size of longitudinal wires in wire fabrics should be either W5 or D5, and the minimum size of transverse wires should be W4 or D4. For steel having a yield strength of 448.5 N/mm², the calculated area of longitudinal steel should be not less than 0.05% of the cross-sectional area of the slab, and the percentage should be revised proportionally upward for steels with lower yield strengths. Also, the slab length should not be higher than 22.8 m. It is also recommended that for this computation, the allowable tensile stress in the steel should be taken as two-thirds of the yield strength of the steel.

EXAMPLE 7.16

Estimating the Required Temperature Steel for a Jointed Reinforced Concrete Airport Runway Pavement

A 305 mm thick airport rigid pavement is to be constructed as a jointed reinforced pavement with transverse joints spaced at 9 m intervals and a paving lane width of 7.5 m. Determine

(a) The required cross-sectional area of the longitudinal steel/m width of slab
(b) The required cross-sectional area of the transverse steel/m length of slab. Assume that the yield strength of steel is 448.5 N/mm²

Solution

(a) Use Equation 7.17 to determine required temperature steel in the longitudinal direction:

$$A_s = 2.0141 \times 10^{-3} \times \frac{L\sqrt{Lt}}{f_s}$$

$$= 2.0141 \times 10^{-3} \times \frac{9 \times 10^3 \sqrt{9 \times 10^3 \times 305}}{\frac{2}{3} \times 448.5}$$

$$= 100.4 \text{ mm}^2/\text{m}$$

Recommended minimum longitudinal steel = 0.05% of cross sectional area = $0.0005 \times 305 \times 100 = 152.5$ mm²/m, which should be used.

(b) Use Equation 7.17 to determine required temperature steel in the transverse direction:

$$A_s = 2.0141 \times 10^{-3} \times \frac{7.5 \times 10^3 \sqrt{7.5 \times 10^3 \times 305}}{\frac{2}{3} \times 448.5}$$

$$= 76.4 \text{ mm}^2/\text{m}$$

Area and Spacing of Longitudinal Reinforcement in Continuously Reinforced Highway Pavements AASHTO stipulates three conditions that should be satisfied in determining the amount of longitudinal reinforcement required in a continuously reinforced concrete pavement:

1. Maximum and minimum spacing between cracks
2. Maximum crack width
3. Maximum steel stress

AASHTO recommends that in order to minimize crack spalling, consecutive cracks should be spaced at 8 ft (2.44 m) or less from each other, and in order to minimize the potential for punchouts, the spacing between cracks should be not less than 3.5 ft (1.06 m). In order to reduce crack spalling and the potential of water penetrating through the pavement, AASHTO also recommends a maximum crack width of 0.04″ (1.01 mm). AASHTO also recommends that in determining the longitudinal steel percentage, consideration should be given to the use of a higher longitudinal steel percentage or smaller-diameter reinforcing bars, as this will result in a smaller crack width. The criteria placed on the maximum steel stress are to ensure that the steel does not fracture or suffer excessive permanent deformation. A maximum steel stress of 75% of the ultimate tensile strength of the steel is used to satisfy these criteria.

The required longitudinal steel percentage is determined through the following steps:

(i) Determine the wheel load tensile strength using Figure 7.41.

(ii) Determine the maximum required steel percentage (p_{max}) to satisfy the minimum spacing (3.5 ft = 1.06 m) between cracks using either the chart or the expression given on Figure 7.42.

(iii) Determine the minimum required steel percentage to satisfy the maximum spacing between cracks (8 ft = 2.44 m) using the chart or the expression shown on Figure 7.42.

(iv) Determine the minimum percent longitudinal reinforcement to satisfy the maximum crack width (0.04″ = 1.01 m) criterion using the chart or expression given in Figure 7.43.

(v) Determine the minimum percent longitudinal reinforcement to satisfy steel stress criteria using the chart or expression given in Figure 7.44.

(vi) Select the largest percentage of the value obtained in steps (iii), (iv), and (v) as the minimum percentage (p_{min})

(vii) Compare (p_{max}) and (p_{min})

If (p_{max}) $\geqslant$ (p_{min}), continue to step (vii)

If (p_{max}) $<$ (p_{min}), review the design input values and make suitable changes for these input values and repeat steps (i) through (vii) until (p_{max}) $\geqslant$ (p_{min}). Also, check the computations for the subbase and slab thicknesses to ensure that the changes made in the design input values have not resulted in required changes in these thicknesses.

(viii) Determine the maximum and minimum numbers of the reinforcing bars or wires using Equations 7.18 and 7.19:

$$N_{max} = 0.01273 \times p_{max} \times W_s \times D/(\varphi^2) \tag{7.18}$$

$$N_{min} = 0.01273 \times p_{min} \times W_s \times D/(\varphi^2) \tag{7.19}$$

where

N_{max} = maximum number of reinforcing bars or wires
N_{min} = minimum number of reinforcing bars or wires
p_{max} = maximum required percent steel

FIGURE 7.41
Chart for estimating wheel load tensile stress.

Source: AASHTO Guide for Design of Pavement Structures, American Association of State Highway and Transportation Officials, Washington, D.C., 1993. Used by permission.

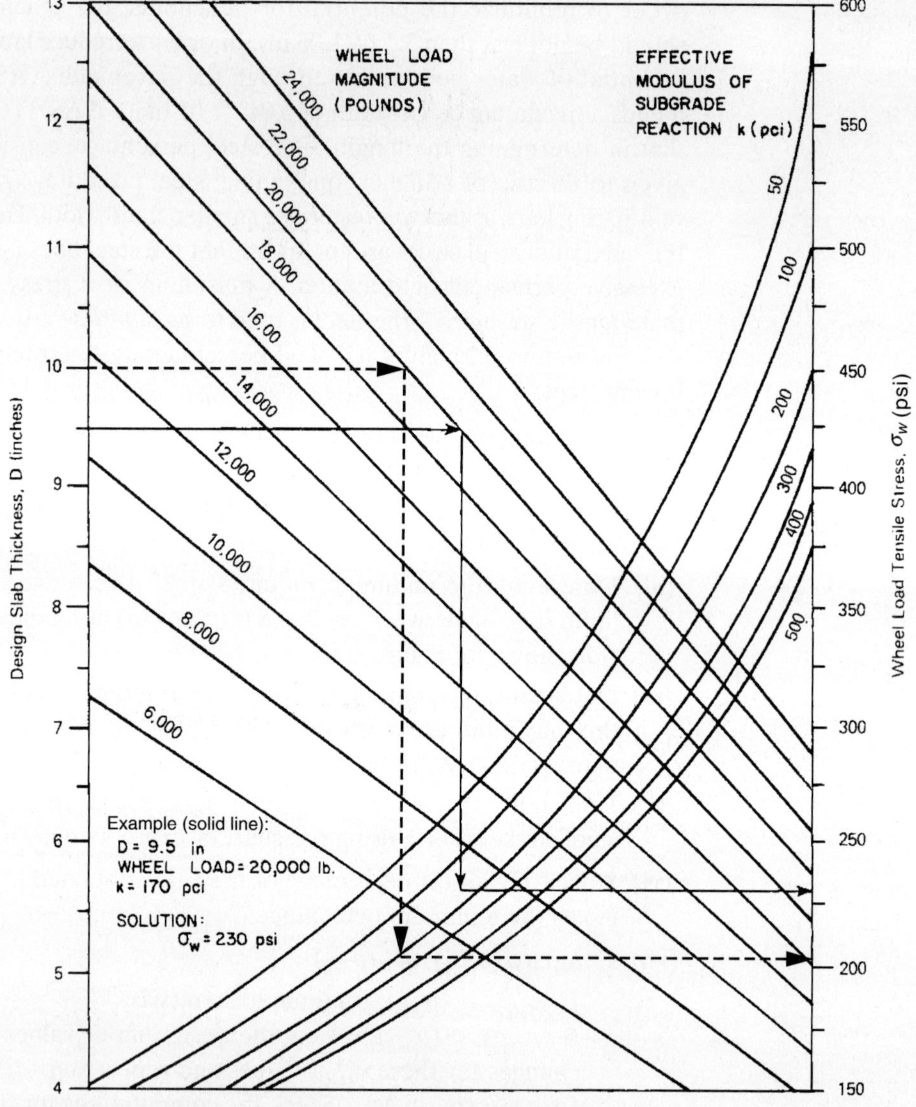

Note: 1 inch = 25.4 mm, 1 lb = 4.5 N, 1000 psi = 6.9 N/mm^2
1000 pci = 0.272 N/mm^3

p_{min} = minimum required percent steel
W_s = total width of pavement section, inches
D = concrete slab thickness, inches
φ = reinforcing bar or wire diameter, inches

(ix) Select the number N_{design} (whole integer) of reinforcing bars or wires such that

$$N_{min} < N_{design} < N_{max}$$

FIGURE 7.42

Percent of longitudinal reinforcement to satisfy crack spacing criteria.

Source: AASHTO Guide for Design of Pavement Structures, American Association of State Highway and Transportation Officials, Washington, D.C., 1993. Used by permission.

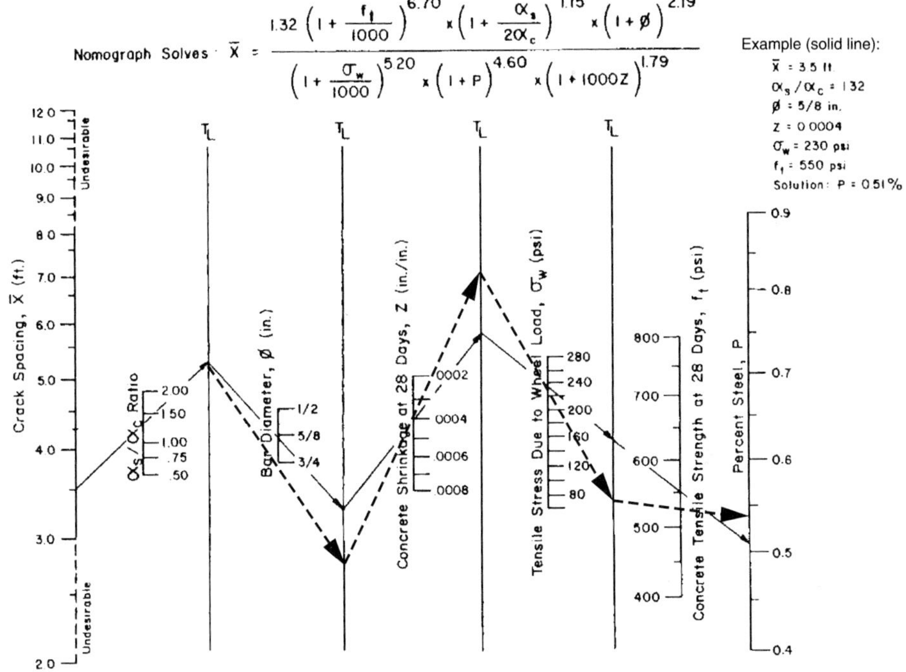

Note: 1 inch = 25.4 mm, 1 ft = 304.8 mm, 1000 psi = 6.9 N/mm²

FIGURE 7.43

Minimum percent of longitudinal reinforcement to satisfy crack width criteria.

Source: AASHTO Guide for Design of Pavement Structures, American Association of State Highway and Transportation Officials, Washington, D.C., 1993. Used by permission.

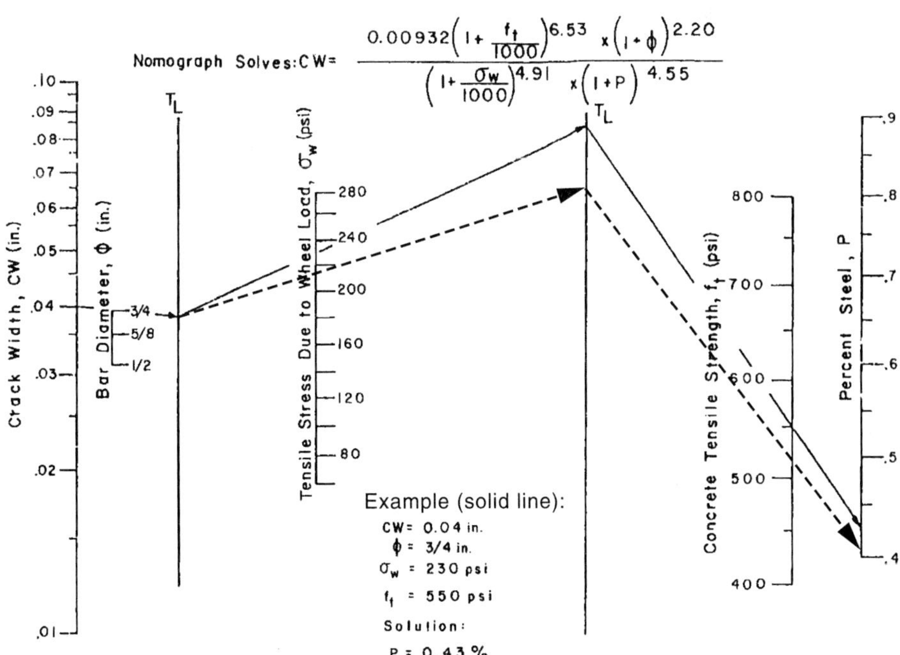

Note: 1 inch = 25.4 mm, 1 ft = 304.8 mm, 1000 psi = 6.9 N/mm²

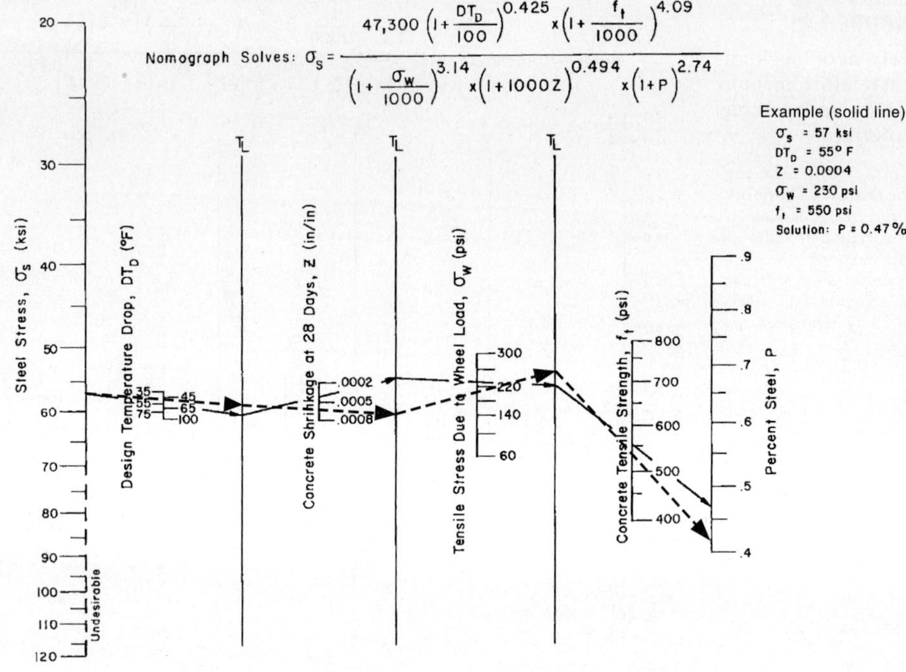

FIGURE 7.44

Minimum percent of longitudinal reinforcement to satisfy steel stress criteria.

Source: AASHTO Guide for Design of Pavement Structures, American Association of State Highway and Transportation Officials, Washington, D.C., 1993. Used by permission.

This selected number of bars or wires can be converted to percent steel, which can then be used to estimate the resultant crack spacing, crack width, and steel stress by working backward using the appropriate charts.

EXAMPLE 7.17

Estimating the Area and Spacing of Longitudinal Steel in Continuously Reinforced Highway Concrete Pavements

A 10-"(254 mm) concrete slab is to be used to construct a continuously reinforced highway concrete pavement. Determine the number of 190.5 mm (#6) bars that should be adequate in the longitudinal direction for the following input values:

Wheel load = 20,000 lb (90.7 kN)
Effective modulus of subgrade reaction (k) = 185 lb/in^3 (0.0503 N/mm^3)
Concrete tensile strength f_t = 525 lb/in^2 (3.622 N/mm^2)
Concrete shrinkage at 22 days = 0.0004
Thermal coefficient of steel $\alpha_s = 5 \times 10^{-6}$

Thermal coefficient of concrete $\alpha_c = 3.8 \times 10^{-6}$
Design temperature drop $DT_D = 50°F$ (maximum temperature of 80°F, and minimum temperature of 30°F)
Ultimate steel tensile stress $\sigma_s = 76 \times 10^3$ lb/in^2 (524.4 N/mm^2)
Width of lane = 12 ft (3.65 m)

Solution

Determine the wheel load tensile stress (σ_w) using Figure 7.41 (step i) (dashed lines):

$$\sigma_w = 215 \text{ lb/in}^2 \text{ (1.483 N/mm}^2\text{)}$$

Determine the maximum required steel percentage (p_{max}) to satisfy the minimum spacing between cracks (3.5 ft) using Figure 7.42 (step ii) (dashed lines):

$$\alpha_s/\alpha_c = \frac{5 \times 10^6}{3.8 \times 10^6} = 1.32$$

$$p_{max} = 0.54\%$$

Determine minimum steel percentage: required steel percentage to satisfy the maximum spacing (8 ft) between cracks = 0.40% from Figure 7.42 (step iii): minimum percent longitudinal reinforcement to satisfy the maximum crack width 0.04″, criterion using Figure 7.43 (step iv): = 0.42% minimum percent longitudinal reinforcement to satisfy steel stress criteria using Figure 7.44 (step v):

$$p = 0.420\%$$

Minimum required steel percentage $p_{min} = 0.42$ (step vi).

Note: Maximum allowable steel stress is 75% of ultimate tensile stress, $0.75 \times 76 = 57 \times 10^3$ lb/in^2 (393 N/mm^2).

Compare p_{max} and p_{min}

$$p_{max} > p_{min}, \text{ go to (step vii)}$$

Determine the maximum and minimum numbers of the reinforcing bars or wires using Equations 7.18 and 7.19 (step viii):

Note: $W_s = 12 \times 12 = 144$

$$\begin{aligned} N_{max} &= 0.01273 \times p_{max} \times W_s \times D/(\varphi^2) \\ &= 0.01273 \times 0.54 \times 144 \times 10/\left(\tfrac{3}{4}\right)^2 \\ &= 17.6 \end{aligned}$$

$$\begin{aligned} N_{min} &= 0.01273 \times p_{min} \times W_s \times D/(\varphi^2) \\ &= 0.01273 \times 0.42 \times 144 \times \left[10 \div \left(\tfrac{3}{4}\right)^2\right] \\ &= 13.69 \end{aligned}$$

Select the number N_{design} (whole integer) of reinforcing bars or wires such that

$N_{min} < N_{design} < N_{max}$ (step ix)

Select $N_{design} = 15$

Area and Spacing of Longitudinal Steel in Continuously Reinforced Airport Runway Concrete (CRCP) Pavements The Federal Aviation Administration stipulates that the longitudinal steel reinforcement in airport runway CRCP pavements should satisfy the following three design conditions:

(1) Minimum steel to resist subgrade restraint
(2) Minimum steel to resist temperature effects
(3) Concrete to steel strength ratio

The minimum steel required to resist subgrade restraint could be obtained by using the chart shown in Figure 7.45. It is based on the tensile strength of the concrete, the allowable steel strength, and the friction factor of the subbase. However, in no case should the longitudinal steel percentage be less than 0.5% of the cross-sectional area of the slab. FAA recommends that the allowable stress of the steel should be 75% of the specified minimum yield strength, the allowable tensile stress of the concrete should be 67% of its flexural strength, and the friction factor of a stabilized base is 1.8. Recommended friction factors for unbound fine-grained and coarse-grained soils are 1.0 and 1.5, respectively. However, the FAA does not recommend the use of these soils as subbase materials in CRCPs.

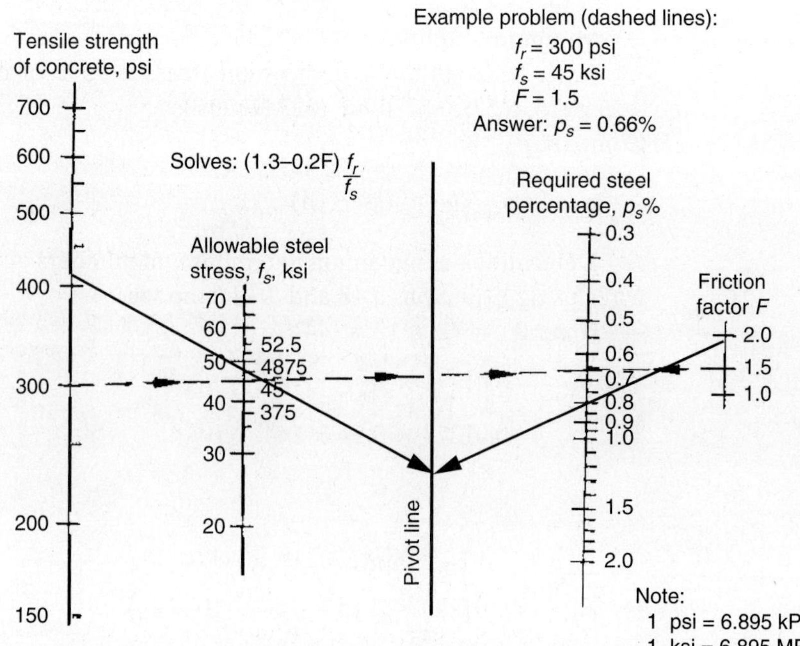

FIGURE 7.45
CRCP-longitudinal steel requirement for resisting subgrade restraint.

Source: Airport Pavement Design and Evaluation, Advisory Circular AC 150/5320-6D (Incorporating Changes 1 through 5), Federal Aviation Administration, Department of Transportation, Washington, D.C., April, 2004.

The minimum steel to resist temperature effects must be capable to withstand the forces due to the expansion and contraction of the slab resulting from temperature changes. It is based on the tensile strength of the concrete, the working strength of the steel, and the maximum seasonal temperature differential in the pavement. It is obtained from Equation 7.20:

$$P_{tc} = \frac{50 \times \frac{1000}{6.9} f_t}{\frac{1000}{6.9} f_s - 195\left(\frac{9}{5}T + 32\right)} \tag{7.20}$$

where

P_{tc} = steel reinforcement to resist temperature crack in percent of cross-sectional area of slab
f_t = tensile strength of concrete, N/mm^2
f_s = working stress of steel usually taken as 75% of specified minimum strength, N/mm^2
T = maximum seasonal temperature differential for pavement in degrees Fahrenheit (°C)

The concrete to steel strength criterion stipulates that the steel reinforcement in percent of the cross-sectional area of the pavement should be not less than the ratio of the concrete strength to the yield strength of the steel multiplied by 100 and is given as

$$P_{c/s} = 100 \frac{f_t}{f_y} \tag{7.21}$$

where

$P_{c/s}$ = steel reinforcement to satisfy concrete to steel strength criterion in percent of the cross-sectional area of the pavement
f_t = tensile strength of the concrete
f_y = minimum yield strength of the steel

Adequate transverse steel reinforcement must also be provided in airport runway CRC pavements to control the longitudinal cracks that may sometimes occur. The transverse steel also helps to support the longitudinal steel during construction. The minimum steel requirement in the transverse direction of airport runway CRCPs as recommended by the FAA can be obtained from either Equation 7.22 or the chart in Figure 7.46.

$$P_{ts} = 1.1314 \times 10^{-3} \frac{W_s F}{f_s} \tag{7.22}$$

where

P_{ts} = transverse steel reinforcement
W_s = width of slab, mm
F = friction factor of subbase
f_s = allowable working stress in steel, N/mm^2

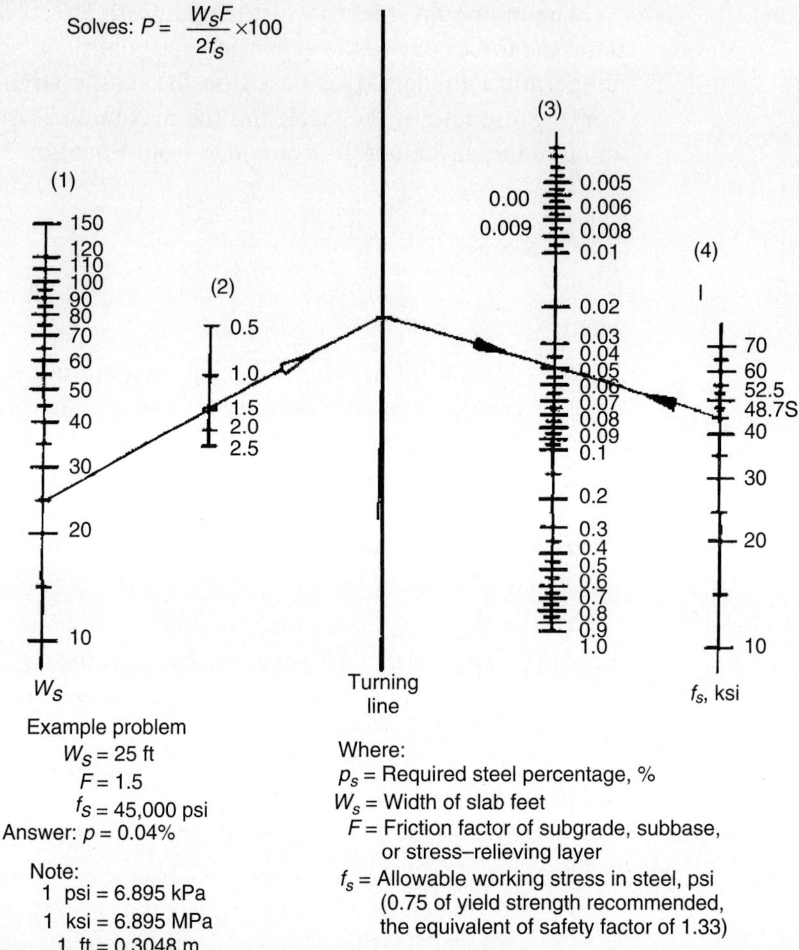

FIGURE 7.46 Continuously reinforced concrete pavement-transverse steel reinforcement.

Source: Airport Pavement Design and Evaluation, Advisory Circular AC 150/5320-6D (Incorporating Changes 1 through 5), Federal Aviation Administration, Department of Transportation, Washington, D.C., April, 2004.

EXAMPLE 7.18

Estimating the Area and Spacing of Longitudinal Steel in Continuously Reinforced Airport Concrete Pavements

Determine the area and spacing of the longitudinal steel for a continuously reinforced airport concrete pavement to meet the design conditions if the concrete flexural strength is 4.14 N/mm^2, the maximum seasonal temperature differential is 35°C, and the subbase is cement stabilized.

Solution

Determine the minimum longitudinal steel to resist subgrade restraint; use Figure 7.45 with the following inputs:

Yield strength of steel = 448.5 N/mm^2

Working stress = 0.75 × 448.5 N/mm^2 = 336 N/mm^2

Friction factor = 1.8

Tensile strength of concrete = 0.67 × 4.14 N/mm^2 = 2.77 N/mm^2

We obtain longitudinal steel percentage = 0.8%.

Determine minimum longitudinal steel to resist the forces generated from seasonal temperature changes; use Equation 7.20:

$$P_{tc} = \frac{50 \times \frac{1000}{6.9} f_t}{\frac{1000}{6.9} f_s - 195\left(\frac{9}{5}T + 32\right)}$$

$$P_{tc} = \frac{50 \times \frac{1000}{6.9} \times 2.77}{\frac{1000}{6.9} 336 - 195\left(\frac{9}{5} 35 + 32\right)}$$

$$= 0.67\%$$

Determine the minimum longitudinal steel to satisfy the concrete to steel strength ratio criterion; use Equation 7.21:

$$P_{c/s} = 100 \frac{f_t}{f_y}$$

$$P_{c/s} = 100 \times \frac{2.77}{448.5}$$

$$= 0.62\%$$

Since the minimum steel to resist subgrade restraint is the maximum, this condition governs.

Area of cross-section of slab/m width of pavement = 305 mm (thickness of slab) × 1000

= 304400 mm^2

Area of steel/m width = (0.8/100)× 304400 mm^2

= 2435 mm^2. This can be provided by using #7 bars at 152 m centers (see Table 7.40b).

EXAMPLE 7.19

Estimating the Area and Spacing of Transverse Steel in Continuously Reinforced Airport Concrete Pavements

Determine the minimum transverse steel required for the slab in Example 7.16 and the input values in Example 7.18.

Solution

Determine the minimum transverse steel from Equation 7.22 using the following input values:

Width of slab = 7.5 m

Friction factor = 1.8

Allowable working stress in steel = 336.4 N/mm^2

Minimum transverse steel percent $= P_{ts} = 1.1314 \times 10^{-3}\frac{W_sF}{f_s}$

(Equation 7.22)

$$= 1.1314 \times 10^{-3} \times \frac{7.5 \times 10^3 \times 1.8}{336.4}$$

$$= 0.046\%$$

Area of steel/m width of slab $= (0.046/100) \times 305$ (thickness of slab) $\times 1000$

$= 140$ mm^2/m width of slab

From Table 7.40b, we can use #3 bars at 457 mm centers.

Effects of Pumping Pumping is another important factor that should be considered in the design of rigid pavements. This is the discharge of water and subgrade (subbase) material through the joints and cracks within the pavement and along the pavement edges. It is primarily caused by the repeated deflection of the pavement slab in the presence of accumulated water beneath it. The water is formed in voids that are created by the intermixing of soft subgrade soils and aggregate base or subbase as a result of repeated loading. A major design consideration for preventing pumping is the reduction or elimination of expansion joints, since it is usually associated with these joints. Pumping can also be eliminated by either chemically or mechanically stabilizing the susceptible soil or by replacing it with a nominal thickness of granular or sandy soils. For example, some highway agencies recommend the use of a 76 to 152 mm layer of granular subbase material at areas along the pavement alignment where the subgrade material is susceptible to pumping. Alternatively, the susceptible material may be stabilized with asphalt material or Portland cement. Also, geotextiles may be used to separate the fine-grained subgrade soil from the overlying pavement aggregates to avoid the mixing of these materials.

The AREMA Method for the Design of Railway Tracks

The basic principle adopted in this procedure is similar to that for highways and airport runways in that the track should be capable of maintaining its functionality and be structurally sound. The functionality refers to the ability of the rail support to ensure a stable wheel–rail interaction, the effective distribution of the applied forces, the damping of rail vibrations, and the ability to minimize frictional movement between wheel and rail. The track's structural ability is its capability to withstand the stresses caused by the dynamic loading applied by the train wheels. The design parameters are the dynamic load applied onto the track by the wheels, the rail support modulus, the maximum allowable tie-ballast bearing pressure, the maximum allowable stress on the subgrade, the contact stress between the tie plate and the tie, and the stresses due to flexure and fatigue on the rail.

The rail support modulus (k_r) is defined by AREMA as the load (in pounds) that causes a 25.4 mm vertical rail deflection/linear inch of track. Factors that influence the value of k_r are the quality of the ties, tie spacing, tie dimensions, the quality of the ballast in terms of its cleanliness, moisture content, temperature, compaction and depth, and the load-carrying capacity of the subgrade.

The rail support modulus may be determined in the field using any available car or locomotive. The wheel loads P of one truck are first determined by placing the loaded car or locomotive on a scale. A measuring stick is vertically attached to the rail web at the location selected for the test. The loaded car or locomotive is then moved at a speed of about 8 km/hr along the track. The deflection of the rail, w_m, when the first wheel is directly above the measuring stick is determined using a level that is located about 18.2 m from the track. The ratio w_m/P is determined and then used in Figure 7.47 to determine the corresponding k_r value for two axle trucks.

AREMA recommends the following maximum allowable stresses:

Maximum allowable tie-ballast bearing pressure $= 0.4485$ N/mm^2

Maximum allowable stress on the subgrade $= 0.1725$ N/mm^2
(it is recommended that lower values be used even for good subgrades but should be definitely reduced for poor-quality subgrades)

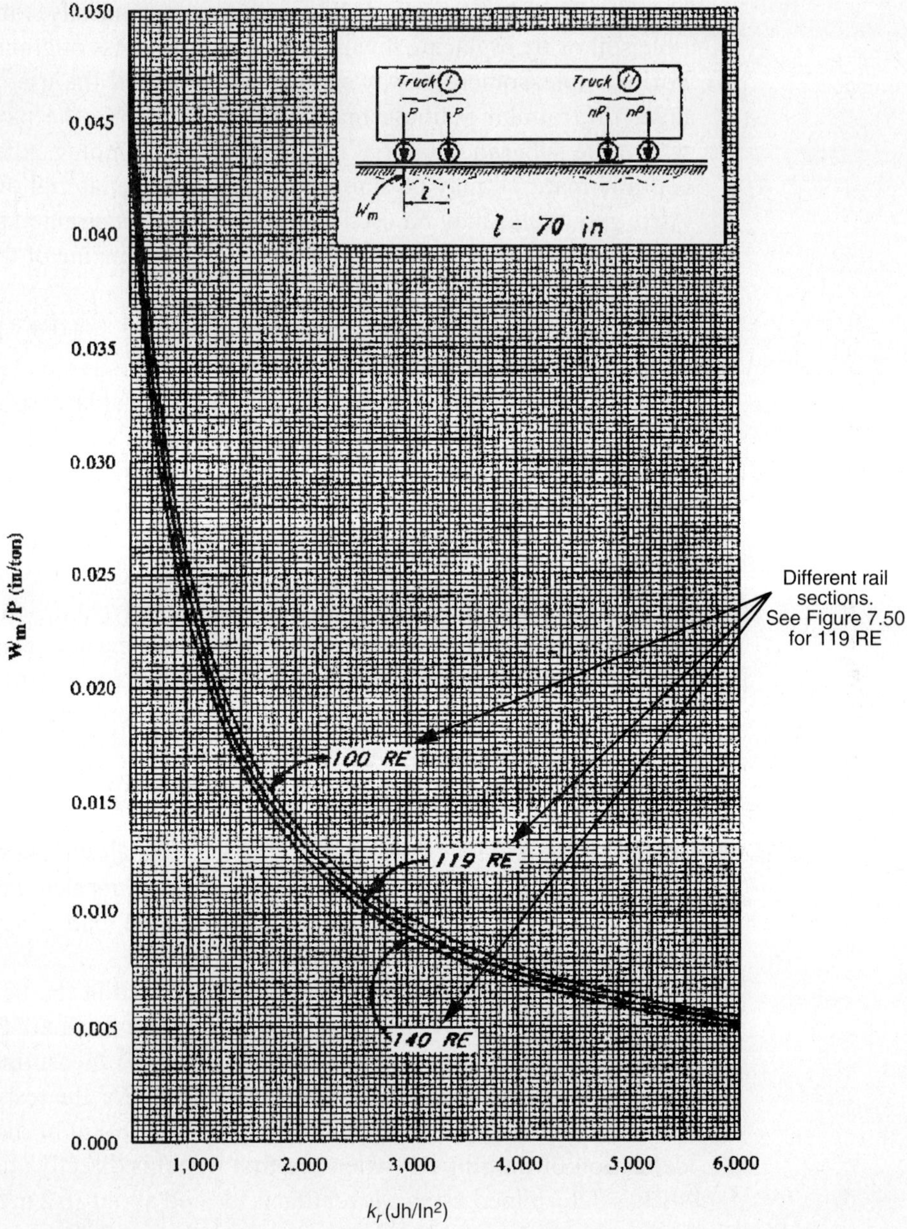

FIGURE 7.47 Master chart for determination of *k*.

Source: Manual for Highway Engineering, American Railway Engineering and Maintenance-of-Way Association, Landover, MD, 2005.

Maximum allowable contact stress between tie plate and tie (for wood) = 1.38 N/mm^2 (since tests have shown that this varies from 2.76 N/mm^2 for hard wood to 1.725 N/mm^2 for softwood)

Maximum allowable bending stress in the rail = 172 .5 N/mm^2

The allowable bending and fatigue stresses in continuous rail steel are obtained from the yield stress and the endurance strength of the rail steel after

TABLE 7.41

AREMA Recommended Reduction Factors to Obtain Permissible Rail Stress for Continuously Welded Rails

Source: Manual for Highway Engineering, American Railway Engineering and Maintenance-of-Way Association, 2005.

Influencing Factor	*Reduction Factor, Severity Assumption (Note 1)*
Lateral bending	20%
Track condition	25%
Rail wear and corrosion	15%
Unbalanced elevation	15%
Temperature stress	20,000 psi

Note1: Actual conditions may be substantially different which require the reduction factors to be modified accordingly.

Note: 1000 psi = 6.9 N/mm²

adjustments have been made for the influencing factors shown in Table 7.41. Table 7.41 also gives AREMA reduction factors for each of the influencing factors and the temperature stress. For example, if the rail has a yield stress of 65,000 lb/in², the permissible rail stress due to flexure and fatigue is given as:

$$\frac{448.5 - 138}{1.2 \times 1.25 \times 1.15 \times 1.15}\text{ N/mm}^2 = 156.5\text{ N/mm}^2$$

Depth of Rail Track Ballast The minimum total depth of the ballasts (ballast and subballast) required below the ties can be determined by using either the Talbot equation, the Boussinesq equation, or Love's formula.

The Talbot equation is given as

$$P_c = 958\frac{P_m}{h^{1.25}} \tag{7.23}$$

where

P_c = maximum intensity of pressure on the subgrade N/mm² (maximum value = 0.1725 N/mm²)
h = depth of ballast below ties, mm
P_m = intensity of pressure on ballast
= $(2q)/A_b$ N/mm²(maximum value = 0.4485 N/mm²)

The Boussinesq equation is given as

$$P_c = \frac{6q_b}{2\pi h^2} \tag{7.24}$$

where

P_c = maximum intensity of pressure on the subgrade (0.1725 N/mm²)
h = depth of ballast below ties, inches
q_b = intensity of pressure on ballast (maximum value = 0.4485 N/mm²)

Love's formula is given as

$$P_c = P_m\left[1 - \left(\frac{1}{1 + \frac{r^2}{h^2}}\right)^{\frac{3}{2}}\right] \tag{7.25}$$

where

P_c = maximum intensity of pressure on the subgrade (0.1725 N/mm^2)
h = depth of ballast below ties
P_m = intensity of pressure on ballast (0.4485 N/mm^2)
r = radius of a uniformly loaded circle whose area equals the effective tie bearing area under one rail seat

EXAMPLE 7.20

Determining the Total Depth of Ballasts Using the Talbot Equation

Using the Talbot equation, determine the required total ballast depth below the base of the wooden ties if the maximum allowable pressure on the ballast is 0.3795 N/mm^2 and the allowable maximum pressure on the subgrade is 0.138 N/mm^2

Solution

Use Equation 7.23 to solve for depth of ballast below ties (h):

$$P_c = 958 \times \frac{P_m}{h^{1.25}}$$

$$0.138 = 958 \times \frac{0.3795}{h^{1.25}}$$

$$h^{1.25} = 2634.5$$

$$h = 545 \text{ mm}$$

Width of Ballast Shoulder at Ends of Ties In order to provide lateral support for the track, the width of the ballasts should be extended beyond the ends of the ties. AREMA has noted that if a tie is buried 102 mm into the ballast with a 152 mm ballast shoulder and carrying no vertical load, a force of approximately 446.4 kg/m will be required to move the tie 25.4 mm. However, the width at curves depends on the lateral force that is produced by the continuously welded rail on the curved track due to temperature changes. This force is given as

$$P_L = 1.1812D(\Delta T) \tag{7.26}$$

where

P_L = lateral force, lb/linear m
D = degree of curve
ΔT = temperature change, °C

The total lateral force acting between ties is therefore given as the lateral force/linear ft, P_L multiplied by the tie spacing (ft). The additional width is obtained by dividing this force by the force that will cause the tie to move 25.4 mm. It should be noted that when nonthermal longitudinal forces are present such as those that occur at grades or when braking or tractive forces are applied, the actual force may be greater than that calculated from Equation 7.26. Also, it is known that because of the rail uplift wave, track buckling frequently occurs immediately ahead of or under a moving train. AREMA therefore suggests that wider ballast shoulders may be required to facilitate adequate lateral stability. Experience and local conditions should be used to determine when this is necessary.

EXAMPLE 7.21

Determining the Width of Ballast Shoulders at Ends of Ties

Determine the minimum width of ballast shoulder that is required at a curved rail road track for the following conditions:

Degree of curvature = 9°

Temperature change = 38.9°C

Tie spacing = 495 mm

Solution

Determine the lateral force, using Equation 7.26:

$$P_L = 1.1812D(\Delta T)$$
$$= 1.1812 \times 9 \times 38.9$$
$$= 4135 \text{ N/linear m}$$

Determine the total force on each tie:

$$= 4135 \times \frac{495}{1000} = 2040 \text{ N}$$

Determine the total width of ballast shoulder (i.e., total for both ends of a tie):

$$= \frac{2040}{446.4} = 0.457 \text{ m}$$

Determine the width of ballast at each end of a tie:

$$= \frac{0.457}{2} = 0.228 \text{ m}$$

Determination of Track Rail Cross-Section A brief discussion of the procedure to determine the track rail cross-section is given. However, specific details relating to the use of the associated equations are beyond the scope of this book, but interested readers may refer to any structural analysis textbook. The cross-section of the rail is selected to ensure that the bending stresses within the rail do not exceed the maximum allowable. These stresses depend on the bending moment and deflection caused by the imposed wheel loads.

The bending moment and deflection are determined from the basic differential equation given as

$$EI\frac{d^4w}{dx^4} + k_r w = q(x) \tag{7.27}$$

where

E = Young's modulus of rail steel
I = moment of inertia of one rail with respect to the horizontal centroidal axis
w = vertical track deflection
q = vertical load distribution (wheel loads) on the rail
x = point on the rail cross-sectional axis
k_R = elastic modulus of rail support for one rail
kw = p = distributed rail-tie contact pressure

Solving Equation 7.27 for the magnitude of the deflection $w(x)$at point x and the bending moment $M(x)$ for a single wheel load gives the following:

$$w(x) = \frac{\beta P^d}{2k_r}e^{-\beta x}[\cos \beta x + \sin \beta x] = \frac{\beta P^d}{2k}\lambda(\beta x) \tag{7.28}$$

$$M(x) = \frac{P^d}{4\beta}e^{-\beta x}[\cos \beta x - \sin \beta x] = \frac{P^d}{4\beta}\mu\beta x \tag{7.29}$$

where

$\beta = \sqrt[4]{\frac{k_R}{4EI}}$ stiffness ratio
$\mu(\beta x)$ and $\lambda(\beta x)$ can be obtained from Figure 7.48
P^d = dynamic wheel load = $(1+\theta)P$
θ = coefficient of impact = 33×0.621 speed (km/hr)/(3.937 $\times$ diameter of wheels (mm)) (see Equation 7.4)
P = static load

Since the maximum deflection and therefore the maximum bending stress occurs at the top of the rail that is immediately underneath the wheel, the maximum deflection and maximum bending moment occur at $x = 0$, which gives

$$w_{\max} = w(x = 0) = \frac{\beta P^d}{2k} \tag{7.30}$$

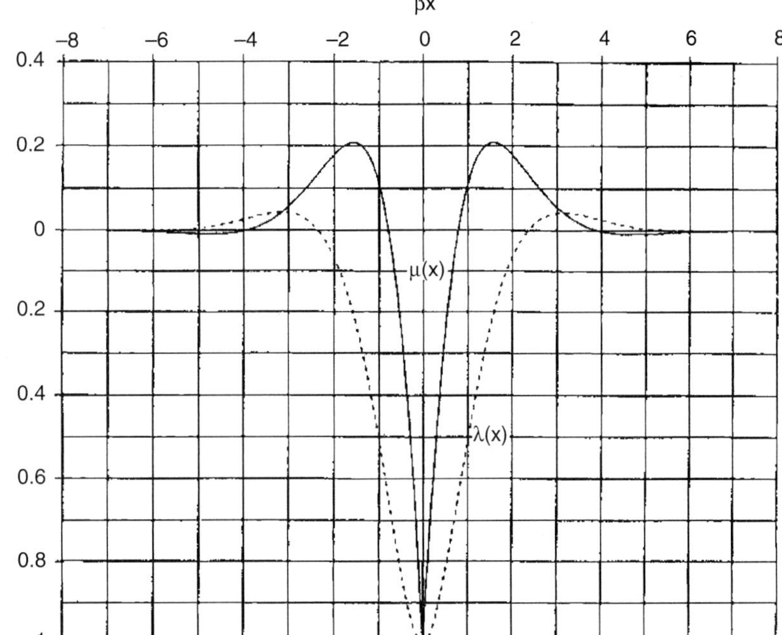

FIGURE 7.48
Influence curves.

Source: Manual for Highway Engineering, American Railway Engineering and Maintenance-of-Way Association, Landover, MD, 2005.

$$M_{max} = M(x = 0) = \frac{P^d}{4\beta} \tag{7.31}$$

The maximum dynamic bending stress in the rail is given as

$$\sigma^d_{max} = \frac{cM^d_{max}}{I} = \frac{M^d_{max}}{Z_b} \tag{7.32}$$

and for design purposes the section modulus of a suitable rail cross-section is obtained from

$$Z_{req} \geq \frac{M^d_{max}}{\sigma^d_{all}} \tag{7.33}$$

where

c = distance from the neutral axis to the rail base
Z_b = section modulus for the rail base
I = moment of inertia of one rail with respect to the horizontal centroid axis

Examples of rail sections recommended by AREMA are given in Figures 7.49 to 7.51. Note that when the axles of a rail truck are closely spaced, more than one wheel load may simultaneously cause deflection and bending at a section of the rail. The combined effect of all wheel loads should be considered in determining w_{max} and M_{max} by using the influence curve given in Figure 7.48. This procedure is beyond the scope of this book.

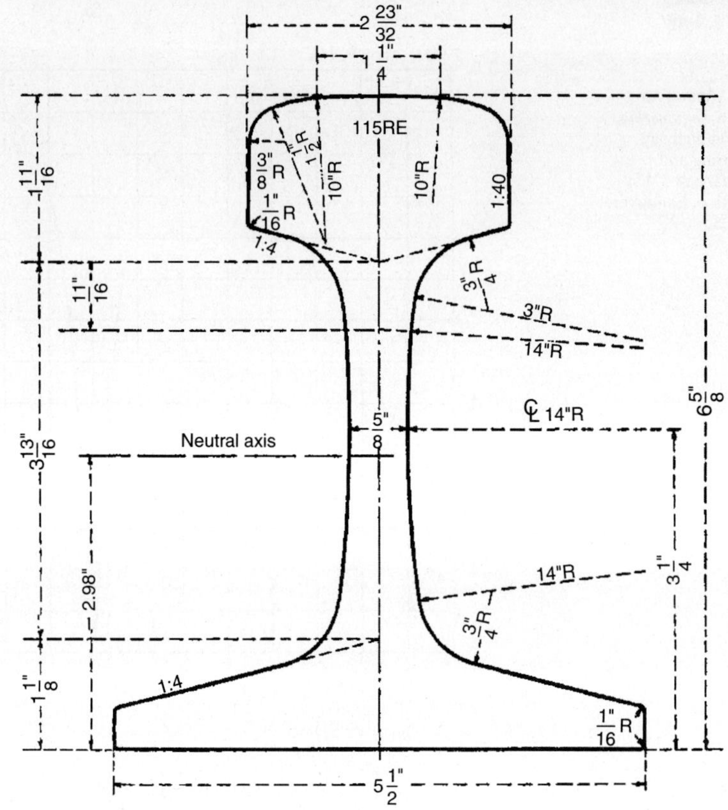

FIGURE 7.49

115RE rail section.

Source: Manual for Highway Engineering, American Railway Engineering and Maintenance-of-Way Association, Landover, MD, 2005.

1. Rail area (square inch)	
Head	3.9156
Web	3.0363
Base	4.2947
Whole rail	11.2465
2. Rail weight (lb/yd) (based on specific gravity of rail steel = 7.84)	114.6758
3. Moment of inertia about the neutral axis	65.9
4. Section modulus of the head	18.1
Section modulus of the base	22.0
5. Height of neutral axis above base	3.00
6. Lateral moment of inertia	10.7
7. Lateral section modulus of the head	7.90
Lateral section modulus of the base	3.90
8. Height of shear center above base	1.45
9. Torsional rigidity is 'KG' where G is the modulus of rigidity and K = (error for K greater than 10%)	4.69

Note: 1″ = 25.4 mm

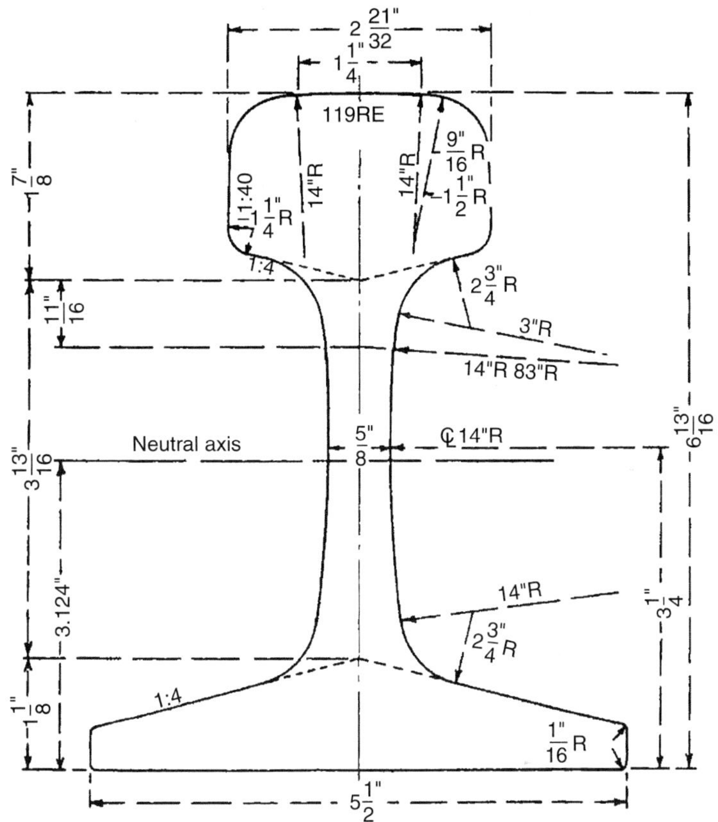

FIGURE 7.50
119RE rail section.

Source: Manual for Highway Engineering, American Railway Engineering and Maintenance-of-Way Association, Landover, MD, 2005.

1. Rail area (square inch)	
Head	4.3068
Web	3.0363
Base	4.2946
Whole rail	11.6378
2. Rail weight (lb/yd) (based on specific gravity of rail steel = 7.84)	118.6657
3. Moment of inertia about the neutral axis	71.4
4. Section modulus of the head	19.4
Section modulus of the base	22.8
5. Height of neutral axis above base	3.13
6. Lateral moment of inertia	10.8
7. Lateral section modulus of the head	8.16
Lateral section modulus of the base	3.94
8. Height of shear center above base	1.51
9. Torsional rigidity is 'KG' where G is the modulus of rigidity and K = (error for K greater than 10%)	5.11

Note: 1″ = 25.4 mm

FIGURE 7.51

132RE rail section.

Source: Manual for Highway Engineering, American Railway Engineering and Maintenance-of-Way Association, Landover, MD, 2005.

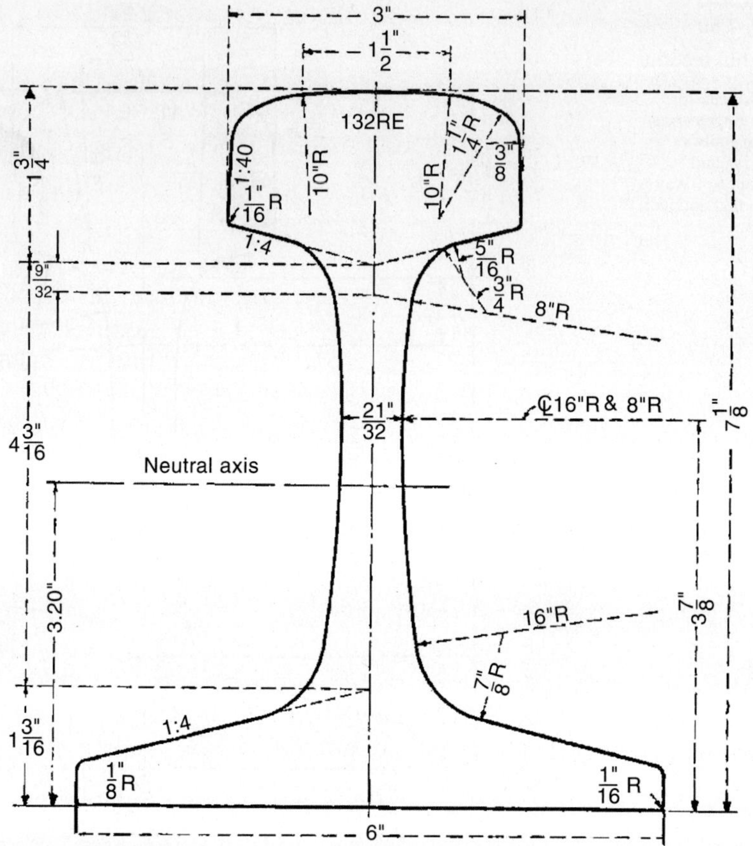

1. Rail area (square inch)	
Head	4.4274
Web	3.6149
Base	4.8701
Whole rail	12.9124
2. Rail weight (lb/yd) (based on specific gravity of rail steel = 7.84)	131.6622
3. Moment of inertia about the neutral axis	87.9
4. Section modulus of the head	22.4
Section modulus of the base	27.4
5. Height of neutral axis above base	3.20
6. Lateral moment of inertia	14.4
7. Lateral section modulus of the head	9.57
Lateral section modulus of the base	4.79
8. Height of shear center above base	1.57
9. Torsional rigidity is 'KG' where G is the modulus of rigidity and K = (error for K greater than 10%)	5.31

Note: 1″ = 25.4 mm

EXAMPLE 7.22

Determining a Suitable Track Rail Cross-Section

Determine a suitable track rail cross-section for a railroad track supporting trains traveling at 88 kmph with a single static wheel load of 16330 kg and wheel diameter of 915 mm. Assume that Young's modulus of rail steel is 2×10^5 N/mm^2 and the elastic modulus of the rail support for one rail (k_R) is 20.7 N/mm^2

Solution

Determine the dynamic wheel load; use Equation 7.4:

$$P^d = (1 + \theta)P$$

where

P^d = dynamic wheel load
P = static wheel load
θ = coefficient of impact = $33 \times 0.621v/(3.937D)$
u = dominant train speed, mi/hr
D = diameter of vehicle wheels, mm

$$\theta = 33 \times \frac{0.621 \times 88}{3.937 \times 915}$$

$$\theta = 0.5$$

$$P^d = (1 + 0.5) \times 16330$$

$$P^d = 244940 \text{ N}$$

Determine the maximum dynamic bending moment; use Equation 7.31:

$$M_{\max} = M(x = 0) = \frac{P^d}{4\beta}$$

$$\beta = \sqrt[4]{\frac{k_R}{4EI}}$$

Since the moment of inertia (I) is required, let us assume a rail section and determine whether it is adequate. Assume a 119RE rail section as shown in Figure 7.50:

$$I = 71.4 \text{ in}^4 = 2.9718 \times 10^7 \text{ mm}^4$$

$$\beta = \sqrt{\frac{20.7}{4 \times 2 \times 10^5 \times 2.9718 \times 10^7}} = 9.6597 \times 10^{-4}\text{/mm}$$

$$M_{\max} = 244940 \times 9.81/(4 \times 9.6597 \times 10^{-4}) = 6.218 \times 10^7 \text{ N mm}$$

Determine the maximum stress in the rail; use Equation 7.32:

$$\sigma^d_{max} = \frac{cM^d_{max}}{I} = \frac{M^d_{max}}{Z_b} \; (Z_B = 22.8 \text{ in}^3 = 3.736 \times 10^5 \text{ mm}^3 \text{ section modulus for railbase from Figure 7.50})$$

$$= \frac{6.218 \times 10^7}{3.736 \times 10^5} = 166.43 \text{ N/mm}^2$$

Maximum allowable bending stress in rail = 172.5 N/mm²

Maximum stress in rail < maximum allowable

Therefore, 119RE rail section can be used.

Determination of Tie Plate Size It is necessary to determine the required size of the tie plate to ensure that the contact stress between the tie plate and tie is not higher than the specified maximum value. The area of the tie plate is obtained from Equation 7.34:

$$A_{req} \geq \frac{F^d_{max}}{\sigma_{allowed}} \tag{7.34}$$

where

A_{req} = area of tie plate
F^d_{max} = rail seat load
$\sigma_{allwood}$ = allowable contact stress between tie plate and tie AREMA recommends the use of 1.38 N/mm² for design analysis.

The rail seat load is a function of the intensity of the continuously distributed load, p, against the underside of the rail and the deflection at the point that is the maximum deflection, and the elastic modulus of rail support for one rail. When there is more than one wheel load, the influence curve in Figure 7.48 is used to obtain p_{max}. For a single wheel load, it is given as

$$F^d_{max} = p_{max}\,(a) \tag{7.35}$$

where

a = tie spacing

$$p_{max} = \text{rail base pressure} = k_R w_{max} = \frac{k_R \beta P^d}{2k_R} = \frac{\beta P^d}{2} \tag{7.36}$$

$$\beta = \sqrt[4]{\frac{k_R}{4EI}}$$

P^d = dynamic wheel load

TABLE 7.42

Design of Tie Plates for Use with AREMA Rail Sections

Source: Manual for Highway Engineering, American Railway Engineering and Maintenance-of-Way Association, 2005.

Rail		Plate	
AREMA Rail Section	**Base Widths**	**Width Inches**	**Length Inches**
140RE, 136RE, 133RE, 132RE	6 inch	8	18
		7 3/4	16
		7 3/4	14 3/4
		7 3/4	14
		7 3/4	13
119RE, 115RE	5 inch	7 3/4	15
		7 3/4	14
		7 3/4	13
		7 3/4	12
100RE	5⅜ inch	7 3/4	12
		7 1/2	11
90RA-A	5⅛ inch	7 1/2	11
		7 1/2	10

Note 1: All tie plate sections canted 1:40.
All tie plate sections have inclined ends
1 inch = 25.4 mm

Table 7.42 gives recommended sizes of tie plates that should be used for different AREMA rail sections. Detailed designs of these tie plates are given in the AREMA *Manual for Railway Engineering.*

EXAMPLE 7.23

Determining the Tie Plate Size

Determine the size of the tie plate that will be required for the rail obtained in Example 7.22. The ties are spaced at 610 mm centers.

Solution

Determine the rail base pressure, use Equation 7.36:

$$p_{\max} = \text{rail base pressure} = k_R w_{\max} = \frac{k_R \beta P^d}{2k_R} = \frac{\beta P^d}{2}$$

$$= \frac{9.6597 \times 10^{-4} \times 244940 \times 9.81}{2}$$

$$= 116 \text{ N/mm}$$

Note: $\beta = 9.6597 \times 10^{-4}$/mm (from Example 7.22)
$P^d = 244940$ N (from Example 7.22)

Determine the tie plate size; use Equation 7.34:

$$A_{req} \geq \frac{F^d_{max}}{\sigma_{allwood}} \geq \frac{P^d a}{\sigma_{allwood}} \geq \frac{116 \times 610}{1.38} = 5.127 \times 10^4 \text{ mm}^2$$

Select AREMA 304.8 mm × 196.8 mm (12″ × $7\frac{3}{4}$″) plate, which is suitable for 119RE rail. This provides an area of 93 in^2.

Determination of Effective Bearing Area of Tie It is also necessary to determine the minimum effective bearing area of the tie to ensure that the contact pressure between tie and ballast is not greater than the maximum allowable. The maximum contact pressure occurs at the rail seat and the minimum is at the tie center. In order to simplify the calculations, the pressure distribution along the tie length shown in Figure 7.52 is assumed. The effective length (L_{eff}) of the tie is therefore considered to be one-third of its length (L) and the effective bearing area (A_b) of the tie is given as

$$A_b = b \times L_{eff} = (b \times L)/3 \tag{7.37}$$

where

b = width of the tie at its base

The corresponding tie-ballast bearing pressure is given as

$$\sigma_{tb} = \frac{3F^d_{max}}{bL} \leq 65 \text{ lb/in}^2 \ (0.448 \text{ N/mm}^2) \tag{7.38}$$

where

σ_{tb} = tie-ballast bearing pressure

$$F^d_{max} = p_{max}(a) \tag{7.40}$$

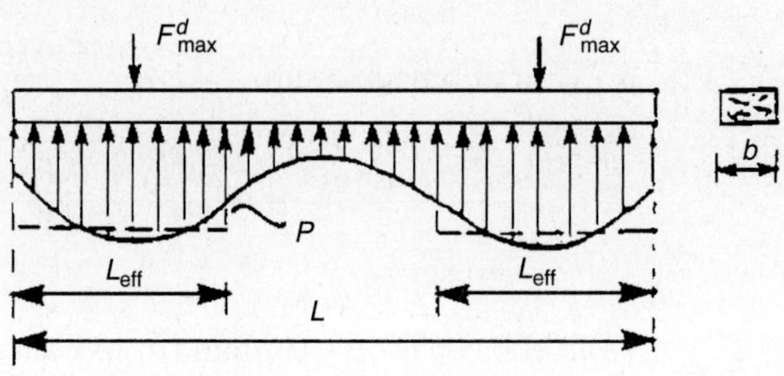

FIGURE 7.52
Pressure distribution along length of tie.

Source: Manual for Highway Engineering, American Railway Engineering and Maintenance-of-Way Association, Landover, MD, 2005.

where

a = tie spacing

p_{max} = rail base pressure = $k_R w_{max} = \dfrac{k_R \beta P^d}{2k_R} = \dfrac{\beta P^d}{2}$

$\beta = \sqrt[4]{\dfrac{k_R}{4EI}}$

P^d = dynamic wheel load

EXAMPLE 7.24

Determining Tie Ballast Bearing Pressure

Determine the bearing pressure imposed by the ties selected for the track in Example 7.23 if they are spaced at 610 mm centers. Assume that the length of each tie is 2590 mm in and the width is 203.2 mm.

Solution

Determine the dynamic force (F^d_{max}) imposed by each tie on the ballast using Equation 7.40:

$$F^d_{max} = p_{max}(a)$$

$$p_{max} = 116 \text{ N/mm (see Example 7.23)}$$

$$F^d_{max} = 116 \times 610$$

$$= 7.076 \times 10^4 \text{ N}$$

Determine the tie ballast bearing pressure; use Equation 7.38:

$$\sigma_{tb} = \frac{3F^d_{max}}{bL} \leq 0.448$$

$$= 3 \times 7.076 \times 10^4/(2590 \times 203.2) = 0.404 \text{ N/mm}^2 \leq 0.448 \text{ N/mm}^2.$$

SUMMARY

This chapter presents the basic principles used in the structural design of travelways for the highway, airport, and rail modes. It is clear that, regardless of the mode being considered, the basic principles used in the design are the same, although the application of these principles may be different from one mode to another. For example, the identification of a suitable subgrade material for the highway pavement, the airport pavement, or the rail track is based mainly on the classification of the subgrade soil with respect to its grain size distribution and its plastic characteristics. The specific classification system used for a given mode may, however, be different from that used for another mode. Similarly, the travelway of each mode is designed so that the stress on the subgrade due to the imposed vehicle load does not cause excessive or permanent deformation of the subgrade. Each mode uses the basic principle of transmitting the imposed wheel load through a number of structural components that make up the travelway. The structural components for the highway and airport modes are the surface, base, and subbase. Those for the rail mode are the rails, ties, ballast, and subballast. The fundamental principle used in designing these structural components is that each component should be capable of withstanding the stress imposed on it by the vehicles using the travelway. Different methodologies that illustrate this fundamental principle are presented. It should be noted, however, that the chapter does not fully cover all the factors that may be considered in the actual design of these structural components, as some of these are beyond the scope of this book.

PROBLEMS

7.1 Compare and contrast the material characteristics of the different structural components of the highway pavement, the rail track and the airport pavement.

7.2 What is the basic principle used to identify suitable soil materials for the subgrade of a travelway? Describe how this principle is used to identify suitable subgrade materials for the airport pavement, the highway pavement and the rail track.

7.3 Describe the three design methods that are used in the FAA design methodology for airport pavements to compensate for soils that are susceptible to frost penetration.

7.4 The characteristics of a soil sample are given below. Determine whether this soil is suitable for use as:

i. subgrade material for a highway pavement
ii. subgrade material for an airport pavement
iii. subgrade material for a rail track

Sieve Analysis: % finer by weight:

4 – 53 %
10 – 42%
40 – 40%
200 – 25%

Liquid Limit = 20%

Plastic Limit = 12%

Coefficient of curvature $C_c = 7$

Coefficient of uniformity $C_u = 2.5$

7.5 An existing 4-lane (two lanes in each direction) primary highway has a flexible pavement and is carrying a current AADT of 6500 vehicles in one direction. If the road is to be re-constructed to meet with interstate highways standards, and the re-construction is expected to be completed three years from now, determine the design ESAL for a design life of 20 years. The vehicle mix and axle loads are given below and the growth rate for all vehicles is 4% per annum.

Passenger cars (1000 lb/axle) = 60%
2-axle single unit trucks (8000 lb/axle) = 30%

3-axle single unit trucks (12,000 lb/axle = 10%

$P_t = 2.5$

Assume Structural Number, SN = 4

7.6 Determine the equivalent annual departures and design load for an airport pavement if the average annual departures and maximum take-off weight of each aircraft type expected to use the runway is given in the table below. Assume that the 737-200 is the design aircraft:

Aircraft	*Gear Type*	*Average Annual Departures*	*Max Take-Off Weight (N)*
727-100	Dual	2500	589670
727-200	Dual	3500	612350
707- 320B	Dual tandem	2000	1247370
DC-10-30	Dual	4800	476270
737-200	Dual	15,350	568350

7.7 An existing end-level straight section of a rail track was designed to carry a single static wheel load of 90720 N for a train having a wheel diameter of 0.762 m and a dominant train speed of 33.5 m/s. The rail authority is considering the use of a different type of train that will be expected to travel at a dominant speed of 38.20 m/s, with a wheel diameter of 0.914 m and a static wheel load also of 90720 N. Determine whether this can be done without checking the size of thc rails. Assume all other conditions remain the same.

7.8 The subgrade M_r values for a proposed flexible pavement road are 138, 138, 34.5, 34.5, 34.5, 62.1, 62.1, 62.1, 65.55, 65.55, 51.75 and 138 N/mm^2 for each month from January through December respectively. Determine the effective modulus of the subgrade that is equivalent to the combined effect of the different seasonal moduli.

7.9 The first year AADT on a six-lane interstate highway located in an urban area is expected to be 10,500 in one direction. The growth rate of two-axle single unit trucks 45360 N/axle is expected to be 5% per annum during the first five years of the pavement life and will increase to 6% per annum for the remaining life of the pavement while the growth rate for all other vehicles is expected to be 4% per annum throughout the life of the pavement. Determine the design ESAL for a 20-year design life.

The projected vehicle mix during the first year of operation is:

Passenger cars (45360 N/axle) = 83%

Two-axle single-unit trucks (45360 N/axle) = 10%

Two-axle single-unit trucks (5443 Kg/axle) = 5%

Three-axle single-unit trucks (6350 Kg/axle) = 2%

$P_i = 3.5$

$P_t = 2.5$

$f_d = 0.7$

Assume SN =4

7.10 The effective resilient modulus M_r of the subgrade of the pavement in Problem 103.5 N/mm^2. Using the AASHTO method, determine whether the assumption of SN = 4 is correct. If this assumption is wrong, what action should the designer take? Use overall standard deviation of 0.40, a reliability level R of 90%, an initial serviceability index of 3.5, and a terminal serviceability index of 2.5.

7.11 A flexible pavement is to be designed to carry the design ESAL obtained in Problem 7.9. The effective resilient modulus M_r of the subgrade of the pavement is 103.5 N/mm^2, the subbase layer is an untreated sandy soil with an effective M_r of 120.75 N/mm^2 and the base material is an untreated granular material with M_r of 186.3. The pavement structure will be exposed to moisture levels approaching saturation 20% of the time and it will take about 1 week to drain the base layer to 50% saturation. Using an SN of 4 obtained in Problem 7.10, determine appropriate depths for the subbase, base, and asphalt pavement layers.

Elastic modulus E_{AC} of the asphalt concrete at 20°C is 3105 N/mm^2

7.12 The flexible pavement of a collector road located in a rural area is being designed to carry a design ESAL of 0.55 × 10^6. The CBR of the subgrade is 8. Select appropriate subbase and base materials, and determine the depth of each layer of the pavement. The M_r of the asphalt surface material is 2760 N/mm^2.

7.13 An airport flexible pavement is being designed to carry 15,000 equivalent annual departures for the A-300 Model B2 aircraft with a gross weight of 907180 N. If the only subbase material available at the vicinity of the site has a CBR value of 12 and the engineer wishes to use the minimum specified thickness for the hot-mix asphalt surfacing, determine the depths of the base and subbase layers. The subgrade has a CBR of 6.

7.14 Determine whether the assumption that the 737-200 aircraft is the design aircraft in Problem 7.6 is correct.

7.15 Using the data given in Problem 7.6 and your answer for Problem 7.14, determine the depth of each structural component for a flexible runway pavement if it consists of a hot-mix asphalt concrete surface, a base course and a subbase course. The CBR values for the subgrade and subbase are 8 and 15, respectively.

7.16 Repeat Problem 7.15 if the CBR of the subgrade is 5, that of the subbase is 12 and the depth for the base course is restricted to maximum 381 mm because of scarcity of the material.

7.17 Briefly describe the four general types of rigid pavements.

7.18 Using the AASHTO rigid pavement design method and the input variables given below, determine the required thickness of a concrete highway pavement to carry an accumulative equivalent single-axle load of 2.0×10^6. The subbase is an 203.2 mm layer cement-treated granular material, and the seasonal values for the roadbed resilient modulus and the subbase elastic modulus are given in the following table:

Loss of support (LS) = 1

Elastic modulus of concrete (E_c) = 34500 N/mm^2

Modulus of rupture of concrete to be used in construction (S_c) = 4.485 N/mm^2

Load transfer coefficient (J) = 3.2

Drainage coefficient (C_d) = 1.0

Overall standard deviation (S_o) = 2.9

Reliability level = 95% (Z_R = 1.645)

Initial serviceability index (P_i) = 4.5

Terminal serviceability index (P_t) = 2.5

(1) Month	(2) Roadbed Modulus M_r (N/mm^2)	(3) Subbase Modulus E_{SB} (N/mm^2)
January	124.2	310.5
	124.2	310.5
February	124.2	310.5
	124.2	310.5
March	27.6	124.2
	27.6	124.2
April	34.5	138
	34.5	138
May	27.6	124.2
	27.6	124.2
June	55.2	172.5
	55.2	172.5
July	55.2	172.5
	55.2	172.5
August	55.2	172.5
	55.2	172.5
September	55.2	172.5
	55.2	172.5
October	55.2	172.5
	55.2	172.5
November	55.2	172.5
	55.2	172.5
December	124.2	310.5
	124.2	310.5

7.19 Determine the longitudinal and transverse steel that will be required for the slab in Problem 7.18 if the slab is a jointed reinforced pavement with a joint spacing of 13.7 m and width of 7.3 m. The yield strength of the steel is 414 N/mm^2.

7.20 Repeat Problem 7.19 for a continuously reinforced pavement using 5/8-″ (#5) bars and the following input data:

Wheel load = 81650 N

Effective modulus of subgrade reaction (k) = 0.051 N/mm^3

Concrete tensile strength, f_t = 3.45 N/mm^2

Concrete shrinkage = 0.0004

Thermal coefficient of steel $\alpha_s = 5 \times 10^{-6}$

Thermal coefficient of concrete $\alpha_c = 3.8 \times 10^{-6}$

Design temperature drop DT_D = 10°C (maximum temperature of 26.67°C and minimum temperature of −1.11°C)

Allowable steel stress σ_s = 0.414 N/mm^2

Width of lane = 3.66 m

7.21 An airport pavement is being designed to carry 22,000 equivalent annual departures for the A-300 Model B2 aircraft with a maximum wheel load of 1020580 N. If the subbase consists of 152.4 mm stabilized material and the modulus k of the subgrade is 0.0136 N/mm^3, determine the required depth of the concrete pavement. The concrete flexural strength is 4.480 N/mm.^{2}

7.22 Determine the longitudinal and transverse steel that will be required for the slab in Problem 7.20 if the slab is a jointed reinforced pavement with a joint spacing of 10.67 m and the paving lane width is 7.62 m. The yield strength of the steel is 414 N/mm^2

7.23 Determine the longitudinal steel that will be required for the slab in Problem 7.21 for a continuously reinforced pavement if the maximum seasonal temperature differential is 29.44°C, yield strength of steel is 448.5 N/mm^2, and width of lane is 3.66 m.

7.24 Using the Talbot equation, determine the required total ballast depth below the base of the wooden ties if the maximum allowable pressure on the ballast is 0.414 N/mm^2 and the allowable maximum pressure on the subgrade is 0.1242 N/mm^2.

7.25 Determine the minimum width of ballast shoulder that is required at a curved rail section with a 10° curvature if the ties are spaced at 508 mm intervals and the temperature change is 44.44°C.

7.26 A rail track is being designed to support trains traveling at 96 kmph with a single static load of 158750 N and a wheel diameter of 915 mm. Determine a suitable rail cross-section if the Young's modulus of the rail steel is 2×10^5 N/mm^2 and the elastic modulus of the rail support for one rail is 20.7 N/mm^2.

7.27 Determine the size of the tie plate that will be required for the rail obtained in Problem 7.26 if the ties are spaced at 610 mm centers. Also determine the bearing pressure imposed by the ties on the ballast if the ties are 2.59 m long and 203 mm wide.

References

1. *Airport Master Plans,* Advisory Circular AC 150/5070-6B, Federal Aviation Administration, Washington, D.C., 2005.
2. *AASHTO Guide for the Design of Pavement Structures,* American Association of State Highway and Transportation Officials, Washington, D.C., 1993.
3. *Airport Pavement Design and Evaluation,* Advisory Circular AC No. 150/5320-6D, Federal Aviation Administration, U.S. Department of Transportation, Incorporating Changes 1 through 5, Washington, D.C., April 2004.
4. *Annual Book of ASTM Standards*, Section 4, Vol 04.03, *Road and Paving Materials*; Pavement Management Technologies, American Society for Testing and Materials, Philadelphia, PA, 2003.

5. *Aerospace Forecasts, Fiscal Years 2006–2017*, U.S. Department of Transportation, Federal Aviation Administration, Office of Policy and Plans, Washington D.C., http:faa.gov/data_statistics.
6. *Manual for Railway Engineering,* American Railway Engineering and Maintenance-of-Way Association, Landover, MD, 2005.
7. *Standards for Specifying Construction of Airports,* Advisory Circular, AC No. 150/5370-10A, Incorporating Changes 1 through 14, Federal Aviation Administration, U.S. Department of Transportation, Federal Aviation Administration, Washington, D.C., 2004.
8. *Standards for Specifying Construction of Airports*, Advisory Circular 150/5370-10B, U.S. Department of Transportation, Federal Aviation Administration Washington, D.C., 2005.
9. *Standard Specifications for Transportation Materials and Methods of Testing,* 20th ed., The American Association of State Highway and Transportation Officials, Washington D.C., 2000.
10. *Superpave Mix Design,* Asphalt Institute, Superpave Series No. 2 (SP-2), Asphalt Institute, Lexington, KY, 2000.
11. *Traffic and Highway Engineering,* 3rd ed., Nicholas J. Garber and Lester A. Hoel, Brooks/Cole, Thompson Learning, 2002.

CHAPTER 8

Transportation Safety

The United States has developed an extensive transportation system that is unsurpassed in the world. The system has provided unprecedented mobility for all its citizens by combining a vast road and highway network with air, rail, and urban transit services. Freight moves from one corner of the globe to another via an intermodal network of shipping, ports, rail, and highway freight carriers. This impressive system is not without its failings, however, and perhaps the most critical problem facing the transportation industry today is to assure a safe environment for vehicle operators and passengers. This chapter discusses causes of safety problems, remedies for these problems, and programs for improving the safety performance of the nation's transportation system. Where statistics are cited, they are intended to provide context regarding the magnitude of the safety problem, and can be found in the references and Web sites furnished at the end of this chapter.

It is estimated that approximately 1.2 million people worldwide are killed and 50 million are injured on roadways throughout the world annually. As developing countries become motorized, as is the case in China, Thailand, and India, it can be expected that these numbers will increase significantly in the future. In the United States, over 40,000 people are killed in motor vehicle crashes each year, and many of these deaths, sadly, represent a youthful and vigorous segment of the population.

Aviation, which is considered to be a very safe mode of transportation, experiences approximately 10 to 15 crashes per 10 million flights worldwide. However, with the expansion of air transportation in ever more crowded and congested conditions, it is estimated that over time the industry can expect the permanent loss of as much as one aircraft per week worldwide as the result of an air accident. Surprisingly, in recent years, there have been very few, if any, fatalities on U.S. scheduled airlines per year, although fatalities are not uncommon in the sector known as general aviation. The contrast between highway and aviation safety performance has confounded transportation experts, since the numbers of highway fatalities each year is so much greater than in the airways. It is believed that commercial airlines are safer because of the industry's emphasis on crash avoidance and the competence of airline pilots.

Rail and bus transit modes are considered to be relatively safe. Travel, on average, is considered to be about two to three times safer on a bus or train than in an airplane and about 40 times safer than in an automobile. Nevertheless, crashes do occur on buses and trains, and these often involve large numbers of passengers. Head-on collisions between passenger railroad trains are rare events, and infrequent but tragic bus–railroad grade crossing crashes have occurred that have resulted in both property damage and loss of life. Freight accidents have occurred involving hazardous material spills that spread over towns and communities. When crashes occur involving collisions between passenger vehicles and trucks or trains, the outcome is either fatal or injurious to the auto occupants.

ISSUES INVOLVED IN TRANSPORTATION SAFETY

Are they Crashes or Accidents?

The term *accident* is commonly accepted as an occurrence involving one or more transportation vehicles in a collision that results in property damage, injury, or death. The term *accident* implies a random event that occurs for no apparent reason other than "it just happened." Have you ever been in a situation where something happened that was unintended? Your immediate reaction might have been, "Sorry, it was just an accident."

In recent years, the National Highway Traffic Safety Administration (NHTSA) has suggested replacing the word *accident* with the word *crash*. Why is this so? Simply because the word *crash* is results oriented, implying that the vehicle collision may have been caused by any number of events. The crash could have been prevented or its effect minimized in several ways. Among the options are modifying driver behavior, improving vehicle design (called "crashworthiness"), modifying roadway geometry, and enhancing the traveling environment. *Crash* is not the term used by all transportation modes, and its most common usage is in the context of highway and traffic incidents. Both terms—*crash* and *accident*—are applied to nonhighway modes, and therefore the word *accident* is a commonly accepted descriptor of a crash.

What Are the Causes of Transportation Crashes?

The occurrence of a transportation crash presents a challenge to safety investigators. In every instance, the question arises, "What sequence of events or circumstances contributed to the incident that resulted in injury, loss of life, or property damage?" In some cases, the answer may be a simple one. For example, the cause of a single-car crash may be that the driver fell asleep at the wheel, crossed the highway shoulder, and crashed into a tree. In other cases, the answer may be complex involving many factors that, acting together, caused the crash.

One of the most noted disasters occurred in 1912 when the *Titanic*, an "unsinkable" ocean liner, went to the bottom of the sea with nearly 1200 passengers and crew off the coast of Nova Scotia. The general belief among most people who are interested in this story is that the cause of this tragedy was that the ship struck an iceberg and sank. In reality, the reason is much more complex and involved many contributing factors. Among these were the shortage of lifeboats to carry passengers from the sinking ship; a lack of wireless information regarding ice fields since the transmitter had been shut down for the evening; poor judgment by the ship's captain in informing passengers and crew of an impending disaster; an ambitious ship owner who wanted to claim the record for the shortest Atlantic crossing time; an inadequate on-board warning system and inadequate passenger drills prior to sailing; an overconfidence in the technology of a ship thought to be invincible; and flaws in the rivets that fastened the ship's steel plates and caused them to break apart. As a result of this horrible disaster, a Congressional investigation identified many of the causes and passed laws regarding ocean travel to assure that what happened to the *Titanic* would not happen again.

Air crashes, when they occur, attract media attention and public concern. Experts from the National Transportation Safety Board (NTSB) are dispatched to the scene to begin their investigation. Data are retrieved from the flight data and voice recorders, sections of the doomed plane are assembled, and interviews from witnesses and survivors are secured. The results of the investigation, which could take months or even years to complete, often provide information regarding the probable cause of the accident and may result in changes in procedures and design specifications that may help to prevent future similar occurrences. The dramatic hijackings on September 11, 2001, when terrorists stormed the cockpit of four separate aircraft, subduing the crew, and crashing the aircraft, resulted in many changes to air transportation. Two results of this horrific event are that the doors to the cockpit are now "hardened" to prevent unauthorized entry and passenger screening procedures have been improved.

Causes of transportation accidents may also involve poor coordination among institutions and organizations. For example, a head-on collision of two commuter passenger trains in central London occurred during the morning rush hour, and some experts attributed the crash to the recent privatization of the railroad system. Since no single organization was in charge (the two railroad trains were owned by different companies), it was argued that a lack of communication among the various parts of the system was a causal factor. As a result, the two trains proceeded toward each other on the same track. Another version of the cause of this crash, in which over 60 passengers were killed, is more simple and direct. One of the train operators failed to stop at a red signal and continued ahead at high speed. Nevertheless, privatization was viewed as a secondary cause of the crash since the railroad had been unwilling to install a device (due to cost factors) that would have warned that a train had proceeded through the red signal and would have caused the train to come to a halt.

The examples cited illustrate types of transportation accidents and their causes. Based on these illustrations and other similar cases, it is possible to construct a general list of the categories of circumstances that influence the occurrence of transportation crashes. When the factors that contribute to crash events are identified, it is possible to modify and improve the transportation system. Then, with the reduction or elimination of the crash-causing factor, a safer transportation system is likely to result.

As an example, crash data have shown conclusively that there is a strong correlation between highway deaths and the use of drugs or alcohol by the driver. Armed with these findings, organizations such as Mothers Against Drunk Driving (MADD) have successfully lobbied for laws that control drug and alcohol use while driving. Over time, limits have been placed on a permissible alcohol content in the bloodstream. In some states, the limit is "zero tolerance," with an increase in fines and the imposition of jail terms. The result of this action has been a significant reduction in the number of highway crashes due to drinking and driving.

What Are the Major Factors Involved in Transportation Crashes?

While the causes of accidents are usually complex and may involve several factors, they can be considered within four separate categories: actions by the driver, condition of the vehicle, geometric characteristics of the travelway, and the physical or climatic environment in which the vehicle operates. These factors will be reviewed in the following section.

The major contributing factor in most accident situations is considered to be the performance of the driver of one or both (in multiple vehicle crashes) of the vehicles involved. Driver error can occur in many ways, including inattention to the roadway and surrounding traffic, failure to yield the right of way, and disobedience of traffic laws. These "failures" can occur due to unfamiliarity with roadway conditions, traveling at high speeds, drowsiness, drinking, using a cell phone, or dealing with other distractions within the vehicle. Figure 8.1 illustrates a crash due to driver error.

The mechanical condition of a vehicle can also be a cause of transportation crashes. If an aircraft runs out of fuel and crashes, the reason may be that the

FIGURE 8.1

Driver error contributes to the majority of highway crashes.

Source: Transportation Research Board. National Academies, Washington, D.C.

fuel gauge was not in proper working order. Faulty brakes in heavy trucks, rail cars, and airplanes have caused crashes. Other reasons are the electrical system, worn tires, and the location of the vehicle's center of gravity. Figure 8.2 illustrates a crash effect that could be the result of vehicle failure. Off-road crashes of this type require that rescue teams be promptly notified, especially in rural areas in order to assure immediate medical attention.

The condition of the travelway, which includes the road, intersections, and the traffic control system, can be a factor in the creation of a transportation crash. Highways must be designed to provide adequate sight distance at the design speed or motorists will be unable to take remedial action to avoid a crash. Traffic signals must provide adequate decision sight distance when the signal goes from green to red. Railroad grade crossings must be designed to operate safely and thus minimize crashes between highway traffic and rail cars. Rail tracks must be carefully aligned to assure that a fast-moving train does not "jump the tracks." The superelevation of highway and railroad curves must be carefully laid out with the correct radius and the appropriate transition sections to assure that vehicles can negotiate curves safely. A travelway failure of a train derailment is illustrated in Figure 8.3.

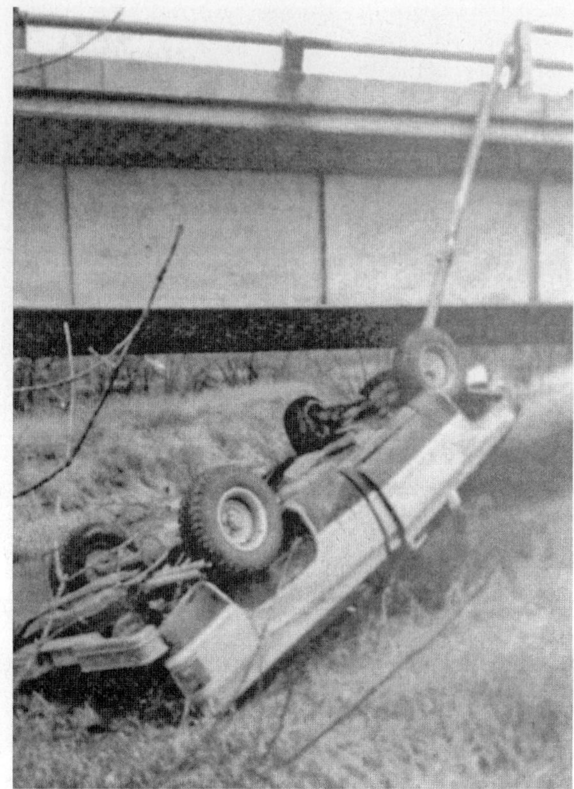

FIGURE 8.2

Off-road crashes require prompt notification.

Source: Transportation Research Board. National Academies, Washington, D.C.

The physical and climatic environment surrounding a transportation vehicle can also be a factor in the occurrence of transportation crashes. The most common cause of crashes is the weather. Transportation systems function at their best when the weather is sunny and the skies are clear. Air transportation is significantly affected by weather, and most travelers can recall a trip by air that was delayed or canceled because of stormy, foggy, windy, or snowy weather at either the origin or destination airport or en route. Weather also affects ships at sea, especially in stormy periods often caused by hurricanes. Great sea sagas have been written about the heroism of sailors trying to survive during a storm.

Water on roads can contribute to highway crashes. For example, a wet pavement reduces stopping friction and can cause vehicles to hydroplane. Many severe crashes have occurred due to fog. Vehicles traveling at high speed are unable to see other vehicles ahead that may have stopped or slowed down, creating a multivehicle pile-up. Geography is another environmental cause of transportation crashes. Mountain ranges have been the site of air crashes. Flooded river plains, swollen rivers, and mudslides on the pavement have been the cause of railroad and highway crashes.

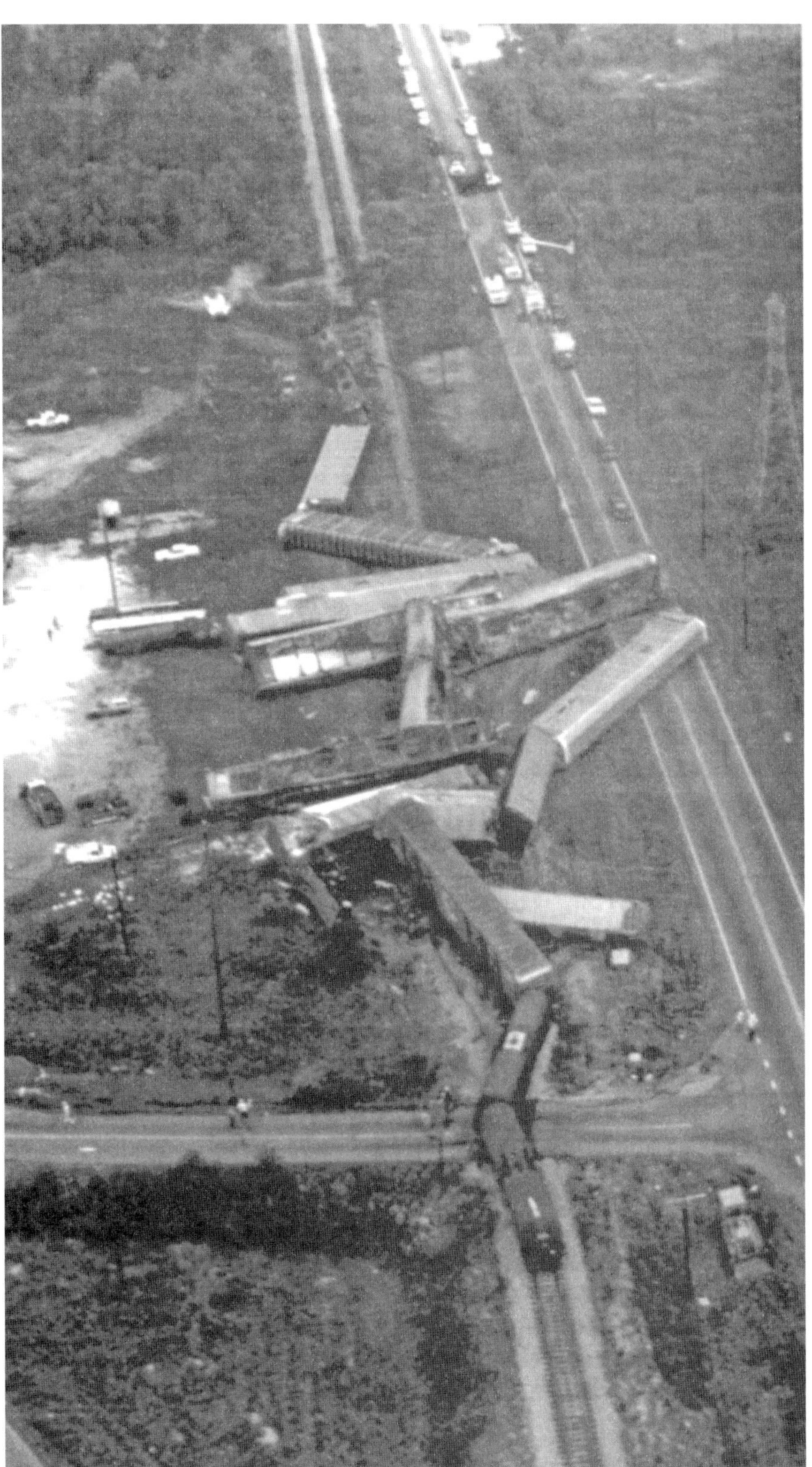

FIGURE 8.3

Rail track must be aligned to avoid derailments.

Source: Transportation Research Board. National Academies, Washington, D.C.

What Are the Ways to Improve Transportation Safety?

Safety improvement measures can be classified as laws and government regulations, enforcement, education, and engineering. Each of these actions is necessary if transportation safety is to be improved. However, in order to effectively apply a particular safety improvement, it is necessary first to determine the intended outcome of each measure, which could be either to prevent the crash or to minimize the effects of the crash once it happens.

The primary goal of any transportation safety program is to prevent crashes. Clearly, if a crash is prevented, it never happens. No one is injured or killed. Prevention is seen as the primary objective by the air transportation industry, and, in the United States, much has been achieved. Passengers know that if an airline crash should occur, they are likely to be killed. Thus, the best option is to prevent crashes, although precautions must be in place to minimize the effect should a crash occur. For example, with the heightened emphasis on security, passengers willingly submit to personal searches intended to assure that a terrorist cannot board the aircraft.

To accomplish air crash prevention, legislation is needed that authorizes public officials to regulate and enforce laws regarding the safe movement of people and goods. For example, the U.S. Department of Transportation is authorized by Congress to regulate the movement of hazardous materials and to certify that commercial transportation meets standards of maintenance. Safety also requires enforcement and education. For example, after laws are passed setting maximum speed and weight limits for heavy trucks, the operators must be certified through education and violators must be apprehended and punished. Finally, engineering plays a major role in accident prevention by assuring that the vehicle and travelway are designed so that driving is as safe as possible.

The second approach is to design the vehicle-travelway system for safety in such a manner that, if a crash occurred, the effect on the occupants would be minimized. Minimizing the effects of crashes is a strategy that is effectively used in highway transportation, with its large annual death and injury rate and extensive property damage. Rather than developing stringent measures to prevent crashes, the highway industry views the problem as saving the public from itself; in other words making an implicit assumption that the number of highway crashes could remain constant or even increase. Thus, the design engineer seeks to eliminate obstacles near the travelway so that, if the vehicle cannot be controlled, it will have a clear zone free of obstacles or, if impact occurs, the occupant's injuries will be minimized.

For example, there have been several situations when an aircraft skidded off the runway and was unable to stop before crashing into fixed objects causing the loss of life. Had the runway area been clear of signs and structures, or provided space to reduce speed, the aircraft and its passengers might have been saved. Similarly, many highway signs are constructed such that, if struck by a motor vehicle, they will break away at the base. Vehicle design is also aimed at minimizing crash effects by installing energy-absorbing bumpers, air bags, and seat belts. Motorcycle helmet laws are intended to reduce head injuries should a crash occur. Only four states (Colorado, Illinois, Iowa, and New Hampshire) do not have laws governing motorcycle helmet use.

In contrast to commercial transportation (air, rail, and water), highway transportation faces formidable barriers to developing a rational basis for safety improvements. Commercial modes can control operator performance and prevent crashes because they desire to avoid loss of revenue and to limit the increase in operating costs.

Furthermore, the U.S. highway system is fragmented and decentralized—consisting of literally millions of drivers and multiple political jurisdictions. Many individual cities, towns, counties, and states administer and enforce traffic safety programs and, at the federal level, multiple agencies and the U.S. Congress promulgate laws, regulations, and design standards assuring a wide variability in safety management approaches. For example, pole-mounted cameras intended to snare "red-light running" and thus reduce fatalities and injury are legal in some states and municipalities but not in others. Similar anomalies exist with respect to seat belt laws, speed limits, cell phone use, and alcohol blood levels.

Other barriers to effective highway crash protection are perceptions by the motoring public that highways are safer than air travel. There is little outcry from the public, the press, or Congress regarding the number of highway deaths per year, while a single air crash creates enormous press coverage and concern. In addition, the public and press have not been willing to accept an economic basis for safety improvements that requires a valuation on the worth of a human life. Thus, advocates urge, for example, seat belts on buses, when other more cost-effective means are available, and approve large investments in structures to reduce railroad-crossing crashes, when this category represents only about 1% of all fatal crashes.

Finally, in contrast to air travel, where passengers willingly submit to laws and regulations regarding passenger safety, the motoring public is not as compliant and views traffic laws as optional or to be obeyed only when there is a chance of being caught. Thus, travelers exceed the speed limit, use cell phones while driving, drink and drive, park illegally, and demonstrate aggressive behavior. Furthermore, the public (and their elected officials) are disinterested in new technology that monitors their driving behavior, and many screening techniques are deemed illegal due to privacy concerns and constitutional issues regarding securing of evidence about an individual's behavior by remote means.

Collection and Analysis of Crash Data

After a crash has occurred, an accident investigation is conducted that seeks to understand what may have been the cause. Data are collected that will be useful in reconstructing the crash event and may lead to a determination of a possible solution. In addition, crash data are assembled over time to determine trends and to evaluate statistically how elements of the overall transportation system are performing. For example, if the number of crashes at one intersection is considerably greater than for other similar locations within a state, it would be beneficial to analyze this site, identify possible causes, and suggest steps to improve its safety.

Crash data are obtained from federal, state, and local transportation officials or police agencies. Soon after a crash occurs, medical emergency assistance is dispatched to the scene to assist the injured. Then crash investigators are assigned to record relevant information about the crash event. Among the data collected regarding the situation at the scene are location of the crash, time of occurrence, environmental conditions, type and number of vehicles involved, path trajectory, and final resting location of each vehicle. The crash site may also be photographed and videotaped. The accident record becomes the basic source of information for further analysis. The data may be used to produce accident reconstructions, to support legal or insurance claims, to establish statistical trends, and to improve knowledge about the factors that cause crashes. Finally, the data can assist in evaluating the effectiveness of improvements (or countermeasures) for reducing fatalities or injuries.

The first step in collecting crash data is the completion of a crash report form at the site of the crash. Information on a reported highway crash is recorded on a crash form by the police officer investigating the crash. Since each state maintains its own reporting form, the format may be different from state to state, but the information recorded is similar. This includes the date and time of the crash, the types of vehicles involved, the severity of any injury that occurs as a result of the crash, and a short description of the crash. Figure 8.4 shows the Virginia police crash report form. Similar forms are used for railroad and airline crashes, as shown in Figures 8.5 and 8.6, respectively. The railroad companies are primarily responsible for collecting and reporting on the crashes involving their trains, while data on civil aviation crashes are usually collected by the NTSB. The NTSB is an independent federal agency mandated by law to investigate and determine the probable cause of all civil aviation crashes in the United States and significant crashes in other modes of transportation. Significant crashes are as follows:

- Selected highway crashes
- Railroad crashes involving passenger trains

FIGURE 8.4a
Virginia police crash report form.

Commonwealth of Virginia • Department of Motor Vehicles

Page ___ of ___ pages ☐ Supplemental report

Police Crash Report

DMV Copy FR300P (Rev 9/03)

Crash date MM / DD / YYYY | Day of week | Military time (24 hr. clock) | County of crash | Official DMV use

☐ City of ☐ Town of | Landmark at scene | GPS Lat.

Location of crash (route/street) | Railroad crossing ID no. (if within 150 ft.) | GPS Long. | Mile marker number | Local case number

☐ at intersection with or ___ ☐ miles ☐ feet N S E W ☐☐☐☐ of | Location of crash (route/street) | Number of vehicles

Vehicle No. 1 | **Vehicle No. 2 (or pedestrian ☐)**

Driver's name (last, first, middle) | Driver fled scene ☐ | Yrs. dr. experience

Address (street and no.)

City | State | ZIP

Birth date MM / DD / YYYY | Gender | Driver's license number | ☐ DL ☐ CDL | State

Vehicle owner's name (last, first, middle) or Commercial motor carrier | ☐ same as driver

Address (street and no.)

City | State | ZIP

A Veh. type | Veh. year | Veh. make | Veh. model | CMV ☐ | Towed ☐

Vehicle plate number | State | B EMV type | C EMV in service | Approximate repair cost

VIN

CMV only: U.S. DOT no. or VA no. | Placard no. and class or name

No. of axles | Truck cover Y☐ N☐ | GVWR ☐ 10,000 and under ☐ 10,001 to 26,000 ☐ over 26,000 | ☐ HAZMAT ☐ Oversize | ☐ Cargo spill ☐ Override ☐ Underride

Vehicle no. 1 damage | Name of insurance company (not agent) | Name of insurance company (not agent) | **Vehicle no. 2 damage**

Check impact area(s)
Circle initial impact
12, 1, 2, 3, 4, 5, 6, 7, 8, 9, 10, 11, 13

See back of FR300T

Crash Diagram

Speed: Before crash | Limit | Max safe | Lane dir.

Passengers age count: Less 6 | 6-17 | 18-21 | Over 21

Indicate north by arrow

Damage to property other than vehicles | Approximate repair cost | Object struck (tree, fence, etc.) | Property owner's name (last, first, middle) and address

Crash description

Offenses charged driver

All injured

12	13	14	15	16	17	18	19	20	Names of injured (if deceased give date of death)	EMS transport	Date of death MM/DD/YYYY

Investigating officer | Badge/code no. | Agency/department name and code no. | Reviewing officer | Report file date

(a) Data recording form

Note: 1 ft = 0.3 m

- Railroad crashes that result in one or more fatalities or major property damage, regardless of whether the train or trains involved is a passenger train or not
- Major marine crashes

FIGURE 8.4b
Virginia police crash report form.

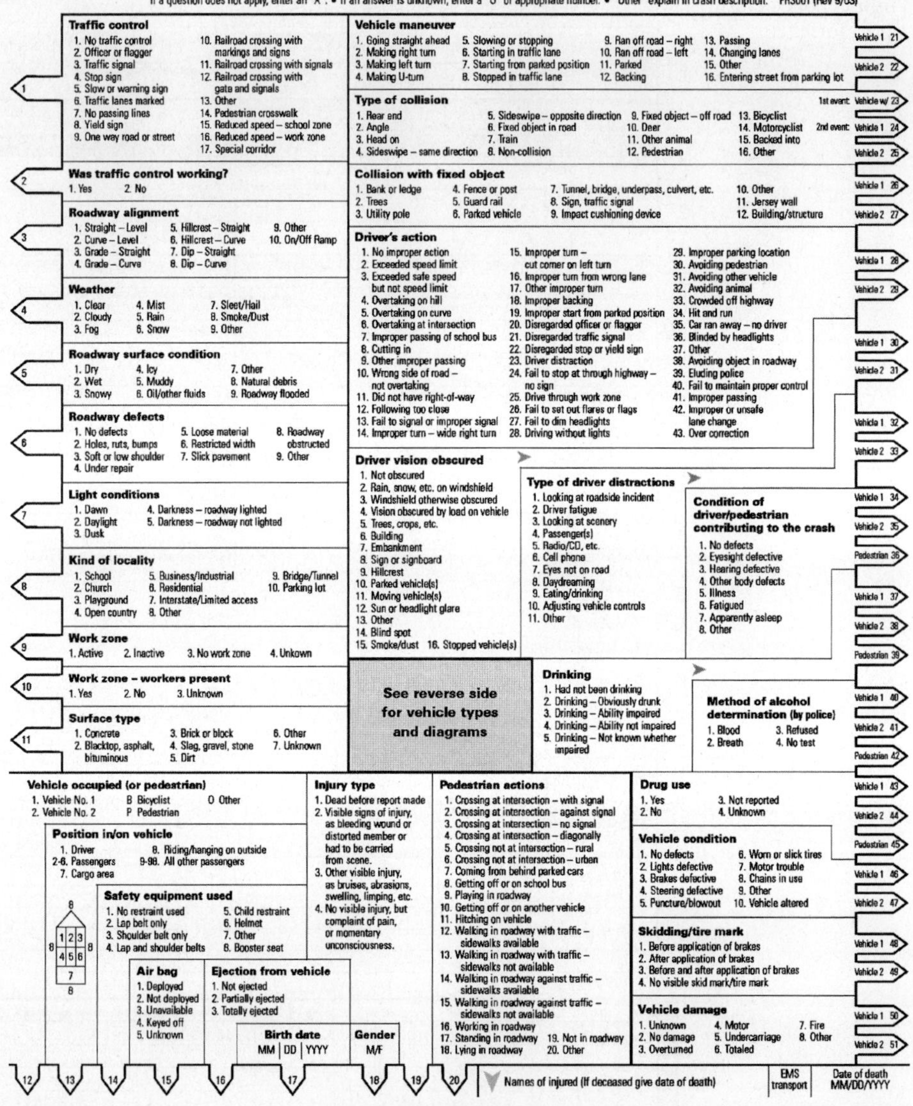

Commonwealth of Virginia • Department of Motor Vehicles

Police Crash Report

If a question does not apply, enter an "X". • If an answer is unknown, enter a "U" or appropriate number. • "Other" explain in crash description. FR300T (Rev 9/03)

1 **Traffic control**
1. No traffic control
2. Officer or flagger
3. Traffic signal
4. Stop sign
5. Slow or warning sign
6. Traffic lanes marked
7. No passing lines
8. Yield sign
9. One way road or street
10. Railroad crossing with markings and signs
11. Railroad crossing with signals
12. Railroad crossing with gate and signals
13. Other
14. Pedestrian crosswalk
15. Reduced speed – school zone
16. Reduced speed – work zone
17. Special corridor

2 **Was traffic control working?**
1. Yes 2. No

3 **Roadway alignment**
1. Straight – Level
2. Curve – Level
3. Grade – Straight
4. Grade – Curve
5. Hillcrest – Straight
6. Hillcrest – Curve
7. Dip – Straight
8. Dip – Curve
9. Other
10. On/Off Ramp

4 **Weather**
1. Clear
2. Cloudy
3. Fog
4. Mist
5. Rain
6. Snow
7. Sleet/Hail
8. Smoke/Dust
9. Other

5 **Roadway surface condition**
1. Dry
2. Wet
3. Snowy
4. Icy
5. Muddy
6. Oil/other fluids
7. Other
8. Natural debris
9. Roadway flooded

6 **Roadway defects**
1. No defects
2. Holes, ruts, bumps
3. Soft or low shoulder
4. Under repair
5. Loose material
6. Restricted width
7. Slick pavement
8. Roadway obstructed
9. Other

7 **Light conditions**
1. Dawn
2. Daylight
3. Dusk
4. Darkness – roadway lighted
5. Darkness – roadway not lighted

8 **Kind of locality**
1. School
2. Church
3. Playground
4. Open country
5. Business/Industrial
6. Residential
7. Interstate/Limited access
8. Other
9. Bridge/Tunnel
10. Parking lot

9 **Work zone**
1. Active 2. Inactive 3. No work zone 4. Unkown

10 **Work zone – workers present**
1. Yes 2. No 3. Unknown

11 **Surface type**
1. Concrete
2. Blacktop, asphalt, bituminous
3. Brick or block
4. Slag, gravel, stone
5. Dirt
6. Other
7. Unknown

Vehicle maneuver (Vehicle 1 21, Vehicle 2 22)
1. Going straight ahead
2. Making right turn
3. Making left turn
4. Making U-turn
5. Slowing or stopping
6. Starting in traffic lane
7. Starting from parked position
8. Stopped in traffic lane
9. Ran off road – right
10. Ran off road – left
11. Parked
12. Backing
13. Passing
14. Changing lanes
15. Other
16. Entering street from parking lot

Type of collision (1st event: Vehicle w/ 23; 2nd event: Vehicle 1 24, Vehicle 2 25)
1. Rear end
2. Angle
3. Head on
4. Sideswipe – same direction
5. Sideswipe – opposite direction
6. Fixed object in road
7. Train
8. Non-collision
9. Fixed object – off road
10. Deer
11. Other animal
12. Pedestrian
13. Bicyclist
14. Motorcyclist
15. Backed into
16. Other

Collision with fixed object (Vehicle 1 26, Vehicle 2 27)
1. Bank or ledge
2. Trees
3. Utility pole
4. Fence or post
5. Guard rail
6. Parked vehicle
7. Tunnel, bridge, underpass, culvert, etc.
8. Sign, traffic signal
9. Impact cushioning device
10. Other
11. Jersey wall
12. Building/structure

Driver's action (Vehicle 1 28, Vehicle 2 29; Vehicle 1 30, Vehicle 2 31; Vehicle 1 32, Vehicle 2 33)
1. No improper action
2. Exceeded speed limit
3. Exceeded safe speed but not speed limit
4. Overtaking on hill
5. Overtaking on curve
6. Overtaking at intersection
7. Improper passing of school bus
8. Cutting in
9. Other improper passing
10. Wrong side of road – not overtaking
11. Did not have right-of-way
12. Following too close
13. Fail to signal or improper signal
14. Improper turn – wide right turn
15. Improper turn – cut corner on left turn
16. Improper turn from wrong lane
17. Other improper turn
18. Improper backing
19. Improper start from parked position
20. Disregarded officer or flagger
21. Disregarded traffic signal
22. Disregarded stop or yield sign
23. Driver distraction
24. Fail to stop at through highway – no sign
25. Drive through work zone
26. Fail to set out flares or flags
27. Fail to dim headlights
28. Driving without lights
29. Improper parking location
30. Avoiding pedestrian
31. Avoiding other vehicle
32. Avoiding animal
33. Crowded off highway
34. Hit and run
35. Car ran away – no driver
36. Blinded by headlights
37. Other
38. Avoiding object in roadway
39. Eluding police
40. Fail to maintain proper control
41. Improper passing
42. Improper or unsafe lane change
43. Over correction

Driver vision obscured
1. Not obscured
2. Rain, snow, etc. on windshield
3. Windshield otherwise obscured
4. Vision obscured by load on vehicle
5. Trees, crops, etc.
6. Building
7. Embankment
8. Sign or signboard
9. Hillcrest
10. Parked vehicle(s)
11. Moving vehicle(s)
12. Sun or headlight glare
13. Other
14. Blind spot
15. Smoke/dust
16. Stopped vehicle(s)

Type of driver distractions
1. Looking at roadside incident
2. Driver fatigue
3. Looking at scenery
4. Passenger(s)
5. Radio/CD, etc.
6. Cell phone
7. Eyes not on road
8. Daydreaming
9. Eating/drinking
10. Adjusting vehicle controls
11. Other

Condition of driver/pedestrian contributing to the crash (Vehicle 1 34, Vehicle 2 35, Pedestrian 36; Vehicle 1 37, Vehicle 2 38, Pedestrian 39)
1. No defects
2. Eyesight defective
3. Hearing defective
4. Other body defects
5. Illness
6. Fatigued
7. Apparently asleep
8. Other

See reverse side for vehicle types and diagrams

Drinking
1. Had not been drinking
2. Drinking – Obviously drunk
3. Drinking – Ability impaired
4. Drinking – Ability not impaired
5. Drinking – Not known whether impaired

Method of alcohol determination (by police) (Vehicle 1 40, Vehicle 2 41, Pedestrian 42)
1. Blood
2. Breath
3. Refused
4. No test

Vehicle occupied (or pedestrian)
1. Vehicle No. 1
2. Vehicle No. 2
B Bicyclist
P Pedestrian
O Other

Position in/on vehicle
1. Driver
2-6. Passengers
7. Cargo area
8. Riding/hanging on outside
9-98. All other passengers

8 / 1 2 3 / 8 4 5 6 8 / 7 / 8

Safety equipment used
1. No restraint used
2. Lap belt only
3. Shoulder belt only
4. Lap and shoulder belts
5. Child restraint
6. Helmet
7. Other
8. Booster seat

Air bag
1. Deployed
2. Not deployed
3. Unavailable
4. Keyed off
5. Unknown

Ejection from vehicle
1. Not ejected
2. Partially ejected
3. Totally ejected

Birth date
MM | DD | YYYY

Gender
M/F

Injury type
1. Dead before report made
2. Visible signs of injury, as bleeding wound or distorted member or had to be carried from scene.
3. Other visible injury, as bruises, abrasions, swelling, limping, etc.
4. No visible injury, but complaint of pain, or momentary unconsciousness.

Pedestrian actions
1. Crossing at intersection – with signal
2. Crossing at intersection – against signal
3. Crossing at intersection – no signal
4. Crossing at intersection – diagonally
5. Crossing not at intersection – rural
6. Crossing not at intersection – urban
7. Coming from behind parked cars
8. Getting off or on school bus
9. Playing in roadway
10. Getting off or on another vehicle
11. Hitching on vehicle
12. Walking in roadway with traffic – sidewalks available
13. Walking in roadway with traffic – sidewalks not available
14. Walking in roadway against traffic – sidewalks available
15. Walking in roadway against traffic – sidewalks not available
16. Working in roadway
17. Standing in roadway
18. Lying in roadway
19. Not in roadway
20. Other

Drug use (Vehicle 1 43, Vehicle 2 44, Pedestrian 45)
1. Yes
2. No
3. Not reported
4. Unknown

Vehicle condition (Vehicle 1 46, Vehicle 2 47)
1. No defects
2. Lights defective
3. Brakes defective
4. Steering defective
5. Puncture/blowout
6. Worn or slick tires
7. Motor trouble
8. Chains in use
9. Other
10. Vehicle altered

Skidding/tire mark (Vehicle 1 48, Vehicle 2 49)
1. Before application of brakes
2. After application of brakes
3. Before and after application of brakes
4. No visible skid mark/tire mark

Vehicle damage (Vehicle 1 50, Vehicle 2 51)
1. Unknown
2. No damage
3. Overturned
4. Motor
5. Undercarriage
6. Totaled
7. Fire
8. Other

12 13 14 15 16 17 18 19 20 Names of injured (If deceased give date of death) | EMS transport | Date of death MM/DD/YYYY

(b) Code definitions form

- All marine crashes involving a public and nonpublic vessel
- Pipeline crashes involving a fatality or substantial property damage
- All crashes in all modes of transportation resulting in the release of hazardous material
- Selected transportation crashes involving problems of a recurring nature

FIGURE 8.4c
Virginia police crash report form.

Vehicle type (put in box A)

1. Passenger car
2. Truck – pick-up/passenger truck
3. Van
4. Truck – straight truck (2-axle), flat bed, dump truck, wrecker, tractor truck
5. Truck – tractor trailer
6. Truck – tractor twin-trailer
7. Motor home, recreational vehicle
8. Special vehicle – oversized vehicle/earthmover/road equipment
9. Bicycle
10. Moped
11. Motorcycle
12. Emergency vehicle
13. Bus – school bus
14. Bus – city transit bus/privately-owned church bus
15. Bus – commercial passenger bus (seats 9 –15 people, including driver)
16. Other
17. Bus – commercial passenger bus (seats 15+ people, including driver)
18. Special vehicle – farm equip, go-cart, hearse, bookmobile
19. Special vehicle – ATV
20. Special vehicle – golf cart
21. Special vehicle – low-speed vehicle
22. Truck – sport utility vehicle
23. Truck – straight truck (3 or more axles)
24. Truck – tractor triple-trailer
25. Truck – truck tractor (bobtail – no trailer)

Emergency vehicle type (put in box B)		**Emergency vehicle status** (put in box C)	
1. Not applicable	5. Tow truck	1. Yes, in emergency	3. Not applicable
2. Police	6. Military	2. No, not in emergency	4. Unknown
3. Fire	7. Maintenance		
4. Ambulance	8. Other		

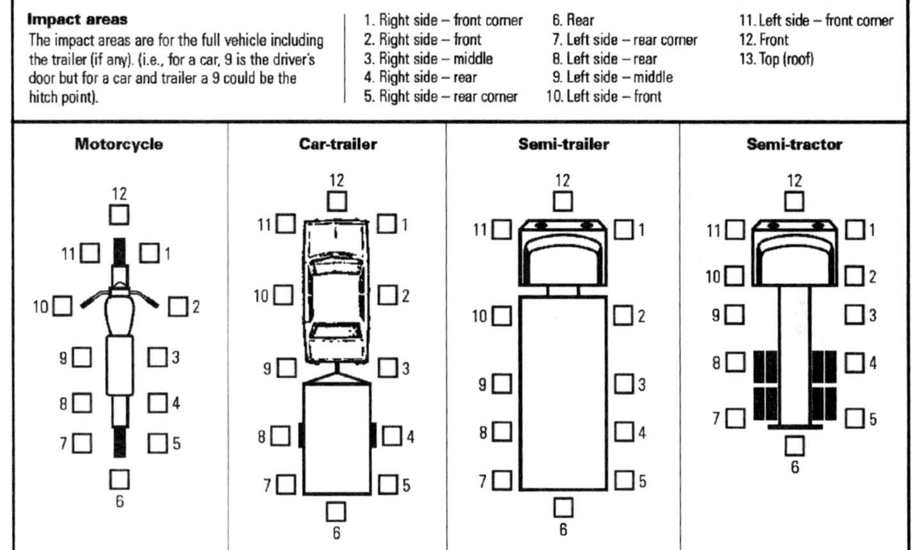

(c) Code definitions form part 2

Because the NTSB investigates selected highway crashes, only a few of the highway crashes that occur in a given year are investigated by the Board, resulting in a much lower proportion of total crashes than those for aviation and railroad. The national statistics on highway crashes are therefore mainly dependent on the information recorded by the investigating police officer of each crash.

National crash databases are regularly provided by different federal transportation agencies. For example, NHTSA, in its annual report *Traffic Safety Facts 2004: A Compilation of Motor Vehicle Crash Data from the Fatality*

FIGURE 8.5a
Highway–rail grade crossing accident/incident report.

DEPARTMENT OF TRANSPORTATION
FEDERAL RAILROAD ADMINISTRATION (FRA)

HIGHWAY-RAIL GRADE CROSSING ACCIDENT/INCIDENT REPORT

OMB Approval No.: 2130-0500

1. Name of Reporting Railroad | 1a. Alphabetic Code | 1b. Railroad Accident/Incident No.

2. Name of Other Railroad Involved in Train Accident/Incident | 2a. Alphabetic Code | 2b. Railroad Accident/Incident No.

3. Name of Railroad Responsible for Track Maintenance *(single entry)* | 3a. Alphabetic Code | 3b. Railroad Accident/Incident No.

4. U. S. DOT Grade Crossing Identification Number | 5. Date of Accident/Incident month day year | 6. Time of Accident/Incident AM ☐ PM ☐

7. Nearest Railroad Station | 8. Division | 9. County | 10. State Abbr. Code

11. City *(if in a city)* | 12. Highway Name or Number Public ☐ Private ☐

Highway User Involved | Rail Equipment Involved

13. Type A. Auto B. Truck C. Truck-trailer D. Pick-up truck E. Van F. Bus G. School bus H. Motorcycle J. Other motor vehicle K. Pedestrian M. Other *(specify)* Code

17. Equipment 1. Train *(units pulling)* 2. Train *(units pushing)* 3. Train *(standing)* 4. Car(s) *(moving)* 5. Car(s) *(standing)* 6. Light loco(s) *(moving)* 7. Light loco(s) *(standing)* 8. Other *(specify)* A. Train pulling- RCL B. Train pushing- RCL C. Train standing- RCL Code

14. Vehicle Speed *(est. mph at impact)* | 15. Direction *(geographical)* 1. North 2. South 3. East 4. West Code | 18. Position of Car Unit in Train

16. Position 1. Stalled on crossing 2. Stopped on crossing 3. Moving over crossing 4. Trapped Code | 19. Circumstance 1. Rail equipment struck highway user 2. Rail equipment struck by highway user Code

20a. Was the highway user and/or rail equipment involved in the impact transporting hazardous materials? 1. Highway user 2. Rail equipment 3. Both 4. Neither Code | 20b. Was there a hazardous materials release by 1. Highway user 2. Rail equipment 3. Both 4. Neither Code

20c. State here the name and quantity of the hazardous material released, if any.

21. Temperature *(specify if minus)* °F | 22. Visibility *(single entry)* 1. Dawn 2. Day 3. Dusk 4. Dark Code | 23. Weather *(single entry)* 1. Clear 2. Cloudy 3. Rain 4. Fog 5. Sleet 6. Snow Code

24. Type of Equipment Consist *(single entry)* 1. Freight train 2. Passenger train 3. Commuter train 4. Work train 5. Single car 6. Cut of cars 7. Yard/switching 8. Light loco(s). 9. Maint./inspect. car A. Spec. MoW Equip. Code | 25. Track Type Used by Rail Equipment Involved 1. Main 2. Yard 3. Siding 4. Industry Code | 26. Track Number or Name

27. FRA Track Class (1-9, X) | 28. Number of Locomotive Units | 29. Number of Cars | 30. Consist Speed R - Recorded E - Estimated *(Recorded speed, if available)* MPH Code | 31. Time Table Direction 1. North 2. South 3. East 4. West Code

32. Type of Crossing Warning 1. Gates 2. Cantilever FLS 3. Standard FLS 4. Wig wags 5. Hwy. traffic signals 6. Audible 7. Crossbucks 8. Stop signs 9. Watchman 10. Flagged by crew 11. Other *(specify)* 12. None Code(s) | 33. Signaled Crossing Warning *(See reverse side for instructions and codes)* Code | 34. Whistle Ban 1. Yes 2. No 3. Unknown Code

35. Location of Warning 1. Both sides 2. Side of vehicle approach 3. Opposite side of vehicle approach Code | 36. Crossing Warning Interconnected with Highway Signals 1. Yes 2. No 3. Unknown Code | 37. Crossing Illuminated by Street Lights or Special Lights 1. Yes 2. No 3. Unknown Code

38. Driver's Age | 39. Driver's Gender 1. Male 2. Female Code | 40. Driver Drove Behind or in Front of Train and Struck or was Struck by Second Train 1. Yes 2. No 3. Unknown Code | 41. Driver 1. Drove around or thru the gate 2. Stopped and then proceeded 3. Did not stop 4. Stopped on crossing 5. Other *(specify)* Code

42. Driver Passed Standing Highway Vehicle 1. Yes 2. No 3. Unknown Code | 43. View of Track Obscured by *(primary obstruction)* 1. Permanent structure 2. Standing railroad equipment 3. Passing train 4. Topography 5. Vegetation 6. Highway vehicles 7. Other *(specify)* 8. Not obstructed Code

Casualties to:	Killed	Injured
46. Highway-Rail Crossing Users		
49. Railroad Employees		
52. Passengers on Train		

44. Driver was 1. Killed 2. Injured 3. Uninjured Code | 45. Was Driver in the Vehicle? 1. Yes 2. No Code

47. Highway Vehicle Property Damage *(est. dollar damage)* | 48. Total Number of Highway-Rail Crossing Users *(include driver)*

50. Total Number of People on Train *(include passengers and train crew)* | 51. Is a Rail Equipment Accident/ Incident Report Being Filed? 1. Yes 2. No Code

53a. Special Study Block | 53b. Special Study Block

54. Narrative Description *(Be specific, and continue on separate sheet if necessary)*

55. Typed Name and Title | 56. Signature | 57. Date

NOTE: This report is part of the reporting railroad's accident report pursuant to the accident reports statute and, as such shall not " be admitted as evidence or used for any purpose in any suit or action for damages growing out of any matter mentioned in said report . . . ." 49 U.S.C. 20903. See 49 C.F.R. 225.7 (b).

FORM FRA F 6180.57 (Revised March 2003)

* NOTE THAT ALL CASUALTIES MUST BE REPORTED ON FORM FRA F 6180.55A

(a) Data recording form

Analysis System and the General Estimate System, gives information on traffic crashes of all severities. Similarly, the Federal Railroad Administration, in its *Railroad Safety Statistics Annual Report*, gives statistical data, tables, and charts describing the nature and causes of many rail-related crashes and incidents. The

FIGURE 8.5b
Railroad crash report form.

INSTRUCTIONS FOR COMPLETING BLOCK 33

Only if Types 1 - 6, Item 32 are indicated, mark in Block 33 the status of the warning devices at the crossing at the time of the accident, using the following codes:

1. Provided minimum 20-second warning.
2. Alleged warning time greater than 60 seconds.
3. Alleged warning time less than 20 seconds.
4. Alleged no warning.
5. Confirmed warning time greater than 60 seconds.
6. Confirmed warning time less than 20 seconds.
7. Confirmed no warning.

If status code 5, 6, or 7 was entered, also enter a letter code explanation from the list below:

A. Insulated rail vehicle.

B. Storm/lightning damage.

C. Vandalism.

D. No power/batteries dead.

E. Devices down for repair.

F. Devices out of service

G. Warning time greater than 60 seconds attributed to accident-involved train stopping short of the crossing, but within track circuit limits, while warning devices remain continuously active with no other in-motion train present.

H. Warning time greater than 60 seconds attributed to track circuit failure (e.g., insulated rail joint or rail bonding failure, track or ballast fouled, etc.).

J. Warning time greater than 60 seconds attributed to other train/equipment within track circuit limits.

K. Warning time less than 20 seconds attributed to signals timing out before train's arrival at the crossing/island circuit.

L. Warning time less than 20 seconds attributed to train operating counter to track circuit design direction.

M. Warning time less than 20 seconds attributed to train speed in excess of track circuit's design speed.

N. Warning time less than 20 seconds attributed to signal system's failure to detect train approach.

P. Warning time less than 20 seconds attributed to violation of special train operating instructions.

R. No warning attributed to signal system's failure to detect the train.

S. Other cause(s). Explain in Narrative Description.

(b) Instructions form

NTSB also publishes the *Aviation Accident Database*, which contains information on the operations, personnel, environmental conditions, consequences, probable causes, and contributing factors of civil aviation crashes. Examples of the type of information that can be obtained from these databases are given in Figures 8.7 and 8.8 for highway and air carrier crashes, respectively, and in Table 8.1 for railroad crashes.

FIGURE 8.6a

Federal Aviation Administration crash report form.

FAA ACCIDENT / INCIDENT REPORT

1. ACCIDENT ☐ INCIDENT ☐

3. DATE OF EVENT MO DA YR

4. FAA OFFICE REGION OFFICE NUMBER

5. NTSB ID

6. LOCATION-CITY/STATE/ZIP

7. OPERATOR NAME FOUR LETTER IDENTIFIER

8. AIRPORT (IF APPLICABLE) 3- OR 4- LETTER ID

9. LOCAL TIME 24- HOUR CLOCK

10A. LATITUDE

10B. LONGITUDE

2. AMENDED DATE MO ☐ DA ☐ YR ☐

13. AIRCRAFT

REGISTRATION

MAKE/MODEL

SERIAL NO.

YEAR OF MANUFACTURE

TOTAL AIRFRAME HRS. (WHOLE HOURS)

AIRFRAME CYCLES (AIR CARRIER ONLY)

16. POWER PLANT MAKE/MODEL/SERIES (IF APPLICABLE)

17. PROPELLER MAKE/MODEL/SERIES (IF APPLICABLE)

18. BIOHAZARD AREA YES ☐ NO ☐

14. FAR PART NUMBER

91	133
103	135 ON DEMAND
105	135 COMMUTER
121	137
125	
129	

15. TYPE OF AIRCRAFT

AIRPLANE

HELICOPTER

GLIDER

BALLOON

DIRIGIBLE

GYROPLANE

HOMEBUILT /AMATEUR/EXP.

ULTRALIGHT

19. TYPE OF LANDING GEAR

CONVENTIONAL	SKIS
TRICYCLE	AMPHIBIOUS
FLOATS	

11. AIRCRAFT DAMAGE

NONE

MINOR

SUBSTANTIAL

DESTROYED

12. COLLISION - BETWEEN TWO AIRCRAFT

YES ☐ AIR ☐

NO ☐ GROUND ☐

REGISTRATION NUMBER SECOND AIRCRAFT

20. INJURY SUMMARY UNKNOWN ☐

	FLT.CREW	CABIN CREW	PASSENGERS	OTHER	TOTAL
NONE					
MINOR					
SERIOUS					
FATAL					
TOTAL					

21. FACTORS - IDENTIFY PRIMARY FACTOR AS A. IDENTIFY SECONDARY FACTORS, IF ANY, AS X.

CHECKING OF FACTORS IS THE OPINION OF THE INVESTIGATOR/INSPECTOR BASED ON THE INVESTIGATION.

21A. TECHNICAL FACTORS

GEAR COLLAPSE	FIRE AFTER LANDING
GEAR UP LANDING	SYSTEM FAILURE
FIRE OR EXPLOSION	COMPONENT FAILURE
FUEL CONTAMINATION	LOST POWER
BLADE/ROTOR FAILURE	FOD
DESIGN OF AIRCRAFT	AUTO/IMPROPER FUEL
METAL FATIGUE	CORROSION
IMPROPER MAINTENANCE	INFLIGHT FIRE
IMPROPER INSTALLATION	SMOKE/FUMES
AD NON-COMPLIANCE	INFLIGHT BREAKUP
DECOMPRESSION	IMPROPER PART
21F. ATA CODE	OTHER

21C. PART NAME 21D. MANUFACTURER 21F PART NUMBER

21B. OPERATIONAL FACTORS

FUEL DEPLETION	SABOTAGE
PILOT INDUCED	PILOT INCAPACITATED
GROUND CREW	PILOT INCP. ALCOHOL
OTHER THAN PILOT	DOWNWIND TAKEOFF
PARACHUTE INCIDENT	CARBURETOR ICE
OVER GROSS WEIGHT	HIT KNOWN OBJECT
CG OUT OF LIMITS	EMERGENCY LANDING
STRUCK ANIMAL	HARD LANDING
BIRD STRIKE	OVERSHOT RUNWAY
PAX DISTURBANCE	UNDERSHOT RUNWAY
STOLEN AIRCRAFT	LOSS OF CONTROL
HIJACK	STALL/SPIN
	MISMANAGED GEAR
	MISMANAGED CONTROLS
	ABORTED TAKEOFF
	AIRFRAME ICE/FROST
	WAKE TURBULENCE
	WEATHER

22. TYPE OF OPERATIONS

PERSONAL

COMMERCIAL

CARGO

INSTRUCTION

CORPORATE

FERRY

AERIAL APPLICATION

AMBULANCE

FIREFIGHTING

BANNER TOW

AIR SHOW

SIGHTSEEING

SKYDIVING

FAR 141 PILOT SCHOOL

MILITARY

FOREIGN

PUBLIC USE

OTHER

23. WX. BRIEFING SOURCE

NOT APPLICABLE/NOT AVAILABLE

NATIONAL WEATHER SERVICE

FLIGHT SERVICE STATION

PATWAS

VOICE RESP. SYSTEM

COMPANY

COMMERCIAL WX. SERVICE

TV/RADIO WEATHER

MILITARY

COMPUTER BRIEFING

24. PRECIPITATION

NOT APPLICABLE/NOT AVAILABLE

RAIN

HAIL

SLEET

SNOW

FREEZING DRIZZLE

FREEZING RAIN

DRIZZLE

OTHER

25. WEATHER FACTORS

NONE / NOT APPLICABLE	THUNDERSTORM
HAZE	CROSSWIND
DUST	TURBULENCE/WINDSTORM
SMOKE	DENSITY ALTITUDE
FOG	LIGHTNING STRIKE
BLOWING DUST	BLOWING SNOW
BLOWING SMOKE	WHITE OUT
ICING CONDITIONS	WIND SHEAR
GUSTY WINDS	OTHER

26. PHASE OF FLIGHT

GROUND	CRUISE	MANEUVER
TAXI	DESCENT	HOVER
TAKEOFF	APPROACH	OTHER
CLIMB	LANDING	

27. ACTUAL WEATHER IMC ☐ VMC ☐ NOT AVAILABLE ☐

28. RUNWAY CONDITIONS NOT APPLICABLE ☐

DRY	SNOW
WET	SLUSH
ICE	STANDING WATER

FAA Form 8020-23 (12-99) SUPERSEDES FAA FORMS 8020-5 and 8020-16 INFORMATION IS PRELIMINARY AND SUBJECT TO CHANGE NSN: 0052-00-923-1000

AFS Electronic Forms System - JetForm FormFlow

(a) Data recording form

The purpose of crash analysis is to identify the existence of patterns in the safety performance of transportation facilities, to determine the most probable causes of the crash, and to identify measures that could be taken to avoid similar crashes in the future.

FIGURE 8.6b
Federal Aviation Administration crash report form.

29 GENERAL AVIATION ACCIDENTS ONLY				30. EVACUATION OVERVIEW (AIR CARRIER ONLY)			
				EVACUATION INITIATED		EVACUATION INJURIES	
DID PILOT ATTEND SAFETY SEMINAR OR CLINIC WITHIN PAST 3 YEARS ?	YES ☐	NO ☐	UNKNOWN ☐	☐	☐	☐	☐
DID PILOT PARTICIPATE IN WINGS PROGRAM WITHIN PAST 3 YEARS ?	YES ☐	NO ☐	UNKNOWN ☐				
DID PILOT ATTEND ANY OTHER RECURRENT TRAINING WITHIN THE PAST 3 YEARS ?	YES ☐	NO ☐	UNKNOWN ☐	YES	NO	YES	NO

31. PILOT INFORMATION NOT APPLICABLE ☐		CERTIFICATE TYPE	SECOND PILOT
NAME		RECREATIONAL ☐	
DATE OF BIRTH	MO DA YR	STUDENT ☐	MO DA YR
DATE HIRED (AIR CARRIER ONLY)	MO DA YR	PRIVATE ☐	MO DA YR
DOMICILE ZIP CODE			
HOURS MAKE AND MODEL		COMMERCIAL ☐	
HOURS LAST 90 DAYS		FLIGHT INST. ☐	
TOTAL HOURS			
CERTIFICATE NO.		ATP ☐	
REGULATORY CHECK RIDE	MO DA YR	NON-PILOT ☐	MO DA YR

32 CORRECTIVE ACTION(S) PLANNED OR INITIATED NONE ☐ 44709 REXAM ☐ EIR ☐ SDR ☐ COUNSELING ☐ M or D ☐ OTHER ☐

33. NARRATIVE (ATTACH ADDITIONAL SHEETS AS NECESSARY)
(ONLY STATE THE FACTS THAT ARE CAUSAL TO THE ACCIDENT/INCIDENT)

CONDUCT OF INVESTIGATION

34. NTSB PARTICIPATION ON-SCENE ☐ LIMITED ☐ | 35 FAA PARTICIPATION ON-SCENE ☐ NOT ON-SCENE ☐ SCENE NOT ACCESSIBLE ☐

36. FAA INITIAL NOTIFICATION	37 FSDO NOTIFICATION	38 FAA IIC ARRIVAL ON SCENE
DATE AND LOCAL TIME	DATE AND LOCAL TIME	DATE AND LOCAL TIME
MO DA YR	MO DA YR	MO DA YR
24 - HOUR CLOCK	24 - HOUR CLOCK	24-HR CLOCK
39. FAA HOURS USED FOR TOTAL INVESTIGATION	40 TOTAL HOURS USED AT ACCIDENT SCENE	41 TOTAL TRAVEL HOURS TO & FROM SCENE

42 FAA NINE RESPONSIBILITIES
IDENTIFICATION OF RESPONSIBILITIES IS THE INVESTIGATORS OPINION BASED ON HIS/HER INVESTIGATION

1. FAA FACILITIES YES ☐ NO ☐	4. AIRMAN/AIR AGENCY COMPETENCE YES ☐ NO ☐	7. SECURITY YES ☐ NO ☐
2. NON FAA FACILITIES YES ☐ NO ☐	5. FAR CHANGE NEEDED YES ☐ NO ☐	8. AIRMAN MEDICAL QUALIF. YES ☐ NO ☐
3. AIRWORTHINESS YES ☐ NO ☐	6. AIRPORT CERTIFICATION YES ☐ NO ☐	9. FAR VIOALATIONS YES ☐ NO ☐

43 BRIEF EXPLANATION OF ISSUES INVOLVED

44 FAA IIC NAME DATE REGION DISTRICT OFFICE

FAA Form 8020-23 (12-99) SUPERSEDES FAA FORMS 8020-5 and 8020-16 INFORMATION IS PRELIMINARY AND SUBJECT TO CHANGE NSN: 0052-00-923-1000
AFS Electronic Forms System - JetForm FormFlow

(b) Data recording form part 2

Crash rates are used in order to facilitate the comparison of crash histories at one location with those of another. Rates for each crash type are typically reported in terms of number of crashes per vehicle or per passenger milc for a given facility type or length of travelway. Rates may also be reported in terms

FIGURE 8.6c

Federal Aviation Administration crash report form.

INSTRUCTIONS FOR ACCIDENT/INCIDENT REPORT

1. **OCCURRENCE INFORMATION:**
THIS FORM IS TO BE FILLED OUT FOR EACH ACCIDENT/INCIDENT AND FORWARDED TO THE REGIONAL FS DIVISION WITHIN 30 DAYS. REGIONAL FS DIVISION WILL FORWARD ORIGINAL FAA ACCIDENT/ INCIDENT REPORT TO AFS-620 AND A COPY OF ACCIDENT REPORTS ONLY TO AAI-220.
2. **AMENDED DATE:**
FOR AMENDED REPORTS FILL IN ITEMS 1, 2, 3, 5, AND 13, REGISTRATION NUMBER ONLY, AND NEW OR CHANGED INFORMATION PERTAINING TO ACCIDENT INVESTIGATION.
3. **DATE OF THE OCCURRENCE:**
MONTH/DAY/YEAR.
4. **FAA (INVESTIGATING OFFICE):**
THE FIRST TWO BLOCKS ARE THE REGION. THE SECOND TWO BLOCKS ARE THE NUMERICAL I.D. OF THE FSDO, E.G., EA 21.
5. **NTSB ID:**
FOR ACCIDENTS ONLY AND SUPPLIED BY THE NTSB OFFICE WITH JURISDICTIONAL RESPONSIBILITY.
6. **LOCATION :**
CITY: NEAREST CITY OR TOWN.
STATE: 2 LETTER IDENTIFIER.
ZIP CODE: SELF- EXPLANATORY.
7. **OPERATOR:**
FOR AIR CARRIER OCCURRENCES ONLY. PROVIDE THE NAME OF THE OPERATOR THAT HAS OPERATIONAL CONTROL. THE 4-LETTER DESIGNATOR IS FROM PTRS.
8. **AIRPORT:**
NAME OF AIRPORT IF OCCURRENCE TOOK PLACE ON AN AIRPORT. AIRPORT DESIGNATOR ACCORDING TO ORDER 7310.1.
9. **TIME:**
LOCAL 24 HOUR CLOCK.
10. **LATITUDE / LONGITUDE:**
SELF-EXPLANATORY. ALASKA ACCIDENTS ONLY.
11. **AIRCRAFT DAMAGE:**
CHECK THE MOST SEVERE DAMAGE.
12. **COLLISION:**
MEANS TWO AIRCRAFT COLLIDED IN THE AIR OR ON THE GROUND. BOTH WERE FLYING OR HAD THE INTENT TO FLY. TWO FORMS REQUIRED IF BOTH AIRCRAFT WERE FLYING OR HAD THE INTENT TO FLY.
13. **AIRCRAFT REGISTRATION NUMBER:**
E.G. N1234M. MAKE/MODEL: MANUFACTURER/MODEL/SERIES, E.G., DC-9-10. SERIAL NUMBER: SELF EXPLANATORY. YEAR OF MANUFACTURE: E.G., 1994 AIRFRAME CYCLES, AIRFRAME HOURS SELF-EXPLANATORY.
14. **FAR PART NUMBER:**
CHECK THE REGULATION THAT THE AIRCRAFT WAS OPERATING UNDER. AN AIR CARRIER DOING POSITIONING, TRAINING, ETC., IS PART 91. PART 135 AIR TAXI OR AIR AMBULANCE IS PART 91 UNTIL PASSENGER PICKUP. MEDICAL PERSONNEL ARE CONSIDERED PART OF THE CREW.
15. **TYPE OF AIRCRAFT:**
SELF- EXPLANATORY (MORE THAN ONE MAY BE CHECKED).
16. **POWERPLANT INFORMATION:**
(ONLY IF CAUSAL TO THE ACCIDENT/INCIDENT)
LIST MAKE/MODEL/SERIES OF ENGINE.
17. **PROPELLER INFORMATION :**
(ONLY IF CAUSAL TO THE ACCIDENT/INCIDENT):
LIST MAKE/MODEL/SERIES OF PROPELLER.
18. **BIOHAZARD AREA:**
CHECK YES IF BODY FLUIDS WERE PRESENT. USE OR NONUSE OF PERSONAL PROTECTIVE EQUIPMENT DOES NOT AFFECT THIS QUESTION.
19. **TYPE OF LANDING GEAR:**
SELF -EXPLANATORY.
20. **INJURY SUMMARY:**
ENTER THE NUMBERS INVOLVED AND ACCOUNT FOR ALL ON BOARD THE AIRCRAFT, AND ACCOUNT FOR THE PERSONNEL INJURED THAT WERE NOT ON THE AIRCRAFT.
21. **FACTORS:**
CHECK THE PRIMARY FACTOR FROM EITHER TECHNICAL OR OPERATIONAL FACTORS BLOCK WHICHEVER IS MOST APPROPRIATE.

21A. **TECHNICAL FACTORS:**
CHECK APPLICABLE BOXES. MORE THAN ONE MAY BE CHECKED. THIS IS THE INSPECTOR/INVESTIGATOR OPINION BASED ON HIS/HER INVESTIGATION.

21B. **OPERATIONAL FACTORS:**
SAME AS 21A.

21C. **PART NAME:**
IDENTIFY THE PART NAME THAT FAILED OR IS SUSPECTED OF FAILURE BY THE PROPER NOMENCLATURE THAT IS DEPICTED IN THE MANUFACTURERS PARTS CATALOGUE.

21D. **MANUFACTURER:**
IDENTIFY THE MANUFACTURER OF THE PART, IF KNOWN.

21E. **PART NUMBER:**
IDENTIFY THE MANUFACTURER PART NUMBER. THIS WOULD BE THE SAME NUMBER NEEDED TO REQUISITION A REPLACEMENT PART.

21F. **ATA CODE:**
REFER TO THE CODE TABLE IN THE FLIGHT STANDARDS GUIDE TITLED: JOINT AIRCRAFT SYSTEM AND COMPONENT CODE TABLE AND DEFINITIONS DATED JANUARY 1996.

22. **TYPE OF OPERATIONS:**
CHECK APPROPRIATE BOXES.
23. **WEATHER BRIEFING SOURCE:**
SAME AS 21A.
24. **PRECIPITATION:**
SAME AS 21A.
25. **WEATHER FACTORS:**
SAME AS 21A.
26. **PHASE OF FLIGHT:**
WHERE ACCIDENT AND INCIDENT SEQUENCE STARTED CHECK APPLICABLE PHASE.
27. **ACTUAL WEATHER CONDITIONS:**
CHECK APPROPRIATE BOX.
28. **RUNWAY CONDITIONS:**
CHECK APPROPRIATE BOX.
29. **GENERAL AVIATION ACCIDENTS ONLY:**
SELF- EXPLANATORY.
30. **EVACUATION OVERVIEW** (AIR CARRIER ONLY)
EVACUATION INITIATED YES/NO.
INJURIES: CHECK YES IF INJURIES ATTRIBUTABLE TO EVACUATION.
31. **PILOT INFORMATION:**
SELF- EXPLANATORY. CHECK THE HIGHEST CERTIFICATE THAT THE PILOT HAS. PIC NAME NOT APPLICABLE IF THE PILOTS ACTIONS OR LACK OF ACTIONS DID NOT CONTRIBUTE TO THE ACCIDENT/INCIDENT. HOWEVER, FOR AIR CARRIER ACCIDENTS, PLEASE PROVIDE PIC DOB, HOURS MAKE AND MODEL, AND TOTAL HOURS.
32. **CORRECTIVE ACTION:**
SELF- EXPLANATORY.
33. **NARRATIVE:**
SELF- EXPLANATORY.
34. **NTSB PARTICIPATION (ACCIDENT ONLY):**
SELF- EXPLANATORY.
35. **FAA PARTICIPATION:**
SELF- EXPLANATORY. ON-SCENE CAN BE CHECKED IF THE INSPECTOR/ INVESTIGATOR PARTICIPATES IN THE INVESTIGATION BEYOND USE OF THE TELEPHONE, I.E., ENGINE TEARDOWN, INTERVIEW,OR WRECKAGE INVESTIGATION NOT AT THE SCENE OF THE ACCIDENT, ETC.
36. **FAA INITIAL NOTIFICATION:**
THIS IS THE TIME THE FIRST FAA PERSON WHO DISCOVERS OR IS NOTIFIED OF THE OCCURRENCE. THIS IS USUALLY AIR TRAFFIC.
37. **FSDO NOTIFICATION:**
THIS IS THE FIRST CALL THAT THE FSDO RECEIVES.
38. **FAA IIC ARRIVAL ON SCENE:**
SELF-EXPLANATORY.
39. **FAA HOURS USED FOR TOTAL INVESTIGATION:**
INCLUDES ON-SCENE, TRAVEL HOURS, AND NON-SCENE ACTIVITIES. WHOLE HOURS ONLY.
40. **TOTAL HOURS USED AT ACCIDENT/INCIDENT SCENE:**
WHOLE HOURS ONLY.
41. **TOTAL TRAVEL HOURS TO & FROM SCENE:**
WHOLE HOURS ONLY
42. **FAA NINE RESPONSIBILITIES (ACCIDENT MANDATORY/INCIDENTS OPTIONAL):**
CHECK WHICH OF THE AREAS OF RESPONSIBILITY WERE INVOLVED. THE DETERMINATION OF RESPONSIBILITIES IS THE OPINION OF THE INSPECTOR/ INVESTIGATOR BASED ON HIS/HER BACKGROUND, TRAINING, SKILL, AND EXPERIENCE. THE ANNOTATION OF ONE OR MORE RESPONSIBILITIES DOES NOT HAVE TO BE JUSTIFIED OR PROVEN. AN AIRMAN WHO MAKES A MISTAKE WHICH RESULTS IN AN ACCIDENT IS ANNOTATED UNDER AIRMAN/AIR AGENCY COMPETENCE. IT IS NOT NECESSARY TO SUBMIT AN EIR BECAUSE OF ANNOTATION OF VIOLATION.
43. **BRIEF EXPLANATION OF ISSUES INVOLVED FOR EACH OF THE NINE RESPONSIBILITIES INVOLVED.**
IF NONE INVOLVED, EXPLAIN WHY. SELF-EXPLANATORY.
44. **FAA IIC NAME:**
PRINT, SIGN, AND DATE.

FAA Form 8020-23 (12-99) SUPERSEDES FAA FORMS 8020-5 and 8020-16 INFORMATION IS PRELIMINARY AND SUBJECT TO CHANGE NSN: 0052-00-923-1000
AFS Electronic Forms System - JetForm FormFlow

(c) Instructions form

of community factors such as number of crashes per registered vehicle or per person.

Crash rates for intersections or grade crossings are typically stated in terms of crashes per million entering vehicles per year expressed as follows:

$$R/MEV/Y = (C_i \times 1{,}000{,}000)/V \quad (8.1)$$

where

$$R/MEV/Y = \text{crash rate per million vehicles entering the crossing/year}$$
$$C_i = \text{number of crashes/year of type } i$$
$$V = \text{annual number of entering vehicles} = (\text{ADT} \times 365)$$
$$\text{ADT} = \text{average daily traffic}$$

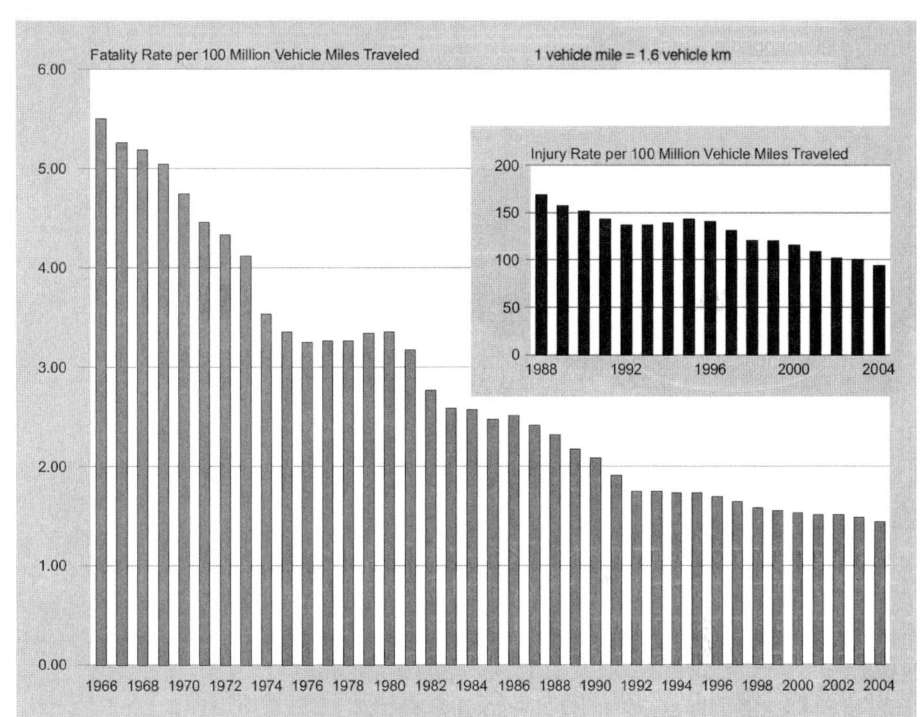

FIGURE 8.7 Motor vehicle fatality and injury rates per 100 million vehicle miles of travel, 1966–2004.

Source: Traffic Safety Facts 2004: A Compilation of Motor Vehicle Crash Data from the Fatal Analysis Reporting System and the General Estimates System, National Center for Statistics and Analysis of the National Highway Traffic Safety Administration, U.S. Department of Transportation, Washington, D.C., 2005.

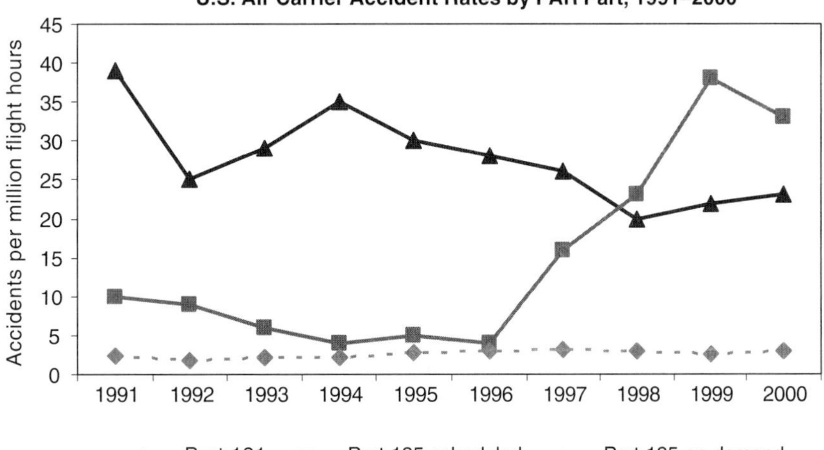

FIGURE 8.8 U.S. carrier accident rates by FAR (Federal Aviation Regulation) part, 1991–2000.

Source: National Transportation Safety Board, GILS: Aviation Accident Data Base, updated February 2006, Office of Aviation Safety, Washington, D.C.

TABLE 8.1
U.S. railroad accident/incident historical summary.

Source: Railroad Safety Statistics 2004 Report, Federal Railroad Administration, U.S. Department of Transportation, Washington, D.C., November 2005.

Category	1993	1994	1995	1996	1997	1998	1999	2000	2001	2002	2003	2004
---GRAND TOTAL---												
Accidents/incidents	24,740	22,465	19,591	17,690	16,699	16,501	16,776	16,918	16,087	14,404	14,279	14,232
Rate[1]	21.82	19.14	16.60	15.05	14.14	13.78	13.72	13.94	13.56	12.18	11.95	11.59
Deaths	1,279	1,226	1,146	1,039	1,063	1,008	932	937	971	951	867	898
Nonfatal conditions	19,121	16,812	14,440	12,558	11,767	11,459	11,700	11,643	10,985	11,103	9,180	8,871
---TRAIN ACCIDENTS-----												
Rate[2]	4.25	3.82	3.67	3.64	3.54	3.77	3.89	4.13	4.25	3.76	4.03	4.28
Total number	2,611	2,504	2,459	2,443	2,397	2,575	2,768	2,983	3,023	2,738	2,997	3,296
Deaths	67	12	14	25	17	4	9	10	6	15	4	13
Injuries	308	262	294	281	183	129	130	275	310	1,884	227	229
Collisions	205	240	235	205	202	168	205	238	220	192	200	237
Derailments	1,930	1,825	1,742	1,816	1,741	1,757	1,961	2,112	2,234	1,989	2,114	2,367
On main line	955	914	912	941	867	934	858	976	1,025	886	962	1,009
On yard track	1,383	1,339	1,279	1,249	1,223	1,306	1,531	1,619	1,569	1,478	1,651	1,860
Yard track rate[3]	15.87	14.91	14.23	14.22	14.41	15.60	17.51	18.21	18.30	18.25	20.21	22.14
Other track rate[4]	2.33	2.06	2.03	2.05	1.98	2.12	1.98	2.15	2.32	1.95	2.03	2.10
Track caused	963	911	856	905	879	900	995	1,035	1,121	941	969	1,010
Track caused rate	1.57	1.39	1.28	1.35	1.30	1.32	1.40	1.43	1.58	1.29	1.30	1.31
Human factor caused	865	911	944	783	855	971	1,031	1,147	1,035	1,050	1,217	1,329
Equipment caused	360	293	279	318	271	307	321	372	427	367	361	416
Signal caused	54	36	27	49	39	38	49	70	42	50	58	65
Equip Dmg (millions $)	121.833	124.850	134.766	160.908	152.092	162.561	164.654	169.172	200.752	173.982	191.411	223.615
Track Dmg (millions $)	48.816	43.899	54.458	51.407	58.637	71.337	80.435	94.040	113.713	92.550	99.118	98.757
Hazmat												
Consists releasing	28	34	26	34	31	42	41	35	32	31	27	29
Cars releasing	57	40	48	69	38	66	75	75	57	56	38	47
People evacuated	3,207	15,336	2,817	8,547	8,812	2,058	996	5,258	52,620	5,438	2,260	5,938
---HIGHWAY-RAIL--------												
Rate[5]	7.97	7.60	6.92	6.34	5.71	5.14	4.90	4.84	4.55	4.22	4.00	3.98
Incidents	4,892	4,979	4,633	4,257	3,865	3,508	3,489	3,502	3,237	3,077	2,977	3,063
Deaths	626	615	579	488	461	431	402	425	421	357	334	368
Injuries	1,837	1,961	1,894	1,610	1,540	1,303	1,396	1,219	1,157	999	1,031	1,081
---OTHER INCIDENTS-----												
Incidents[6]	17,237	14,982	12,499	10,990	10,437	10,418	10,519	10,433	9,827	8,589	8,305	7,873
Deaths	586	599	553	526	585	573	521	502	544	579	529	517
Injuries	16,976	14,589	12,252	10,667	10,044	10,027	10,174	10,149	9,518	8,220	7,922	7,561

[1]*Total accident/incident rate of all reported events × 1,000,000/(train miles + hours).*
[2]*Total train accidents × 1,000,000/total train miles.*
[3]*Accidents on yard track × 1,000,000/yard switching train miles.*
[4]*Accidents on other than yard track × 1,000,0000/(total train miles − yard switching).*
[5]*Total incidents × 1,000,000/total train miles.*
[6]*Other events that cause death, injury to any person; or illness to a railroad employee.*

Note: 1 mile = 1.6 km

Crash rates per travelway segment are typically stated in terms of crashes per 100 million vehicles or passenger kilometers for a given segment length or per million flight kilometers and are expressed as follows:

$$R/HMVM/Y = (C_i \times 100{,}000{,}000)/(VMT) \tag{8.2}$$

where

$R/HMVM/Y$ = crash rate per hundred million vehicle kilometers per year
C_i = number of crashes per year of type i
VMT = number of vehicle miles of travel = (ADT) (365) (travelway length)

(*Note:* Crash rates for passenger kilometers can be obtained by substituting *PMT* for *VMT*.)

EXAMPLE 8.1

Computing Accident Rates at Intersections or Grade Crossings

There are eight crashes per year at a rural railroad grade crossing at which one train passes through per hour. The average 24-hour volume entering the crossing is 5500 vehicles/day. Determine the crash rate per million entering vehicles.

Solution

$$\begin{aligned} R/MEV/Y &= (C_i \times 1{,}000{,}000)/V \\ &= (8 \times 1{,}000{,}000)/(5500 \times 365) \\ &= 3.98 \text{ crashes/million entering vehicles/year} \end{aligned}$$

EXAMPLE 8.2

Computing Crash Rates on Travelway Segments

The number of crashes for a commuter airline with service between two cities located 300 km apart is three in a five-year period. There are seven flights per day with an average passenger load of 29. Compute the crash rate per million vehicle and passenger kilometers.

Solution

$$\begin{aligned} R/MVM/Y &= (C_i \times 1{,}000{,}000)/(VMT) \\ &= (3/5) \times (1{,}000{,}000)/(7)(365)(300) \\ &= 0.78 \text{ crashes per million flight kilometers per year} \end{aligned}$$

$$\begin{aligned} R/MPM/Y &= 3/5(1{,}000{,}000)/(7)(29)(300)(365) \\ &= 0.027 \text{ crashes per million passenger kilometers per year} \end{aligned}$$

Given the multiplicity of reasons that transportation crashes can or have occurred, it is remarkable that the U.S. transportation system is relatively safe. The causes of crashes or accidents described previously are well known to the transportation professional community, and a great deal has been done to assure safety while traveling.

Many improvements have been made in each of the areas that are known to cause crashes. During the past 30 years, transportation accident rates have

FIGURE 8.10

Growth in U.S. transportation travel.

Source: Transportation Research Board. National Academies, Washington, D.C.

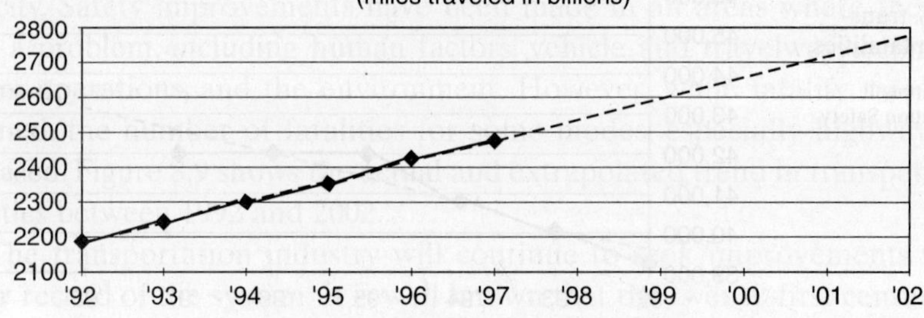

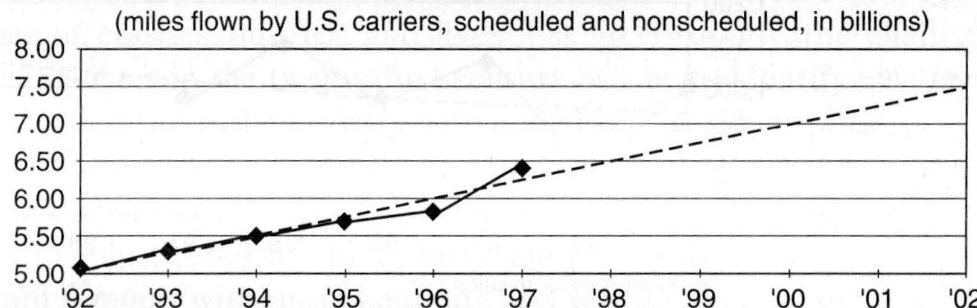

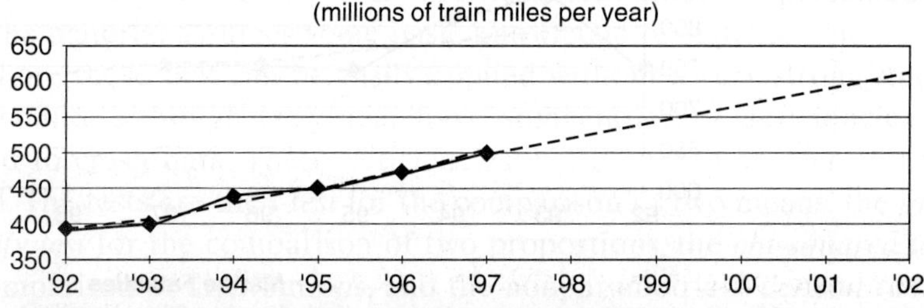

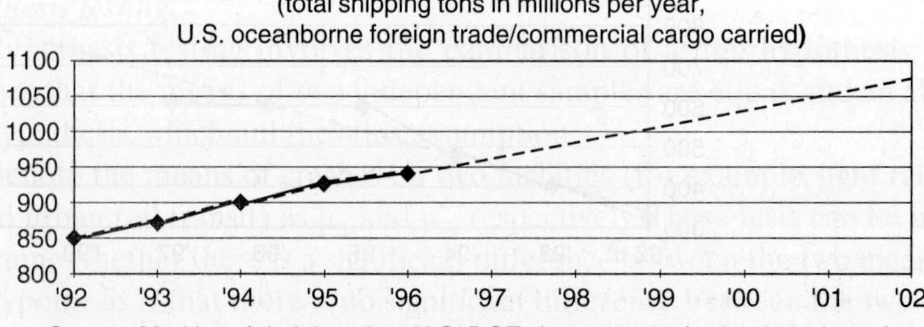

Note: 1 mile = 1.6 km

The alternative hypothesis will depend on the problem statement and can be one of the following:

H_1: $\mu_1 < \mu_2$ (one-tail test)

H_0: $\mu_1 > \mu_2$ (one-tail test)

H_0: $\mu_1 \neq \mu_2$ (two-tail test)

The null hypothesis is tested by computing a test statistic using the estimated mean and/or variances of the populations obtained from the data sets. The test statistic is then compared with a similar value obtained from the theoretical distribution. The theoretical value depends on the degree of freedom for the assumed distribution.

In using the *t-test,* the test statistic is given as

$$t = \frac{\overline{X}_1 - \overline{X}_2}{S\sqrt{\frac{1}{n_1} + \frac{1}{n_2}}} \tag{8.3}$$

where

$\overline{X}_1$ and $\overline{X}_2$ = sample means
n_1 and n_2 = sample sizes
S = square root of the pooled variance given by

$$S^2 = \frac{(n_1 - 1)S_1^2 + (n_2 - 1)S_2^2}{n_1 + n_2 - 2} \tag{8.4}$$

where

S_1 and S_2 = variances of the populations
t has a degree of freedom of $(n_1 + n_2 - 2)$

The theoretical values for t for different levels of significance are given in Appendix A for different levels of confidence α, which is the probability of rejecting the null hypothesis when it is true and is usually referred to as a Type I error. Values normally used in transportation safety are 5% and 10%. The main deficiency in applying the *t-test* to crash data is that the test assumes a normal distribution for the data being tested, while research has shown that the Poisson and negative binomial distributions usually describe the occurrence of crashes. However, large sample sizes tend to diffuse this deficiency. The region of rejection of the null hypothesis is as follows:

If H_1 is	*Then reject H_0 if*
$\mu_1 < \mu_2$	$t \leq -t_\alpha$
$\mu_1 > \mu_2$	$t > t_\alpha$
$\mu_1 \neq \mu_2$	$t < -t_{\alpha/2}$ or $t > t_{\alpha/2}$

EXAMPLE 8.3

Using the *t*-Test for Significant Difference in Crashes

It is required to test whether large trucks are significantly more involved in rear-end crashes on interstate highways with differential speed limits (DSL) for passenger cars and large trucks than on those with a uniform speed limit (USL). Using the *t*-test and data for the same three-year period shown in the following table, determine whether you can conclude that trucks are more involved in rear-end crashes on interstate highways with differential speed limits. Use a 5% significance level. List the deficiencies inherent in this type of analysis.

Interstate Highways with DSL		*Interstate Highways with USL*	
Site No.	*Number of Crashes*	*Site No.*	*Number of Crashes*
1	10	1	8
2	12	2	9
3	9	3	11
4	8	4	12
5	11	5	5
6	6	6	7

Solution

$$\boldsymbol{H_0}: M_{\text{DSL}} = M_{\text{USL}}$$

Rear-end crashes on interstate highways with differential speed limits are the same for interstate highways with uniform speed limits.

$$\boldsymbol{H_A}: M_{\text{DSL}} > M_{\text{USL}}$$

Trucks are more involved in rear-end crashes on interstate highways with differential speed limits.

$$M_{\text{DSL}} = 9.333, S^2_{\text{DSL}} = 4.667$$

$$M_{\text{USL}} = 8.667, S^2_{\text{USL}} = 6.667$$

$$S^2 = \frac{(n_{\text{DSL}} - 1)S^2_{\text{DSL}} + (n_{\text{USL}} - 1)S^2_{\text{USL}}}{n_{\text{DSL}} + n^{-2}_{\text{USL}}}$$

$$= \frac{(6 - 1)4.667 + (6 - 1)6.667}{6 + 6 - 2} = 5.667$$

$$t = \frac{M_{\text{DSL}} - M_{\text{USL}}}{S\sqrt{\frac{1}{n_{\text{DSL}}} + \frac{1}{n_{\text{USL}}}}} = \frac{9.333 - 8.667}{\sqrt{5.667\left(\frac{1}{6} + \frac{1}{6}\right)}} = 0.485$$

At the 95% confidence level and degree of freedom of $t = 6 + 6 - 2 = 10$, $t_\alpha = 1.812$. Since $t < t_\alpha$, we cannot reject the null hypothesis. Thus, it can be concluded that trucks are *not* more involved in rear-end crashes on interstate highways with differential speed limits. The main deficiency in this solution is that the use of the t-test incorrectly assumes that crashes are normally distributed.

The *proportionality test* is used to compare two independent proportions p_1 and p_2. For example, the procedure can be used to compare the proportion of night-landing commercial airline crashes occurring at commercial service–primary airports with those occurring at reliever airports. The null hypothesis is also usually

$$\mathrm{H}_0: p_1 = p_2$$

The alternative hypothesis will depend on the problem statement and can be one of the following:

$$\mathrm{H}_1: p_1 < p_2$$

$$H_1: p_1 > p_2$$

$$H_1: p_1 \neq p_2$$

The test statistic is given as

$$Z = \frac{p_1 - p_2}{\sqrt{p(1-p)\left(\frac{1}{n_1} + \frac{1}{n_2}\right)}} \tag{8.5}$$

where

$$p = \frac{(x_1 + x_2)}{n_1 + n_2}$$

$$p_1 = x_1/n_1, \quad p_2 = x_2/n_2$$

n_i = total number of observations in data set, i
x_i = successful observations in data set, i

The value of Z is then compared with Z_α, the standard normal variant corresponding to a significance level of α.

EXAMPLE 8.4

Using the Proportionality Test for Significant Difference in the Proportions of Serious Crashes at Work Zones and Non–Work Zones

The following table shows Fatality and Injury and Property Damage Only crashes at work zones and non–work zones at six locations on the same interstate highways over the same period. Using the proportionality test, determine whether the probability of large trucks being involved in F&I crashes is significantly different at work zones than at non–work zones at a 5% significance level.

Work Zones		*Non–Work Zones*	
PDO Crashes	*F&I Crashes*	*PDO Crashes*	*F&I Crashes*
8	6	7	3
5	2	8	4
6	4	5	1
10	7	9	4
2	0	10	2
6	1	11	8
Σ 37	20	50	22

$\boldsymbol{H_0}: p_1 = p_2$

The probability of large trucks being involved in F&I crashes at work zones is not higher than at non–work zones.

$\boldsymbol{H_A}: p_1 > p_2$

The probability of large trucks being involved in F&I crashes at work zones is higher than at non–work zones.

$$p_1 = \frac{20}{37 + 20} = 0.351, \qquad p_2 = \frac{22}{50 + 22} = 0.306$$

$$p = \frac{20 + 22}{(37 + 20) + (50 + 22)} = 0.326$$

$$Z = \frac{0.351 - 0.306}{\sqrt{0.326(1 - 0.326)\left(\frac{1}{57} + \frac{1}{72}\right)}} = 0.546$$

At the 95% confidence level, $Z_\alpha = 1.645$; $Z < Z_\alpha$, so we cannot reject the null hypothesis. It can be concluded that the probability of large trucks being involved in F&I crashes is *not* significantly different at work zones from that at non–work zones at a 5% significance level.

The *chi-squared test* can be used to perform a "before and after" evaluation of a countermeasure or safety treatment. This usually entails the use of crash data for a treated site and a control site before and after the implementation of a countermeasure at the treated site, to determine whether the crash frequency at the treated site after the implementation of the countermeasure is significantly different from that before implementation. The null hypothesis is usually that there is no difference between the number of crashes at the treated site before the implementation of the treatment and after the implementation of the treatment. We shall discuss two tests: (1) where the control area is so large that the control ratio can be assumed to be free from error, and (2) the case where no such control ratio exists.

For the case where it is assumed that the control ratio is free from error, it is assumed that crashes at the study site for the before and after periods can be distributed in accordance with those of the control area.

Let

a = accidents after at study site
b = accidents before at study site
A = accidents after at control area
B = accidents before at control area
C = control ratio A/B
$n = a + b$

With this assumption, redistribute the total crashes in a study area in proportion to those in the control area (A and B). The number of expected crashes before the implementation of the countermeasure at the study area is given as

$$\frac{Bn}{A + b} = \frac{n}{1 + C} \tag{8.6}$$

The number of expected crashes after the implementation of the countermeasure is given as

$$\frac{An}{A + B} = \frac{Cn}{1 + C} \tag{8.7}$$

The χ^2 test is based upon a *contingency* table—a table showing both observed and expected values as shown follows:

Original Data Set

	Before	*After*	*Total*
Study site	B	a	$(a + b) = n$
Study site observed	b	a	$(a + b) = n$
Study site expected	$n/(1 + C)$	$Cn/(1 + C)$	N
Control area	B	A	$B + A$

Crashes at Study Site after Redistribution

	Before	*After*	*Total*
Study site observed	b	a	$(a + b) = n$
Study site expected	$n/(1 + C)$	$Cn/(1 + C)$	n

Using χ^2, we have

$$\chi^2 = \sum_{i=1}^{m}\sum_{i=1}^{n}\frac{(O_{ij} - E_{ij})^2}{E_{ij}} \tag{8.8}$$

where

O_{ij} = the observed value in column i and row j
E_{ij} = the expected value in column i and row j
m = the number of rows
n = the number of columns

The computed χ^2 is then compared with the theoretical value of χ^2, for a degree of freedom of $(m - 1)(n - 1)$. If the computed χ^2 is less than the theoretical χ^2 for a selected significance level α, then there is no reason to conclude that there is a significant difference between the observed and expected crashes and the null hypothesis will be accepted. However, if the computed χ^2 is larger than the theoretical value, then the null hypothesis is rejected, and thus there is a significant difference between the actual and expected number of crashes. The table of theoretical chi-squared values is also given in Appendix A.

	Before	*After*	*Total*
Site 1	b	a	$(a + b)$
Site 2	d	c	$(c + d)$
Total	$(b + d)$	$(a + c)$	$(a + b + c + d) = T$

For the case where it cannot be assumed that the control ratio is free from error, both sets of data should be redistributed, giving the following contingency tables:

Expected Crashes at Both Sites after Redistribution

	Before	*After*	*Total*
Site 1 expected	$(b + d)(a + b)/T$	$(a + b)(c + d)/T$	$(a + b)$
Site 2 expected	$(b + d)(c + d)/T$	$(a + c)/T$	$(c + d)$
Total	$(b + d)$	$(a + c)$	T

The χ^2 is computed using both the actual and the expected for both sets of data as

$$\chi^2 = \frac{\left[b - \frac{(b + d)(a + b)}{T}\right]^2}{\frac{(b + d)(a + b)}{T}} + \frac{\left[a - \frac{(a + b)(a + c)}{T}\right]}{\frac{(a + b)(a + c)}{T}}$$

$$+ \frac{\left[d - \frac{(b + d)(c + d)}{T}\right]^2}{\frac{(b + d)(c + d)}{T}} + \frac{\left[c - \frac{(a + c)(c + d)}{T}\right]^2}{\frac{(a + c)(c + d)}{T}}$$

$$\chi^2 = \frac{T(ad - bc)^2}{(a + b)(c + d)(a + c)(b + d)} \tag{8.9}$$

As in the previous case, the calculated χ^2 will be compared with the theoretical value for the appropriate degree of freedom at a significance level of α. The null hypothesis is rejected if the computed value is higher than the calculated value. The main deficiency of this procedure is that exposure is not considered, where exposure refers to the impact of factors that influence the occurrence of crashes at a specific location. An example of exposure is the traffic volume.

EXAMPLE 8.5

Use of Chi-Squared Test for Significant Difference in the Effect of Active Warning Controls at Railroad Crossings

The data in the following table show crashes collected over the same period at sites at which active warning controls at railroad crossings were previously installed and at several other similar sites without the controls. Using the chi-squared test and without assuming that the control ratio is free from error, determine whether it can be concluded that crashes tend to be higher at locations without the active controls. Use a significance level of 5%.

Locations with Active Controls	*Locations without Active Controls*
Before = 31	Before = 98
After = 35	After = 106

H_0: Crashes are the same at sites with active controls and at sites without active controls.

H_A: Crashes tend to be higher at locations without active controls.

$$\chi^2 = \frac{T(ad - bc)^2}{(a + b)(c + d)(a + c)(b + d)}$$

$$= \frac{270(98 \times 35 - 31 \times 106)^2}{66 \times 204 \times 129 \times 141} = 0.0229$$

Degree of freedom = (2 − 1) (2 − 1) = 1

At a 5% confidence level, $\chi_c^2 = 3.841$. $\chi^2 < \chi_c^2$, so we cannot reject the null hypothesis. It can be concluded that crashes are *not* higher at locations without the controls, at a significance level of 0.05.

The methods described thus far require the assumption of some type of distribution of the populations from which the sample data were obtained. As this assumption is often not met, the use of an analysis methodology that does not require the assumption of any distribution may be considered. Techniques that do not require an assumption of some distribution are known as *nonparametric.*

A commonly used nonparametric technique is the *Wilcoxon rank sum test for independent samples.* It can be used to test the null hypothesis that the probability distributions associated with two populations are not significantly different. Consider two sets of data for large truck crashes on secondary roads and primary roads. The procedure involves the following steps:

(i) Rank the two sets of data from the sample observations as if both sets of data come from the same distribution. The ranking starts with the smallest value in the combined data being ranked as 1, with the other values being ranked increasingly to the highest value that will have the rank of the total number of data points in the combined data. If the two populations have the same distribution, these ranks will randomly fall within each data set; that is, both high and low ranks will lie within each data set. However, if the distributions are very different, high ranks will tend to lie within one data set while the lower ranks lie in the other data set. Note that the total sum of all ranks (T) is given as

$$T = T_A + T_B = n(n + 1)/2 \tag{8.9}$$

where

T_A = the rank sum for data set A
T_B = the rank sum for data set B
n = $n_A + n_B$
n_A = number of data points in data set A
n_B = number of data points in data set B

Since T is constant for any two sets of data, a large value of T_A results in a small value of T_B. This implies that there is no evidence that the two data sets are from the same populations. It is also possible that the combined data set will have one or more sets of numbers that have the same value. If the number of these ties is much smaller than the number of data sets, the test is still valid. When this occurs, assume that these values are not tied and use the mean of the ranks that they would have been assigned for each. For example, if the sixth and seventh data points are tied, assign the mean of 6.5 to both.

(ii) Determine the critical values for T_L (the lower boundary region) and T_U (the upper boundary region) associated with the sample that has the fewer number of data sets for the selected significance level α, which are given in Appendix Table A.3. If n_1 is the same as n_2, any of the two rank sums may be used as the test statistic.

(iii) The null hypothesis is rejected if the rank sum of the data set that has the fewer data points does not lie between T_L and T_U; that is, reject the null hypothesis if

$$T_L \geqslant T_i \text{ or } T_u \leqslant T_i$$

where T_i is the rank sum for the data set with the fewer number of data points.

Example 8.6

Using the Wilcoxon Rank Sum Test for Significant Difference in Crashes

Use the *Wilcoxon rank sum test* to solve Example 8.3:

$\boldsymbol{H_0}$: $M_{\text{DSL}} = M_{\text{USL}}$

Trucks are not more involved in rear-end crashes on interstate highways with differential speed limits.

$\boldsymbol{H_A}$: $M_{\text{DSL}} > M_{\text{USL}}$

Trucks are more involved in rear-end crashes on interstate highways with differential speed limits.

Category	*Number of Crashes*	*Rank*
USL	5	1
DSL	6	2
USL	7	3
DSL	8	4.5
USL	8	4.5
DSL	9	6.5
USL	9	6.5
DSL	10	8
DSL	11	9.5
USL	11	9.5
DSL	12	11.5
USL	12	11.5

$$T_{\text{DSL}} = 2 + 4.5 + 6.5 + 8 + 9.5 + 11.5 = 42$$

$$T_{\text{USL}} = (6 + 6)\,(6 + 6 + 1)/2 - 42 = 36$$

At a 5% confidence level and $n_A = n_B = 6$, we obtain $T_U = 50$ and $T_L = 28$. Since both T_{DSL} and T_{USL} are less than T_U and greater than T_L, we cannot reject the null hypothesis. It can be concluded that trucks are not more involved in rear-end crashes on interstate highways with differential speed limits.

In conducting crash analysis, important issues that should be considered include *regression to the mean, crash migration,* and *sample size.* Regression to the mean is the phenomenon of crashes to fluctuate about a mean value over time, particularly on highways. If a site is selected for improvement based on a high crash rate over a short period, it is probable that the reduction in crashes observed soon after the implementation of the improvement may not be due to that improvement, as a lower crash rate might have occurred (regression to the mean) even if the improvement was not undertaken. It is therefore important that this phenomenon of regression to the mean effect be considered in analyzing crashes. *Crash migration* occurs as a result of changes in trip patterns due to implementation of a safety countermeasure. For example, if local streets previously used by commuters are closed to through traffic because of complaints from residents, crashes on these local streets may reduce but may increase on the main highways because of commuters having to use the main highways. The *sample size* selected for any crash analysis is also very important. Crashes occur infrequently and randomly at any one location. It is therefore necessary to have an adequate sample size for any analysis. It is, however, worthwhile to note that there are sophisticated statistical techniques available that overcome the deficiencies associated with the techniques presented here. The *empirical Bayes method,* for example, takes into consideration that crash frequencies are often described by the Poisson or negative binomial distribution and uses regression equations based on crashes that occur at untreated similar sites to estimate the expected crashes on the treated sites after the treatment. The computed number of expected crashes is then compared with the actual crashes that occurred at the treated sites. A detailed discussion of these procedures is beyond the scope of this text, but interested readers may refer to the references given at the end of this chapter.

High-Priority Safety Improvements

Every federal transportation agency has an agenda for the improvement of safety. An examination of the Web sites (shown at the end of the chapter) will provide the most recent information regarding agency priorities. The agencies are the Federal Administrations for Aviation, Highway, Railroads, Transit, and Highway Traffic Safety. From a national perspective, the agency responsible for the investigation of major transportation crashes is the NTSB, an independent federal agency charged by Congress with investigating every civil aviation accident in the United States as well as significant accidents in other modes of transportation. This section illustrates some of the key areas identified by the NTSB for safety improvements at the national and state level in terms of highway, air, rail, maritime, and cross-modal transportation issues.

Transportation safety experts have recognized that, because of many uncontrollable factors, it is impossible to prevent crashes completely. Instead, focus is being placed on creating a safer vehicle, one in which the occupant will have a lower likelihood of an injury or fatality if a crash should occur. This strategy has been successful in saving lives. For example, by adding lap and shoulder belts, installing air bags, increasing the structural strength of the vehicle, and installing energy-absorbing fenders, highway-related deaths and injuries have been reduced.

In keeping with this strategy, the NTSB has identified a high payoff area to be the **protection of vehicle occupants** through increased usage of seat belts and child restraints. While the current rate of seat belt usage is significant, the National

Highway Traffic Safety Administration estimates that, if seat belt usage were to increase to 85% of all drivers, the number of persons killed each year on the nation's highways would decline by over 5000. Unfortunately, some drivers refuse to "buckle up" in spite of the evidence that "the life they save may be their own."

One method to increase seat belt usage is education through driver training programs, public service announcements, and talks to community and civic organizations. A second method is to enact seat belt laws and require primary enforcement. Every state, with the exception of New Hampshire, has laws mandating seat belt usage, but only very few states permit a vehicle to be stopped solely for a seat belt violation (known as primary enforcement). In the remaining states, a seat belt citation can be written only in connection with another moving violation, such as speeding. The percentage of motorists using seat belts in states with primary enforcement for seat belt usage is much greater than where primary enforcement is not required. If primary enforcement were required nationwide and involved penalties and fines, the expectation is that total deaths and injuries from motor crashes would significantly decline.

The protection of children in moving vehicles requires special consideration. Because they are small and unable to help themselves, a parent or guardian must assure that they are seated in a secure and safe position. State laws require that small children should be "locked in" with a secure infant seat located in the back of the vehicle. Letting children sit in the front seat creates a potential hazard if the passenger-side air bag should be actuated. The rear seats in many vehicles do not provide for child safety since the infant seat cannot be fully secured. If automobile manufacturers offered integrated restraint systems, the need for supplementary hardware would be eliminated and the safety of the back seat would be enhanced.

A major highway safety concern is to reduce the crash rate for **young drivers**. It is well known that one of the major causes of death and injury for youth, aged 16–21, is traffic related. There are many reasons for this situation, including inexperience, driving at high speeds, drinking, and recklessness. With age comes experience and knowledge, but at the beginning stages, youngsters are unaware of how the laws of physics apply to driving a motor vehicle. Consequently, many younger drivers exceed safe speeds on curves and drive too fast for the road conditions, resulting in the inability to stop or to control the vehicle. Among the recommended actions to control young drivers are enforcement of tougher drinking and driving laws, nighttime driving restrictions for the novice driver, and provisional licensing. Alcohol sales are now prohibited in all states to persons under the age of 21, which has resulted in fewer crashes and fatalities.

To illustrate the influence that speed and alertness has on safety, consider the basic relationships for stopping sight distance (SSD), which is the minimum distance required for a driver to stop a vehicle after seeing an object in the roadway. The two components of SSD are the distance traveled during perception reaction time (prior to braking) and the distance traveled during deceleration (during braking). Perception reaction time is commonly taken as 2.5 s for normal driving conditions. However, additional time may be required for a driver to detect unexpected situations or environmental conditions, such as fog, darkness, or the multiplicity of roadway signs. Decision sight distance may also be influenced by the condition of the driver, such as age, fatigue, alcohol impairment, or distraction. When these conditions occur, the perception reaction time increases from the AASHTO-recommended value of 2.5 s to values between 5 and 10 s.

The equation for stopping distance is

$$SSD = 0.28\,ut + \frac{u^2}{255(f \pm G)} \tag{8.10}$$

where

u = speed (km/h)
t = perception reaction time, 2.5 s for normal driving
f = coefficient of friction between tire and roadway
f = $a/g = 0.35$
G = grade in decimal (for upgrades use $+G$, for downgrades $-G$)
SSD = distance (m)

Example 8.7

Determining the Effect of Speed and Fatigue on Stopping Distance

An alert motorist driving at the posted speed of 80 km/h on a rural two-lane road with a 3% downgrade requires a perception reaction time of 2.5 s, whereas a fatigued teenage driver who is speeding at 110 km/h requires a perception reaction time twice that of the alert driver. What is the distance that the safe and alert driver will require in order to stop? Compare this stopping distance with that of the unsafe and fatigued driver.

Solution

SSD for the alert driver:

P R time = 2.5 sec, u = 80 km/h

SSD $= 0.28(80)(2.5) + 80^2/255(0.35 - 0.03) = 56 + 78.4 = 134.4$ m

SSD for the speeding and fatigued driver:

P R time = 5.0 sec, u = 110 km/h

SSD $= 0.28(110)(5.0) + 110^2/255(0.35 - 0.03) = 154 + 148.3 = 302.3$ m

Thus the distance that a speeding fatigued driver requires to stop is 168 m greater than that of a driver who is alert and drives at the speed limit.

In most traffic situations, a margin of error as great as 168 m (almost three city blocks) is usually not available and, if the need should arise to brake in order to avoid a collision, the fatigued driver is at risk and a crash could ensue. Thus, the safest strategy for this driver is either to wait until rested, let someone else drive, or drive slowly and with caution.

Air transportation safety involves several elements, including reliability of the aircraft, the air and ground traffic control system, weather, and the aircraft's crew. Overall, air safety is excellent, although occasionally a crash will occur with a heavy death toll. Detailed investigations are made of the cause, followed by recommendations for changes in the air system. Air crashes are so unique and infrequent that the public remembers specific incidents such as Pan Am 106 over Lockerbie, Scotland, TWA Flight 800 over Long Island, and the ValuJet crash over the Florida Everglades.

Less dramatic but nonetheless important are collisions that occur on runways, which may result in damage to the aircraft or injuries to passengers. An infrequent accident is the head-on collision between aircraft that simultaneously take off from opposite ends of the runway. Runway incursions have increased in recent years due in no small part to the heavy traffic of arriving and departing aircraft.

Several action programs relate to three types of aircraft accidents that are caused by airframe structural icing, airport runway incursions, and explosive mixtures in fuel tanks.

Airframe structural icing is caused by the build-up of ice on the wings of an aircraft, and several crashes have occurred during take-off or in flight due to this phenomenon. The NTSB believes that the safety regulations regarding icing should be revised and that new technology is necessary to detect and protect aircraft from the build-up of ice caused by freezing drizzle.

Airport runway incursions are defined as an occurrence at an airport involving an aircraft, vehicle, person, or object on the ground that creates a collision hazard. Runway incursions usually occur at high-volume complex airports and during periods when visibility is impaired. They may be due to faulty judgment by the pilot, operational errors, or lack of attention by ground vehicles or pedestrians. The most dramatic incursion in recent history occurred in 1977 on a runway in the Canary Islands when a Boeing 747, not cleared for take-off, proceeded down the runway in dense fog, colliding with an oncoming aircraft and causing the deaths of 583 passengers and crew. A variety of mitigating actions have been suggested to reduce incursions, including signage and runway markings, pilot and vehicle operator training, and intelligent technologies such as use of inductive loops and runway status lights for entry and take-off.

Explosive mixtures in fuel tanks was a condition identified during the investigation of the TWA Flight 800 crash over Long Island. To assure that the problem was corrected in other Boeing 747s, the NTSB suggested aircraft design modifications using nitrogen-inerting systems and the installation of insulation between heat-generating equipment and fuel tanks. The Federal Aviation Administration has performed safety inspections of aircraft that included manufacturing practices and types of material used for insulation.

One of the highest priorities in **railroad safety** is the development of systems that assure a positive separation between successive trains. Many of the approximately 31 rail collisions each year are the result of operator error in failing to obey signal systems or to drive the train at higher than allowed speeds. The reasons for these errors may be the incapacitation of the train operator, inattention, or lack of training.

Positive train control (PTC) systems have been initiated or demonstrated by various railroads to assess their value. The industry has invested over $200 million to develop PTC technology. The Association of American Railroads, the Federal Railroad Administration, and the state of Illinois have initiated a four-year, $60 million project to build and test a PTC system on a 198-kilometer railroad line between Chicago and St. Louis. The goal of the project is to demonstrate the benefits of safety improvements, system operability, and cost-effectiveness of PTC technology. If PTC systems were federally mandated, the problem of railroad crashes due to operator failure would be diminished.

Recreational boating accounts for the greatest loss of life in marine transportation. A visit to many U.S. lakes or rivers will attest to the utter chaos that exists when recreational boaters are out in numbers. Speeding, reckless maneuvers, and drinking account for the many boat accidents. To reduce the drowning deaths due to boating accidents, a three-part program addresses this problem. Elements of the program are the use of life jackets by all children in a recreational vessel; assurance that the boat operator is certified to operate the craft by requiring demonstrated knowledge of boating rules of the road, navigational skills, and safe boating practices; and mandatory licensure to operate a motorized recreational vessel.

Safety concerns common to all modes are driver fatigue and the securing of crash data through the use of automated data recorders within vehicles. The effect of driver fatigue occurs because human beings are constituted such that they perform at their best when well rested and on a normal time schedule. For the driver of a personal vehicle, no laws exist that mandate stopping periodically to rest or limit driving to a maximum time period. For this reason, some drivers exceed their physical limitations. We can all recall a news story about a vehicle crash that involved a driver who momentarily fell asleep at the wheel and, as a result, caused a vehicle crash. Research has demonstrated conclusively that drivers perform better on a schedule that includes a full night's rest. It is also well known that a lengthy period of driving without a break induces boredom and fatigue.

For commercial vehicles, driver fatigue has been a contributing cause in accidents involving railroads, trucks, aircraft, and marine vehicles. While regulations exist regarding the allowable number of hours of work, there is often a lack of training regarding the importance of including rest schedules in the travel plan. Furthermore, where operator schedules are shifted between daytime and nighttime assignments, safety and operator performance can be further affected. Recommendations on driver fatigue have been implemented by federal agency regulations. Among the requirements are that drivers be medically fit to drive, be allowed short breaks, and be provided with periods of rest.

Automatic information-recording devices are common in certain transportation vehicles, particularly aircraft. The usefulness of the "black box" for re-creating the conditions immediately prior to a crash has been clearly demonstrated. Recording devices are needed on other commercial vehicles as well, particularly large trucks and railroads. The purpose of these devices is to provide information such as speed, direction of the vehicle, operator comments, wind speed, and temperature. In addition to providing information that might be helpful in reconstructing an accident and identifying its probable cause, the data are used to detect unsafe or improper procedures that may help to correct deficiencies before an accident happens.

As the preceding discussion has demonstrated, transportation safety is a national problem that involves four elements: driver, vehicle, travelway, and environment. These problems are typically addressed on a modal basis because the technology, systems, operations, and environment of each mode are different. The changes required to improve safety in the air differ markedly from the approaches used in railroads or highways. Even in the cross-modal areas of fatigue and information, the solutions and regulatory requirements will vary. Regardless of the mode, a safety improvement process is followed that includes collecting crash information and maintaining a safety database, analyzing the data and identifying probable causes of crash incidents, developing appropriate countermeasures to address safety deficiencies, ranking the order of safety countermeasure projects and establishing priorities, implementing safety projects, and monitoring the results.

The AASHTO Comprehensive Approach to Safety

The American Association of State Highway and Transportation Officials (AASHTO) prepared a strategic highway safety plan, which has been implemented, with the goal of reducing highway fatalities by 5000–7000 lives each year and reducing injuries and property damage. In the United States, over 3.5 million motor vehicle injury crashes occur each year as well as 4.5 million crashes that cause property damage. Although crash fatality and injury statistics have remained stable in recent years, motor vehicle fatalities represent the leading cause of non-health-related deaths. Other notable non-health-related deaths are falling, poison, drowning, and fire.

The elements of the AASHTO plan are instructive because the plan places in perspective the major areas of concern in highway safety and, by extension, the safety of other modes. The plan recognizes that improvements in travelway design, systems operations, and infrastructure maintenance, while important, are not sufficient to achieve progress in safety. The nation's highway system has incorporated many design and engineering improvements, with the result that the number of fatalities per year has remained relatively constant. However, if the average crash rate, which is a function of vehicle kilometers traveled, were to remain unchanged, the effect would be that fatalities would increase. To illustrate, with a constant rate over the lifetime of children born in the year 2000, one child in 84 could die in a motor vehicle crash and 6 out of every 10 children would be injured. Thus, a lowering of the annual fatality rate is essential to achieve a reduction in total deaths or injuries. Greater attention must be given to driver behavior, pedestrians, bicycles, motorcycles and trucks, and highway–vehicle interactions, as well as improved decision support and safety management systems.

A focus of the highway safety community is greater attention to driver behavior. Given that roads are generally in good condition and vehicles come equipped with a variety of safety devices, it is the operator of the vehicle that has become the major causal factor of motor crashes. In today's stressful world, with increased congestion, longer driving times, and multiple destinations reached by car, it is not surprising that the phenomenon of "road rage" has developed and cell phone use is comparable to alcohol consumption as a major cause of crashes.

Accordingly, AASHTO has devoted 8 of its 22 safety elements to improving driver performance. These are graduated licensing for younger drivers; reducing the number of drivers on the road whose privileges have been revoked or suspended; improving the safety of older drivers; controlling aggressive driving behavior; reducing or eliminating driving while under the influence of drugs and alcohol; reducing driver fatigue to assure that drivers are alert; increasing the awareness of the motoring public about the importance of highway safety; and increasing the use of seat belts.

The strategies proposed by AASHTO to achieve improvements in driver performance are varied and depend on the element selected. Among these are enacting laws to ensure driver competence; improving driver education and training programs; developing means to identify problem drivers who are repeat offenders; designing highway signs and signals for greater visibility by older drivers; increasing law enforcement of speeders and seat-belt violators; implementing safety-related intelligent transportation system technologies; increasing driver checkpoint programs to identify impaired drivers; retrofitting highway shoulders and bike lanes with rumble strips to alert drivers; revising hours-of-service regulations to reduce truck driver fatigue; and developing national public awareness of safety issues.

Streets and highways are often shared with both motorized and nonmotorized traffic. In crashes involving a vehicle and a pedestrian or bicycle, death or injury is invariably the result for the individual not in the motor vehicle. Each year, approximately 5600 pedestrians are killed on the nation's roadways, and there are approximately 800 bicycle-related deaths and 61,000 injuries per year. One-third of bicycle-related deaths involved children between the ages of 5 and 15 years. Most pedestrian and bicycle accidents occurred because the individual failed to yield or used the roadway inappropriately. Many bicyclists are unaware of (or disregard) the rules of the road and take chances by traveling in a direction opposite moving traffic or disregarding stop signs and signals. Some pedestrians do not understand the laws of physics as they relate to the time required to stop a vehicle and they enter crosswalks when the stopping distance is inadequate.

Strategies recommended to reduce pedestrian and bicycle crashes include developing agreed-upon standards for the provision of pedestrian facilities; improving outreach and safety training programs for pedestrians and bicyclists; improving bicycle–pedestrian facilities at intersections and interchanges; developing measures to increase the use of bicycle helmets; implementing a coordinated program of safety improvements; integrating engineering (intersection design); education (children, seniors, and the impaired); and enforcement (speeding, red-light running, jaywalking).

AASHTO has recognized that there are three classes of vehicles with special safety problems—motorcycles, trucks, and passenger cars. Motorcycles have many desirable characteristics, such as economy, speed, agility, and flexibility. However, the downside is similar to that of the bicycle in that motorcycles lack stability when on curves or rough pavement and in wet weather. Motorcycle crashes account for approximately 2000–3000 deaths per year, with males between the ages of 18 and 27 being the most frequent victim.

Large trucks account for approximately 4000–5000 deaths each year, with predicted rates expected to increase. As with most crash events, the occupant in the larger vehicle is less likely to be injured or killed. In crashes involving trucks and light vehicles, the factor is about 6:1 in favor of the truck. Most auto drivers are aware of this difference and dislike driving close to a large truck. Truck crashes are often attributed to driver fatigue, inadequate awareness of trucks by other drivers, and defects in the truck itself such as tires, brakes, and steering.

The perception that trucks are unsafe has thwarted attempts by trucking interests to increase allowable size and weight. Furthermore, several high-profile crashes and predictions that annual fatalities could increase in coming years have resulted in action by Congress to replace the Office of Motor Carriers in the Federal Highway Administration with the Federal Motor Carrier Safety Administration, thus reflecting the perceived need to monitor, regulate, and enforce truck safety at the highest levels of government.

Passenger vehicles have been continually enhanced in an effort to create a safe environment, and often the traveling public decides if the new technology is cost efficient. It can be expected that a process of continual upgrading of safety features will continue. For example, the popularity of sport utility vehicles is based in part on the perception that they are safer than an ordinary sedan. As other safety features are introduced, the government may regulate their use (for example, requirements of air bags in passenger vehicles) or the driving public may add options such as GPS and emergency road services. Safety improvements in motor vehicles include the following:

Motorcycle safety. Reduce alcohol-related fatalities; increase awareness and safe driving by motor vehicles; expand motorcycle driver training; improve highway design, operations, and maintenance; and increase the use of helmets.

Truck safety. Identify truck firms with poor safety performance; educate commercial and other vehicle drivers; implement traffic control measures and highway design for trucks; and identify and enhance vehicle safety technologies.

Passenger vehicle safety. Educate drivers in the use of antilock brakes; reduce carbon monoxide poisoning which kills vehicle occupants; expand Intelligent Transportation Systems (ITS) crash avoidance research; and improve compatibility between vehicle and roadside design features.

Motor vehicle crashes occur on highways, intersections, interchanges, grade crossings, and work zones. Thus, highway and traffic engineers must develop geometric designs, traffic signals, and markings that will assist motorists to successfully navigate through roadway sections.

One area of concern is **railroad grade crossings**, which each year account for hundreds of vehicular fatalities. Some crashes are the result of reckless drivers who take chances by avoiding gates or warning signals in an unsuccessful attempt to "beat the train." Other crashes are not the fault of the driver and could have been avoided had positive warning devices at the crossing been installed or were in proper working order. Figure 8.11 depicts a truck–train collision at a railroad grade crossing. Strategies to improve safety at railroad grade crossings include improving the effectiveness of passive warning signs at locations where active controls such as gates, flashers, and enforcement are not economically feasible; improving driver training and awareness of the need for caution; replacing unsafe grade-crossings with grade-separated structures; using advanced technology (such as photo radar) to minimize RR crossing violators; and implementing U.S. DOT grade-crossing recommendations.

Off-road crashes represent a highway-related safety problem caused when a single vehicle leaves the roadway and crosses the median or shoulder. These events are not always life-threatening but become so when the vehicle overturns and/or collides with a fixed object or another vehicle. Again, this event could have been purposeful (i.e., trying to avoid an animal in the roadway) but is usually the result of inattention or fatigue. This type of crash is a major cause of traffic deaths, accounting for one-third of all fatalities nationwide and two-thirds of all fatalities in rural areas. Strategies to improve safety due to off-road crashes include improving pavement marking visibility; installing rumble strips along shoulders and on vehicle or bicycle lane delineators; installing safe roadside hardware such as guard and bridge rails, curbs, and drainage gates; where possible, removing poles and trees from the side of the road; improving ditch and side-slope designs to minimize rollovers and impact; installing centerline rumble strips on two-lane highways; and installing median barriers on freeways and arterial roadways with narrow medians.

One of the unfortunate byproducts of the massive highway reconstruction program in the United States is the number of deaths and injuries that occur in **work zone areas** with the result that highway construction employment is considered to

FIGURE 8.11
Truck and train collision.

Source: Transportation Research Board. National Academies, Washington, D.C.

be one of the most dangerous job categories. The problem exists because the work must be performed while traffic continues to move in adjacent lanes and work crews typically must be on the job both during daytime and at night. Work zones present a hazard to motorists because they must negotiate through a constricted area along unfamiliar paths guided by cones, barriers, and temporary directional signs. Work zone safety strategies include developing procedures to reduce the number and duration of work zones; improving traffic control through work zone areas; increasing public awareness of work zone safety through education; and assuring vigorous enforcement and conviction of speed limit violators.

Information is the basic building block necessary to effectively administer and implement safety programs. Thus data must be gathered and analyzed in order to establish priorities for the investment of scarce resources and to assure that the expenditure will achieve results. The recording of relevant information after a crash provides the analyst with a profile of the event, including the circumstances and location of the crash, damage, injuries and fatalities, how and why the crash occurred, and characteristics of those who were involved when the event took place. The resulting information can be utilized in a variety of ways, including reconstructing accidents, developing trends, and identifying hazardous locations. Similarly, traffic safety programs are processes that integrate the results of information systems to develop regional and statewide strategies for improving highway safety on a systemwide basis.

Recommendations related to information and decision support systems include improving the quality of safety data; providing resources for a national clearinghouse of safety data; managing and using highway safety information;

training professionals to be skilled in data analysis; and interpreting and establishing technical standards for highway information systems.

Recommendations for improving safety management systems include identifying and sharing the experiences of successes; promoting cooperation, coordination, and communication of safety initiatives; establishing performance measurement systems of safety investments; and developing a national highway safety agenda.

Implementing AASHTO Recommendations

The Transportation Research Board (TRB) through the *National Cooperative Highway Research Program (NCHRP)* has furnished guidance for the implementation of the AASHTO highway safety plan. A series of reports (called the 500 series) has been published that correspond directly to the key emphasis areas of the AASHTO plan. The titles of the several volumes are provided in the references at the end of this chapter, and expanded versions are available on the AASHTO Web site. Regardless of the problem to be addressed, the process for implementing the strategies recommended by the TRB research studies will be the same. The process is illustrated in Figure 8.12.

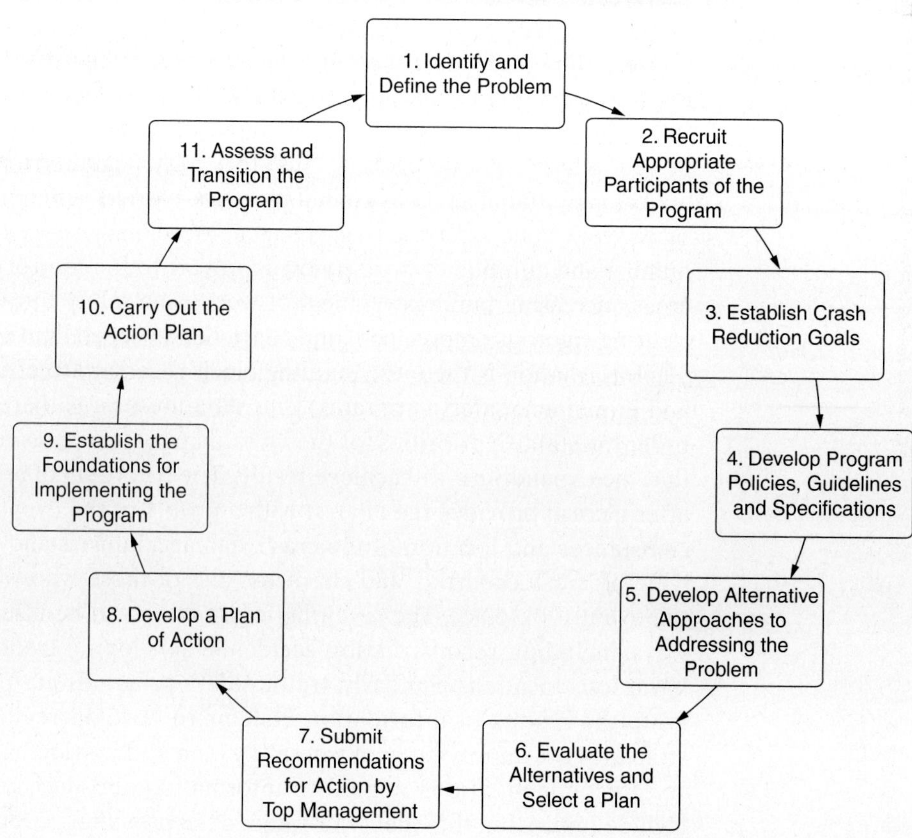

FIGURE 8.12
AASHTO Strategic Highway Safety Plan model implementation process.

Source: NCHRP Report 500, Transportation Research Board, National Academies, Washington, D.C.

EXAMPLE 8.8

Reducing Collisions on Horizontal Curves

According to national statistics, approximately 25% of the 42,815 people killed in 2003 on U.S. highways were involved in crashes on horizontal curves. Further, studies suggest that the accident rate for horizontal curves is about three times that of straight tangent sections. About 75% of curve-related crashes occur in rural areas, predominantly on secondary roads. Over 85% of the crashes either involve single vehicles running off the road or head-on collisions.

A state DOT has allocated funds for the safety improvement of horizontal curves on its secondary road system in rural areas. Using the NCHRP 500 series recommendations, develop a safety program to address this issue.

Solution

Volume 7 of NCHRP Report 500, *A Guide for Reducing Collisions on Highway Curves*, provides an extensive set of strategies listed by time required to implement and project cost. Among the low-cost, short-time (one year) strategies are providing advance warning of unexpected changes in horizontal alignment and installing shoulder rumble strips. Moderate-cost and medium-time (1–2 years) strategies include providing skid-resistant pavement surfaces and lighting in the curve. Higher-cost strategies include modifying the horizontal alignment and designing safer slopes and ditches to prevent rollovers.

The safety design team will evaluate these and other suggested strategies and develop a safety improvement program for those horizontal curves deemed to be most dangerous and within budget limitations. To illustrate, the team is considering the strategy of modifying the horizontal alignment. According to the NCHRP Report Volume 7, the alignment may be modified by increasing the radius of curvature, providing spiral transition curves, and eliminating compound curves.

HIGHWAY SAFETY: WHO IS AT RISK AND WHAT CAN BE DONE?

The previous section described a comprehensive set of strategies and actions to improve highway safety by reducing the number of crashes and the resulting deaths, injuries, and damage to property. These actions are based on statistical data collected over many years regarding the number, type, and characteristics of crashes that have occurred. Thus, it would seem logical that if the causes of crashes can be identified, then remedies can be found that will solve the problem. Unfortunately, there is no "cure" for traffic fatalities and injuries, as there is in medical science for an illness, because the reasons that crashes occur are both complex and personal. This section summarizes the major findings of safety studies and suggests how improvements that have been achieved in the public health sector may shed light on the potential for significant breakthroughs in highway safety.

Traffic crashes are a major public health problem because of the outcome. For example, traffic deaths account for almost 50% of all deaths in teenage years. The total number of preretirement years that are lost due to traffic deaths is approximately the same as that caused by deaths due to cancer and heart disease combined.

Injuries and fatalities from traffic crashes are related to age and sex. Younger drivers are at greater risk exhibiting the highest crash rates of any age group. Older drivers have slower reaction times and other physical impairments, yet they have fewer crashes per person partly due to a compensation of driving less, having more experience, avoiding nighttime driving, driving cautiously, and driving more slowly. Ironically, young drivers are the group with the best driving skills and yet appear to be most vulnerable on the highway. The reason for this paradox is that driving ability, while necessary, is not a sufficient condition to assure safety—rather driving performance is an absolute requirement. Typically, young drivers are more prone to take risks, drive too fast, and disregard laws of nature and society. Driving behavior is correlated with personal behavior. People who are aggressive, stressful, emotionally unstable, and irresponsible in general tend to exhibit similar traits when behind the wheel of a motor vehicle.

The use of alcohol and drugs while driving is a major cause of traffic crashes. It is estimated that if motorists were not driving while under the influence of alcohol, fatalities, injuries, and property damage would diminish considerably. Because there are laws specifying the maximum blood-alcohol level that a body may sustain while driving, the problem has been mitigated except for repeat offenders. In the United States the "per se" limit is typically between 0.08 and 1%, whereas elsewhere, such as Scandinavian countries, the limit is lower and the penalties greater. Other factors that may reduce drinking and driving are minimum drinking age laws and other social norms that promote less alcohol consumption by the general population.

Vehicle characteristics and technology can affect traffic safety. Vehicle mass or size has a significant relationship to the risk of injury or fatality when a crash occurs. Thus it is not surprising that large sport utility vehicles are popular as an alternative to the family sedan. The type of roadway also affects crash rates. For example, two-lane roads have a much higher crash rate per 100 million vehicle kilometers than do limited access divided arterials. In-vehicle devices such as air bags and lap belts improve the chances of surviving a crash compared with driving without restraint protection. However, the availability of devices such as better brakes, enhanced night vision, and warning systems are not guarantees that the expected results will be achieved. If drivers modify their behavior because they perceive that there is no additional risk in driving faster, longer, and more recklessly, then the benefits of advances in technology could be negated with the perverse result of higher crash rates than expected.

Modest improvements can produce significant safety results. Improvements in traffic safety will be based on a wide variety of interventions of which some will cause small reductions and others will be of greater significance. However, because the magnitude of the problem is so large, even small improvements are important. If a safety improvement can reduce the number of fatalities by as little as 2%, then approximately 900 lives would be saved each year. Most interventions are in the low percentage category—for example, if every vehicle were equipped with an air bag, traffic fatalities could be reduced by 5%, and if nonhelmeted bikers used helmets, fatalities could decline by 1%. On the other hand, more dramatic results are possible by reducing the use of alcohol while driving—a measure that has already witnessed reductions in excess of 10%.

Driver responsibility is the key to safety. The analogy to achieving good health and extending life is useful when we consider where further high payoff is likely to be achieved in highway safety. Medical science has concluded that unless the public does its share in maintaining a healthy body, the significant breakthroughs that have been achieved by technology and medications will not produce desired results. Thus, we are admonished to stop smoking, eat a healthy diet, exercise, and limit the use of alcohol. Cardiologists believe that the ultimate success of heart bypass surgery is due mainly to how patients revise their lifestyles after the operation has been performed. Similarly, traffic safety improvements cannot be achieved solely by building safer highways and vehicles or by enacting, regulating, and enforcing laws. Safety requires that the driver takes responsibility for actions on the roadway that will reduce crashes by driving at posted or uniform speeds, minimizing speed differentials, avoiding unsafe maneuvers (such as tailgating and sudden lane changing), and driving only when sober and rested.

To achieve a safety-conditioned society will require that each individual understands how safe driving is directly related to personal health and well-being and the health of passengers and other drivers on the road.

Commercial Transportation Safety: A Team Approach

Commercial transportation differs in one significant way from highway travel in that the vehicle is under the control of a professional operator and the traveler is a passive participant in the trip. Since the general public operates personal vehicles, driver behavior represents a major element in highway safety programs. On the other hand, when travelers board a commercial vehicle, they expect that the carrier, be it an airline, railroad, or shipping line, will assure that the trip is completed in a safe and efficient manner and that the operator is trained and experienced.

The transportation industry is concerned about safety for several reasons. First and foremost is the impact that safety has on its business. If the public perceives that a transport mode or an individual carrier is unsafe, travel demand will decrease. For example, steamships were an early form of transportation but soon were replaced by railroads. In addition to competitive attributes such as speed and cost, railroads were perceived as being safer than steamboats because many ship disasters had occurred due to boiler explosions.

With the advent of air transportation, there was an initial reluctance by the public to embrace this new mode because of the fear of flying and the catastrophic consequences of an air crash. Safety is also a concern of the transportation industry because the loss of life and equipment can be costly. Commercial aircraft, ships, and rail cars are expensive to replace, and where injury or loss of life is the result of negligence by the transportation company, the company may be liable for significant payments to the passengers or their families.

The components of commercial transportation safety are similar to those previously described and include the vehicle, the travelway, and the service provider. In each mode these elements differ and, as such, require coordination among participating organizations. The elements and safety programs for commercial transport modes are described as follows.

In all transportation modes, commercial vehicles are manufactured by companies that specialize in the production of a specific transport mode, such as aircraft, ships, railroad locomotives, and transit vehicles. Manufacturers have the responsibility to produce vehicle designs that include technologies to assure maximum safety during normal operations and in emergencies. They are also responsible for providing training support as well as recommendations for vehicle maintenance.

The provision of the travelway varies from one mode to another. Air transportation travelways are the "open skies" regulated and controlled by the federal government through the air traffic control system operated by the Federal Aviation Administration. Each railroad company owns and operates its own travelway system and, as such, is responsible for the coordination of train schedules and track assignment for both passenger and freight services on its system. Ships travel on the "open sea" but in designated sea lanes and guided through harbors by special pilots familiar with channel locations. Urban transit authorities own and operate light and heavy rail lines but must depend on the street and highway network for provision of bus routes and traffic control.

The operators of transportation modes are private companies or public authorities whose responsibility is the delivery of transportation services to the public. Familiar company names are United, American, or Delta Airlines; CSX, Norfolk Southern, and Burlington Northern Railroad, and Sea Land; and Holland America and Evergreen Shipping Lines. Service providers develop procedures and policies that include safety elements, such as vehicle maintenance schedules, safety inspection programs, and operator training.

In the United States, aviation safety responsibilities are shared among three major groups: manufacturers, operators, and the government. In addition, the media and the public are involved when crash investigations are underway and in pursuing safety legislation and reforms. The result has been impressive in that the chances of being involved in a major commercial accident is only about 1 in 2 million flights. Nonetheless, the industry is continually striving to identify the circumstances under which accidents have occurred and to develop new procedures, strategies, and technology that will result in safe travel by air.

With the growth in commercial jet transportation expected to double in the next 20 years, a hull loss accident per week could occur if the current accident rates were maintained. Thus, as was the case with highway safety, where large increases in travel volumes are expected, lowering the total number of crashes will require reducing crash rates. A further complication in air transportation is its international character and the wide variation in accident rates in other parts of the world. For example, the number of accidents per million departures from the United States and Canada is significantly lower than the corresponding values for Latin America and Africa. These differences suggest that significant improvements are possible worldwide as lessons learned in one region or airline can be transferred elsewhere.

The major differences between U.S. and worldwide accident types are in three areas: loss of control in flight; snow- and ice-related accidents; and runway incursions. In recent years, of the five fatal accidents worldwide caused by ice or snow, three involved U.S. carriers. Of the four fatal runway incursion accidents worldwide, all involved U.S. carriers at U.S. airports. This accident history provides the basis for the NTSB safety priorities described earlier. Other accident types show a similar pattern between the United States and the world. The principal cause of fatalities worldwide is called *controlled flight into terrain (CFIT)* and occurs typically at night and under poor conditions of visibility

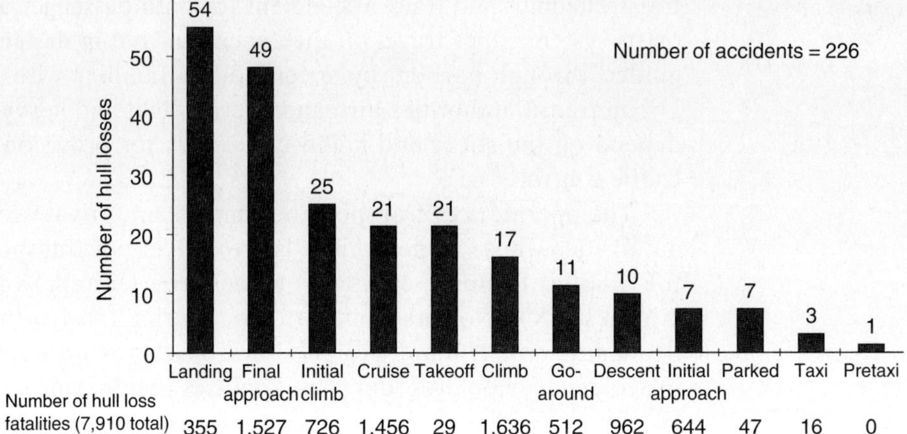

FIGURE 8.13
Airline hull loss accidents worldwide.

Source: Transportation Research Board. National Academies, Washington, D.C.

when a pilot misses a runway or loses orientation and crashes into the ground or sea.

The overall excellent air safety record in the United States is due in no small part to a mature infrastructure with multiple redundancies that have contributed to the lower numbers of CFIT accidents. Technology installed within aircraft and on the ground includes complete radar coverage, minimum safe altitude warning approach radar, and ground proximity warning systems. Further improvements in performance will be made by examining the causes of crashes and incidents worldwide and by segregating hull loss accidents by phase of flight. Figure 8.13 shows that of the 226 hull loss accidents worldwide between 1988 and 1997, the highest number, 54, occurred during landing and the second highest, 49, happened during the final approach.

The improvements to air transport safety have become a national concern. The Federal Aviation Administration has developed an Aviation Safety Plan, which has as its goal the attainment of zero accidents. A White House Commission on Aviation Safety and Security was formed following the crash of TWA Flight 800 over Long Island, a tragic event whose cause would not be known until completing several years of painstaking investigation. At the time of the crash, speculation was rampant regarding its cause with theories ranging from sabotage to a guided missile. As noted earlier, the cause was determined to be ignition of fuel vapors in an empty tank. The White House Commission announced a national goal of an 80% reduction in the fatal accident rate in the United States by 2007 and recommended that cooperation be enhanced among all concerned parties to achieve this result. Congressional action resulted in the creation of the National Civil Aviation Review Commission to advise the FAA regarding improved safety. The Commission recommended that performance measures and milestones be developed to

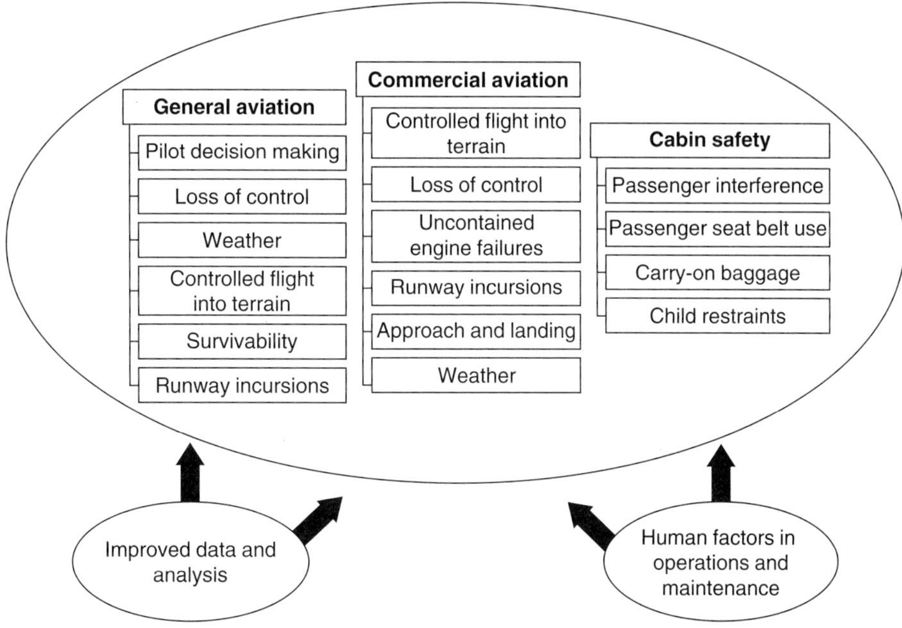

FIGURE 8.14

FAA Safe Skies Agenda.

Source: Federal Aviation Administration.

assess safety progress. In 1998, the FAA developed a Safe Skies Agenda, depicted in Figure 8.14.

An Industry Safety Strategy Team (ISST) was formed comprising the Air Transport Association of America, Boeing Aircraft Company, and the Airline Pilots Association in an effort to coordinate the airline industry's efforts in air safety. The ISST produced a Commercial Aviation Safety Agenda and the group changed its name to Commercial Safety Strategy Team (CSST). Later, it was recognized that the organization should include equal partners from industry and government that would involve most major players from both sectors. The new consortium was changed to Commercial Aviation Safety Team (CAST) with a broad membership, as illustrated in Figure 8.15. The Commercial Aviation Safety Agenda for CAST contains the following elements, which represent a strategic blueprint for safer skies: The program will include continual review of all available accident and incident data and new data from on-board sources; and CAST will assess threats to commercial aviation safety and seek new technologies and operating procedures that will lessen the chances of further accidents and reduce accident rates to meet stated goals.

A major effort is proposed to assure that the flight itself will be as safe as possible. In order to achieve this objective, improved piloting skills and advanced technology will be required. Accordingly, each element of the flight safety agenda is directed to reduce a specific type of air accident by changes in operations or technology.

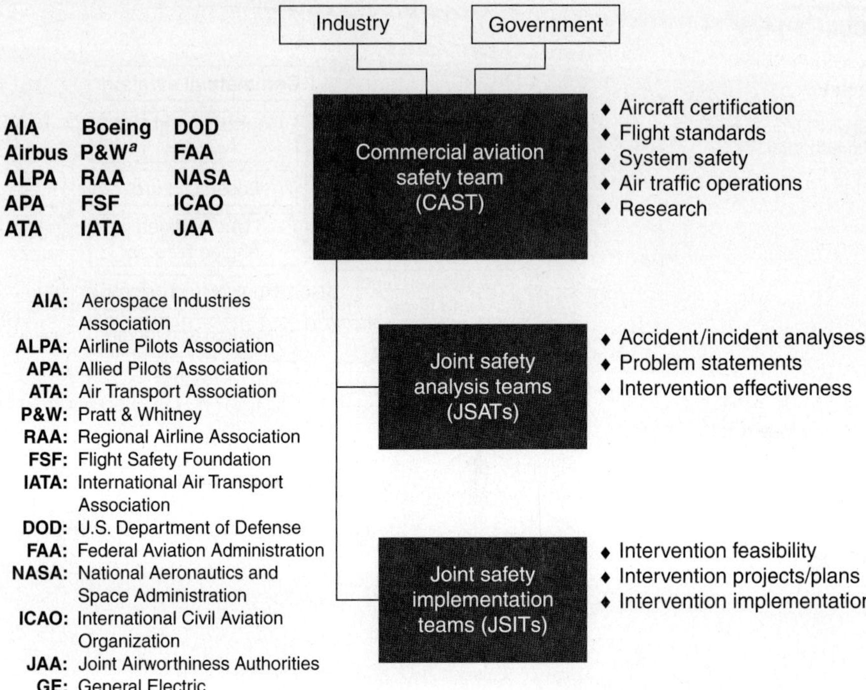

FIGURE 8.15

Membership and composition of Commercial Aviation Safety Team.

Source: Air Transport Association.

Among the items included are reduced controlled-flight-into-terrain mishaps through CFIT training; enhancement of ground proximity warning systems; use of global positioning systems to improve navigational accuracy; reduced loss-of-control accidents by improving flight crew skills and applying new tools and training processes; focus on reducing human errors by improved training and operational procedures; application of crew resource management techniques; development and compliance with standard operating procedures; elimination of inappropriate crew responses to nonnormal situations; enhancement of flight crew-automated cockpit interface; improvement of situational awareness; improvement of simulation fidelity; reduction of approach and landing mishaps by applying stabilized approach procedures and emphasizing the "go-around" option; reduction of weather and turbulence-related accidents by detection caused by wind shear, ice, and deicing/anti-icing on the ground and in flight, wake turbulence, and clear air turbulence; reduction of runway/taxiway incursion accidents by implementing the joint industry–government airport runway incursion action plan; installation of airport-surface detection systems; and installation of airport movement area safety systems.

In many instances where an unusual situation occurs during a routine flight, injuries or deaths to airline passengers have resulted that might have been prevented. Many of these happen when the airplane undergoes turbulence and passengers and objects are thrown about the cabin. Other instances are when an aircraft crashes, usually during landing, and passengers who survive the crash are trapped and killed or injured by the ensuing smoke or fire. Thus, a series of strategies aimed at improving safety of airline passengers and crew include installing fire-resistant or fire-retardant materials that reduce turbulent-related injuries and improving of child seat restraint specifications.

A safe airplane is dependent upon the integrity of the aircraft and its engines as well as the quality and completeness of maintenance. The aircraft fleet throughout the world is aging, and thus inspection and maintenance of aircraft is a key element affecting the industry's safety performance. Implementing maintenance improvements, staggering procedures and standardizing maintenance procedures, and keeping diligent and detailed records and documentation can achieve reduction in maintenance errors.

As noted earlier, the basis for any safety improvement program, regardless of mode, is the acquisition of complete data and its analysis. Data are useful for identifying causes of accidents, identifying mitigation options, and observing trends in safety. The actions regarding data include protecting safety information provided voluntarily to enable sharing and analysis; focusing on accident and incident prevention by data collection and analysis of accidents and incidents; inspecting aging aircraft and aging systems; implementing flight operations quality assurance; and ensuring confidential safety reporting systems.

The safety record of the railroad industry has shown considerable improvement during the past two decades. In 1980, railroads experienced 7.1 accidents per million train-kilometers (mtkm), and in a period of 20 years the accident rate was reduced to 2.2 accidents/mtkm, as shown in Figure 8.16. The industry has invested heavily in track and equipment, and the safety improvements resulting from this investment in equipment and facilities (as well as increased awareness by employees and management) have had dramatic results.

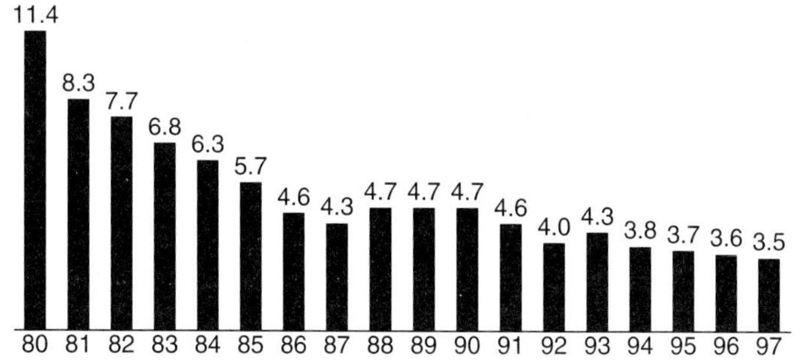

FIGURE 8.16

Total railroad accidents per million train-miles.

Source: Federal Railroad Administration.

The safety programs of the railroad industry represent a cooperative arrangement among the Federal Railroad Administration; the Association of American Railroads (AAR); commuter, regional, and short line railroads; labor unions; AASHTO; transportation equipment suppliers; the Federal Transit Administration (FTA); and the American Public Transit Association (APTA). These groups are assembled through a Railroad Safety Advisory Committee (RSAC) and can comment on proposed safety regulations and suggest areas requiring research.

The FRA and AAR jointly collaborate in railroad safety research through the Transportation Technology Center (TTC), a research facility located near Pueblo, Colorado. The TTC conducts safety research and testing in areas such as heavy axle load technologies, detection of rail and roller bearing defects, track geometry and strength, tank car safety, and transportation of hazardous materials. In addition to conducting railroad safety research and testing, the TTC assists railroads and suppliers to develop improved products, maintenance practices, and employee training. Several of the research projects are illustrated in Figure 8.17.

FIGURE 8.17
Examples of railroad safety research.

Source: Transportation Technology Center.

(a) Track performance under heavy loads

(b) Internal rail defect inspection

(c) Early detection of defective bearings

The railroad industry is large, diverse, and complex. As of 2005, there were only four major freight railroads—Burlington Northern, Santa Fe, CSX, and Norfolk Southern—of approximately equal size. These are augmented by midsized regional railroads, such as the Illinois Central, Kansas City Southern, Wisconsin Central, Florida East Coast, and Soo Line/CP, and short line railroads formed as a result of the selling of unprofitable branch lines. High-speed passenger services are being added in heavily traveled corridors by sharing the same track with freight trains. This growth in demand, diversity of services, co-mingling of freight and passenger trains, and conflicts at rail–highway crossings suggests the need for coordination of safety programs by the federal government.

The railroad safety agenda is driven by the Federal Railroad Administration's goal of zero tolerance for accidents or casualties. A strategic plan for research development and demonstration to achieve this goal contains the following elements: reducing accidents associated with human factors;

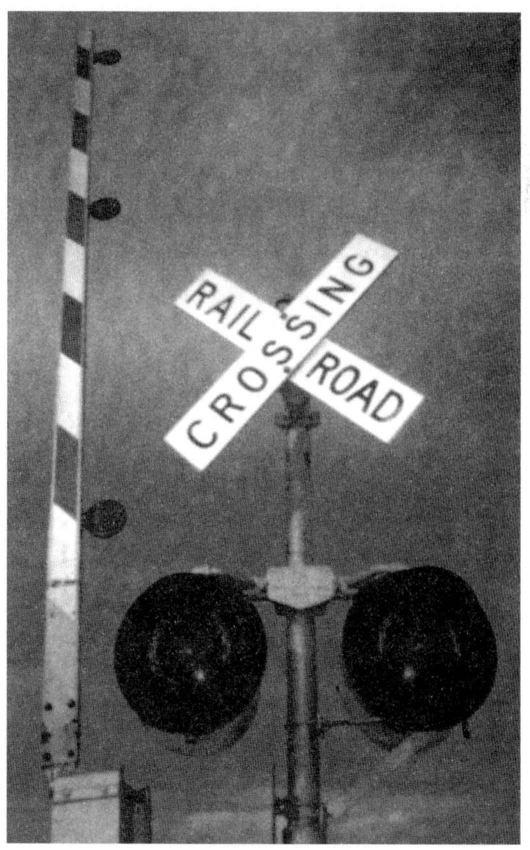

FIGURE 8.18
Railroad grade crossing safety involves many organizations.

Source: Transportation Research Board. National Academies, Washington, D.C.

detecting rolling stock defects and improving performance; detecting and preventing track and structure defects; enhancing track–train interaction safety; preventing train collisions and overspeed accidents; preventing grade crossing accidents; improving the safety of hazardous materials transportation; improving railroad system safety; and improving Research and Development (R&D) facilities and test equipment.

There are many safety railroad elements that interact with other transport modes. One of these—grade crossings—was discussed earlier as a highway issue. Not surprisingly, the FRA and the railroad industry cooperate with AASHTO in developing standards for railroad crossings and incorporating results of intelligent transportation research, as shown in Figure 8.18. The area of human factors is of concern to all modes, as a high proportion of transportation crashes can be attributed to human error. Thus, the National Science and Technology Council coordinates human factors research in transportation through its program Human-Centered Transportation Systems. Another major concern is the transportation of hazardous materials, a safety problem common to all other modes, particularly the trucking and railroad industry.

SUMMARY

Safety is an important element in the design and operations of the nation's transportation system. It is a multidisciplinary activity that involves consideration of the vehicle, travelway, human factors, the environment, and law enforcement. An essential element is the cooperation among industry, government officials, and citizens. The concerted effort underway to reduce or eliminate the number of crashes that occur each year can be expected to continue throughout the twenty-first century. However, accomplishing these goals while demand for travel increases will require that crash rates diminish even further than they have in recent years.

National organizations have taken the lead by developing long-range strategic plans to accomplish safety goals. To succeed will require a team effort of states, localities, vehicle manufacturers, and industry. Transportation safety improvement programs have approached the task from three fronts: preventing crashes, minimizing the effects of crashes when they do occur; and developing data analysis and retrieval systems that will provide information regarding the most probable cause. Each of these phases requires the input of engineering and scientific professionals.

The ultimate effectiveness of any safety strategy depends on actions by the drivers or operators of vehicles in the system. Thus improvements in the nation's safety record involve systems enhancements coupled with driver/operator competency, training, maturity, and experience.

PROBLEMS

8.1 Discuss the difference between a crash and an accident. Which term is more appropriate in transportation safety?

8.2 Select an article from the newspaper or a news Web site regarding a recent transportation crash. Describe the event and its probable cause.

8.3 Compare the articles found by other class members regarding safety-related events. Categorize these by transport mode, whether single or multiple vehicles were involved, and the number of injuries–fatalities. Based on this record, what can you conclude about the relative safety of modes?

8.4 What was the primary cause for the sinking of the ship *Titanic* that resulted in the deaths of 1200 people? What were other factors contributing to the magnitude of this disaster?

8.5 What are major factors that can influence the occurrence of transportation crashes?

8.6 Describe the basic approaches to improving transportation safety.

8.7 The number of crashes per year at a railroad grade crossing with 12 trains per day is five over a three-year period. The average number of vehicles entering the intersection is 2500 per day. Determine the crash rate per million entering vehicles.

8.8 Crash studies were conducted along two segments of railroad with similar characteristics. Segment A is 24 km long, averages 30 trains per day, and experienced seven incidents over a three-year period. Segment B is 34 km long, averages 45 trains per day, and experienced 11 incidents in three years. Determine the crash rate/MVM/Y. Discuss the implications of this result.

8.9 It is required to test whether large trucks are significantly more involved in rear-end crashes on interstate highways with differential speed limits (DSL) for passenger cars and large trucks than on those with a uniform speed limit (USL). Using the t-test and data for the same three-year period shown in the following table, determine whether you can conclude that trucks are more involved in rear-end crashes on interstate highways with differential speed limits. Use a 5% significance level.

Interstate Highways with DSL		*Interstate Highways with USL*	
Site No.	*Number of Crashes*	*Site No.*	*Number of Crashes*
1	14	1	11
2	10	2	10
3	6	3	8
4	9	4	6
5	12	5	12
6	8	6	9

List the inherent deficiencies in this statistical analysis.

8.10 The following table shows F&I and PDO crashes at work zones and non–work zones at six locations on the same interstate highways over the same periods. Using the proportionality test, determine whether the probability of large trucks being involved in F&I crashes is higher at work zones than at non–work zones at a 5% level of significance.

Work Zones		*Non—Work Zones*	
PDO Crashes	***F&I Crashes***	***PDO Crashes***	***F&I Crashes***
9	5	8	5
5	4	6	3
8	6	5	2
6	7	7	3
2	0	10	6
4	2	11	5

8.11 The data below show rear-end crashes collected over the same period at sites with red-light running cameras and at several other similar sites with no cameras. Using the chi-squared test without assuming that the control ratio is free from error, determine whether it can be concluded that rear-end crashes tend to be higher at locations with red-light running cameras. Use a significance level of 5%.

Locations with Red-Light Running Cameras	*Locations without Red-Light Running Cameras*
Before = 45	Before = 102
After = 36	After = 98

8.12 The table below gives large truck crash rates for seven equal periods during peak and nonpeak periods on an interstate highway. Using the Wilcoxon rank sum test, determine whether it can be concluded that truck crash rates during peak and nonpeak periods are similar. Use a 5% significance level.

	Crash Rates (# of crashes/100M VMT)	
Period	***Non-Peak Periods***	***Peak Periods***
1	1.47 (6)	2.21 (13)
2	2.12 (11)	2.18 (12)
3	1.03 (2)	1.34 (5)
4	1.00 (1)	1.82 (9)
5	1.56 (7)	1.21 (3)
6	1.62 (8)	1.31 (4)
7	1.84 (10)	2.24 (24)

8.13 List major safety concerns requiring attention for the following modes: highways, aviation, railroads, maritime.

8.14 What are the two safety issues that are common to all modes?

8.15 List the elements of a safety improvement process that applies to all modes.

8.16 What are the key findings of safety research studies that suggest the "high-payoff" areas to improving highway safety?

8.17 List possible methods to improve safety for drivers, pedestrians and bicycles, motorcycles, and trucks.

8.18 What is the increased stopping sight distance required of a motorist if the perception reaction time increases from 2.5 to 6 sec at a speed of 80 km/h?

8.19 What is the relationship between speed and safety? How much additional distance is required to stop a vehicle on a level travelway at 65 km/h versus 105 km/h once the brakes have been applied?

8.20 Explain the difference between commercial and noncommercial transportation. How does this difference influence safety?

8.21 What are the elements of the Commercial Aviation Safety Agenda?

8.22 The Federal Railroad Administration has established a goal of zero tolerance for accidents or casualties. Explain how the FRA proposes to achieve this goal.

References

1. American Association of State Highway and Transportation Officials, *AASHTO Strategic Highway Safety Plan,* 1998.
2. Bozin, William G., "Commercial Aviation Safety Team: A Unique Government-Industry Partnership," *TRNews*, Number 203, July–August 1990. Transportation Research Board, National Academies, National Research Council.
3. Cole, Thomas B., "Global Road Safety Crisis Remedy Sought," *Journal of the American Medical Association*, Vol. 290, 2004.
4. Ditmeyer, Steven R., "Railroad Safety Research," *TRNews*, Number 203, July–August 1999. Transportation Research Board, National Academies, National Research Council.
5. Evans, Leonard, *Traffic Safety and the Driver*, Van Nostrand Reinhold, New York, 1991.
6. Evans, Leonard, *Traffic Safety, Science Serving Society*, Bloomfield Hills, MI, 2004.
7. Ogden, K. W., *Safer Roads: A Guide to Road Safety Engineering*, Ashgate Publishing Ltd, Burlington, VT, 1996 (reprinted 2002).
8. Skinner, Robert E., Jr., "Policy Making to Improve Road Safety in the United States" paper presented to the Road Safety Congress, Pretoria, South Africa, September 2000.
9. Sweedler, Barry M., "Toward a Safer Future: National Transportation Board Priorities," *TRNews*, Number 201, March–April 1999. Transportation Research Board, National Academies, National Research Council.
10. NCHRP Report 500, *Guidance for Implementation of AASHTO Strategic Highway Safety Program*:
    Volume 1: *A Guide for Addressing Aggressive-Driving Collisions*
    Volume 2: *A Guide for Addressing Collisions Involving Unlicensed Drivers and Drivers with Suspended or Revoked Licenses*

Volume 3: *A Guide for Addressing Collisions with Trees in Hazardous Locations*
Volume 4: *A Guide for Addressing Head-On Collisions*
Volume 5: *A Guide for Addressing Unsignalized Intersection Collisions*
Volume 6: *A Guide for Addressing Run-Off-Road Collisions*
Volume 7: *A Guide for Reducing Collisions on Horizontal Curves*
Volume 8: *A Guide for Reducing Collisions Involving Utility Poles*
Volume 9: *A Guide for Reducing Collisions Involving Older Drivers*
Volume 10: *A Guide for Reducing Collisions Involving Pedestrians*
Volume 11: *A Guide for Increasing Seat Belt Use*
Volume 12: *A Guide for Reducing Collisions at Signalized Intersections*
Volume 13: *A Guide for Reducing Collisions Involving Heavy Trucks*

Web sites Related to Transportation Safety

Air Transport Association: *www.air-transport.org*
American Association of State Highway and Transportation Officials: *www.aashto.org*
American Automobile Association Foundation for Traffic Safety: *www.aaafts.org*
Federal Aviation Administration: *www.faa.gov*
Federal Railway Administration: *www.fra.gov*
Insurance Institute for Highway Safety: *www.hwysafety.org*
National Highway Traffic Safety Administration: *www.nhtsa.gov*
National Transportation Safety Board: *www.ntsb.gov*
Transportation Research Board: *www.trb.org*
United States Coast Guard: *www.uscg.mil*

CHAPTER 9

Intelligent Transportation and Information Technology

Information Technology (IT) has had a dramatic impact on society and transportation. This chapter discusses applications of IT to transportation systems, which is also called the Intelligent Transportation Systems (ITS) program.

ITS refers to the application of information technologies such as computer software, hardware, communications technologies, navigational devices, and electronics to improve the efficiency and safety of the transportation system. ITS offers a modern approach to meeting the challenges of increasing travel demand that replaces the physical construction of additional capacity with optimization of existing capacity. The benefits of ITS include improving traffic flow, reducing delay, and minimizing congestion. ITS improves the level of service and safety by providing timely information, advanced warnings, and efficient commercial vehicle operations.

Under the umbrella of Intelligent Transportation Systems, several applications can be identified. There are many applications of ITS; some are designed to improve the safety and efficiency of passenger transportation, and others focus on freight transportation. ITS applications may reside within the transportation infrastructure and others within the vehicles themselves. Thus, ITS applications may be referred to as Intelligent (or Smart) Roads or Intelligent (or Smart) Vehicles.

This chapter describes infrastructure-based technologies designed to improve the safety and mobility of passenger transportation. Among the areas covered are

1. Freeway and incident management systems (FIMS)
2. Advanced arterial traffic control (AATC)
3. Advanced public transportation systems (APTS)

4. Multimodal traveler information systems (MTIS)
5. Advanced technologies for rail

The topics covered should furnish an understanding of successful ITS applications and issues related to them. For each area, the operational concept will be described and illustrated by a brief description of real-world examples. The chapter will also discuss the modeling and analysis tools that can be used to aid in the planning, design, and analysis of ITS applications.

FREEWAY AND INCIDENT MANAGEMENT SYSTEMS

Freeway and incident management systems (FIMS) are designed to improve the flow of people and goods on limited access facilities. They combine field equipment (such as traffic detectors, variable message signs, and ramp meters), communications networks, and traffic operations centers and operating personnel. These elements assist FIMS to control and manage traffic efficiently and safely by reducing congestion. Figure 9.1 depicts a typical traffic control center.

Congestion on a freeway occurs when demand exceeds capacity. There are generally two types of congestion: recurrent congestion and nonrecurrent congestion. Recurrent congestion occurs on a regular basis usually during peak hours. Nonrecurrent congestion is less predictable since it is caused by occurrences such as traffic accidents, adverse weather conditions, and short-term construction work. These events result in a reduction in the capacity of a freeway segment and an increase in congestion. FIMS can deal with both types of congestion but are more effective in dealing with nonrecurrent congestion.

FIGURE 9.1
Traffic operations center.

FIMS Objectives and Functions

The objectives commonly identified for FIMS are to

- Continuously monitor the status of traffic flow, and to implement appropriate traffic control actions that reduce congestion;
- Minimize the duration and severity of nonrecurrent congestion by restoring capacity to its normal level;
- Reduce the frequency of recurrent congestion and mitigate its adverse impacts;
- Maximize freeway efficiency and improve safety; and
- Provide real-time travel information concerning the status of traffic that assists drivers to alter route plans.

The functions of FIMS are traffic surveillance; incident detection and management; ramp metering; information dissemination and dynamic route guidance; and lane management. Each of these functions is described next.

Traffic Surveillance

Traffic surveillance is the continuous monitoring of the status of the transportation system. This function provides the basis for all the other functions and applications of ITS, because ITS applications rely on the use of real-time information about the state of the system. The traffic surveillance system collects several types of data, among the most important of which are data on the status of traffic operations. Traffic operations are evaluated on the basis of the three fundamental measures of traffic previously mentioned in Chapter 4: flow, speed, and density. These measures constitute an essential component of the data collected by modern surveillance systems, but additional types of data are also captured by surveillance technologies. These are video images of transportation system operations, queue length, travel time between given origin and destination pairs, location of emergency response vehicles, bus or transit vehicle location, and environmental data, including pavement temperature, wind speed, road surface condition information, emission levels, and air quality.

Traffic Surveillance System Components and Technologies

A surveillance system consists of four components: detection methods, hardware, computer software, and communications. Detection methods use technologies such as inductive loops, nonintrusive detection devices, closed circuit TV cameras, probe vehicles, reports from police or citizens, and environmental sensors to monitor weather conditions. Hardware elements include computers, monitors, controllers, and displays. Computer software is used to convert data collected by detection devices and to interface and communicate with the field devices. The communications system connects the components located at the control center with the field devices.

Detection Methods Detection methods include inductive loop and nonintrusive detectors such as microwave, infrared, ultrasonic, and acoustic sensors; closed circuit TV; video image processing (VIP); vehicle probes, police and citizens' reporting, and environmental sensors. These technologies are described next in terms of characteristics, applications, advantages, and disadvantages.

Inductive loop detectors (ILDs) are widely used for vehicle detection. Their main use is at intersections with advanced signal traffic control systems and on freeways for incident detection and traffic monitoring. ILDs are made of insulated wire embedded in the pavement. The loop is connected with lead-in cable to the detector unit, which senses changes in the inductance within the embedded wire when a vehicle passes over the loop (Figure 9.2).

ILDs can operate in either the pulse mode or the presence mode. In the pulse mode, the loop sends a short signal (typically in the order of 0.125 sec) to the detector unit, and the pulse mode serves to detect volume counts. In the presence mode, the signal persists as long as the vehicle occupies the detection area. The presence mode provides volume counts and the time occupied by the vehicle. ILDs also measure speeds (by installing two pulse loops a short distance apart) and can determine vehicle classification. The ILDs, however, are not always reliable and may fail to operate when damaged by heavy traffic. Furthermore, installation and maintenance of loops require lane closure and modifications to the pavement.

Occupancy calculations where presence-type inductive loops provide occupancy measurements, defined as the proportion of time that a detector is "occupied" or covered by a vehicle during a given time period. Occupancy measurements can be used to calculate traffic density, one of the fundamental

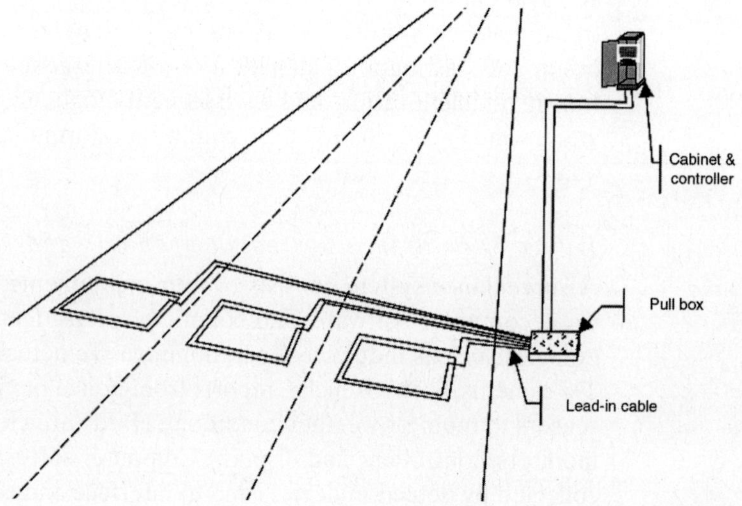

FIGURE 9.2
Inductive loop detectors.

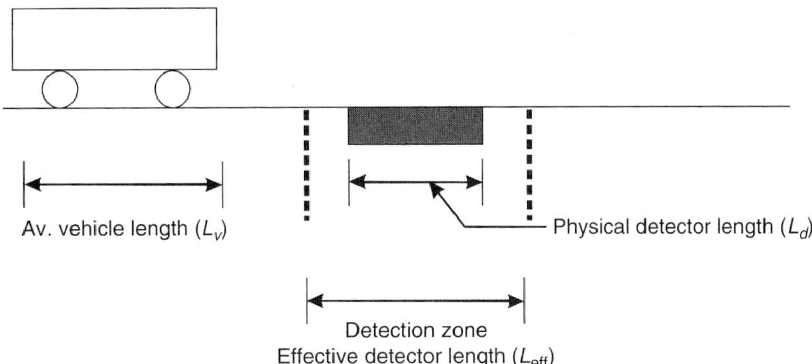

FIGURE 9.3
Occupancy measurements.

measures of traffic flow discussed in Chapter 4, by using an estimate of the average length of vehicles in the traffic stream (L_v) and the effective length of the detector (L_{eff}). The value of L_{eff} is typically greater than the physical length of the loop since vehicles are detected prior to and after they are within the loop (see Figure 9.3).

If the average length of a vehicle (L_v) is known, the following equation can be used to estimate traffic density from occupancy measurements:

$$D = \frac{10 \times Occ}{L_v + L_{\text{eff}}} \tag{9.1}$$

where

D = traffic density in veh/km/lane
Occ = occupancy measurement (percent time occupied)
L_v = average vehicle length in meter
L_{eff} = effective detector length in meter

Note that the vehicle and the detector length are added, since the detector is activated as the front bumper enters the detection zone and is deactivated when the rear bumper clears the zone. Since the occupancy measurement is for a single detector in a designated lane, the density value applies only to that lane.

EXAMPLE 9.1

Calculating Traffic Density from ILD Occupancy Measurements

A freeway detection station in one direction of a six-lane highway (3 lanes/direction) provides the occupancy measurements shown in Table 9.1. The average vehicle length of vehicles is 6 m for lane 1, 5.5 m for lane 2, and 5 m for lane 3. The effective length of each loop detector is 2.5 m. Determine the traffic density for (a) each lane, and (b) for the freeway.

TABLE 9.1
Occupancy Measurements

Lane No.	Occupancy (%)
Lane 1	22
Lane 2	15
Lane 3	12

Solution

(a) Estimate the density for each lane using Equation 9.1. The results are shown in Table 9.2.

TABLE 9.2
Per Lane Density

Lane No.	Occupancy (%)	Av. Veh. Length (ft)	Density (veh/km/ln)
Lane 1	22.00	6	25.9
Lane 2	15.00	5.5	18.8
Lane 3	12.00	5	16.0

(b) The overall density for the freeway direction measured is the sum of the per lane density for each lane:

$$\text{Overall density} = 25.9 + 18.8 + 16.0 = 60.7 \text{ veh/km}$$

Microwave radar detectors are nonintrusive devices whose installation and maintenance does not require lane closure and pavement modifications since they are mounted on a structure over or to the side of the road (Figure 9.4). The types of data collected by the sensor depend on the form of electromagnetic wave transmitted. Sensors that transmit a continuous wave are designed to sense vehicle speeds by measuring the Doppler shift in the returned wave. They cannot detect stationary vehicles and thus cannot function as a presence-type detector. Microwave sensors that use a frequency-modulated continuous wave (FMCW) can measure speed and detect vehicles. The presence of a vehicle is detected by measuring the change in range when a vehicle enters the field of detection.

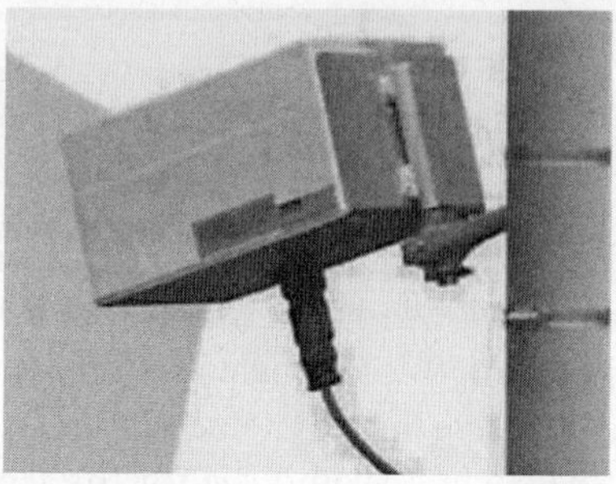

FIGURE 9.4
Nonintrusive traffic detector.

A major advantage of microwave sensors is their ability to function under all weather conditions. Since these sensors are installed above the pavement surface, they are not subject to the effects of ice and snow plowing. Microwave sensors can be expected to function adequately under rain, fog, snow, and wind.

Infrared sensors are nonintrusive detectors that can be either passive or active. Passive sensors do not transmit energy but detect the energy that is emitted or reflected from vehicles, road surfaces, and other objects. The amount of transmitted energy is a function of surface temperature, size, and structure. When a vehicle enters a detection zone, it creates an increase in the transmitted energy compared with a static road surface. Passive infrared detectors can measure speed, vehicle length, vehicle volume, and occupancy. Since their accuracy is affected by adverse weather conditions, they are not always reliable.

Active infrared detectors are similar to microwave radar detectors since they direct a narrow beam of energy toward the roadway surface. The beam is then directed back to the sensors, and vehicles are detected by noting changes in the round-trip propagation time of the infrared beam.

Active infrared detectors measure vehicle passage, presence, and speed information. Speed is measured by noting the time it takes for a vehicle to cross two infrared beams that are scanned across the road surface a known distance apart. Some active detectors have the ability to classify vehicles by measuring and identifying their profiles. Accuracy may be compromised by weather conditions such as fog and precipitation.

Ultrasonic detectors are similar to microwave detectors in that they actively transmit pressure waves, at frequencies above the human audible range. The waves can either be continuous or pulse. Detectors that use continuous waves sense vehicles by using the Doppler principle and measure volume, occupancy, and speed. Pulse wave detectors can also determine classification and presence. Since ultrasonic sensors are sensitive to environmental conditions, they require a high level of maintenance.

Acoustic detectors measure acoustic energy or audible sound from a variety of sources both within the vehicle and from the interaction of the tires with the road surface. Acoustic detectors use an array of acoustic microphones to detect these sounds from a single lane on a roadway. When a vehicle passes through the detection zone, a signal-processing algorithm detects an increase in sound energy and a vehicle presence signal is generated. When the vehicle leaves the detection zone, the sound energy decreases below the detection threshold and the vehicle presence signal is terminated.

Acoustic sensors can be used to measure speed, volume, occupancy, and presence. Vehicle classification can also be obtained by matching the sonic signature of a vehicle against a database of sonic signatures for different vehicles. Speed is measured by using an array of microphones such that the time delay of sound arrival will vary for each microphone. The advantage of acoustic sensors is their ability to function under all lighting and weather conditions.

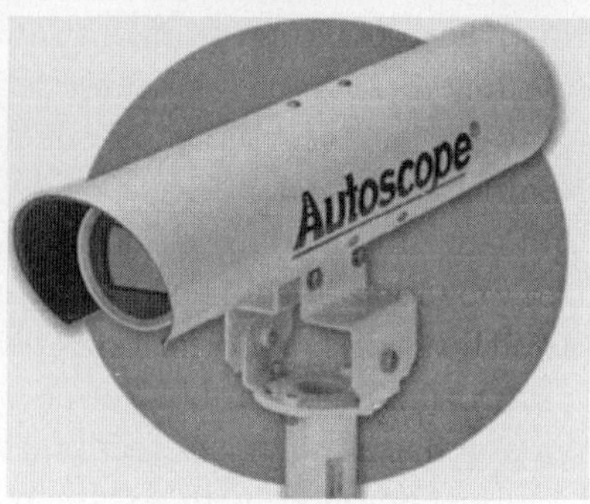

FIGURE 9.5

Integrated video camera/image processing system.

Source: Autoscope Web site, http://www.autoscope.com/.

Video image processing (VIP) is a traffic detection technique that purports to meet the needs of traffic management and control. VIP detectors identify vehicles and traffic flow parameters by analyzing imagery supplied by video cameras. The images are digitized and passed through a series of algorithms that identify changes in the image background. New designs include an integrated machine vision processor, color camera, and zoom lens (Figure 9.5).

An advantage of VIP systems stems from their ability to provide wide area detection across several lanes and in multiple zones within the lane. The user can modify the detection zones through the graphical interface, without the need for pavement excavation or traffic lane closures. The performance of VIP systems can be compromised by poor lighting, shadows, and inclement weather.

VIP and closed circuit TV (CCTV) can be combined to provide an excellent detection tool, particularly for incident detection and verification purposes. When an incident occurs, the user can switch from the VIP mode to the standard CCTV mode and verify the occurrence of the incident via pan/tilt/zoom controls.

Vehicle probes involve tracking vehicles, using positioning and communications technologies, and communicating the location of the vehicle to a central computer where data from different sources are combined to determine the status of traffic flow over the measured transportation system. Vehicle probes can provide useful information not available from other detection techniques. These include link travel times, average speeds, and origin–destination information. Three different technologies that use vehicles as probes are automatic vehicle identification (AVI); automatic vehicle location (AVL); and anonymous mobile call sampling (AMCS).

Automatic vehicle identification (AVI) can identify vehicles as they pass through a detection zone. A transponder (or tag) mounted on the vehicle is read by a roadside device, using dedicated short-range communications (DSRG). The information is then transmitted to a central computer. The most common application of AVI technologies is in conjunction with automatic toll collection systems. The toll charge is automatically deducted from a driver's account when a vehicle enters the toll plaza. This technology can also be used for detection purposes by determining the average travel time on freeways between roadside readers.

Automatic vehicle location (AVL) determines the location of vehicles as they travel over the network. AVL can locate and dispatch emergency vehicles, locate buses in real time, and determine their expected arrival time at bus stops. Several technologies are used for AVL, including dead reckoning, ground-based radio, signpost and odometer, and the global positioning system (GPS), which is the currently most commonly used technology for location identification and navigation. To operate, GPS relies on signals transmitted from 24 satellites orbiting the earth at an altitude of 20,200 km. GPS receivers calculate the location of a point using the time it takes for those electromagnetic signals to travel from the satellites to the GPS receiver.

Anonymous mobile call sampling uses triangulation techniques to determine a vehicle's position by measuring signals emanating from a cellular phone within the vehicle. This concept provides a wealth of information at a relatively low cost. Anonymous mobile call sampling requires two elements: a geo-location control system (GCS), and a traffic information center. The GCS provides the latitude and longitude of cellular probes, which is communicated to the traffic information center, where the information is fused and analyzed. This concept was first tested in the Washington, D.C., area in the mid-1990s.

Mobile reports constitute another source of freeway surveillance information. In many cases, reports of incidents from citizens and the police can provide system monitoring information, at a lower cost than other surveillance technologies. Mobile reports provide event information at unpredictable intervals that could be useful for traffic management purposes. In particular, mobile reports are effective for incident detection. Examples of mobile reporting methods include cellular phones and freeway service patrols.

Cellular phones can serve as an effective tool for incident detection. Many jurisdictions around the country have established an incident reporting hotline to encourage citizens to report traffic incidents. This method has the advantage of low start-up costs. A freeway service patrol consists of a team of trained drivers who are responsible for covering a given segment of the freeway. The freeway service patrol vehicle is equipped to help stranded motorists and to clear an incident site. Examples of the supplies used by a service patrol vehicle include gasoline, water, jumper cables, vehicle repair tools, first-aid kit, push bumpers, and warning lights. Service patrols can then detect incidents and perform the entire incident management process of detecting and clearing an incident.

Environmental sensors are used to detect adverse weather conditions, such as icy or slippery conditions. This information can then be used to alert drivers via variable message signs (VMS) and can be used by maintenance personnel for optimizing maintenance operations. Environmental sensors can be divided into road condition sensors, which measure surface temperature, surface moisture,

FIGURE 9.6
Environmental station.
Source: Nu-Metrics Web site.

and presence of snow accumulations; visibility sensors, which detect fog, smog, heavy rain, and snowstorms; and thermal mapping sensors, which can be used to detect the presence of ice. In addition, many manufacturers currently provide complete weather stations that are capable of monitoring a wide range of environmental and surface conditions. Figure 9.6 shows one example of a weather station.

Hardware Computer hardware is the second component of a traffic surveillance system. Computers receive information from field devices and sensors; communicate data from the control center to field devices (e.g., control data to provide for pan/tilt/zoom of a field CCTV camera); process data to derive meaningful traffic parameters from real-time data collected by sensors; and archive the data collected.

In addition to computer hardware, a surveillance system typically includes graphical displays in the control center to provide for a visual description of the transportation system operations as obtained from the field cameras. Graphics can be provided on the workstations' monitors or on large-screen graphics displays. These displays take the form of an array of video screens (Figure 9.1).

Software Computer software constitutes the third component of a traffic surveillance system. Examples of traffic surveillance system software include incident detection algorithms, decision support systems (DSS) for incident management, and software for controlling field devices. Following sections discuss incident detection algorithms and decision support systems for incident management.

Communications System The communications system is needed to provide communications among the components of a control center and between the control center and the field devices. Communications within the center is accomplished via a local area network (LAN). Between the center and the field devices, a wire-line communication system (e.g., fiber optic, coaxial cable, twisted pair) or wireless system is used. The choice of the communications medium (for example, fiber optic versus coaxial cable) depends on the bandwidth requirements of the data transmitted. For example, video requires a wide bandwidth that can best be met using fiber optic cables.

Incident Management

The second function provided by a FIMS is incident management. Congestion on freeways can be either recurrent or nonrecurrent. Incident management systems are designed primarily to deal with nonrecurrent congestion conditions. Incident management is defined as "a coordinated and planned approach for restoring traffic to its normal operations after an incident has occurred." An incident can be a random event (a freeway accident or a stalled vehicle) or a planned or scheduled event (such as work zone lane closure). In both cases, the goal is to systematically utilize both human and mechanical resources to achieve the following:

- Quickly detect and verify the occurrence of an incident.
- Assess the severity of the situation and identify the resources needed to deal with the situation.
- Determine the most appropriate response plan that will restore the facility to normal operation.

The process of incident management can be conceptually viewed as consisting of the following four sequential stages:

- Detection and verification
- Response
- Clearance
- Recovery

The goal of the incident management process is to reduce the time needed to complete each stage and to restore normal operations. A brief discussion of these four stages follows:

Incident Detection and Verification

Incident detection is the identification of an incident. Verification is the acquisition of information about the incident, such as its location, severity, and extent. Verification provides the information used to devise an appropriate response plan. Incident detection and verification have always been the responsibility of state and local police. Technologies now available for incident detection and verification augment these functions and are either nonautomated or automated. Nonautomated detection techniques include cellular phone calls to a 911 or 511 number, an incident reporting hotline, freeway service patrols, citizen-band radio monitoring, motorist call boxes, and fleet operators. These techniques often serve an important role in the incident management process as a supplement to automated surveillance technologies. Automated incident detection methods will be discussed in a following section.

Incident Response

With an incident detected and verified, the next step in the incident management process is that of incident response. Incident response involves the activation, coordination, and management of personnel and equipment to clear the incident. Traffic incident response can be divided into two stages. Stage 1 is concerned with identifying the closest incident response agencies required to clear the incident, communicating with those agencies, coordinating their activities, and suggesting the required resources to deal with the incident effectively. Stage 2 involves traffic management and control activities aimed at reducing the adverse impacts of the incident. These include informing the public about the incident using variable message signs (VMS) or other information dissemination devices, implementing ramp metering and traffic diversion strategies, and coordinating corridor-wide traffic control strategies.

The primary goal of incident response technologies is to optimize resource allocation and minimize response time. The three elements of response time are verification of the occurrence and location of an incident; dispatch of an incident response team; and travel time of the incident response team. A number of techniques and technologies are available to reduce response time, including incident response manuals, tow truck contracts, techniques to improve emergency vehicle access, and improved traffic flow through alternative route planning.

Incident Clearance

Incident clearance refers to the safe and timely removal of an incident. There are several technologies for improving the efficiency of incident clearance. Inflatable air bag systems are one example of such technologies. The main purpose of these systems is to restore an overturned vehicle to an upright position. The system consists of rubber inflatable cylinders having various heights. These cylinders are placed underneath the overturned vehicle and inflated until the task is complete.

Incident Recovery

This stage refers to the time taken by traffic to return to normal flow conditions after the incident has been cleared. The goal is to use appropriate traffic management techniques to restore normal operations and to prevent the effect of congestion from spreading elsewhere.

Automatic Incident Detection Methods

Automatic incident detection (AID) uses algorithms to detect incidents in real time using data collected from traffic detectors. The development of AID algorithms began in the 1970s, and since then many algorithms have been developed. Evaluation of AID algorithms is based on the detection rate (DR); the false alarm rate (FAR); and time to detect (TTD).

Detection rate (DR) is a measure of how effective an AID algorithm detects incidents. It is the ratio of the number of incidents that the algorithm detects and the total number of incidents that occurred. Values range from 0 to 100% and the closer to 100, the more effective the algorithm.

False alarm rate (FAR) is the ratio of the number of false detections and the total number of observations. Most algorithms observe incidents at regular time intervals, such as every 30 seconds or every minute. The FAR results are a percentage for each detector station or simply the total number of false reports over the time period observed.

Time to detect (TTD) is the time difference between the moment an incident was detected and when the incident occurred. The *mean time to detect* (MTTD) is the average TTD over a specified number of incidents.

The three parameters are correlated. For example, increasing the value of the DR results in a corresponding increase in the FAR. If the algorithm time to detection is increased (TDD), both the DR and FAR values would improve. Experience with deployed AID has not always been favorable. In many cases, the number of false alarms that AID algorithms produce has become so unreliable that several traffic operations centers have stopped using them.

EXAMPLE 9.2

Calculating the Detection and False Alarm Rates for an AID Algorithm

A certain automatic incident detection (AID) algorithm is used in a traffic management center. The algorithm is applied every 30 s. To evaluate the performance of the algorithm, traffic was observed over a period of 30 days in which a total of 57 incidents took place. Of this number, the algorithm correctly detected a total of 49 incidents. The algorithm provided 1000 false alarms during the observation period. Determine (a) the DR and (b) FAR for this algorithm.

Solution

(a) DR:

The DR is the ratio of the number of incidents detected to the total number of incidents occurring. Thus

$$DR = \frac{49}{57} \times 100 = 86\%$$

(b) FAR is the ratio of the number of incorrect detections to the total number of times that the algorithm was applied. Therefore, it is necessary to first determine the number of times the algorithm was applied. Since the algorithm is applied every 30 s, and the observation period was 30 days long, the number of times the algorithm was applied is

30 days × 24 hours × 60 minutes × 2 applications/minute = 86,400 times

Thus

$$\text{FAR} = \frac{1000}{86400} \times 100 = 1.16\%$$

Although the FAR rate is relatively low (only about 1% in this problem), the absolute number of false alarms (1000) is high, which could become quite annoying to operators in traffic centers.

Comparing the Performance of Incident Detection Algorithms

The performance index, PI, is a measure that is used to compare different AID algorithms. This measure can also be used to calibrate the algorithms for a given location. The performance index is defined in Equation 9.2, with lower values of PI indicating better performance of the algorithm:

$$PI = \left[\frac{(100 - DR)}{100}\right]^m \times FAR^n \times MTTD^p \tag{9.2}$$

where

DR and FAR = detection and false alarm rates, respectively
$MTTD$ = mean time to detect in minutes
m, n and p = coefficients that can be used to emphasize or weigh how the three performance measures are used in evaluating an algorithm's performance (e.g., using higher values for the coefficient m compared to n and p would emphasize the role of the detection rate for the algorithm in judging its performance)

EXAMPLE 9.3

Comparing the Performance of AID Algorithms

The performance of seven AID algorithms was evaluated by recording the DR, FAR, and MTTD for each algorithm. The results are shown in Table 9.3. Values of the coefficients m, n, and p are all equal to 1.0. Using Equation 9.2, determine how each AID algorithm performs. Which of these is preferred?

TABLE 9.3
Comparing the Performance of AID Algorithms

	DR (%)	*FAR (%)*	*MTTD (min)*
AID1	82	1.73	0.85
AID2	67	0.134	2.91
AID3	68	0.177	3.04
AID4	86	0.05	2.5
AID5	80	0.3	4
AID6	92	1.5	0.4
AID7	92	1.87	0.7

Solution

To solve this problem, the PI is computed for each AID. The results are shown in Table 9.4. From Table 9.4, it can be seen that AID4 has the lowest PI value and is therefore the algorithm with the best performance.

TABLE 9.4
PI Calculations for $m = 1; n = 1;$ and $p = 1$

	DR (%)	*FAR (%)*	*MTTD (min)*	*PI*
AID1	82	1.73	0.85	0.265
AID2	67	0.134	2.91	0.129
AID3	68	0.177	3.04	0.172
AID4	**86**	**0.05**	**2.5**	**0.018**
AID5	80	0.3	4	0.240
AID6	92	1.5	0.4	0.048
AID7	92	1.87	0.7	0.105

EXAMPLE 9.4

Detection Time of AID Algorithms

In the previous example, the traffic engineer is interested in emphasizing the importance of the quick detection of incidents. Given this, the engineer decides to rerun the analysis but doubling the value for the p coefficient (i.e., $p = 2$). Which algorithm would be judged to have the best performance in this case?

Solution

The analysis is rerun with $m = 1, n = 1$, and $p = 2$. The results are shown in Table 9.5.

In this case, AID6, which has a mean detection time of only 0.4 min, is judged to be the best algorithm.

TABLE 9.5
PI Calculations

	DR (%)	*FAR (%)*	*MTTD (min)*	*PI*
AID1	82	1.73	0.85	0.225
AID2	67	0.134	2.91	0.374
AID3	68	0.177	3.04	0.523
AID4	86	0.05	2.5	0.044
AID5	80	0.3	4	0.960
AID6	**92**	**1.5**	**0.4**	**0.019**
AID7	92	1.87	0.7	0.073

Types of AID Algorithms

AID algorithms can be broadly divided into four groups based upon the principle behind the algorithm's operation: (1) comparative-type or pattern recognition algorithms; (2) catastrophe theory algorithms; (3) statistical-based algorithms; and (4) artificial intelligence–based algorithms. This section describes the comparative-type or pattern recognition algorithms as these form the basis for other applications. Other algorithms are briefly discussed.

Comparative-type or pattern recognition algorithms These are among the most commonly used AID algorithms. These algorithms are based on the premise that the occurrence of an incident results in an increase in the density of upstream traffic detectors and in a decrease in density for the downstream detectors. Figure 9.7 illustrates this phenomenon.

Comparative-type algorithms attempt to distinguish between "usual" and "unusual" traffic patterns by comparing the values of the traffic volumes, densities, and speeds at upstream and downstream detector stations with preestablished thresholds. If the field-observed values exceed established thresholds, an alarm is triggered indicating that an incident may have occurred. The most challenging part in implementing comparative-type algorithms involves establishing values for preestablished thresholds since values differ for specific freeway locations. A few examples of comparative-type algorithms are given below.

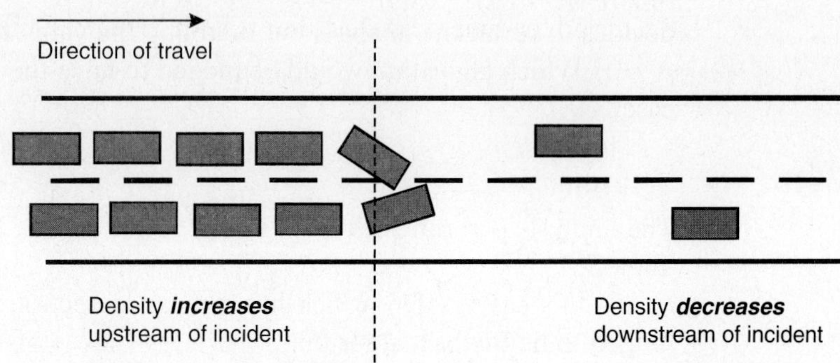

FIGURE 9.7
Occupancy changes as a result of an incident.

The California algorithm is one of the earliest comparative-type AID algorithms to be developed and is often used for comparisons and benchmarking. The algorithm tests for an incident by comparing occupancy (density) values from two adjacent detection stations, according to the following logic:

Step 1: The difference between the upstream station occupancy (OCC_{up}) and the downstream station occupancy (OCC_{down}) is compared against threshold T_1. If the threshold value is exceeded, then the algorithm proceeds to step 2.

Step 2: The ratio of the difference in the upstream and downstream occupancies to the upstream station occupancy ($OCC_{up} - OCC_{down}$)/OCC_{up}) is checked against threshold T_2. If this threshold is exceeded, the algorithm proceeds to step 3.

Step 3: The ratio of the difference in the upstream and downstream occupancies to the downstream station occupancy ($OCC_{up} - OCC_{down}$)/OCC_{down} is checked against threshold T_3. If this threshold is exceeded, a potential incident is indicated. No alarm is indicated, yet step 2 is repeated for the following time interval. If thresholds T_2 and T_3 are again exceeded, a potential incident is assumed.

An incident state is terminated when threshold T_2 is no longer exceeded. The thresholds are calibrated from empirical data. The application of the California algorithm is quite straightforward, yet it is challenging to determine appropriate values of the algorithm's thresholds (T_1, T_2, and T_3) for each location.

EXAMPLE 9.5

Applying the California Algorithm for Incident Detection

Table 9.6 provides occupancy readings for two detection stations along a freeway equipped with a California-type AID algorithm. The algorithm is applied at regular time intervals of 30 s. Based on off-line calibration, three threshold values T_1, T_2, and T_3 were determined to be equal to 20, 0.25, and 0.50. Apply the California algorithm logic to determine the time step when an incident alarm would be triggered, and the time step when the incident state would be terminated.

Solution

For each time step, compute the values for the following three quantities:

Step 1: $(Occ_{up} - Occ_{down})$
Step 2: $(Occ_{up} - Occ_{down})/Occ_{up}$
Step 3: $(Occ_{up} - Occ_{down})/Occ_{down}$

The calculations are shown in Table 9.7 in columns 4 through 6.

TABLE 9.6
Detection Stations Readings

Time Step	Occ_{up} (%)	Occ_{down} (%)
1	60	10
2	62	15
3	59	17
4	65	14
5	67	22
6	64	19
7	59	22
8	48	27
9	37	29
10	32	29
11	30	28
12	32	31

TABLE 9.7
California Algorithm Calculations

Column [1]	*Column [2]*	*Column [3]*	*Column [4]*	*Column [5]*	*Column [6]*
Time Step	Occ_{up} (%)	Occ_{down} (%)			
1	60	10	50	0.83	5.00
2	62	15	47	0.76	3.13
3	59	17	42	0.71	2.47
4	65	14	51	0.78	3.64
5	67	22	45	0.67	2.05
6	64	19	45	0.70	2.37
7	59	22	37	0.63	1.68
8	48	27	21	0.44	0.78
9	37	29	8	0.22	0.28
10	32	29	3	0.09	0.10
11	30	28	2	0.07	0.07
12	32	31	1	0.03	0.03

Column [4] = $(Occ_{up} - Occ_{down})$ = Column [2] − Column [3]
Column [5] = $(Occ_{up} - Occ_{down})/Occ_{up}$ = (Column [2] − Column [3])/Column [2]
Column [6] = $(Occ_{up} - Occ_{down})/Occ_{down}$ = (Column [2] − Column [3])/Column [3]

The values in columns 4 through 6 are then compared to the three threshold values, T_1, T_2, and T_3, respectively, to determine whether the threshold values are exceeded. The results are shown in Table 9.8.

It can be seen that an alarm would be triggered after time step 2, since the algorithm needs two time steps where the thresholds are exceeded before an alarm is triggered. The incident state would then be terminated after time step 9.

TABLE 9.8
California Algorithm Results

Column [1] Time Step	*Column [2] Occ_{up} (%)*	*Column [3] Occ_{down} (%)*	*[4] > T_1*	*[5] > T_2*	*[6] > T_3*
1	60	10	YES	YES	YES
2	***62***	***15***	***YES***	***YES***	***YES***
3	59	17	YES	YES	YES
4	65	14	YES	YES	YES
5	67	22	YES	YES	YES
6	64	19	YES	YES	YES
7	59	22	YES	YES	YES
8	48	27	YES	YES	YES
9	***37***	***29***	***NO***	***NO***	***NO***
10	32	29	NO	NO	NO
11	30	28	NO	NO	NO
12	32	31	NO	NO	NO

Since the original California algorithm was first developed, refinements have been made on its performance. At least 10 new algorithms have been produced, of which algorithms 7 and 8 are the most successful. The TSC 7 algorithm represents an attempt to reduce the false alarm rate of the original algorithm. To do this, the algorithm requires that traffic discontinuities continue for a specified period of time before an incident is declared. The TSC 8 algorithm provides a repetitive test for the propagation of congestion effects upstream of the incident. It also categorizes traffic volumes into different states, which require that more parameters be calibrated. The TSC 8 algorithm can be regarded as the most complex algorithm to emerge from the modified California series, but also the best performer.

Catastrophe Theory Algorithms Catastrophe theory derives its name from sudden changes that take place in one variable that is being monitored, while other related variables under investigation show smooth and continuous changes. For incident detection, catastrophe theory algorithms monitor the three fundamental variables of traffic flow—namely, speed, flow, and occupancy (density). When the algorithm detects a drastic drop in speed, without an immediate corresponding change in occupancy and flow, this is an indication that an incident has probably occurred. This is because incidents typically cause a queue to form suddenly. The advantage of catastrophe theory algorithms as compared to the comparative-type is that catastrophe theory algorithms use multiple variables and compare these to previous trends of the data, whereas the comparative type typically uses a single variable and compares it to a preestablished threshold. By using more than one variable, catastrophe theory algorithms are better at distinguishing between nonrecurrent and recurrent congestion. The McMaster algorithm, developed at McMaster University in Canada, is a good example of an algorithm that is based on this idea.

Statistical-Based Algorithms The idea behind these algorithms is the use of statistical and time series methods to forecast future traffic states or conditions. By comparing real-time observed traffic data with data forecasts, unexpected changes are then classified as incidents. An example of these algorithms is the auto-regressive integrated moving-average (ARIMA) time series algorithm. In this algorithm, ARIMA, a time series technique, is used to provide short-term forecasts of traffic occupancies based upon observed data from three previous time intervals. The algorithm also computes the 95% confidence intervals. If observations fall outside the 95% range as predicted by the model, an incident is assumed to have occurred.

Artificial Intelligence–Based Algorithms A number of the computational paradigms using artificial intelligence (AI) have been applied to problems in transportation engineering and planning. Automatic incident detection is one application. The incident detection problem is a good example of a group of problems called pattern recognition or classification problems. Several AI paradigms are available to solve classification problems, of which neural networks (NNs) are among the most effective.

NNs are biologically inspired systems consisting of a massively connected network of computational "neurons," organized in layers (Figure 9.8). By adjusting the weights of the network connections, connecting the neurons in the different layers of the network, NNs can be "trained" to approximate virtually any nonlinear function to a required degree of accuracy. NNs typically learn by providing the network with a set of input and output exemplars. A learning algorithm would then be used to adjust the weights of the network so that the network would give the desired output, in a type of learning commonly called supervised learning. Once trained, the NN can be used to predict the likely output for new cases.

Over the years, several NN types and architectures were developed. The type used for incident detection is the multilayer perceptron (MLP) neural network,

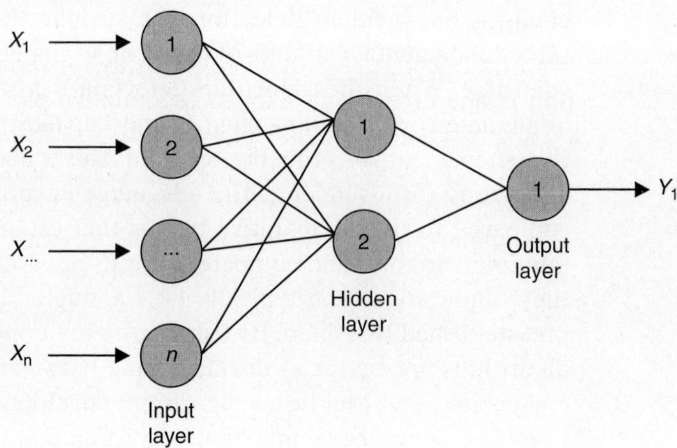

FIGURE 9.8
A multilayer perceptron neural network.

which is among the most widely used of NN architectures. As seen in Figure 9.8, MLPs typically consist of three layers: (1) the input layer; (2) the hidden layer(s); and (3) the output layer. The input layer takes data from loop detectors, the intermediate layer processes data, and the output layer gives an incident or an incident-free signal. Training is performed by presenting the network with a set of both incident and incident-free traffic scenarios. The training helps the network adjust its weights so as to be able to distinguish between traffic states that are incident-free and those that point to the occurrence of an incident.

Estimating the Benefits of Incident Management Systems

A major benefit of incident management systems is the reduction in the duration of an incident. The components of incident duration reduction are reductions in the time to detect and verify the occurrence of an incident; respond to the incident; and clear the incident. Incident management systems have been known to reduce the duration of an incident by up to 55%. The reduction in an incident duration resulting from the deployment of incident management systems can be used to estimate the likely benefits of their deployment, as illustrated by the following example.

EXAMPLE 9.6

Estimating the Benefits of Incident Management Systems Deployment

A six-lane freeway system (three lanes in each direction) carries an estimated 4200 veh/h during the peak hour in the peak direction. The capacity of the freeway is 2000 veh/h/lane. An incident occurs of 60 min duration that blocks 50% of the freeway capacity. Determine the time savings possible if an incident management system is implemented such that the duration is reduced to 30 minutes.

Solution

We first calculate the total vehicle delay for the case when the incident duration is one hour. To do this, the cumulative plot method is used as shown in Figure 9.9 (a similar problem was solved in Chapter 2, Example 2.4).

The arrival rate is 4200 veh/h. The departure rate is 3000 veh/h for a duration of 60 min and, when the incident clears, departure is 6000 veh/h. The total vehicle delay is computed as the area of the triangle between the arrival and departure curves.

(a) Solve for X, the time required for the queue to dissipate:

$$(3000)(1) + (6000)(X) = 4200(1 + X)$$

$$3000 + 6000X = 4200 + 4200X$$

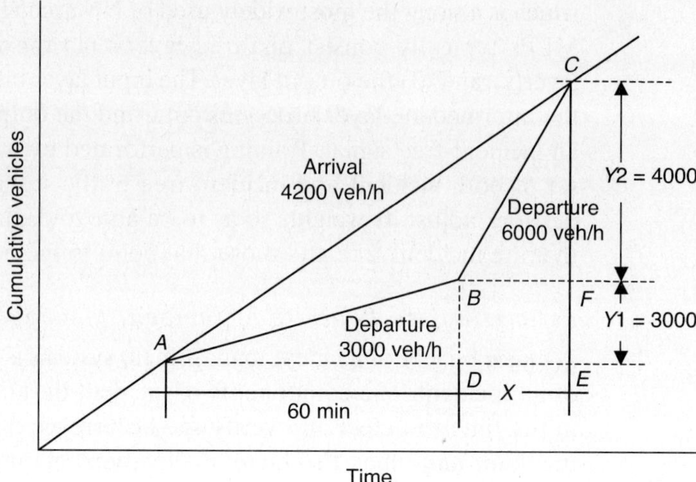

FIGURE 9.9 Cumulative vehicle arrivals and departures.

$$1800X = 1200$$

$$X = 0.667 \text{ h}$$

(b) Determine the cumulative number of vehicles, indicated by the vertical distances $Y1$ and $Y2$, as follows:

$$Y1 = 3000 \times 1 = 3000 \text{ vehicles}$$

$$Y2 = 6000 \times 0.6667 = 4000 \text{ vehicles}$$

(c) Determine the total delay in veh.hours due to congestion by calculating the area of the triangle, ABC, between the arrival and departure curves. The area of triangle ABC is determined by first calculating the area of triangle AEC and then subtracting the area of triangle ABD, the area of rectangle $BDEF$, and the area of triangle CBF, as follows:

$$(\tfrac{1}{2})(7000)(1.67) - (\tfrac{1}{2})(1)(3000) - (3000)(0.67) - (\tfrac{1}{2})(0.67)(4000)$$
$$= 1000 \text{ veh.hours}$$

Next, we consider the case when the incident duration is reduced to 30 min. Figure 9.10 develops the cumulative plot for this case. The total delay is calculated in a manner similar to that described previously, as follows:

(a) Solve for X, the time required for the queue to dissipate:

$$(3000)(0.5) + (6000)(X) = (4200)(0.5 + X)$$

$$1500 + 6000X = 2100 + 4200X$$

$$1800X = 600$$

$$X = 0.333 \text{ h}$$

(b) Determine the cumulative vehicles $Y1$ and $Y2$:

$$Y1 = 3000 \times 0.5 = 1500 \text{ vehicles}$$

$$Y2 = 6000 \times 0.333 = 2000 \text{ vehicles}$$

(c) Determine the total delay in veh.hours due to congestion:

$$(\tfrac{1}{2})(3500)(0.833) - (\tfrac{1}{2})(0.5)(1500) + (1500)(0.33) + (\tfrac{1}{2})(0.33)(2000) = 250 \text{ veh.hours}$$

The deployment of the incident management system has reduced delay from 1000 to 250 veh.hours or a reduction of 75%.

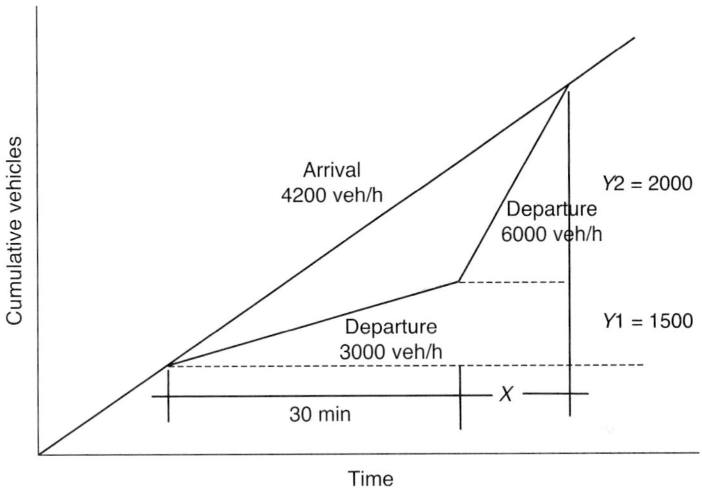

FIGURE 9.10
Cumulative vehicles arrivals and departures.

Ramp Metering

Ramp metering is the third function of a FIMS. It involves the regulation of vehicle entry to a freeway system by traffic signals at the entrance ramps. Ramp metering systems are intended to reduce recurrent congestion during peak hour periods as well as to improve safety when some geometric deficiencies exist.

Ramp metering is not a new strategy and existed in the early 1950s and 1960s. Ramp control systems operate in many areas, including Minneapolis/St. Paul, Minnesota, Seattle, Washington, and Austin, Texas. Most of these systems have achieved the goals of reducing delay and improving safety. The focus of this section is on various types of ramp control systems and their operational concepts. The main system components and technologies used by ramp control systems are described followed by examples of successful ramp metering projects.

Ramp Metering Control Philosophy

As was discussed in Chapter 4, as traffic flow (q) increases, traffic density (k) increases, reaching an optimum k_o at maximum capacity ($q_{\max}$). At density levels greater than k_o, traffic flow conditions deteriorate and flow conditions change from stable to unstable. Ramp metering is designed to prevent unstable flow. Ramp metering controls the amount of traffic entering the freeway in an attempt to maintain densities at or below optimal (k_o) and to assure that traffic does not transition to an unstable or congested condition.

Benefits of Ramp Metering

Ramp metering is designed to achieve the following improvements in traffic operations:

Improved system operation: The main objective of ramp metering is to reduce congestion on a facility by controlling the number of entering vehicles. It is important, however, to make sure that congestion is not moved to surface streets. Vehicle queues on the ramps should not exceed the ramp length. Ramp metering systems can also minimize turbulence caused by merging at the junction of the ramp and the main line lanes, shortening platoons of entering vehicles such that they join the main line flow stream one or two vehicles at a time.

Improved safety: Many freeway accidents occur near on-ramps as the intensity of the merge maneuver increases and large platoons of vehicles arrive. By breaking up platoons of merging vehicles and smoothing the merge operation, ramp metering systems improve the safety of traffic operations. Also, by reducing stop-and-go conditions, ramp metering systems improve safety of traffic operations on a freeway facility.

Reduction in vehicle emissions and fuel consumption: There is a direct relationship between improved traffic operations and the reduction in harmful vehicle emissions and fuel consumption. Thus, ramp metering can improve air quality and energy consumption.

Promotion of demand management strategies: Ramp metering can be designed to encourage demand management and reduction strategies. For example, ramp metering systems can be designed to provide high-occupancy and transit vehicles with preferential treatment by adding a separate lane at the ramp entrance that allows these vehicles to bypass the ramp metering signal. Thus, ramp metering can contribute to strategies aimed at reducing single occupancy vehicles.

Classification of Ramp Metering Strategies

Ramp metering strategies can be classified as restrictive and nonrestrictive metering, and local versus systemwide metering.

Restrictive and nonrestrictive metering: Restrictive ramp metering sets the metering rate at a level lower than the nonmetered ramp volume. As a result, restrictive metering results in queue buildup on ramps and causes drivers to use alternative surface streets. Nonrestrictive metering sets the metering rate equal to or even greater than the average arriving volume. As a result, smaller queues form and diversion to surface streets is reduced. Nonrestrictive metering is often used for the purposes of improving the safety of operations at the vicinity of the ramp by breaking up ramp platoons. It also helps to delay the onset of congestion by smoothing the merge process.

Local versus systemwide ramp metering: Local ramp metering rates are determined based on traffic conditions in the vicinity of the ramp. Local metering is used when traffic congestion can be reduced by metering a single ramp or when several nonmetered ramps are near metered ramps. Systemwide ramp metering rates are implemented at more than one ramp along a freeway section in an integrated fashion and are more effective overall than local ramp metering.

Metering Rate Strategies

Ramp metering success depends on the metering rate selected that permits vehicles to enter the system. The metering rate for single lane ramps is between 240 and 900 veh/h. Ramp metering rates are either pretimed or traffic responsive. Pretimed strategies maintain a constant metering rate for a given time period regardless of the actual traffic volumes on the freeway. Traffic-responsive strategies vary the metering rates based on actual traffic volumes. Traffic-responsive strategies may be *local*, based upon the local traffic conditions detected at the vicinity of the ramp, or *systemwide*, where several ramps are controlled together as a part of an integrated system and metering rates are determined based upon traffic measurements over a large segment of the freeway. Types of ramp metering strategies are described as follows:

Pretimed Metering Pretimed metering rates are determined based on historical observations. Rates are specified for different time periods within a typical day. The metering rate selected depends on the objective to be achieved; that is, whether metering is designed to reduce congestion or to improve safety.

If the system is intended to relieve congestion, the rates are determined to ensure that main line traffic flow is less than the capacity. Thus, the metering rate will be a function of the upstream traffic flow, the ramp volume, and the downstream capacity. The metering should satisfy Equation 9.3 as depicted in Figure 9.11:

$$\text{Metering rate} + \text{upstream volume} \leq \text{downstream capacity} \tag{9.3}$$

Other factors to be considered in setting the metering rate are the availability of adequate storage on the ramp to accommodate queuing and adequate capacity throughout the corridor to accommodate vehicles that may be diverted.

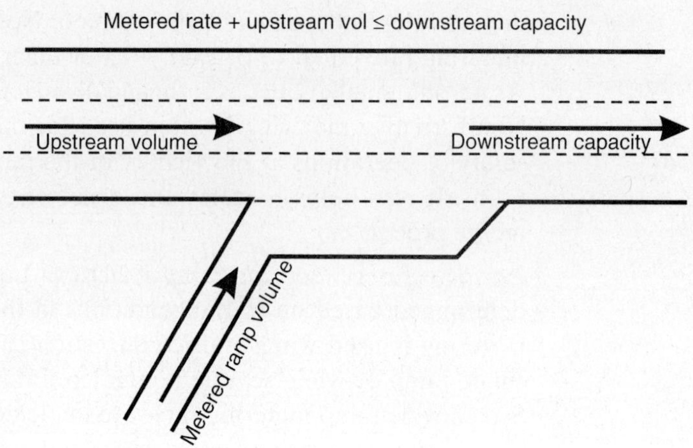

FIGURE 9.11 Pretimed ramp metering.

If the system is intended to improve safety, the metering rate is selected based on merging conditions at the end of the ramp. At ramps and junctions, rear-end and lane-change collisions can occur when vehicle platoons attempt to merge with main line traffic. Ramp metering can alleviate this situation by reducing the number of vehicles in a platoon. The metering rate depends on the geometrics of the ramp and the availability of acceptable gaps in the freeway traffic stream.

With pretimed ramp control, the ramp signal operates according to a predefined plan for the period under consideration. The timing of the red, yellow, and green intervals differs for single-entry, platoon metering, or two-abreast metering, as discussed next.

Single Entry Only one vehicle is allowed per green interval. The green interval (or the green-plus-yellow) is thus typically in the order of 1.5–2.0 s to ensure that only one vehicle enters per green interval. The red interval duration depends on the metering rate in effect.

EXAMPLE 9.7

Designing a Pretimed Single-Entry Ramp Meter

Design a pretimed single-entry ramp metering system on a four-lane freeway. The upstream traffic volume is equal to 3400 veh/h/direction, and the freeway capacity is 2000 veh/h/lane. Green interval is 2 s.

Solution

The downstream capacity in one direction is calculated as (2)(2000) = 4000 veh/h.

The ramp metering rate can be calculated using Equation 9.3:

Metering rate + 3400 = 4000

Metering rate = 4000 − 3400 = 600 veh/h

Since the green interval is 2 sec, the red interval is (cycle length) − (2.0):

Cycle length = 3600/600 = 6 s

Thus, the red interval is (6.0 − 2.0) = 4.0 s/cycle, and the ramp meter signal cycle is green for 2 s and red for 4 s.

PLATOON METERING For metering rates greater than 900 veh/hr, platoon metering is used, where two or more vehicles per cycle enter the freeway. The minimum length of the green interval must be sufficient to allow the platoon to pass.

EXAMPLE 9.8

Designing a Pretimed Platoon Metering System

Design a signal plan for a ramp metering system, given the following information:

Upstream volume = 4800 veh/h

Number of lanes/direction = 3 lanes

Capacity = 2000 veh/h/lane

Solution

Calculate the metering rate using Equation 9.3 using a downstream capacity of 3 × 2000 = 6000 veh/h.

Metering rate = 6000 − 4800 = 1200 veh/h

Since the metering rate is greater than 900 veh/h, platoon metering is required. The metering rate is 1200/60 = 20 veh/min.

If two vehicles enter on the green signal, 10 cycles/min are required (i.e., (2)(10) = 20). The cycle length is 60/10 = 6 s and the green interval is 4 s for 2 s per vehicle. The red interval is 6 – 4 = 2 s.

TWO-ABREAST METERING Two vehicles are released side by side (on a two-lane ramp). Vehicles are released alternately, and the green interval is set to allow the release of one veh/cycle. With two-abreast metering, as many as 1700 veh/h can be accommodated.

Local Traffic Responsive Metering Traffic-responsive metering rates are not prefixed. Rather, they are determined in real time, based on traffic measurements. Local traffic responsive metering rates are selected based on real-time measurements of traffic conditions in the vicinity of the ramp. Traffic-responsive ramp metering systems use traffic flow models that include the variables of flow rate (q), speed (u), and density (k); see Figure 4.2 of Chapter 4. The basic strategy of traffic-responsive metering is to

- Obtain real-time measurements of current traffic flow parameters.
- Determine the current state of traffic flow from traffic flow models.
- Determine the maximum ramp metering rate that would ensure that the flow is kept within the uncongested portion of the fundamental traffic flow diagram (see Figure 9.12).

Ramp metering strategies differ from one another based upon which traffic flow parameters they use in order to determine the appropriate ramp metering rate. Two of the most widely used traffic-responsive ramp metering strategies are demand-capacity control and occupancy control.

DEMAND-CAPACITY CONTROL Metering rates are based on real-time comparisons of upstream traffic *volumes* against downstream capacity. The upstream volume is measured in real time; the downstream capacity is determined either on the basis of historical data or computed in real time based upon downstream volume measurements. The ramp metering rate for the next control period (typically 1 min) is computed as the difference between the downstream capacity and the upstream volume to ensure that downstream capacity is not exceeded. For example, if at a certain control interval the upstream volume equals 3000 veh/h (i.e., 50 veh/min) and the downstream capacity equals 3600 veh/h (i.e., 60 veh/min), a metering rate up to $(60 - 50) = 10$ veh/min could be accommodated.

A problem with using *volume* alone as the traffic flow performance measure is that low volume values may be associated with free-flow conditions as well as with congested conditions, depending upon whether traffic density is less or greater than the density at capacity. As can be seen from Figure 9.12, corresponding to a volume value, $V1$, there are two possible density values, one corresponding to uncongested conditions and the other to congested conditions. To overcome this problem and be able to distinguish between congested and uncongested conditions, occupancy measurements are required.

OCCUPANCY CONTROL Metering rates are selected based upon *occupancy* measurements upstream. There are two types of occupancy control: open-loop occupancy control and closed-loop occupancy control.

Open-loop occupancy control provides a schedule of metering rates. Based upon occupancy measurements upstream the metered ramp, one out of several predefined metering rates is selected for the next control period. The predefined

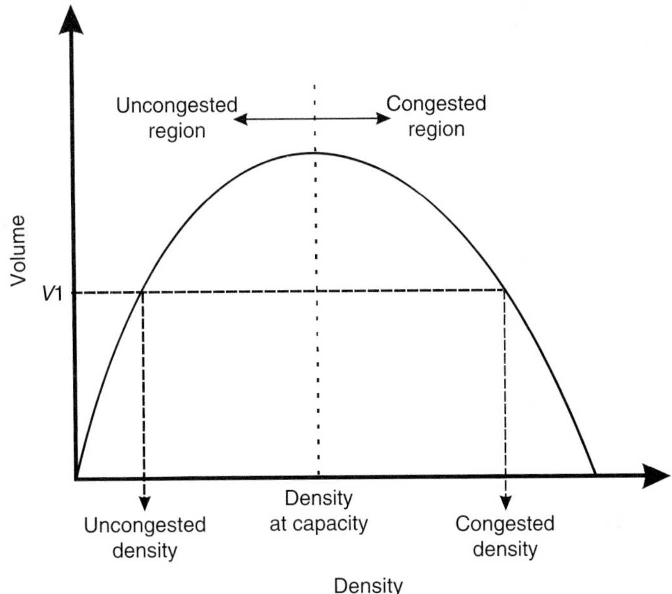

FIGURE 9.12 A typical volume–density plot.

metering rates are determined from a study of a plot of the relationship between volumes and occupancy for the facility of interest. Using this plot, for each level of occupancy, a metering rate can be established that corresponds to the difference between the predetermined estimate of capacity and the real-time estimate of the volume that corresponds to the occupancy measured. The volume can be estimated using a volume–occupancy plot as shown in Figure 9.13, which determines an approximate relationship between occupancy and volume.

Table 9.9 can be used to determine appropriate local traffic-responsive metering rates as a function of the measured mainline upstream capacity. As can

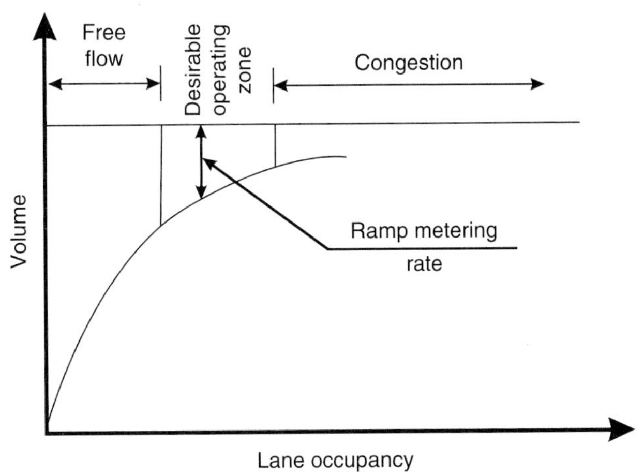

FIGURE 9.13 Calculating ramp metering rates based upon volume–occupancy plots.

TABLE 9.9
Metering Rates as a Function of Upstream Occupancy

Occupancy (%)	*Metering Rate (Veh/Min)*
≤10	12
11–16	10
17–22	8
23–28	6
29–34	4
>34	3

be seen in Table 9.9, if the measured occupancy exceeds the preset capacity occupancy (i.e., 34% in this case), a minimum metering rate of 3 veh/min is selected.

The type of control described is termed *open loop* because it controls the flow rate based upon preset values and does not check the impact of the control action on the controlled environment. It does not control the flow rate to explicitly achieve a parameter value sensed by detectors, such as the downstream occupancy, as does closed-loop control.

EXAMPLE 9.9

Determining Metering Rates for an Open-Loop Ramp Meter

Given the occupancy measurements at a local traffic-responsive ramp metering site shown in Table 9.10, determine the metering rates for the different control periods.

TABLE 9.10
Data for Example 9.9

Control Period	*Measured Occupancy (%)*
1	23%
2	25%
3	29%
4	21%
5	18%

Solution

This problem can be solved using Table 9.9 to determine the appropriate metering rates for each occupancy level. The solution is given in Table 9.11.

TABLE 9.11
Metering Rates for Example 9.9

Control Period	*Measured Occupancy (%)*	*Metering Rate (Veh/min)*
1	23	6
2	25	6
3	29	4
4	21	8
5	18	8

Closed-loop occupancy control explicitly controls the downstream occupancy to conform with the desired occupancy value. Occupancy values measured downstream of the ramp are fed back to the controller in order to determine the metering rate that would bring the downstream capacity to its desired value. One of the more well-known closed-loop ramp metering control algorithms is called ALINEA. The algorithm is designed to operate with a main line detector station that measures occupancy values downstream of the ramp. The metering rate for a given control period, i, is then calculated using the following equation:

$$r(i) = r(i-1) + K_R(o_s - o_{\text{out}}(i)) \tag{9.4}$$

where

$r(i)$ = metering rate for the control interval i
$r(i-1)$ = metering rate during the previous control interval $(i-1)$
o_s = preset or desired value for the downstream occupancy
$o_{\text{out}}(i)$ = measured downstream occupancy for the control interval i

K_R is a coefficient that is commonly referred to as the gain coefficient. The value of K_R affects the sensitivity of the controller and how quickly it reacts to changes in the controller input. The higher the value of K_R, the faster the controller reacts to changes. At the same time, however, high values of K_R tend to make the control more oscillatory and more sensitive to errors in the measured occupancy.

For inductive loop detectors, the occupancy set point (o_s) is typically established in order to ensure that the level of service (LOS) on the freeway does not fall below a specified LOS (e.g., LOS D or E). The calculation proceeds by first looking up the upper value density for the specified LOS from the *Highway Capacity Manual* (HCM) curves or tables. With this determined, Equation 9.1, which relates the density and occupancy values, can be used to calculate the corresponding occupancy set point. The following example illustrates the procedure.

EXAMPLE 9.10

Determining the Occupancy Set Point for a Traffic-Responsive Ramp Meter

A closed-loop, occupancy control ramp meter works by measuring the downstream occupancy using an inductive loop detector and then determining the metering rate using the ALINEA algorithm. It is desired to establish the occupancy set point for the ALINEA algorithm so that the LOS on the freeway will be LOS E. Determine this set point, given the following information:

Average passenger car length = 5.4 m

Average commercial vehicle length = 8.1 m

Percentage of commercial vehicles in traffic stream = 4%

Effective length of the detector = 2.4 m

Upper density level corresponding to LOS E = 28 passenger cars/km/lane

Solution

Calculate the average vehicle length with 4% commercial vehicles (i.e., 96% passenger cars). The average vehicle length, L_v, is

$$L_v = 5.4 \times (1 - 0.04) + 8.1 \times (0.04) = 5.51 \text{ m}$$

The occupancy set point is calculated using Equation 9.1:

$$D = \frac{10 \times Occ}{L_v + L_{\text{eff}}}$$

$$28 = \frac{10 \times Occ}{5.51 + 2.4}$$

$$\text{Occupancy} = \frac{28 \times 7.91}{10} = 22.15\%$$

Systemwide Traffic-Responsive Metering This is the application of metering strategies to a series of ramps. For each control interval, real-time traffic measurements are made of parameters such as volume and/or occupancy, which define demand capacity conditions at each ramp. Ramp metering rates are determined for the entire system as well as for individual ramp meters. Appropriate algorithms will include stored pretimed metering rates. The system will typically use the more restrictive of the traffic responsive and pretimed rates.

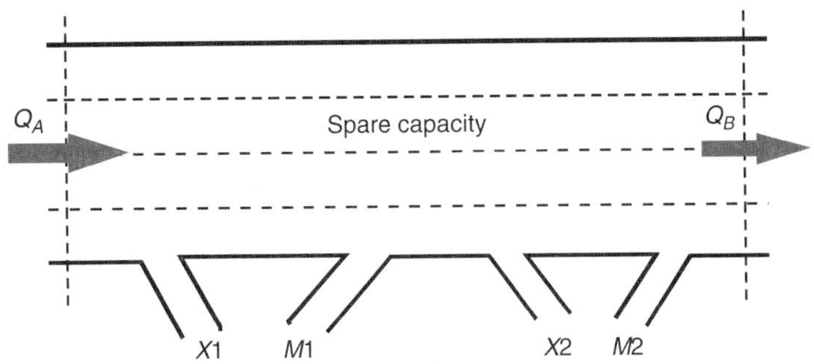

FIGURE 9.14
Volumes entering and leaving a freeway zone.

Most systemwide traffic-responsive metering algorithms start by dividing the freeway into a number of zones. For each zone, the algorithm calculates the number of excess vehicles based on direct mainline measurements. The metering rates of the ramps within that zone are then established based upon the number of excess vehicles.

The Minnesota algorithm serves to illustrate the process. The algorithm regulates traffic within freeway zones by ensuring that the total number of vehicles leaving each zone is greater than the number entering. As shown in Figure 9.14, each freeway zone has three input variables (representing vehicles entering the zone) and three output variables (representing vehicles leaving the zone).

The input variables are

Q_A = main line upstream volume entering the zone, determined from an upstream detection station

M = total ramp volume entering the zone through the metered on-ramps. In Figure 9.14, $M = M1 + M2$

U = total unmetered ramp volume entering the zone

The output variables are

Q_B = main line downstream volume leaving the zone

X = total volume exiting through the zone's off-ramps. In Figure 9.14, $X = X1 + X2$;

S = spare capacity or the additional volume that can enter the zone without causing congestion; calculated based on measured main line speed and volume data

The Minnesota algorithm can be expressed as follows:

$$Q_B + X + S \geq Q_A + M + U \tag{9.5}$$

Therefore

$$M \leq Q_B + X + S - Q_A - U \tag{9.6}$$

Equation 9.6 is the maximum number of vehicles that can pass through all the ramp meters in a given freeway zone. The volume M is then dispersed throughout the zone in proportion to the demand (D) on the metered entrance ramps, using Equation 9.7:

$$R_n = M \times (D_n/D) \tag{9.7}$$

where

R_n = metering rate for on-ramp, n
D_n = demand at ramp, n
D = total demand on all metered ramps within the zone

Example 9.11

Determining Metering Rates for a Systemwide Ramp Metering System

Determine appropriate metering rates for on-ramps A and B of the freeway zone shown in Figure 9.15. The projected demand rates for ramps A and B are $D_A = 550$ and $D_B = 700$ veh/h. Traffic is flowing smoothly within the zone, and spare capacity is 1000 veh/h.

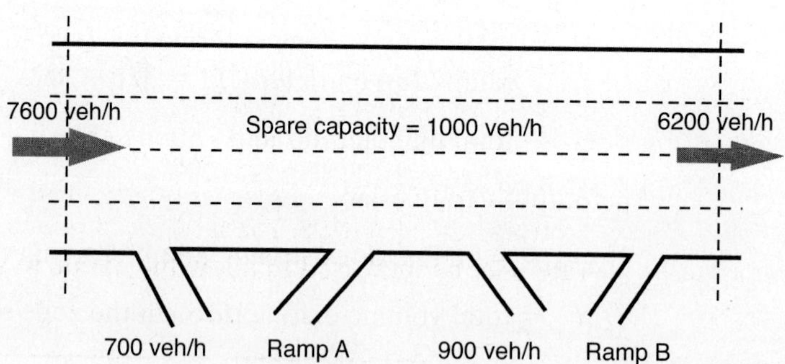

FIGURE 9.15
Traffic volumes for Example 9.11.

Solution

Calculate the total number of vehicles that can pass through the metered ramps A and B. Use Equation 9.6:

$$M = Q_B + X + S - Q_A - U$$

$$= 6200 + (700 + 900) + 1000 - 7600 - 0 = 1200 \text{ veh/h}$$

(Note that U is equal to 0, since all on-ramps within the zone are metered.)

The metering rate for ramps A and B can then be determined using Equation 9.7 as follows:

$$R_1 = 1200 \times \frac{550}{(550 + 700)} = 528 \text{ veh/h (answer)}$$

$$R_2 = 1200 \times \frac{700}{(550 + 700)} = 672 \text{ veh/h (answer)}$$

Ramp Metering System Layout

The typical components of a metering system are shown in Figure 9.16. The system consists of the following elements:

- Ramp metering signal, either a red-yellow-green signal or just a red-green signal;
- Local controller, similar to those used at signalized intersections;
- Advance ramp control sign to inform drivers that the ramp is being metered;
- Vehicle detectors, devices that establish conditions within the ramp area. There are five types of detectors at ramp metering systems, as described next.

  Check-in detectors: Ramp signal remains red until a vehicle is detected. A minimum metering rate, however, is used to avoid problems caused by possible detector failure or a vehicle not stopping near enough to the line to actuate the detector.

  Check-out detectors: These ensure single-vehicle entry. When a vehicle is allowed to pass the ramp, it is detected by the check-out detector and the green is terminated. This ensures that the green interval is sufficient for the passage of one vehicle only.

  Queue detectors: Queue detectors sense the backing or spillback of ramp traffic onto the surface roads. When a queue is sensed, the ramp metering rate may be increased to allow the queue to shorten.

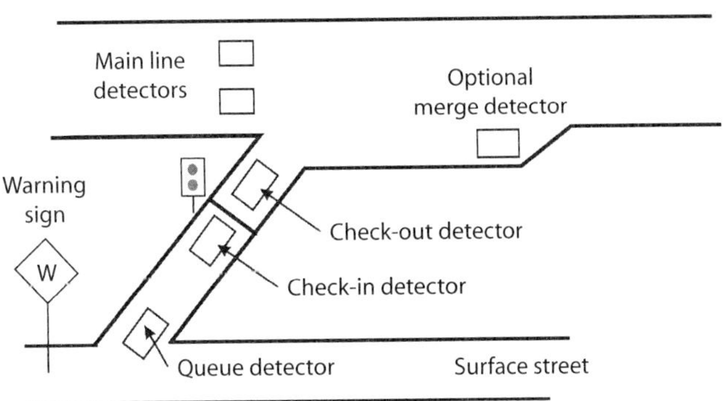

FIGURE 9.16
Ramp metering system layout.

Merge detectors: Merge detectors may be used to detect the presence of vehicles in the merge area. When a vehicle is blocking the merge area, the ramp signal remains red until the detected vehicle merges with freeway traffic.

Main line detectors: Main line detectors sense traffic volumes upstream of the merge area and may be either single or multilane. These detectors provide the input information for the ramp meter control algorithm.

Ramp Storage Requirements

Adequate storage space is required at ramps in order to avoid queues from the ramp backing onto the street network. Storage requirements for ramps can be calculated using the principles of queuing theory as described in Chapter 2.

As discussed in Chapter 2, queuing systems are classified based on the way in which customers arrive and depart. For ramp meters, both the interarrival as well as the service times are best described by the negative exponential distribution. Thus, the *M*/*M*/1 queuing model, discussed in Chapter 2, can be used to solve ramp metering queuing problems. The following example illustrates the procedure.

EXAMPLE 9.12

Ramp Meters Queue Analysis

Traffic is to be regulated on an on-ramp leading to a freeway using a traffic-responsive ramp meter. The ramp has adequate storage space for eight vehicles. During the peak hour, it is estimated that the metering rate will not exceed 600 veh/h. The average volume on the ramp during a typical peak hour is 480 veh/h. Using queuing theory, determine (1) the average queue length on the ramp; (2) the average delay for vehicles at the ramp meter; and (3) the probability that the ramp will be full.

Solution

(1) We first calculate the ratio of the arrival to the service flow rate, ρ, for the ramp meter described in the problem as follows:

$$\rho = \lambda/\mu = 480/600 = 0.80$$

As discussed in Chapter 2, the average queue length, $\overline{Q}$, for an *M*/*M*/1 queue is given by Equation 2.28 as

$$\overline{Q} = \frac{\rho^2}{(1 - \rho)}$$

Therefore,

$$\overline{Q} = \frac{0.80^2}{(1 - 0.80)} = 3.2 \text{ vehicles}$$

(2) Also from Chapter 2, the average delay, $\overline{W}$, for an $M/M/1$ queue is given by Equation 2.29 as follows:

$$\overline{W} = \frac{\lambda}{\mu(\mu - \lambda)}$$

where

λ = the arrival rate (customers/time)
μ = the service rate (customers/time)

In this example,

$$\lambda = 480 \text{ veh/h} = 480/60 = 8 \text{ veh/min}$$

$$\mu = 600 \text{ veh/h} = 600/60 = 10 \text{ veh/min}$$

Therefore

$$\overline{W} = \frac{\lambda}{\mu(\mu - \lambda)} = \frac{8}{10(10 - 8)} = 0.40 \text{ min/veh or } 24 \text{ s/veh}$$

(3) The ramp will be full when we have more than eight vehicles in the queue. For $M/M/1$ queues, the probability of having exactly n customers, p_n, in the queue is given by Equation 2.31 as follows:

$$P_n = (1 - \rho)\rho^n$$

The probability of $n > 8$ can be expressed as follows:

$$p(n > 8) = 1.0 - p(n \leq 8)$$

That is,

$$p(n > 8) = 1 - p(0) - p(1) - p(2) - p(3) - p(4) - p(5) - p(6) - p(7) - p(8)$$

The calculations can be easily performed using Excel, as shown in Figure 9.17. The probability that the ramp will be full is equal to 0.1342.

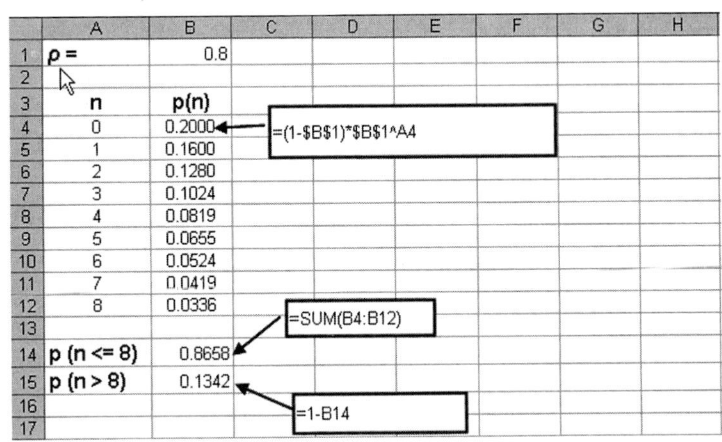

	A	B	C	D	E	F	G	H
1	ρ =	0.8						
2								
3	n	p(n)						
4	0	0.2000	=(1-$B$1)*$B$1^A4					
5	1	0.1600						
6	2	0.1280						
7	3	0.1024						
8	4	0.0819						
9	5	0.0655						
10	6	0.0524						
11	7	0.0419						
12	8	0.0336						
13			=SUM(B4:B12)					
14	p (n <= 8)	0.8658						
15	p (n > 8)	0.1342						
16			=1-B14					
17								

FIGURE 9.17 Probability calculations for Example 9.12.

Information Dissemination

Information dissemination is another function that FIMS are designed to provide. Effective communication with drivers is an essential component of the freeway management process. Freeway management systems utilize several travel information dissemination devices to keep drivers informed about current as well as expected travel conditions on the freeway. Travel information dissemination is divided into pretrip information dissemination and en route information dissemination.

Pretrip travel information concerns providing travelers with information before they start their trips. Examples of pretrip travel information includes current or expected traffic conditions, current and expected weather conditions, and bus schedules and fares. The information is typically provided via devices such as cable TV and the Internet and is intended to enable travelers to make informed route/mode/time of departure decisions. Such informed decisions are likely to improve the overall level of service of the transportation network.

En route travel information dissemination involves providing travelers with information while they are en route. Dissemination devices in this case include dynamic message signs (DMS) (Figure 9.18), Highway Advisory Radio (HAR), low-power FM radio, cell phones, and in-vehicle display devices. En route travel information systems provide information about current and expected traffic and weather conditions, incidents, and alternative routes. Dynamic route guidance (DRG) uses real-time information about traffic flow conditions to reroute drivers around congested areas or incident locations.

FIGURE 9.18
Dynamic message signs (DMS).

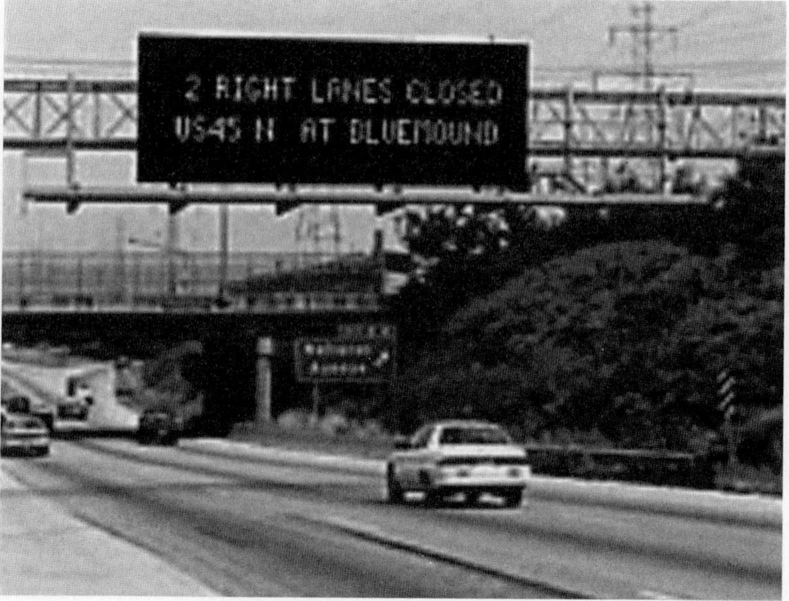

Dynamic Route Guidance

Closely associated with the information dissemination function of a FIMS is the notion of dynamic route guidance (DRG). Travelers typically select the shortest route to their destination while taking account of congestion, if possible. For individual travelers, it is difficult to know in advance the congestion level on the route they plan to use. This is especially true in cases where unexpected incidents and accidents occur on the transportation network. The idea behind DRG is to take advantage of the information provided by the advanced monitoring and surveillance equipment of an intelligent transportation infrastructure and use this information to develop an optimal way to assign or distribute traffic on the network in real time. Routing recommendations are then communicated to drivers via DMS (Figure 9.18) or in-vehicle display devices.

While developing optimal routes, DRG algorithms consider real-time traffic and congestion levels. These algorithms are therefore called dynamic traffic assignment (DTA) algorithms as opposed to the static assignment techniques previously discussed in relation to transportation planning, which focus on average, steady-state conditions. The following section describes the difference between the dynamic and static assignment problems in some detail.

Dynamic versus Static Traffic Assignment

The general traffic assignment problem includes a network and a set of ordered pairs of points on the network where trips originate and end. For each origin–destination pair, a function $R(t)$, $0 \leq t \leq T$, where T is the planning horizon, is given that defines the rate at which vehicles leave the origin at time t, destined to a particular destination. This function yields what is called the origin–destination (O–D) matrix, as shown in Figure 9.19. In addition, the capacity of each link

FIGURE 9.19
The traffic assignment problem.

Origin–Destination Matrix

Zone	1	2	3	4	5
1	0	1000	2000	900	0
2	500	0	1200	1700	700
3	1200	900	0	1100	1500
4	800	700	1500	0	2000
5	1100	750	1150	1500	0

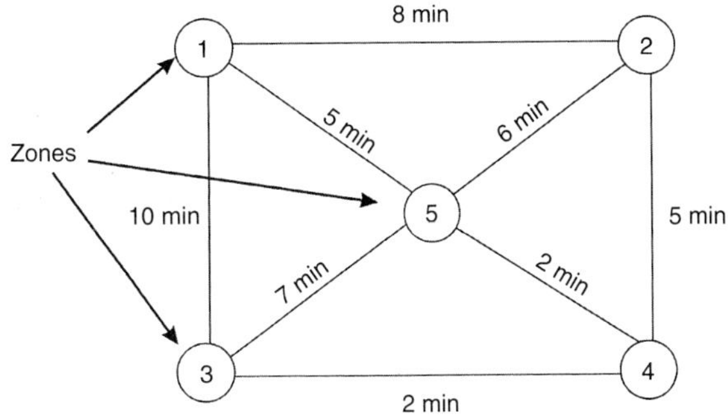

(roadway segment) in the network is provided, $Cap(t)$. The assignment problem is to determine the traffic pattern or flows on the links of the network satisfying certain optimality or equilibrium conditions.

When both $R(t)$ and $Cap(t)$ are constant over time, the problem reduces to the static traffic assignment problem. While this assumption may be reasonable for transportation planning applications, it is not very realistic for real-time modeling and control of transportation networks. The assumption of constant demand and supply does not hold for many realistic traffic situations. Peak-hour conditions, for example, are typically characterized by variations in traffic demand. The occurrence of incidents affects the capacity (i.e., the supply side) of the network. For such conditions of variable demand and/or supply, a dynamic traffic assignment (DTA) problem formulation is needed. This is the formulation required for optimally routing drivers in real-time in the DRG problem.

Mathematical Formulation of the Dynamic Route Guidance or Traffic Assignment Problem

The DRG or DTA problem can be formulated as a mathematical program. For this problem, the decision variables are the time-varying traffic splits at each diversion point that optimize network performance (e.g., minimize total travel time). This defines how traffic should be distributed over the network. The objective function expresses the measure of the highway network's performance to be optimized (such as the total travel time for all vehicles), and the set of constraints attempt to model traffic flow in the region and ensure flow conservation at the nodes and along the links of the network. The formulated model is solved to determine the routing strategy that will optimize the objective function.

Challenges of the DRG The DRG problem is a challenging problem. For realistic transportation networks, with hundreds and even thousands of nodes, links, and alternative routes, the computation effort required to solve the problem is intensive. This is especially true given the fact that the recommended routing strategies need to be developed in real time. As soon as traffic conditions change, such as when an incident occurs, routing strategies must be revised to address the new situation. Second, the problem formulation discussed previously assumes that travel demand and travelers' origins and destinations are known. In practice, predicting travelers' origins and destinations is far from being a straightforward problem. One also needs to be able to predict how drivers will respond to the routing recommendations generated. Finally, there is the problem of missing or incomplete information, since the surveillance system will cover only a subset of the network. Moreover, sensor malfunctions are a common occurrence in the harsh freeway environment.

Real-Time Execution Requirement of the DRG Problem

Real-time execution, in the context of incident traffic flow management, refers to the immediate, online response to an incident, through the implementation of a

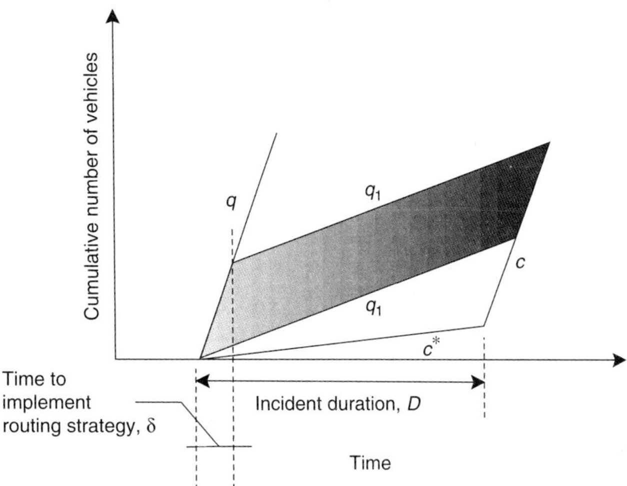

FIGURE 9.20
Incident queuing diagram.

routing strategy, so as to minimize resulting delays. Delay in the implementation of the routing strategy results in incurring additional delays. To illustrate this, we will look at an example next that uses cumulative plots. Consider Figure 9.20, which depicts a queuing diagram for cumulative vehicle arrivals and vehicle departures during a particular incident scenario of duration, D, minutes, similar to the cumulative plots we developed in Chapter 2.

The traffic arrival rate before a routing strategy is implemented is denoted by q (veh/h) and is represented by the slope of the cumulative arrival function. Similarly, q_1 (veh/h) denotes the reduced traffic arrival rate after a routing strategy is implemented. The reduced capacity of the segment caused by the occurrence of the incident is denoted by c^* (veh/h), whereas the normal capacity on the absence of incidents is denoted by c (veh/h). The capacities, c^* and c, are represented by the slope of the cumulative departure curves.

As Figure 9.20 illustrates, waiting for a period, say δ minutes, to implement the routing strategy results in incurring additional delay costs, as indicated by the shaded region of Figure 9.20. Geometrically, the area of the shaded region can be shown to be equal to

$$(q - q_1) \times \delta \times \left[\frac{\delta(q - c) + 2D(c - c^*)}{120(c - q_1)}\right] \text{veh.min} \tag{9.8}$$

where

q = traffic arrival rate before routing (veh/h)
q_1 = reduced traffic arrival rate after routing (veh/h)
δ = waiting time before implementing a routing strategy in minutes
c = normal capacity of segment without any incidents (veh/h)
c^* = reduced capacity of segment as a result of an incident (veh/h)
D = incident duration in minutes

The units for the resulting extra delay incurred will be veh.min.

EXAMPLE 9.13

Extra Delay Resulting from Waiting to Implement Routing Strategies

A six-lane freeway segment, whose unrestricted capacity is 2200 veh/h/lane, carries an average volume of 6000 veh/h. An incident occurs that results in a 60% reduction in the capacity of the segment. The incident lasts for 45 min. To relieve congestion during the incident, traffic routing is implemented which reduces the traffic volume on the segment down to 3600 veh/h. What would be the extra delay incurred if it would take 5 min to implement the routing strategy versus only 30 s?

Solution

To calculate the extra delay incurred as a result of waiting for a period of δ minutes to implement routing strategies, we use Equation 9.8.

For the 30-s case:

$q = 6000$ veh/h

$q_1 = 3600$ veh/h

$\delta = 0.5$ min

$c = 6600$ veh/h

$c^* = 0.4 \times 6600 = 2640$ veh/h

$D = 45$ min

Substituting in Equation 9.8 gives

$$\text{Extra delay} = (q - q_1) \times \delta \times \left[\frac{\delta(q - c) + 2D(c - c^*)}{120(c - q_1)}\right]$$

$$= (6000 - 3600) \times 0.5 \times \left[\frac{0.5(6000 - 6600) + 2 \times 45 \times (6600 - 2640)}{120(6600 - 3600)}\right]$$

$= 1187$ veh.min

For the 5-min case, substituting in Equation 9.8 gives

$$\text{Extra delay} = (q - q_1) \times \delta \times \left[\frac{\delta(q - c) + 2D(c - c^*)}{120(c - q_1)}\right]$$

$$= (6000 - 3600) \times 5 \times \left[\frac{5(6000 - 6600) + 2 \times 45 \times (6600 - 2640)}{120(6600 - 3600)}\right]$$

$$= 11780 \text{ veh.min}$$

Therefore, the extra delay incurred as a result of taking 5 min to implement the routing strategy versus only 30 s is equal to 11,780 − 1187 = 10,593 veh.min.

Lane Management

The lane management function of a FIMS attempts to maximize the utilization of the available lane capacity of the freeway. One major application of the lane management function involves the use of reversible-lane flows, which change the directional capacity of a freeway to accommodate peak directional traffic demands. The use of reversible lanes is warranted when traffic flow exhibits significant directional imbalance (e.g., when there is more than 70% of the two-directional traffic volume in the peak direction). In such cases, the use of reversible lanes allows for using the existing capacity in a more efficient way. Reversible lanes or contraflow lanes are also very useful during some incident management scenarios and for emergency evacuation.

The use of reversible lanes, however, raises some safety concerns, and appropriate measures would need to be implemented to ensure safe operations. These include the use of barrier gates to prevent vehicles from entering in the wrong direction; pop-up lane delimiters; video cameras for vehicle detection; and dynamic message signs (DMS) to inform motorists of the current operating direction.

Real-World Examples of Freeway and Incident Management Systems and Their Benefits

Real-world freeway and incident management systems can now be found throughout the United States, as well as around the world. Examples in the United States include systems in Atlanta, Houston, Seattle, Minneapolis–St. Paul, New York, Chicago, Milwaukee, Los Angeles, San Diego, and Northern Virginia, among others. Freeway and incident management has been proven to be quite effective in alleviating recurrent and nonrecurrent congestion. The San Antonio's TransGuide freeway management system in Texas, for example, helped reduce accidents by 15% and emergency response time by 20%. Ramp metering was shown to help increase throughput by 30% in the Minneapolis–St. Paul metro area, with peak hour speeds increasing by 60%. Ramp meters in Seattle, Washington, are credited with a 52% decrease in travel time and a 39% decrease in accidents. The evaluation of the initial operation of the Maryland CHART program showed a benefit/cost ratio of 5.6:1, with most of the benefits resulting from a 5% decrease (which amounted to around 2 million veh/h/yr) in delays from nonrecurrent congestion.

Advanced Arterial Traffic Control (AATC) Systems

Signalized intersections play a major role in determining the overall performance of arterial networks and many other types of transportation facilities. They are the points where conflicting traffic streams meet and compete for the same physical space, creating many potential conflicts. For a long time, transportation practitioners have been thinking of ways to make signalized intersections more efficient, and a key tool that they have been trying to take advantage of in this regard is information technology (IT). To a large extent, improving the performance of signalized intersections through the use of IT has centered on two simple ideas.

The first idea attempts to make the traffic signal more intelligent and more responsive to actual traffic demands. The concept is to use traffic sensors or loop detectors, similar to those described in relation to FIMS, at the approaches to the intersection. These sensors would detect the presence or passage of vehicles and would communicate this information to the signal controller. Based on this information, the signal controller would attempt to optimize the signal plan so as to minimize the vehicle's delay at the intersection. These signals are commonly referred to as actuated traffic signals.

The second idea involves controlling a group of signals that lie along a major corridor in an integrated or, to use signal control terminology, in a coordinated way. This means that the signal plans of the individual intersections would be coordinated in such a fashion that a platoon of vehicles discharged from one intersection will not be stopped right away at a downstream intersection, but instead would proceed through a sequence of coordinated intersections without stopping. In addition to actuated and coordinated signals, AATC applications include adaptive traffic control and signal preemption to allow emergency vehicles to safely reach their destination as soon as possible. The following sections will describe these applications in more detail.

Actuated Traffic Signals

Actuated signal control could be regarded as one of the very first applications of IT to transportation problems, which predates the term *ITS* by several years. As opposed to pretimed signals, actuated signals have the capability to revise their timing plans based upon actual traffic demands obtained from traffic detectors. The idea behind the use of actuated controllers is to have an adaptive type of control that is responsive to the continuously changing traffic conditions. For pretimed controllers, the implemented signal plan is only optimal for the volumes assumed in developing the "offline" plan. These volumes could be very different from the actual volumes, particularly if signal plans are not updated regularly, which is often the case. Actuated controllers are capable of optimizing the allocation of time based upon real traffic volumes.

To understand the basic concept of operations for actuated controllers, we first need to define the following three parameters:

Minimum green. Each signal phase of an actuated controller is assigned a minimum green time. This time is typically taken to be equal to the time it takes a queue of vehicles potentially stored between the stop line and the approach detector location to enter the intersection.

Passage time interval. The passage time interval is the time it takes a vehicle to travel from the detector location to the stop line. The passage time also defines the maximum gap, which is the maximum time period allowed between vehicles' arrivals at the detector for the approach to retain the green. If a time period equal to the passage time interval elapses without vehicle actuations at the detector, the green for that approach is terminated, and another approach, with a call for service waiting, gets the green. In such a case, the terminating phase is said to have "gapped out."

Maximum green time. In addition to assigning each phase a minimum green, each phase is assigned a maximum green. If the demand on one approach is sufficient to retain the green until this limit (i.e., vehicles keep arriving before the maximum gap expires), the phase is terminated after the maximum green time is exceeded. In this case, the terminating phase is said to have "maxed out."

Figure 9.21 shows the operational concept of an actuated controller. When a certain phase becomes active, the minimum green is displayed first. Following this, the green is extended by the vehicle passage time. Depending upon vehicle actuations, the minimum green is extended by the passage time interval for each vehicle actuation. If a subsequent actuation occurs within one passage time interval, another passage time interval is added (measured from the time of the new actuation, and not from the end of the previous passage time interval). Finally, the green is terminated according to one of two mechanisms: A passage time elapses without a vehicle actuation (the phase gaps out); or the maximum green time for that phase is exceeded (the phase gaps out).

Readers interested in learning the details of actuated signal controller design are referred to appropriate references in traffic and highway engineering, including *Traffic and Highway Engineering* by Garber and Hoel.

Signal Coordination

When a number of traffic signals are located close to one another along a major corridor, one simple idea to improve the efficiency of the transportation system

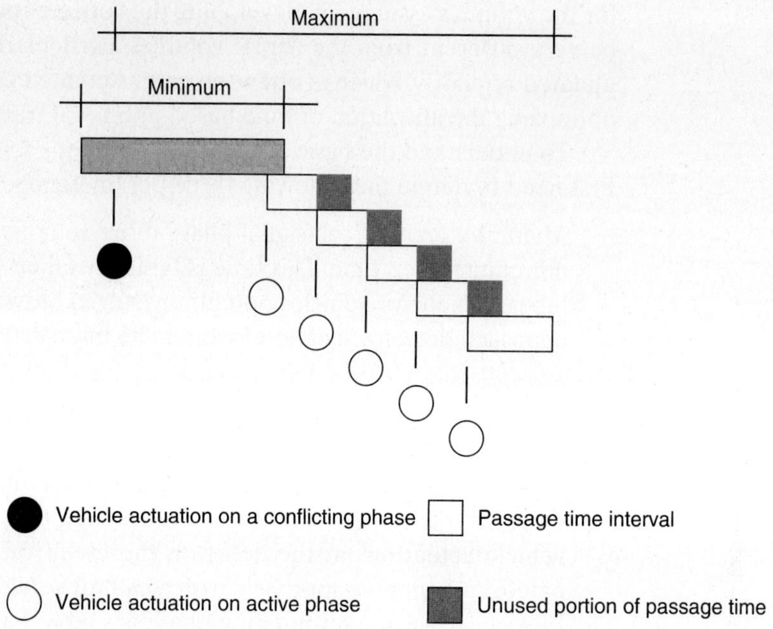

FIGURE 9.21
Actuated control operational concept.

is to coordinate the start of the green for those signals. By carefully setting the time difference between the start of green at successive intersections (i.e., this difference is commonly referred to as the signal offset, as will be explained later), it may be possible to create a "green wave" along the corridor that would allow drivers to go through those signals without having to stop at each and every intersection.

A key requirement for coordinating signals is that successive signals are close enough to one another, thereby allowing vehicles to arrive at intersections in the form of platoons (i.e., a group of vehicles closely spaced to one another). Intersections that are far apart from one another are not good candidates for coordination because vehicles, after traveling for long distances in between intersections, tend to disperse, and the platoonlike structure of the traffic stream is destroyed. In these cases, intersections can be regarded as if they were isolated intersections, and vehicles' arrival patterns at those intersections tend to become random.

To allow for coordination, all signals along a coordinated system have to have the same cycle length (in some cases, however, an intersection with exceptionally high volumes may be allowed to have double the cycle length). A common cycle length is needed so that the start of the green would occur at the same time relative to the nearby intersections. While the cycle length has to be the same, the length of the green at different intersections can vary. Given the requirement for common cycle lengths, most signals along coordinated systems are set to operate on a pretimed basis. It is still possible to coordinate traffic-actuated signals, but they would have to have a common background cycle length.

Coordinated actuated controllers are therefore often of the semiactuated type, which allow for varying the green given to the side streets from one cycle to the next.

For signal coordination, individual controllers need to be interconnected in order to achieve the necessary synchronization. Typically, in a coordinated system, a master controller would send coordination pulses to all other controllers within the coordinated system (these are typically referred to as local controllers). Direct communication could be established using hard-wired cable, telephone lines, coaxial cable, fiber optic cable, or radio communications. In addition, indirect communication could be established using time-based coordinators.

Time–Space Diagram and Signal Coordination

A powerful tool that has historically been used to design coordinated signal systems is the time–space diagram introduced earlier in Chapter 2. At the present time, the use of the time–space diagram for the design of coordinated signal plans has largely been replaced by more powerful traffic simulation software and optimization algorithms. Nevertheless, the diagram is still quite useful in illustrating the concepts, factors, and challenges of signal coordination.

Figure 9.22 shows a typical time–space diagram for a signal coordination problem. To the left of the y-axis of the diagram, which represents the distance, we draw to scale a plan of the corridor or street along which the signals are to

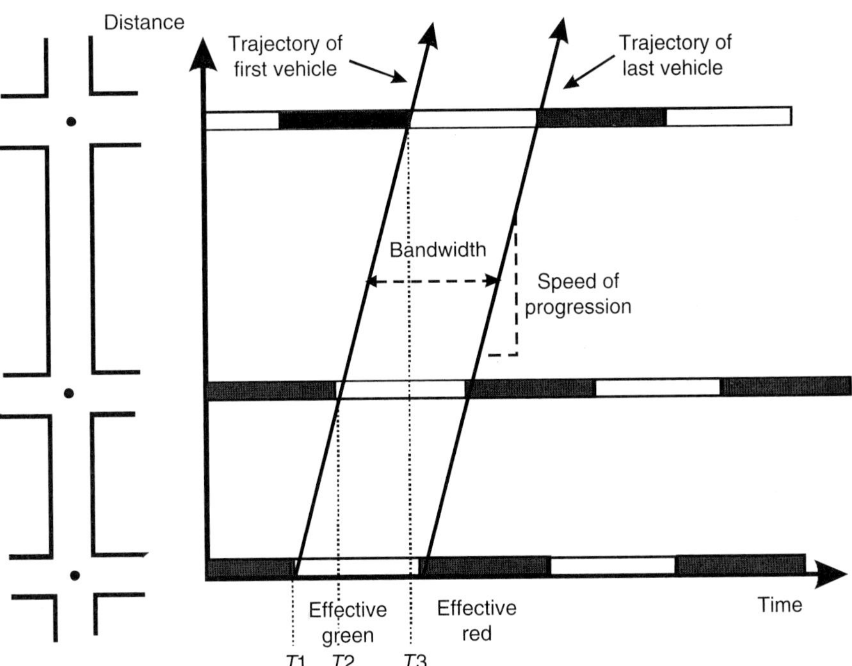

FIGURE 9.22
Signal coordination on a time–space diagram.

be coordinated. We then focus on a given direction (say the NB direction in this example), and at the location of each intersection and along the x-axis, we draw a schematic representation of the phase sequences for the selected direction at that particular intersection. To make things simpler, we typically only plot the duration of the effective green (i.e., the green + the yellow) as a blank line, and the effective red as a solid line. When representing the signal plan for each intersection, it is important to correctly record the beginning of the green for each signal. Vehicles' trajectories could then be drawn and their interactions with the signal plan studied.

As can be seen from Figure 9.22, the first signal turns green at time, $T1$, followed by the second signal at time, $T2$, and the third at time $T3$. The difference between the time when an upstream signal turns green and when the downstream turns green is referred to as the signal offset. Typically the offset is defined as $(T2 - T1)$ or $(T3 - T2)$ and therefore is usually a positive number between 0 and the common cycle length for the coordinated system of signals. Also shown in Figure 9.22 is the concept of bandwidth. This is the amount of green that can be used by a platoon of vehicles moving through a series of intersections without having to stop at any of these intersections.

Determining "Ideal" Offsets

If we focus on one direction (such as the NB direction in Figure 9.22), determining the values for the "ideal" offsets is straightforward. If the offset for a given signal is to be related to the signal directly upstream from that signal, the ideal offset can be easily computed as follows:

$$O_{\text{ideal}} = L/S \tag{9.9}$$

where

L = distance between signalized intersections
S = average vehicle speed

The calculations are illustrated by the following example.

EXAMPLE 9.14

Calculating Ideal Offsets for Signal Coordination

It is required to coordinate signals along the one-way corridor shown in Figure 9.23. All signals shown have a common cycle length of 80 s, and the effective green for the direction to be coordinated for all signals is about 60% of the cycle length. Given that the average vehicle speed along the corridor is 55 km/h and the distances between intersections are shown in Figure 9.23, calculate the ideal offsets for the signals.

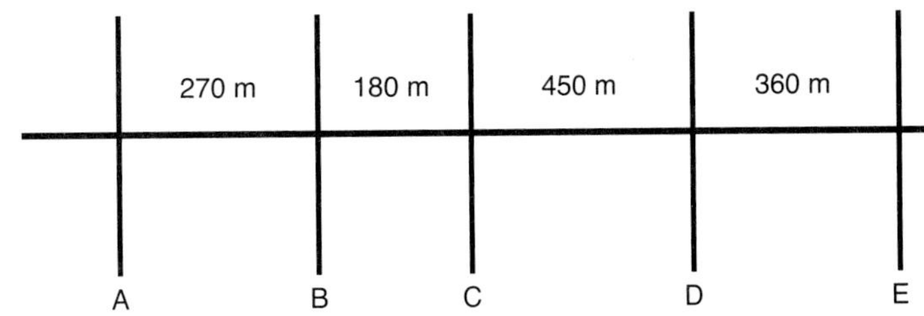

FIGURE 9.23 Calculation of ideal offsets for one-way progression.

Solution

We first convert the speed given in km/h into the equivalent value in m/s as follows:

$$\text{Average speed} = 55 \text{ km/h} = 55 \times 1000/3600 = 15.3 \text{ m/s}$$

We then apply Equation 9.9 to calculate the offsets as shown in Table 9.12. The offset for a given signal is calculated relative to the one right upstream of that signal.

TABLE 9.12 Calculating the Ideal Offsets for the Corridor of Figure 9.23

Signal	*Offset Calculated Relative to Signal*	*Ideal Offset (s)*
B	A	270/15.3 = 17.6 s
C	B	180/15.3 = 11.8 s
D	C	450/15.3 = 29.4 s
E	D	360/15.3 = 23.5 s

Bandwidth Concept

As previously mentioned with reference to Figure 9.22, the bandwidth can be defined as the time difference, in seconds, between the trajectories of the first and last vehicle in a platoon of vehicles capable of moving through a series of intersections without having to stop at any of these intersections. *Bandwidth efficiency* gives an indication of the efficiency of the coordination scheme. It is generally defined as the ratio of the bandwidth to the cycle length, as given by Equation 9.19.

$$\text{Bandwidth efficiency} = \left(\frac{BW}{C}\right) \times 100 \tag{9.10}$$

where

BW = bandwidth, in second
C = cycle length, in second

In general, a bandwidth around 50% is regarded as an indication for good coordination.

Bandwidth capacity gives the number of veh/h that can go through the coordinated system without stopping. The bandwidth capacity can be easily computed by first determining the number of vehicles per traffic lane that go without stopping in each traffic signal cycle. This can be done by dividing the bandwidth in seconds by the discharge or saturation headway, which is typically around 2 s/veh (see Chapter 4). This number is then multiplied by the number of signal cycles/h, and by the number of traffic lanes, as shown in Equation 9.11.

$$\text{Bandwidth capacity (in veh/h)} = \frac{3600 \times BW \times N}{C \times h} \tag{9.11}$$

where

BW = bandwidth, s
N = number of through lanes in the indicated direction
C = cycle length, s
h = discharge or saturation headway, s

Determining the bandwidth for a given coordinated system can be estimated graphically from a time–space diagram similar to the one shown in Figure 9.22. Once this is determined, the efficiency and capacity of the bandwidth can be calculated. The following example illustrates the procedure.

EXAMPLE 9.15

Calculating Bandwidth, Bandwidth Efficiency and Bandwidth Capacity

Figure 9.24 shows a set of three traffic signals along an arterial with two lanes in each direction. The signals are coordinated primarily for the NB direction. The cycle length, the duration of the green for the N-S phase, and the offset for each of the three signals (A, B, and C) are as shown in Table 9.13.

TABLE 9.13
Signal Data for Example 9.15

Signal	*Cycle Length*	*Green for N-S Phase*	*Offset Relative to Upstream Signal*
Signal A	80 s	35 s	0 s
Signal B	80 s	45 s	20 s
Signal C	80 s	40 s	15 s

Given that the average vehicle speed along the corridor is 66 km/h,

1. Draw a time–space diagram for the coordinated system.
2. Determine the bandwidth efficiency and the bandwidth capacity for the NB direction.
3. Determine the bandwidth efficiency and capacity for the SB direction.

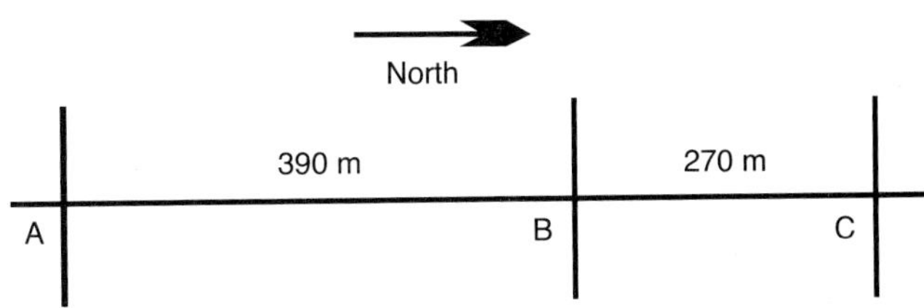

FIGURE 9.24
Roadway sketch for Example 9.15.

Solution

(1) The first step in solving this problem is to draw the time–space diagram for the coordinated system as shown in Figure 9.25.

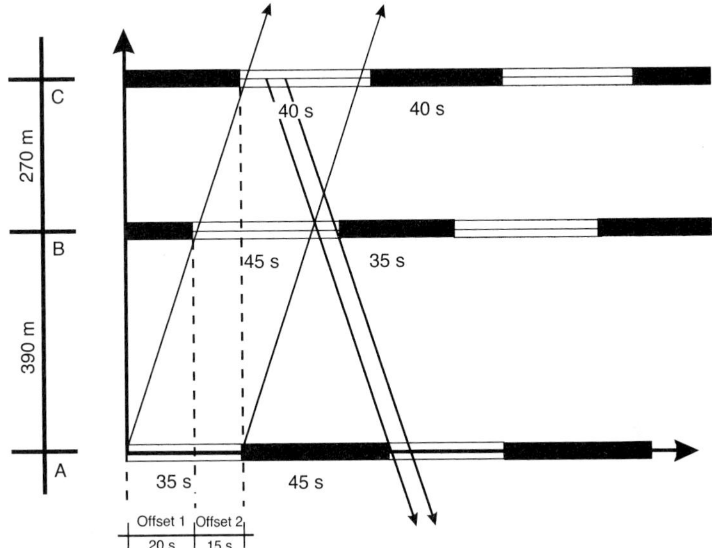

FIGURE 9.25
Time–space diagram for the northbound direction.

The road with the three signals was first drawn to scale along the y-axis of the time–space diagram. Next, the signal timings for each of the three signals A, B, and C were sketched along the x-axis. For signal A and for the N-S direction, we have 35 s of green followed by 45 s of red (to make the 80 s cycle). Signal B has 45 s of green followed by 35 s of red. Since the offset for signal B is 20 s, the green of signal B is drawn so as to start 20 s after the start of the green for signal A. Finally, signal C gets 40 s of green and 40 s of red. The green for signal C starts 15 s after the green for signal B.

We then draw the vehicle trajectories. The average vehicle speed along the corridor is 66 km/h, which is equivalent to $66 \times 1000/3600 = 18$ m/s. Vehicle trajectories are therefore represented by straight lines having a slope of 18 m/s, as shown in Figure 9.25, for both the NB and SB directions.

(2) As can be seen from Figure 9.25, for the NB direction, the bandwidth is equal to 35 s. Given this, the bandwidth efficiency can be easily calculated from Equation 9.10 as follows:

$$\text{Bandwidth efficiency} = \left(\frac{BW}{C}\right) \times 100 = \left(\frac{35}{80}\right) \times 100$$
$$= 43.75\%$$

Bandwidth capacity can be calculated from Equation 9.11 as

$$\text{Bandwidth capacity} = \frac{3600 \times BW \times N}{C \times h} = \frac{3600 \times 35 \times 2}{80 \times 2}$$
$$= 1575 \text{ veh/h}$$

(3) For the SB direction, as can be clearly seen for Figure 9.25, the bandwidth is much smaller, only about 6 s. With the bandwidth determined, the bandwidth efficiency and safety can be easily calculated from Equations 9.10 and 9.11 as follows:

$$\text{Bandwidth efficiency} = \left(\frac{BW}{C}\right) \times 100 = \left(\frac{6}{80}\right) \times 100$$
$$= 7.5\%$$

Bandwidth capacity can be calculated from Equation 9.11 as

$$\text{Bandwidth capacity} = \frac{3600 \times BW \times N}{C \times h} = \frac{3600 \times 6 \times 2}{80 \times 2}$$
$$= 270 \text{ veh/h}$$

Challenges in Coordinating Signals

While signal coordination on one-way streets is straightforward, this is not the case for two-way streets and grid networks. The complexity arises from the fact that, for a two-way street, once the offsets are determined for a given direction (based upon the needs of that direction), the offsets for the other direction are fixed (see Figure 9.26). These offsets (for the other direction) may be inappropriate for the needs of that other direction, as Figure 9.26 and Example 9.15 illustrate.

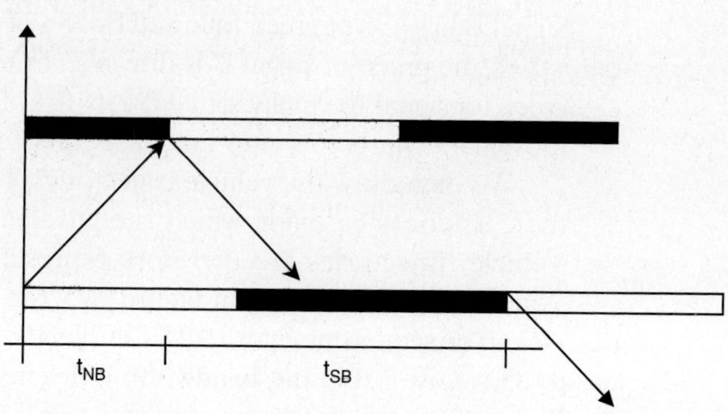

FIGURE 9.26 Relationship between offsets on two-way streets.

Determining the offsets for a two-way street starts with the realization that the offsets in the two directions add to one cycle length, or to an integer multiple of cycle lengths for longer block lengths (see Figure 9.26). Therefore, with reference to Figure 9.26, we could say that

$$t_{NB} + t_{SB} = C \tag{9.12}$$

where

t_{NB} = offset in the northbound direction
t_{SB} = offset in the southbound direction
C = cycle length

The actual offset, which has to satisfy Equation 9.12, can then be expressed as

$$t_{actual} = t_{ideal} + e \tag{9.13}$$

The goal of most signal optimization programs is to minimize the weighted sum of the difference between the actual and ideal offsets.

A number of computer programs are currently available to help in the design of optimal signal timing plans for coordinated systems. The idea behind these programs is to find a set of timing parameters (such as offsets, cycle lengths, and phase intervals) that would minimize a given performance measure (such as total delay or total number of stops) while satisfying the different constraints (such as the constraint defined in Equation 9.12). Among the most famous of these computer programs are TRANSYT-7F and SYNCHRO.

TRANSYT (TRAffic Network StudY Tool) was first developed by the U.K.'s Transport Road Research Laboratories in the late 1960s and has undergone several revisions since then. TRANSYT Version 7 was Americanized for the Federal Highway Administration (FHWA) in the late 1970s and early 1980s; hence the name TRANSYT-7F. Currently, TRANSYT-7F is one of the most widely used computer programs for developing optimal timing plans for corridors and networks. To develop optimal coordinated signal plans, TRANSYT varies the cycle length, the phase lengths, and the signal offsets until the plan that optimizes a user-defined objective function is identified.

SYNCHRO is another signal timing software that could be used to generate optimal signal plans (cycle length, phase lengths, and offsets). For optimization, SYNCHRO uses an objective function that attempts to minimize a combination of delay, number of stops, and number of vehicles in queue. A unique advantage of SYNCHRO is its ability to accurately model the operation of actuated controllers within a coordinated system.

Adaptive Traffic Control Systems

Adaptive or computer traffic control refers to the use of a digital computer to control the operation of a group or system of signals. Adaptive traffic control systems combine the concept of actuated or computer control with the concept of signal coordination. These systems can therefore be regarded as the next step in the evolution of traffic signal control systems. The idea behind adaptive or computer traffic control systems is to take advantage of the power of digital computers to control many signals, along an arterial or in a network, from one central location. Computer traffic control systems predate ITS by several decades; the first installation of these systems took place in the early 1960s. These systems, however, have undergone continuous refinement since that time. In the following paragraphs, we give some historical perspective on the development of these systems.

The most basic type of computer signal control system first appeared in the 1960s. The idea was for a computer to control a series of controllers, but with no "feedback" of information from the field detectors to the computers. In such a system, the traffic plans implemented are not responsive to the actual traffic demand. Instead, the plans are developed "offline" from historical traffic counts and implemented based upon the time of day and day of the week. While this system appears rather simplistic, it offers several advantages, including the ability to update signal plans from a central location, the ability to store a large number of signal plans, and the automatic detection of malfunctioning equipment.

The next development was to have signal control systems in which information from the field traffic detectors is fed back into the central computer. The computer would then use this information to select the signal plan to be implemented. Plan selection is conducted according to one of the following methods.

Select plan from a library of predeveloped plans. In this method, the system has access to a database (library) that stores a large number of different traffic patterns along with the "optimal" signal plans for each pattern (these plans are developed offline). Based upon information from the traffic detector, the computer matches the observed traffic pattern against the patterns stored in the library and identifies the closest one. The plan associated with the identified pattern is then implemented. This type of adaptive traffic control system is often referred to as a first-generation system. The distinguishing feature of these systems is that the plans, while responsive to traffic conditions, are still developed offline. Typically, the frequency of signal update is every 15 min. First-generation systems do not generally have traffic prediction capabilities.

Develop plan online. In this method, the "optimal" signal plan is computed and implemented in real time. This requires enough computational power to do the necessary computations online. Systems that develop plans online are classified as either second-generation or third-generation systems. These systems typically have a much shorter plan update frequency compared to first-generation systems. In addition, signal plans are computed in real time based on forecasts of traffic conditions obtained from feeding the detector's information into a short-term traffic-forecasting algorithm. For second-generation systems, the plan update frequency is every 5 min, whereas third-generation systems have an update interval ranging from 3 to 5 min. The next section will describe some examples of these systems that are in use throughout the world.

Adaptive Traffic Control Algorithms

A number of adaptive traffic control algorithms are currently available. Among the most widely accepted of these are SCOOT, and SCATS. SCOOT (Split, Cycle, Offset Optimization Technique) is an adaptive traffic control system developed by the U.K.'s Transport Research Laboratory (TRL) in the early 1980s. In 1996, the system was in operation in more than 130 towns and cities around the world. SCOOT operates by attempting to minimize a performance index (PI), which is typically taken as the sum of the average queue length and the number of stops at all approaches in the network. In order to do this, SCOOT modifies the cycle lengths, offsets, and splits at each signal in real time in response to the information provided by the vehicle detectors.

The operation of SCOOT is based upon cyclic flow profiles (CFPs), which are histograms of traffic flow variation over a cycle, measured by loops and detectors, placed midblock on every significant link in the network. Using CFPs, the offset optimizer calculates the queues at the stop line. The optimal splits and cycle length are then computed. In recent years, a number of additional features have been added to SCOOT to improve its effectiveness and flexibility. These include the ability to provide preferential treatment or signal priority for transit vehicles; the ability to automatically detect the occurrence of incidents; and the addition of an automatic traffic information database that feeds historical data into SCOOT, allowing the model to run even if there are faulty detectors.

The Sydney Co-ordinated Adaptive Traffic System (SCATS) was developed in the late 1970s by the Roads and Traffic Authority of New South Wales in Australia. For operation, SCATS requires only stop line traffic detectors, and not midblock traffic detection as does SCOOT. This is definitely an advantage since the majority of existing signal systems are equipped with sensors only at stop lines. SCATS is a distributed intelligence, hierarchical system that optimizes cycle length, phase intervals (splits), and offsets in response to detected volumes. For control, the whole signal system is divided into a large number of smaller subsystems ranging from 1 to 10 intersections each. The subsystems run individually unless traffic conditions require the "marriage" or integration of the individual subsystems.

In developing real-time signal plans, SCATS' objective is generally to equalize the saturation flow ratio of the conflicting approaches. Consequently, the system in many cases does not minimize delays on major arterials, which may actually exhibit deterioration in their level of service particularly during peak periods. This was evident in the FAST-TRAC ITS field test in Oakland County, Michigan. In that project, video detection was used to feed a SCATS system, which then developed timing plans in real time.

Signal Preemption and Priority

Advanced arterial traffic control (AATC) systems often include signal preemption and priority capabilities. These capabilities allow controllers to detect vehicles approaching signalized intersections and to provide them with some type of preferential treatment. There are several instances where such systems could be used. For example, signal preemption could be used to provide an approaching emergency vehicle with the green light, an act that could save the lives of those in an emergency situation. Signal preemption could be used at highway–rail grade crossings to prevent a vehicle from being trapped on the railway track. They could also be used to provide transit vehicles with some preferential treatment that may involve extending the green at an intersection for an approaching bus in order to allow buses to stay on their schedule.

Historically, the term *signal preemption* has been used to refer to highway–rail grade crossing systems, emergency vehicles systems, and transit systems. More recently, the use of the term *signal priority* is preferred to reflect the fact that there is a need to assign different priorities to different requests. For example, a highway–rail crossing typically is assigned the highest priority, which would typically involve instantaneous response from the controller to avoid trapping vehicles on the railway track. Emergency vehicles are typically assigned a slightly lower priority to allow a signal from a highway–rail grade crossing to override the emergency vehicle request, if a need arises to do so. Finally, transit vehicles are assigned an even lower priority. Such requests received from transit vehicles typically do not cause major disruptions of the phase sequence but may extend a green split by a prescribed amount, allowing the bus to pass the signal.

There are several control strategies that could be used to grant transit vehicles preferential treatment at signalized intersections. In this section, however, we focus primarily on emergency vehicle systems. In our discussion, we will use the term *preemption* since this is the term that is most often used today to refer to emergency vehicle systems. Signal preemption is usually designed to give the green light in the direction of the oncoming emergency vehicle while posting a red light for all other approaches (Figure 9.27). The other, less frequently used, alternative is for preemption to result in all approaches turning red. There are basically two different approaches for signal preemption—the first is based on local communication between the vehicle and the controller. In this case, the controller identifies approaching vehicles using acoustic, optic, or special loop technology.

Under the second approach, the right-of-way is granted based upon requests from an emergency management center to a traffic management center. This approach requires a highly integrated ITS system whereby the emergency management center would track its vehicles in real time, using global positioning systems technology, and would send signal preemption requests to the traffic management center. The center would then grant the emergency or transit vehicle

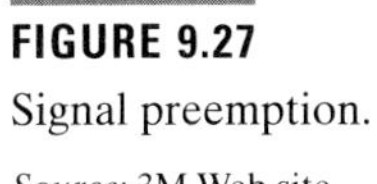

FIGURE 9.27
Signal preemption.
Source: 3M Web site.

the right-of-way. The approach allows for the development of more sophisticated signal coordination strategies compared to the local signal preemption approach, strategies that would anticipate the vehicle's turning movements and would minimize overall system disruption. However, it is much more complex and more expensive than the local signal preemption approach.

Advanced Arterial Traffic Control Systems Benefits

The benefits to be expected from advanced arterial traffic control systems include travel time reduction benefits; environmental benefits resulting from improved traffic flow conditions, lower emission rates, and less fuel consumption; and safety-related benefits resulting from lower accident rates under improved travel conditions. Following is a brief discussion of each of these benefits.

Travel Time Reduction Benefits

Evaluation studies conducted throughout the United States indicate that advanced signal control systems could result in travel time reduction in the range of 8 to 25%. The exact value of time reduction will depend upon a number of factors, including the variability of travel demand; the overall level of congestion; the time interval between signal timing plan modifications; and the density of traffic signals.

Environmental Benefits

Studies show that advanced signal control systems could result in a reduction in air pollutants (such as hydrocarbons and carbon monoxide) that range between 16 and 19%. They could also result in a 4 to 12% reduction in fuel consumption.

Safety Benefits

Some studies show that advanced signal control systems could also result in a reduction in the frequency of injury-related accidents ranging from 6 to 27%.

As an example of how the benefits of an advanced arterial traffic control system may be calculated, consider a segment on an arterial carrying an average annual daily traffic (AADT) of about 20,000 veh/day. Assuming that the average trip length for this segment is about 10 min, and using the conservative estimate of a 10% reduction in trip travel time (as discussed, studies show a reduction in the range 8 to 25%), the time savings resulting from deploying the ITS system could be estimated at ($0.10 \times 10 = 1.0$ min/veh/day). Assuming that the value of time is equal to \$8.90/h, the benefits could be computed as follows:

$$\text{Benefits} = (\text{\# of vehicles}) \times (\text{time saved}) \times (\text{time value}) \times 365$$

$$= 20{,}000 \times (1.0/60) \times (\$8.90) \times 365 = \$\,1{,}082{,}833/\text{year}$$

Advanced Public Transportation Systems

Advanced public transportation systems attempt to improve the efficiency, productivity, and safety of transit systems. They also strive to increase ridership levels and customer satisfaction. In this section, we describe some examples of advanced public transportation systems. The examples described can be categorized under the following four categories: automatic vehicle location (AVL) systems; transit operations software; transit information; and electronic fare payment systems.

Automatic Vehicle Location (AVL) Systems

AVL systems are designed to allow for tracking the location of transit systems in real time. These systems work by measuring the actual real-time position of each vehicle and communicating the information to a central location. This information can then be used to increase dispatching and operating efficiency; allow for quicker response to service disruptions; provide input to transit information systems; and increase driver and passenger safety and security.

While a number of technologies are available for AVL systems, including dead reckoning, ground-based radio, signpost and odometer, and GPS, most agencies now are choosing GPS-based systems. GPS is a navigational and positioning system that relies upon signals transmitted from satellites for its operation. In 1996, there were 86 transit agencies across the country operating, implementing, or planning AVL systems; over 80% of these were using GPS technologies. The following section describes some real-world implementations of AVL systems.

Real-World Examples of Transit AVL Systems

In Atlanta, Georgia, about 250 buses of the Metropolitan Atlanta Rapid Transit Authority (MARTA)'s 750-vehicle fleet are equipped with AVL. The system is linked to the Georgia Department of Transportation traffic management center. There are also electronic signs at a few bus stops for displaying information to passengers waiting at these stops. The system was shown to yield concrete benefits, including improved on-time performance as well as increased safety.

The Tri-County Metropolitan Transportation District of Oregon (Tri-Met) has recently completed the implementation of a GPS AVL system for 640 fixed-route vehicles and 140 paratransit vehicles. The AVL is being employed as part of a regional ITS system, whereby the buses will be used as probes for traffic monitoring, as discussed previously.

The Milwaukee Transit System (MTS) completed the installation of a GPS AVL on 543 buses and 60 support vehicles. Preliminary results indicate a 28% decrease in the number of buses more than one minute behind schedule.

Transit Operations Software

Transit operations software allow for automating, streamlining, and integrating many transit functions. This includes applications such as computer-aided dispatching (CAD), service monitoring, supervisory control, and data acquisition. The use of operations software can improve the effectiveness of operations dispatching, scheduling, planning, customer service, and other agency functions. Operations software is available for fixed-route bus operations as well as for paratransit or demand-responsive operations.

Operations software for demand response transit implement new scheduling and dispatching software for improved performance and increased passenger-carrying capability of the vehicles. Systems vary widely in their capabilities; the high-end systems have integrated automated scheduling and dispatching software with AVL, geographic information systems, and advanced communications systems. These systems provide dispatchers with the capability to view maps of the service area with the locations of all the vehicles in real time. Drivers have mobile data terminals displaying the next hour's pickups and dropoffs.

Real-World Examples of Transit Operations Software Implementations

Kansas City, Missouri, was able to reduce up to 10% of the equipment required for bus routes using an AVL/CAD system. This allowed Kansas City to recover its investment in the system within two years. On-time performance was improved by 12% in the first year of operating the AVL system. In *Ann Arbor, Michigan*, the city's paratransit service (A-Ride) has implemented computer-aided dispatch, automated scheduling, and advanced communications for eight AVL-equipped paratransit vehicles. This system is able to provide service 24 hours a day, with the services of a dispatcher needed only to take reservations and cancellations from callers and to confirm rides.

Transit Information Systems

Transit information systems implement traveler information systems that provide travelers with transit-related information. Three types of such systems can be identified: pretrip systems; in-terminal/wayside systems; and in-vehicle transit information systems.

Pretrip information systems provide travelers with accurate and timely information before starting their trips to allow them to make informed decisions about modes as well as routes and departure times. The information provided can cover a wide range of categories, including transit routes, maps, schedules, fares, park-and-ride locations, points of interest, and weather. In addition, these systems often support itinerary planning. Methods of obtaining pretrip information include touch-tone phones, pagers, kiosks, the Internet, fax machines, and cable TV.

In-terminal/wayside systems provide information to transit riders who are already en route. This information is typically communicated using electronic signs, interactive information kiosks, and CCTV monitors. The overall goal is to provide real time bus and train arrival and departure times, reduce waiting anxiety, and increase customer satisfaction. In-vehicle information systems provide en route information for travelers on board the vehicle. The major impetus behind such systems is to comply with the applicable provisions of the American Disabilities Act of 1991.

Real-World Examples of Transit Information Systems

A good example of pretrip transit information systems can be found in Seattle, Washington, where a key product of Seattle Metro's transit information system is a Web site whereby travelers can obtain information on transit schedules and fares, van and carpooling, ferries, and park-and-ride facilities. This site also provides assistance to transit users in planning their trips. In addition, the University of Washington has developed a Java applet to allow users to view the locations of all buses traveling throughout the Metro System. The University has also developed Web pages to help travelers predict the time of arrival of buses at different bus stops.

Examples of in-terminal and in-vehicle information systems can be found in Ann Arbor, Michigan, where a pair of 79 cm video monitors is used to display real time data generated by the AVL system to inform passengers at the downtown transit center (in-terminal) about arrival status, delays, and departure times. The Ann Arbor system also includes in-vehicle annunciators and displays from which passengers will receive next-stop and transfer information. The latter will take the form of announcements that identify valid bus transfers at upcoming stops.

Electronic Fare Payment Systems

The idea behind electronic fare payment systems is to facilitate the collection and management of transit fare payments by using electronic media rather than cash or paper transfers. These systems consist of two main components: a card and a card reader. Cards could be of the magnetic-strip type, where the reader does most of the data processing. They could also be equipped with a microprocessor (smart cards), and, in this case, data processing could occur on the card itself. Electronic fare payment systems offer a number of advantages: they offer convenience to vehicle operators by eliminating the need for any actions to be taken on their part; they eliminate the need for a passenger to worry about having exact change for the bus fare; they facilitate the collection and processing of fares; and they allow for adopting more complex and effective fare structures.

There are two types of electronic fare payment systems: (a) closed systems, and (b) open systems. Closed systems are limited to one main purpose (i.e., paying transit fares) or to a few other applications, such as paying parking fees. However, the value stored on the card cannot be used outside the defined set of activities; hence the name *closed system*. Open systems can be used outside the transit system. A prime example of an open system is a credit card, which naturally can be used with multiple merchants.

Transit AVL and Operations Software Benefits

Studies show that the deployment of transit AVL and operations software could result in both capital as well as operating cost savings to the operating agency. The default values are a reduction between 1 and 2% in fleet size and a reduction in the range of 5 to 8% in operating costs. To illustrate how the benefits from deploying such systems could be estimated, consider the case of a transit agency whose annual capital and operating costs are \$2,000,000 and \$1,500,000, respectively. Assuming a cost saving of 1.5% for capital costs and 6% saving for operating costs, the annual savings are equal to

$$2{,}000{,}000 \times 1.5/100 + 1{,}500{,}000 \times 6/100 = \$120{,}000/\text{yr}$$

Multimodal Traveler Information Systems

Multimodal traveler information systems are designed to provide static, as well as real-time, travel information over a variety of transportation modes (e.g., highways, transit, ferries, etc.). In essence, these systems integrate the traffic information dissemination functions of freeway and incident management systems with the functions of transit information systems. They then add more information from sources such as Yellow Pages, tourist organizations, and weather services. Traveler information can be provided before or during a trip (pretrip and en route traveler information). Table 9.14 shows an extensive list of data of potential interest to the traveler that could be part of the traveler information system. Information is classified as either static or real-time information.

Following the collection and processing of data, telecommunications technologies, including voice, data, and video transmission over wire-line and wireless channels, are then used to disseminate the information to the public. Among

TABLE 9.14 Potential Contents of a Multimodal Traveler Information System

Static information: Known in advance, changes infrequently	Planned construction and maintenance activities
	Special events, such as state fairs and sporting events
	Transit fares, schedules, and routes
	Intermodal connections (e.g., ferry schedules along Lake Champlain)
	Commercial vehicle regulations (such as Hazmat (hazardous material) and height and weight restrictions)
	Parking locations and costs
	Business listings, such as hotels and gas stations
	Tourist destinations
	Navigational instructions
Real-time information: Changes frequently	Roadway conditions, including congestion and incident information
	Alternate routes
	Road weather conditions, such as snow and fog
	Transit schedule adherence
	Travel time

the means of information dissemination are the Internet, cable TV, radio, phone systems, kiosks, pagers, personal digital assistants (PDAs), and in-vehicle display devices. In addition, efforts are currently underway to implement a national number, 511, which would provide travelers in the United States with real-time multimodal travel information.

Benefits of Multimodal Traveler Information Systems

Multimodal traveler information services allow users to make more informed decisions regarding time of departure, routes, and mode of travel. They have been shown to increase transit usage and reduce congestion when travelers choose to defer or postpone trips or select alternate routes. A good example of a regional multimodal traveler information system is given by the Smart Trek Model Deployment Initiative in Seattle, Washington. At the center of this system is a set of protocols and paradigms designed to collect and fuse data from a variety of sources; process the data to derive useful information; disseminate the information derived to independent information service providers; and warehouse the data for the purposes of long-range planning.

Challenges Facing Multimodal Traveler Information Systems

Although multimodal traveler information systems have the potential to yield significant benefits for both travelers and system operators, the demand for products has been slow to materialize. The size of the market for traveler information systems has been modest to date. Several reasons could be provided for the slow growth of the traveler information market. First, consumer awareness of traveler information products is currently quite low. Second, the price of some products, especially in-vehicle display devices, is still high. Finally, the quality of the information and the extent of coverage need to be increased.

Advanced Technologies for Rail

The rail industry has also been quite active in applying IT to rail transportation applications. While there are numerous examples of the application of advanced technologies to improve the safety and efficiency of rail transportation, we limit our discussion here to just two representative examples, namely, positive train control (PTC) systems, and (2) intelligent rail intersections.

Positive Train Control

Positive train control (PTC) systems are designed to allow for controlling train movements with safety, security, and efficiency. PTC systems integrate digital data communications networks, GPS navigation systems, computers on board trains, in-cab displays, and control center computers and displays. PTC systems will allow personnel in the control center to track the location of trains and maintenance crews in real time, and will control train movements so as to achieve optimum speeds and hence maximum track capacities. PTC will also allow a control center to stop a train should the locomotive crew be incapacitated, thereby providing a greater level of safety and security. Demonstration projects of PTC systems are currently underway in the United States, and wide-scale implementation is expected to begin shortly.

Intelligent Highway–Rail Intersections (HRIs)

Intelligent HRI systems are designed to eliminate accidents at highway–rail grade crossings. Active warning systems at intersections (such as flashing lights and gates) are activated as an approaching train is detected. The equipment at the HRI may also be connected to the adjacent signal system. In case a trapped vehicle on the railway track is detected, the system would immediately preempt the signal and simultaneously warn the locomotive engineer. Intelligent HRI also continuously monitors the health of the detection and warning systems and reports any detected malfunctioning to the appropriate authorities.

Summary

In this chapter, we have discussed some of the applications of information technology to improve the efficiency and safety of the transportation system. As was discussed, this effort is often referred to as Intelligent Transportation Systems or ITS applications. The chapter focused primarily on five key ITS applications: (1) freeway and incident management systems; (2) advanced traffic signal systems;

(3) advanced public transportation systems; (4) multimodal traveler information systems; and (5) advanced technologies for rail. The operational concept behind each application was discussed, as well as a description of the likely benefits of these systems. A brief mention was also made of advanced technologies applications to rail transportation. The application of advanced technologies to transportation is still an evolving field, and new ideas are proposed every day.

PROBLEMS

9.1 Select an ITS project in your state that you are familiar with, and briefly describe its basic concept of operation. What are the likely benefits of that project?

9.2 List the primary objectives of a freeway and incident management system (FIMS).

9.3 What is the difference between *recurrent* and *nonrecurrent* congestion?

9.4 Describe the four basic components of a traffic surveillance system.

9.5 Select four different methods for traffic detection, and briefly discuss the advantages and disadvantages of each method.

9.6 A freeway detection station on an eight-lane freeway gives the occupancy measurements shown below. The average length for vehicles is 6.25 m for lane 1, 5.75 m for lane 2, 5.25 m for lane 3, and 5 m for lane 4. Assuming that the effective length for the loop detectors is 2.5 m, determine the traffic density for each lane and for the freeway direction.

Lane No.	*Occupancy (%)*
Lane 1	24%
Lane 2	17%
Lane 3	14%
Lane 4	12%

9.7 In the context of traffic monitoring, explain what is meant by "probe vehicles." Discuss the different technologies that can be used to implement the concept.

9.8 Briefly discuss the four different stages of the incident management process.

9.9 Describe the parameters used to evaluate the performance of automatic incident detection (AID) algorithms.

9.10 To evaluate the performance of an AID algorithm, its performance was observed over a period of 45 days. During this period, a total of 80 incidents occurred, out of which the algorithm managed to detect 63 incidents. At the same time, the algorithm gave a total of 1300 false alarms. If the algorithm is applied every 30 seconds, determine the detection rate (DR) and the false alarm rate (FAR) for the algorithm.

9.11 A transportation agency desires to compare the performance of five different AID algorithms in order to select one algorithm for implementation in its traffic operations center. To do this, the agency looks at some historical data that give the detection rate (DR), false alarm rate (FAR), and Mean Time to Detect (MTDD) for the five algorithms. The data compiled are shown in the following table. Assuming that the agency would like to place equal emphasis on the three performance measures, which algorithm should the agency select?

	DR (%)	*FAR (%)*	*MTTD (min)*
AID1	83	0.37	2.1
AID2	92	0.86	1.2
AID3	87	0.03	0.4
AID4	95	1.24	0.7
AID5	72	0.73	0.2

9.12 In Problem 9.11, assuming that the agency would like to lay twice as much emphasis on the time to detect incidents as on either the detection rate, or false alarm rate, which algorithm should the agency choose?

9.13 Pick three different types of AID algorithms, and briefly outline how they work.

9.14 Two detection stations on a freeway equipped with a California-type AID algorithm give the following readings. The calibration process for the AID algorithm shows that the values for the algorithm's three thresholds (T_1, T_2, and T_3) are equal to 25%, 0.30, and 0.45, respectively. Determine the time step when an incident alarm would be triggered and the time step when the incident state would be terminated.

Time Step	Occ_{up} (%)	Occ_{down} (%)
1	55	18
2	67	17
3	72	15
4	70	14
5	67	13
6	69	14
7	74	9
8	65	17
9	60	24
10	42	30
11	39	34
12	37	33

9.15 A four-lane freeway system carries an estimated volume of 3400 veh/h during the peak hour in the main direction. An incident occurs resulting in a loss of about 60% of the original freeway capacity. Without an incident management system, the incident is likely to last for a period of 1 hour. Determine the time savings possible if an incident management system is implemented such that the duration is reduced to only 20 mins. Assume the capacity of a freeway lane is equal to 2200 veh/h/lane.

9.16 An accident occurs on a six-lane freeway carrying a peak hourly volume of 4800 veh/h in the main direction. The accident blocks two of the freeway's three lanes, resulting in a significant reduction of the freeway capacity to a value of only 2000 veh/h. Assuming a lane capacity of 2100 veh/h/lane, compare the maximum length of queue, the maximum delay incurred by a vehicle, and the total vehicle delay for the following two cases:

(a) Without an incident management system, the incident lasts for a duration of 75 minutes.

(b) With an incident management system in place, the incident duration is reduced to only 30 minutes.

9.17 A four-lane freeway system carries an estimated 2600 veh/lane during the peak hour in the peak direction. Studies have shown that the maximum capacity of the freeway is 2000 veh/h/lane. For an incident with a duration of 90 minutes that would block 50% of the capacity of the freeway, what are the time savings resulting from implementing an incident management system that would cut the incident duration in half?

9.18 Briefly discuss the likely benefits of ramp metering.

9.19 Distinguish between restrictive and nonrestrictive ramp metering.

9.20 Briefly describe the different types of ramp metering strategies.

9.21 Briefly describe the difference between open-loop and closed-loop control for ramp metering systems.

9.22 Design a pretimed single entry ramp metering system on a four-lane freeway. The upstream traffic volume is 3800 veh/h/direction, and the freeway lane capacity is equal to 2300 veh/h/lane. Assume the green interval is equal to 2 s.

9.23 Design a signal plan for a ramp metering system on a six-lane freeway that carries a total volume of 5220 veh/h in the peak direction. Assume the freeway lane capacity is equal to 2100 veh/h/lane.

9.24 The occupancy measurements at a local traffic-responsive ramp metering station are as shown in the following table. Determine the metering rates for the different control periods.

Control Period	*Measured Occupancy (%)*
1	12
2	18
3	17
4	24

9.25 Determine the set point for an ALINEA-type control ramp meter given the following:

Average passenger car length = 5.25 m
Average truck length = 8.5 m
Percentage of trucks in traffic stream = 8%
Effective detector length = 2.5 m
Upper density level that corresponds to LOS E = 28 passenger car/km/lane

9.26 Determine the appropriate metering rates for the two on-ramps A and B shown if the projected demand for ramp A is 900 veh/h and that for ramp B is 700 veh/h.

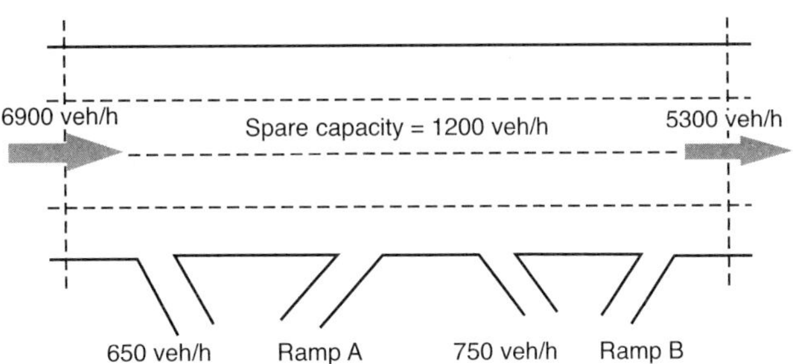

9.27 A traffic-responsive ramp has adequate storage space for 10 vehicles. During the peak hour, it is estimated that the metering rate will not exceed 750 veh/h, whereas the average traffic demand is 620 veh/h. Using queuing theory, determine the average number of vehicles on the ramp, the average delay for the on-ramp vehicles, and the probability that the ramp will be full.

9.28 Explain the difference between pretrip and en route travel information user services.

9.29 Discuss the difference between dynamic versus static traffic assignment.

9.30 As was discussed earlier, delay in implementing dynamic traffic routing strategies results in incurring additional delays. The additional delay can be calculated using Equation 9.8 as follows:

Additional delay in veh/minutes =

$$(q - q_1) \times \delta \times \left[\frac{\delta(q - c) + 2D(c - c^*)}{120(c - q_1)}\right]$$

where

q = traffic arrival rate before routing (veh/h)
q_1 = reduced traffic arrival rate after routing (veh/h)
δ = waiting time before implementing a routing strategy in minutes
c = normal capacity of segment without any incidents (veh/h)
c^* = reduced capacity of segment as a result of an incident (veh/h)
D = incident duration in minutes

Use a cumulative plot to confirm the validity of the preceding equation.

9.31 A four-lane freeway segment carries an average volume of 3800 veh/h. The unrestricted capacity of a freeway lane can be assumed to be equal to 2300 veh/h/lane. An incident occurs that results in a 65% reduction in the capacity of the segment. The incident lasts for 60 mins. To relieve congestion, traffic routing is implemented, which reduces the traffic volume on the segment to 2200 veh/h. What would be the extra delay incurred if it took 4 mins to implement the routing strategy versus only 20 s?

9.32 Briefly define the following terms in relation to actuated signals: (1) minimum green; (2) passage time; and (3) maximum green.

9.33 Briefly describe the operational concept for an actuated controller.

9.34 It is required to coordinate the signals along the two-lane, one-way corridor shown:

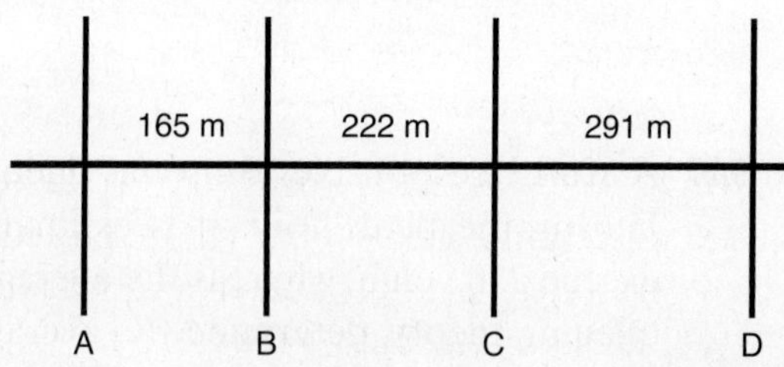

All signals have a common cycle of 90 s, and the effective green for the direction to be coordinated is 66.67% of the cycle length. Given that the average speed along the corridor is 63.4 km/h, calculate the following:

(a) The ideal offsets
(b) The bandwidth efficiency
(c) The bandwidth capacity

9.35 A north-south arterial with two lanes in each direction has four of its signals coordinated for the northbound direction, as shown. The average vehicle speed along the corridor is 55.5 km/h.

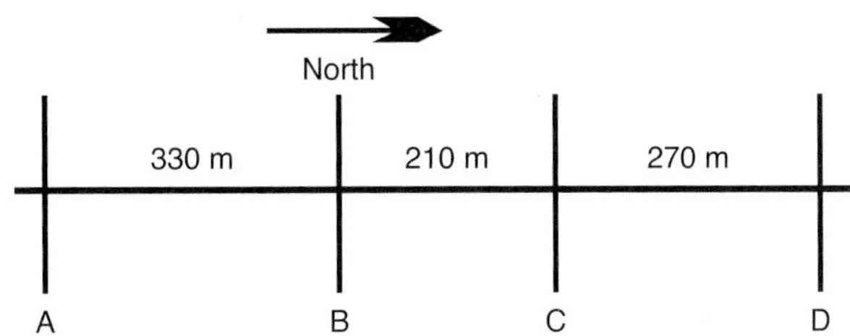

The cycle length, the duration of the green for the N-S phase, and the offset for each of the four signals (A, B, C, and D) are as shown:

Signal	*Cycle Length (s)*	*Green for N-S Phase*	*Offset Relative to Upstream Signal*
A	100	55	0
B	100	60	21
C	100	45	14
D	100	50	18

(a) Draw a time–space diagram for the coordinated system.

(b) Determine the bandwidth efficiency and bandwidth capacity for the NB direction.

(c) Determine the bandwidth efficiency and bandwidth capacity for the SB direction.

9.36 Briefly trace the development of adaptive traffic control systems since the early 1960s.

9.37 Discuss the difference between signal preemption and signal priority.

9.38 Determine the benefits to be expected from the deployment of an advanced traffic signal system on an arterial with an average ADT of 30,000 veh/day. The average trip length on the segment where the system is to be deployed is 15 mins. Assume that the value of time is equal to $10.00/h.

9.39 Give some examples for (1) real-world transit tracking systems; and (2) transit information systems.

9.40 A transit agency has an annual capital and operating cost of $3,000,000 and $2,500,000, respectively. Determine the benefits to be anticipated from deploying an AVL and transit operations software system.

9.41 Describe two applications of advanced technologies in rail transportation.

References

1. Bishop, R., *Intelligent Vehicles Technology and Trends*, Artech House, Inc., Norwood, MA, 2005.
2. Bretherton, D., "Current Developments in SCOOT: Version 3," in *Transportation Research Record 1554,* TRB, National Research Council, Washington, D.C., 1996.
3. Cambridge Systematics and ITT Industries, *ITS Deployment Analysis System User's Manual.* Cambridge, MA, 2000.
4. Chowdhury, M. A., and Sadek, A., *Fundamentals of Intelligent Transportation Systems Planning*, Artech House, Inc., Norwood, MA, 2003.
5. Cheu, R. L., and Ritchie, S. G., "Automated Detection of Lane-Blocking Freeway Incidents Using Artificial Neural Networks," *Transportation Research C,* Vol. 3(6), pp. 371–388.
6. Dailey, D. J., Smart Trek: A Model Deployment Initiative. U.S. Department of Transportation, 2001. Available at http://www.its.washington.edu/pubs/smart_trek_report.pdf
7. Garber, N. J. and Hoel, L. A., *Traffic & Highway Engineering,* Brooks/Cole, Pacific Grove, CA, 2002.
8. Hansen, B. G., Martin, P. T., and Perrin, H. Joseph, Jr., "SCOOT Real-Time Adaptive Control in a CORSIM Simulation Environment," in *Transportation Research Record 1727,* TRB, National Research Council, Washington, D.C., 2000.
9. Head, K. L., Mirchandani, P. B., and Sheppard, D., "Hierarchical Framework for Real Time Traffic Control," in *Transportation Research Record 1360,* TRB, National Research Council, Washington, D.C., 1992.
10. Institute of Transportation Engineers, *Intelligent Transportation Systems Primer*, Washington, D.C., 2001.
11. Lucas, D. E., Mirchandani, P. B., and Head, K. L., "Remote Simulation to Evaluate Real Time Traffic Control Strategies," in *Transportation Research Record 1727,* TRB, National Research Council, Washington, D.C., 2000.
12. Michalopoulos, P. G., Jacobson, R. D., Anderson, C. A., and Barbaresso, J. C., "Field Deployment of Machine Vision in the Oakland County ATMS/ATIS Project," *Proceedings,* IVHS America 1994 Annual Meeting, Atlanta, GA, April 1994, pp. 335–342.
13. Mitretek Systems, Inc., *ITS Benefits: 1999 Update*, Report No. FHWA-OP-99-012, Federal Highway Administration, U.S. Department of Transportation, Washington, D.C., 1999.
14. Neudorff, L. G., Randall, J. E., Reiss, R., and Gordon, R., *Freeway Management and Operations Handbook*, Report No. FHWA-OP-04-003, Federal Highway Administration, U.S. Department of Transportation, Washington, D.C., 2003.
15. Payne, H. J., and Tignor, S. C., "Freeway Incident Detection Algorithms Based on Decision Trees with States," in *Transportation Research Record 682*, TRB, National Research Council, Washington, D.C., 1978, pp. 30–37.
16. Persaud, B., and Hall, F. L., "Catastrophe Theory and Pattern in 30-Second Freeway Traffic Data—Implication for Incident Detection," *Transportation Research A*, Vol. 23(2), pp. 103–113, 1989.
17. Smith, Brian L., Pack, Michael L., Lovell, David J., and Sermons, M. William, "Transportation Management Applications of Anonymous Mobile Call Sampling," *Proceedings of the 11th Annual Meeting of ITS America*, Miami, FL, 2001.
18. U.S. Department of Transportation, Federal Highway Administration, *The National Intelligent Transportation Systems Architecture, Version 5.1*, 2005. Available from http://www.iteris.com/itsarch.
19. U.S. Department of Transportation, Federal Highway Administration, *Intelligent Transportation Systems, Compendium of Field Operational Test—Executive Summaries*, Washington, D.C., 1998.
20. U.S. Department of Transportation, Federal Highway Administration, *Developing Traveler Information Systems Using the National ITS Architecture*, Washington, D.C., 1998.
21. U.S. Department of Transportation, *Advanced Public Transportation Systems: The State of the Art—1998 Update*, Report No. FTA-MA-26-7007-98-1, Federal Transit Administration, Washington, D.C., 1998.
22. U.S. Department of Transportation, Federal Highway Administration, *The National Intelligent Transportation Systems Program Plan*, Washington, D.C., 1995.

APPENDIX A

TABLE A.1

Critical values of t

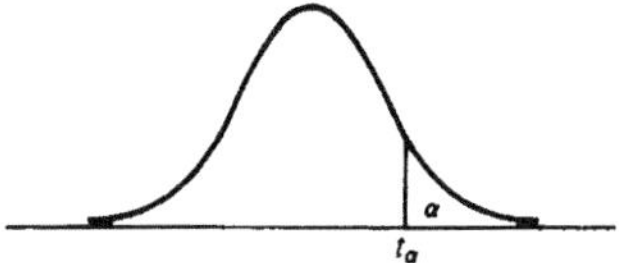

Degrees of Freedom	$t_{0.100}$	$t_{0.050}$	$t_{0.025}$	$t_{0.010}$	$t_{0.005}$
1	3.078	6.314	12.706	31.821	63.657
2	1.886	2.920	4.303	6.965	9.925
3	1.638	2.353	3.182	4.541	5.841
4	1.533	2.132	2.776	3.747	4.604
5	1.476	2.015	2.571	3.365	4.032
6	1.440	1.943	2.447	3.143	3.707
7	1.415	1.895	2.365	2.998	3.499
8	1.397	1.860	2.306	2.896	3.355
9	1.383	1.833	2.262	2.821	3.250
10	1.372	1.812	2.228	2.764	3.169
11	1.363	1.796	2.201	2.718	3.106
12	1.356	1.782	2.179	2.681	3.055
13	1.350	1.771	2.160	2.650	3.012
14	1.345	1.761	2.145	2.624	2.977
15	1.341	1.753	2.131	2.602	2.947
16	1.337	1.746	2.120	2.583	2.921
17	1.333	1.740	2.110	2.567	2.898
18	1.330	1.734	2.101	2.552	2.878
19	1.328	1.729	2.093	2.539	2.861
20	1.325	1.725	2.086	2.528	2.845
21	1.323	1.721	2.080	2.518	2.831
22	1.321	1.717	2.074	2.508	2.819
23	1.319	1.714	2.069	2.500	2.807
24	1.318	1.711	2.064	2.492	2.797
25	1.316	1.708	2.060	2.485	2.787
26	1.315	1.706	2.056	2.479	2.779
27	1.314	1.703	2.052	2.473	2.771
28	1.313	1.701	2.048	2.467	2.763
29	1.311	1.699	2.045	2.462	2.756
∞	1.282	1.645	1.960	2.326	2.576

Source: From M. Merrington, "Table of Percentage Points of the t-Distribution," *Biometrika*, 1941, 32, 300. Reproduced by permission of the *Biometrika* Trustees.

TABLE A.2
Critical values of χ^2

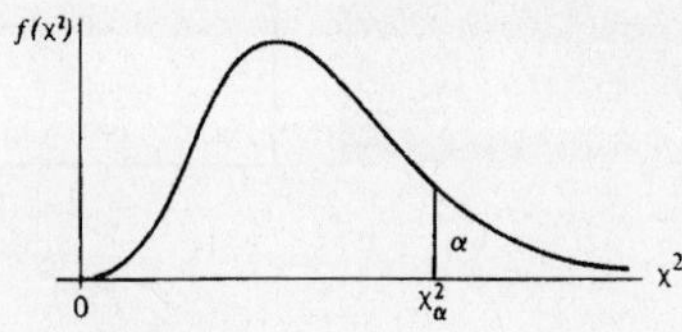

Degrees of Freedom	$\chi^2_{0.995}$	$\chi^2_{0.990}$	$\chi^2_{0.975}$	$\chi^2_{0.950}$	$\chi^2_{0.900}$
1	0.0000393	0.0001571	0.0009821	0.0039321	0.0157908
2	0.0100251	0.0201007	0.0506356	0.102587	0.210720
3	0.0717212	0.114832	0.215795	0.351846	0.584375
4	0.206990	0.297110	0.484419	0.710721	1.063623
5	0.411740	0.554300	0.831211	1.145476	1.61031
6	0.675727	0.872085	1.237347	1.63539	2.20413
7	0.989265	1.239043	1.68987	2.16735	2.83311
8	1.344419	1.646482	2.17973	2.73264	3.48954
9	1.734926	2.087912	2.70039	3.32511	4.16816
10	2.15585	2.55821	3.24697	3.94030	4.86518
11	2.60321	3.05347	3.81575	4.57481	5.57779
12	3.07382	3.57056	4.40379	5.22603	6.30380
13	3.56503	4.10691	5.00874	5.89186	7.04150
14	4.07468	4.66043	5.62872	6.57063	7.78953
15	4.60094	5.22935	6.26214	7.26094	8.54675
16	5.14224	5.81221	6.90766	7.96164	9.31223
17	5.69724	6.40776	7.56418	8.67176	10.0852
18	6.26481	7.01491	8.23075	9.39046	10.8649
19	6.84398	7.63273	8.90655	10.1170	11.6509
20	7.43386	8.26040	9.59083	10.8508	12.4426
21	8.03366	8.89720	10.28293	11.5913	13.2396
22	8.64272	9.54249	10.9823	12.3380	14.0415
23	9.26042	10.19567	11.6885	13.0905	14.8479
24	9.88623	10.8564	12.4011	13.8484	15.6587
25	10.5197	11.5240	13.1197	14.6114	16.4734
26	11.1603	12.1981	13.8439	15.3791	17.2919
27	11.8076	12.8786	14.5733	16.1513	18.1138
28	12.4613	13.5648	15.3079	16.9279	18.9392
29	13.1211	14.2565	16.0471	17.7083	19.7677
30	13.7867	14.9535	16.7908	18.4926	20.5992
40	20.7065	22.1643	24.4331	26.5093	29.0505
50	27.9907	29.7067	32.3574	34.7642	37.6886
60	35.5346	37.4848	40.4817	43.1879	46.4589
70	43.2752	45.4418	48.7576	51.7393	55.3290
80	51.1720	53.5400	57.1532	60.3915	64.2778
90	59.1963	61.7541	65.6466	69.1260	73.2912
100	67.3276	70.0648	74.2219	77.9295	82.3581

Source: From C. M. Thompson, "Tables of the Percentage Points of the χ^2-Distribution." *Biometrika*, 1941, 32, 188–189. Reproduced by permission of the *Biometrika* Trustees.

TABLE A.3

Critical values of T_L and T_U Wilcoxon Rank Sum Test: independent samples

Test statistics is rank sum associated with smaller sample (if equal sample sizes, either rank sum can be used)

(a) $\alpha = 0.025$ one-tailed; $\alpha = 0.05$ two-tailed

n_2 \ n_1	3		4		5		6		7		8		9		10	
	T_L	T_U	T_L	T_U	T_L	T_U	T_L	T_U	T_L	T_U	T_L	T_U	T_L	T_U	T_L	T_U
3	5	16	6	18	6	21	7	23	7	26	8	28	8	31	9	33
4	6	18	11	25	12	28	12	32	13	35	14	38	15	41	16	44
5	6	21	12	28	18	37	19	41	20	45	21	49	22	53	24	56
6	7	23	12	32	19	41	26	52	28	56	29	61	31	65	32	70
7	7	26	13	35	20	45	28	56	37	68	39	73	41	78	43	83
8	8	28	14	38	21	49	29	61	39	73	49	87	51	93	54	98
9	8	31	15	41	22	53	31	65	41	78	51	93	63	108	66	114
10	9	33	16	44	24	56	32	70	43	83	54	98	66	114	79	131

(b) $\alpha = 0.05$ one-tailed; $\alpha = 0.10$ two-tailed

n_2 \ n_1	3		4		5		6		7		8		9		10	
	T_L	T_U	T_L	T_U	T_L	T_U	T_L	T_U	T_L	T_U	T_L	T_U	T_L	T_U	T_L	T_U
3	6	15	7	17	7	20	8	22	9	24	9	27	10	29	11	31
4	7	17	12	24	13	27	14	30	15	33	16	36	17	39	18	42
5	7	20	13	27	19	36	20	40	22	43	24	46	25	50	26	54
6	8	22	14	30	20	40	28	50	30	54	32	58	33	63	35	67
7	9	24	15	33	22	43	30	54	39	66	41	71	43	76	46	80
8	9	27	16	36	24	46	32	58	41	71	52	84	54	90	57	95
9	10	29	17	39	25	50	33	63	43	76	54	90	66	105	69	111
10	11	31	18	42	26	54	35	67	46	80	57	95	69	111	83	127

Source: From F. Wilcoxon and R. A. Wilcox, "Some Rapid Approximate Statistical Procedures," 1964, 20–23. Reproduced with the permission of American Cyanamid Company

F

G

H

I

J

Q

R

S

U

V

PRINCIPAL UNITS USED IN MECHANICS

Quantity	International System (SI)			U.S. Customary System (USCS)		
	Unit	Symbol	Formula	Unit	Symbol	Formula
Acceleration (angular)	radian per second squared		rad/s^2	radian per second squared		rad/s^2
Acceleration (linear)	meter per second squared		m/s^2	foot per second squared		ft/s^2
Area	square meter		m^2	square foot		ft^2
Density (mass) (Specific mass)	kilogram per cubic meter		kg/m^3	slug per cubic foot		$slug/ft^3$
Density (weight) (Specific weight)	newton per cubic meter		N/m^3	pound per cubic foot	pcf	lb/ft^3
Energy; work	joule	J	N·m	foot-pound		ft-lb
Force	newton	N	$kg{\cdot}m/s^2$	pound	lb	(base unit)
Force per unit length (Intensity of force)	newton per meter		N/m	pound per foot		lb/ft
Frequency	hertz	Hz	s^{-1}	hertz	Hz	s^{-1}
Length	meter	m	(base unit)	foot	ft	(base unit)
Mass	kilogram	kg	(base unit)	slug		$lb\text{-}s^2/ft$
Moment of a force; torque	newton meter		N·m	pound-foot		lb-ft
Moment of inertia (area)	meter to fourth power		m^4	inch to fourth power		$in.^4$
Moment of inertia (mass)	kilogram meter squared		$kg{\cdot}m^2$	slug foot squared		$slug\text{-}ft^2$
Power	watt	W	J/s (N·m/s)	foot-pound per second		ft-lb/s
Pressure	pascal	Pa	N/m^2	pound per square foot	psf	lb/ft^2
Section modulus	meter to third power		m^3	inch to third power		$in.^3$
Stress	pascal	Pa	N/m^2	pound per square inch	psi	$lb/in.^2$
Time	second	s	(base unit)	second	s	(base unit)
Velocity (angular)	radian per second		rad/s	radian per second		rad/s
Velocity (linear)	meter per second		m/s	foot per second	fps	ft/s
Volume (liquids)	liter	L	$10^{-3}\ m^3$	gallon	gal.	$231\ in.^3$
Volume (solids)	cubic meter		m^3	cubic foot	cf	ft^3